Waves in Nonlinear Layered Metamaterials, Gyrotropic and Plasma Media

Online at: https://doi.org/10.1088/978-0-7503-2336-9

Waves in Nonlinear Layered Metamaterials, Gyrotropic and Plasma Media

Yuriy Rapoport
Taras Shevchenko National University of Kyiv, Kyiv, Ukraine
and
University of Warmia and Mazury in Olsztyn, Poland

Vladimir Grimalsky
Autonomous University of State Morelos, Cuernavaca, Morelos, Mexico

IOP Publishing, Bristol, UK

ISBN 978-0-7503-2336-9 (ebook)
ISBN 978-0-7503-2334-5 (print)
ISBN 978-0-7503-2337-6 (myPrint)
ISBN 978-0-7503-2335-2 (mobi)

DOI 10.1088/978-0-7503-2336-9

Version: 20221201

IOP ebooks

British Library Cataloguing-in-Publication Data: A catalogue record for this book is available from the British Library.

Published by IOP Publishing, wholly owned by The Institute of Physics, London

IOP Publishing, No.2 The Distillery, Glassfields, Avon Street, Bristol, BS2 0GR, UK

US Office: IOP Publishing, Inc., 190 North Independence Mall West, Suite 601, Philadelphia, PA 19106, USA

Contents

Preface

The present book is devoted to the nonlinear wave processes in metamaterials (MMs), gyrotropic and plasma media. The fundamental concept of metamaterials is based on the programmed compositions of various materials to produce new desired qualities, especially to linear and nonlinear wave propagation. The modern nano- and meta-photonic technologies make it possible to create desired construction and composition of MMs for providing a specified response of any system to external electromagnetic (EM) and acoustic fields, which is usually impossible in the natural media. In MM the set of specific features can be realized jointly in the same structure, for instance specific wave resonances, the extreme field concentration, the resonant nonlinearity, bi-anisotropy, the dispersion and diffraction wave management etc.

In distinction to other books devoted to this subject, our book is concentrated on the following main issues.

(1) It includes the general approach to the layered nonlinear and active metamaterials, gyrotropic and plasma media that is called the 'metamaterial approach'. This term was used already earlier. Nevertheless we are developing such an approach much farther and in a more general way. The applicability of the general approach is possible to very different media and scales, from 0.01 Hz, 1000 km in the natural geophysical systems up to 10^{15} Hz and several nanometers in artificial MMs. This has caused a positive effect in investigations of an extremely wide class of materials and their use.

(2) The book covers the MMs, including the contemporary ones, and also relative gyrotropic, plasmonic and hyperbolic materials. This is rather a relevant choice because: (a) plasmonic materials form now an inherent part of the MMs, from THz to optical range; (b) gyrotropic materials may be used as an integrated part of the MMs, providing, in particular, a negative magnetic permeability at the microwave range and controllability by means of external fields in, for example, the optical range. Then the processes with weak, moderate, and strong nonlinearity are included. Because the most interesting features of scientific and practical applications of the MMs exploit their resonant behavior, the losses are inevitable problems for the MMs. The method of overcoming losses is also addressed in the book. The other examples of active media include interacting waves with parametric coupling in layered MMs and ferrite films, amplification of EM waves in active MMs with metaparticles loaded on the active elements like diodes, while the microwave range is considered, or hyperbolic MMs with active quantum molecules with inversed energy layers in optical range. Naturally, if amplification is applied or an input pulse amplitude in nonlinear media increases, then finally the regime of weak and moderate nonlinearity should be replaced by a regime of the strong nonlinearity. Respectively, the approximation of slowly varying

amplitudes should be replaced by other, more accurate methods. The set of such methods is described in the book, like a very general method NEELS (nonlinear evolution equation for layered structures) for the weak and moderate nonlinearity and specific methods for strong nonlinearities and other cases, where slowly varying amplitude approximation fails.

(3) At the same time, we develop the general method for the derivation of nonlinear evolution equations (NEELS) for envelopes of pulses in layered waveguiding structures and for a moderate nonlinearity and spectrum width. This method is effective and useful from the practical point of view. It allows us to get corresponding nonlinear evolution equations regardless of the type of nonlinear layered media and with accounting for both volume and surface nonlinearities, higher-order nonlinear effects, bi-anisotropy, parametric interactions and other active interactions/wave amplification, the nonlinear spatial dispersion including corresponding nonlinear auxiliary boundary conditions. Formation of such remarkable nonlinear quasi-2D structures in nonlinear layered and active structures, as vortices and bullets is addressed. A description of the nonlinear pulses adopts effects of controllability, such as the magnetooptic control of nonlinear beams/pulses in the layered MM waveguiding structures with magnetooptic effects. Using the same method, the magnetic bullets, a stabilization of their collapse and new nonlinear structures based on their parametric coupling, as well as magnetooptic control over solitons with higher-order nonlinear effects have been obtained, first. Then a unified method is put forward 'from properties of individual particle to the nonlinear waves in layered bi-anisotropic structures' including: (1) modeling separate chiral or Ω-particles; (2) the new method of nonlinear homogenization for bi-anisotropic media; and (3) generalization of the NEELS method for the waves in bi-anisotropic waveguiding media including their parametric coupling.

(4) It is shown first using NEELS method that under a condition of temporal resonance of the second harmonic, when the main harmonic pulse has a carrier frequency close to half of plasma frequency, the surface nonlinear plasmons demonstrate the giant second harmonic generation in the 'epsilon-near-zero' regime, with a maximum amplitude, larger than the amplitude of the main harmonic, while the surface nonlinearity prevails over the volume one both quantitatively and quantitatively.

(5) Important scientific and practical results are obtained for the active systems, while a condition of spatial amplification/convective instability and absence of the generation (absolute instability) of EM waves are determined. It is shown for the first time that in order to provide the spatial amplification of EM waves and negative phase behavior in MMs with metaparticles loaded by active elements, the band of frequencies, where the corresponding elements are active, must be finite. On the basis of the analysis of three main possible sources of the absolute instability, the conditions of spatial amplification with the simultaneous negative phase

behavior of MMs with metaparticles loaded by active elements with frequency-dependent conductors are found. The generalized theoretical scheme of the construction of such an active MM, which provides the spatial amplification of EM waves in a negative phase MM environment, is proposed. The possibility is mentioned of significant and non-trivial influence of higher order nonlinear effects, both nonlinear dispersion and diffraction, on the instability of bullets in a nonlinear waveguide. In particular, it is shown that the higher order effects may lead to a splitting of bullets in the direction of propagation. It is shown for the first time that with the use of diffraction management, the bullets in the active waveguide in the presence of these effects can be amplified with the preservation of the form at a greater distance than without the diffraction management. The conclusion is done that the influence of the higher order effects are nontrivial and essential for the formation, distribution and control, with the help of external fields, of nonlinear wave structures in layered complex media of different physical nature.

(6) The new effects in the systems with strong nonlinearities and in the regimes, for which the slowly varying amplitude approximation fails, are presented. For the first time, the existence of strong nonlinearity effects in a multilayered MM dielectric-graphene structure is shown, in particular nonlinear threshold switching of THz pulses from reflection mode to transmission mode. The nonlinear MM field concentrator, or the nonlinear EM 'black hole', is proposed. For a case of the strong nonlinearity, when the NEELS method is inapplicable, a new method is proposed and developed. This method combines a new variant of complex geometric optics in a linear region of the concentrator, where the wavelength is much smaller than the characteristic length of the inhomogeneity, and a full-wave solution of the Maxwell equations in the nonlinearity region, in accordance with the new proposed boundary conditions for the matching of both solutions. The method has theoretically revealed a new physical effect associated with strong nonlinearity in active MMs in the IR range, namely the threshold jump of the focus point and the formation of a self-consistent highly localized nonlinear resonator, so-called the hot spot, on the boundary of the linear and nonlinear regions in the field concentrator. It is proven that the jump of the point of focus is reached when the threshold of the field is exceeded, regardless of the physical method of such an excess (increasing the amplitude of the incident pulse, or linear amplification, or using an additional weak beam).

(7) Giant double-resonant second-harmonic generation in multilayered dielectric-graphene MMs will be formulated. The equations for the second-harmonic generation will be derived comparing graphene-based, hydrodynamic and kinetic approaches. The numerical modeling supports all the conclusions.

(8) The synergetic approach to the wave processes is nicely discussed for active quantum hyperbolic MMs. The points addressed include: (a) the system of Maxwell–Bloch equations for nonlinear waves in active quantum

nonlinear hyperbolic MMs (b) Ginzburg–Landau equations for nonlinear waves in active nonlinear hyperbolic MMs propagating under a finite angle to optic axis. The evolution Ginzburg–Landau equations for resonant active MM in the resonance regime are derived with complex coefficients. A choice of the materials formed for active hyperbolic structures with different frequency ranges is given together with numerical modeling based on the spectral-finite difference method, including the method of splitting by the directions and by physical factors, for the pulse propagation. There was accounted for the finite length of the MM slabs and proper EM boundary conditions both on the MM boundary and between the elementary layers inside the MM; moreover, the condition of the necessity of the mesoscale approach to a nonlinear active MM, which goes beyond the effective media approximation, in spite of the satisfaction of a formal MM approximation, i.e. the elementary cell length is much less than the wavelength, is presented.

(9) The following additional problems are included, namely the rogue waves in hyperbolic MMs, in particular the wave excitation corresponding to rogue NSE breather solutions in transparent double-negative MMs, wherein higher-order dispersive or nonlinear effects are included. The impact of the self-steepening will be investigated connecting to both Peregrine and Akhmediev breathers. The propagation of the rogue waves in MMs under magnetooptic influence is demonstrated.

(10) In the conclusion chapter of the book, the main results obtained in the previous chapters, and conclusions are summarized. Then the list of the future tasks is presented, which is a subject of personal preferences of the author(s). Note that the following is also mentioned in the conclusion chapter of the book, which follows from a possibility to apply the methods, developed for MM, also to the natural geophysical systems. Taking into account the above properties of MMs and the deep physical analogy between the characteristics of wave processes occurring in layered gyrotropic, chiral, hyperbolic and nonlinear artificial MMs and in a layered nonlinear gyrotropic and anisotropic active natural system, it is suitable to use a unified MM approach to study the mechanisms of wave coupling in the geophysical systems.

The authors are grateful to our colleagues and co-authors of the scientific works Prof. S Koshevaya, Prof. V Ivchenko, Prof. G Milinevsky, Prof. O Cheremnykh, Prof. E Gutierrez, Prof. I Nefedov, Prof. A Lavrynenko, Dr J McNiff, Dr R C Mitchell-Thomas, Dr L Velasco for the pleasant and inspiring time of scientific collaboration. Also the authors are grateful to Prof. V Shalaev, Prof. M Noginov, Prof. S Nikitov, Prof. Yu Kivshar, Prof. K Simovsky, Prof. V Tretyakov, Prof. V Yampol'skii for their outstanding contribution into the science of nonlinear MMs that helped us to get the results on this subject and to write this book.

We are very grateful to the late Professor Allan Boardman for his cooperation, mentoring, co-creation, collaboration in the development of the scientific direction reflected in this book, and for involving us in its development when it first began in the world.

One of the authors (YuR) is grateful to Prof. Allan Boardman and Dr J McNiff for the invitation to participate in a number of projects on nonlinear wave processes in composites and MMs, as well as for the organization of scientific visits to the University of Salford, UK; and Prof. Sergei Tretyakov and Professors Craig Zaspel and Andrei Slavin for their joint work on nonlinear wave processes in metamaterials and solitons in ferrite films, respectively, and for an invitation to participate in scientific projects as a visiting Professor at Aalto University, Finland and the University of Montana, USA, respectively. This communication and fruitful scientific collaboration also greatly contributed to the writing of this book.

This book and the table of contents in a preliminary form were conceived during 2016–18, and most of the articles that formed its basis were written during 2002–18, together with Professor A D Boardman, a brilliant physicist and a remarkable and witty person, one of the world's leading experts in nonlinear nanophotonics, a pioneer in many fields of which he was. These areas include, for example, magnetooptical control of optical solitons, highly nonlinear surface waves in layered solid-state structures, including dielectric-metamaterial-dielectric structures, diffraction management for layered nanometaphotonic materials, and many others. YuR thanks the staff of scientific project 20BF051-02/0120U102178, in particular Y K Gorban, N R Hensitskyi, S S Petrishevskii, T V Volkova, who helped in the technical preparation of a number of chapters of the book, as well as other employees of Taras Shevchenko Naional University of Kyiv, Ukraine, its Physics Department, the Department of Astronomy and Space Physics and the Laboratory of Space Physics, who created such a benevolent and creative atmosphere that had a beneficial effect on implementation of the opportunity to successfully work on this book.

The authors are grateful to the team of publishers of this book, led by Ashley Gasque, for the tremendous organizational work that helped us a lot in preparing the book.

The authors are very grateful to their families for their patience, great moral support and understanding in preparing this book. One of the authors (Yu R) is especially grateful to his wife Svetlana for her love and support, as well as for the tremendous work at all stages of the technical preparation of the book, without which this book would hardly have been born.

This book is dedicated to the memory of the late Professor Boardman, who did not have time to take part in the writing of its text, but with whom the physical meaning of most of the physical effects described in the book was discussed and together with whom its concept was developed. Professor Boardman has a dedicated page in this book.

YuR and all his family are very grateful to University of Warmia and Mazury (UWM) in Olsztyn, Poland and its administration and to Prof. Andrzej Krankowski, Head of Space Radio-Diagnostic Research Centre in UWM, and his strong and friendly team for their great help, hospitality and patience. Without all of this support after February 24, 2022, when the full-scale Russian invasion of

Ukraine started, it would be very hard to finish this book up to now. YuR is also grateful to the Ministry of Education and Science of Ukraine, grant 20БФ051-02 [0120U102178] 'Wave processes and effects in active resonant layered plasma media and metamaterials', for the financial support during the three-year period of the book preparation and to the Kosciuszko Foundation Program for Ukrainian Scientists 'Freedom starts with your mind' (Poland) and National Science Centre, Poland, grant No. 2022/01/3/ST10/00072, for the financial support, very important for finishing this book.

Author biographies

Yuriy Rapoport

Yuriy Rapoport (YuR) is PhD (1986) and Dr of Phys.-Math. Science (2017), Leading research fellow, from Taras Shevchenko National University of Kyiv, Physics Faculty, Space Physics Laboratory, Kyiv, Ukraine, and professor in the University of Warmia and Mazury, in Olsztyn, Poland, leader and executor of more than 10 Ukrainian and international scientific projects. He graduated from Physical-Mathematical school No 145 (now Physics-Mathematical Lyceum) in Kyiv, and then from Taras Shevchenko State University, Kyiv, Ukraine, Department of Theoretical Physics in 1978. YuR is author and co-author of pioneer works on superheterodyne amplification of magnetoacoustic wave in ferrite film, on energetic method for nonlinear waves of different physical nature in the layered media and then on magnetostatic solitons and bullets in ferrite films, magnetooptic control over spatial and temporal solitons in metamaterial waveguides, theory of absolute and convective instability in active metamaterials, strong nonlinearities and formation of hot spots in active nonlinear hyperbolic metamaterial planar structures and cylindrical wave concentrator, controllable nonlinear wave modulation in multi-layered grapheme-dielectric metamaterials etc. YuR took part in the NSF project on magnetostatic solitons in ferrite films in USA in 1999–2000, project on the determination of the characteristics of an artificial composite material in microwave ranges in UK in 2001–02 and in the first project of EPSRC on the nonlinear wave phenomena in metamaterials led by Prof. Allan Boardman in UK in 2007–10. He collaborates with colleagues from Ukraine, Mexico, Japan, Finland, Denmark, USA, Poland, Italy, etc. Current research interests/new possible scientific projects with involvement of post-graduate students and university students include: **I.** theory and modeling effects of the controllable nonlinear scattering on the strongly resonant layered metamaterial active structures; **II.** nonlinear singular and topological GHz/THz metaphotonics including amplitude and phase singular structures (vortices) using metamaterials and metasurfaces; propagation of GHz/THz electromagnetic vortices and their robustness in the atmosphere; **III.** generation and amplification of plasmon-polaritons by drift of carriers in graphene; nonlinear modulation and generation of ultrashort pulses in layered metamaterial structures in THz range; **IV.** strongly nonlinear controllable wave structures in highly resonant active metamaterials and on metasurfaces; **V.** methods of control over quantum waves of electron states (QWES) in 2D electron gas metamaterials; **VI.** VLF (very low-frequency) waves in the system Lithosphere (Earth)–Atmosphere–Ionosphere–Magnetosphere (LEAIM); Ionosphere as a sensitive indicator of the influences from ‘below’ (the strongest earthquakes, hurricanes, etc) and ‘above’ (magnetic storms). In his Dr Sci. Thesis on Theoretical Physics, YuR developed a metamaterial approach to the wave phenomena in both artificial and natural layered media with volume and surface nonlinearities and elaborated the world-recognised method of modeling nonlinear pulses in such a system called method NEELS (nonlinear evolution equations for wave processes in layered structures). He has about 200 scientific publications with more than 1000 citations and is a Leader of Research Group at Taras Shevchenko National University of Kyiv, Ukraine (SCOPUS data base: https://www.scopus.com/authid/detail.uri?authorId=7005316101). He has presented around 10 plenary and invited papers at international conferences and was a co-author of a set of invited papers and book (monograph) chapters.

Volodymyr Grimalsky

Volodymyr (Vladimir) Grimalsky graduated from T. Shevchenko Kiev State University (KSU), former USSR (now T. Shevchenko National University, Ukraine), in 1982 with the honorous diploma on the theoretical physics. He obtained his PhD in physics in 1986 at the same university. V Grimalsky worked as the researcher and the senior researcher in the Radiophysical Faculty of KSU and in the Institute for Space Research, Kiev. Since 1999 he has worked in Mexico in the National Institute for Astrophysics, Optics, and Electronics (INAOE), Puebla, and then in the Autonomous University of State Morelos (UAEM) as the professor-investigator.

V Grimalsky is the author of more than 200 papers in refereed journals, conference proceedings, and book chapters. His scientific interests are with the linear and nonlinear propagation and interaction of electromagnetic and acoustic waves in complex media including metamaterials; physical fundamentals of semiconductor devices of high frequencies including terahertz range; and investigation of the physical fundamentals of electromagnetic and acoustic processes in the ionosphere connected with seismic and volcanic activity phenomena.

Page dedicated to Professor Allan D Boardman

Professor Allan D Boardman

The last appointment: Professorial Research Fellow and Professor of Applied Physics

The late Professor Allan Boardman was from the UK University of Salford, and was a worldwide expert on the global revolution known as metamaterials that is now transforming science. These are artificial, very specialised, materials beyond the control of Nature. There are many conferences on metamaterials and the very important SPIE Photonics Europe meeting was held during April 2018 in Strasbourg. It was a focus of global leaders in this field. Allan was made co-Chair of this huge metamaterial conference under the Photonics Europe heading. He held a Doctor of Science degree from the University of Durham and was responsible for 328 peer-reviewed and other publications, generating 5432 citations. He was Topical Editor for the *Journal of the Optical Society of America B* for Metamaterials and Photonic Structures and for the UK Institute of Physics *Journal of Optics*. Professor Boardman was a Fellow of the SPIE and a Fellow of the Optical Society of America. He was also a Fellow of the UK Institute of Physics and the Institute of Mathematics and its Applications. He is well-known both in the UK, and globally, for his work on surface plasmons and guided wave optics, especially nonlinear waves and solitons. He was the leading theorist looking at nonlinear guided magnetooptic waves in metamaterials, which is why he was the co-Chair of the globally leading SPIE Photonics Europe held biannually in Europe. This was in addition to being, for some time, the co-Chair of SPIE Annual Optics and Photonics meeting held annually in San Diego, USA. He organized many conferences and has been a Director of a number of NATO Advanced Study Institutes. His principal research activities cover linear and nonlinear surface waves, covering both electromagnetics and acoustics. His particular emphasis was on negative index and the more modern hyperbolic complex metamaterials. His work maps beautifully onto, solitons in nonlinear gyroelectric media, nonlinear electromagnetic modelling, nonlinear microwave magnetics, complex microwave materials, magneto-photonic band gap phenomena, dissipative solitons and electromagnetic modelling for mm-wave devices. All of this metamaterial work maps perfectly onto making cloaking devices that nicely hide objects that must not be seen by certain people. It also maps onto biomedical applications that will transform tumour control by hiding them from dangerous actions. Some dramatic breakthroughs for his work involved the publication on Trapped Rainbows with a citation of 201 and his co-authorship of the Space-time History Editor paper, which was voted by *Physics World* to be the third out of 10 most important scientific breakthroughs of 2011. Finally, he was establishing new pathways for ***rogue*** waves controlled by metamaterials, and was an eminent scientist in the field of nonlinear metaphotonics. Professor Allan Boardman educated a galaxy of students and graduate students who became doctors of science. Professor Allan Boardman made a great contribution to the papers on which this book is based and selection the material for the book (contents) and the discussion of the ideas of the future book. To our deep regret, Professor Boardman died before this book began to be written as such. This book/monograph is dedicated to the blessed memory of Professor Allan Boardman.

Selected publications of the authors

Journal papers

[1] Buttner O, Bauer M, Demokritov S O, Hillebrands B, Kivshar Y S, Grimalsky V V, Rapoport Y and Slavin A N 2000 Linear and nonlinear diffraction of dipolar spin waves in yttrium iron garnet films observed by space- and time-resolved Brillouin light scattering *Phys. Rev. B* **61** 11576–87

[2] Boardman A D, Hess O, Mitchell-Thomas R C, Rapoport Y G and Velasco L 2010 Temporal solitons in magnetooptic and metamaterial waveguides *Photon. Nanostruct. – Fundam. Appl.* **8** 228–43

[3] Boardman A D, Grimalsky V V, Yu Kivshar, Koshevaya S V, Lapine M, Litchinitser M, Malnev V N, Noginov M, Rapoport Y G and Shalaev V M 2011 Active and tunable metamaterials *Laser Photonics Rev.* **5** 287–307

[4] Boardman A D, Rapoport Y G, Grimalsky V V, Ivanov B A, Koshevaya S V, Velasco L and Zaspel C E 2005 Excitation of vortices using linear and nonlinear magnetostatic waves *Phys. Rev. E* **71** 026614

[5] Rapoport Y G, Boardman A D, Grimalsky V V, Ivchenko V M and Kalinich N 2014 Strong nonlinear focusing of light in nonlinearly controlled electromagnetic active metamaterial field concentrators *J. Opt.* **16** 055202

[6] Rapoport Y *et al* 2020 Model of propagation of VLF beams in the waveguide Earth–Ionosphere. Principles of tensor impedance method in multilayered gyrotropic waveguides *Ann. Geophys.* **38** 207–30

[7] Boardman A D, Alberucci A, Assanto G, Grimalsky V V, Kibler B, McNiff J, Nefedov I S, Rapoport Y G and Valagiannopoulos C A 2017 Waves in hyperbolic and double negative metamaterials including rogues and solitons *Nanotechnology* **28** 444001

[8] Rapoport Y, Grimalsky V, Lavrinenko A V and Boardman A 2017 Double resonant excitation of the second harmonic of terahertz radiation in dielectric-graphene layered metamaterials *J. Opt.* **19** 095104

[9] Rapoport Y G *et al* 2017 Ground-based acoustic parametric generator impact on the atmosphere and ionosphere in an active experiment *Ann. Geophys.* **35** 53–70

[10] Rapoport Y, Selivanov Y, Ivchenko V, Grimalsky V, Tkachenko E, Rozhnoi A and Fedun V 2014 Excitation of planetary electromagnetic waves in the inhomogeneous ionosphere *Ann. Geophys.* **32** 449–63

[11] Rapoport Y, Grimalsky V, Krankowski A, Pulinets S, Fedorenko A and Petrishchevskiii S 2020 Algorithm for modeling electromagnetic channel of seismo-ionospheric coupling (SIC) and the variations in the electron concentration *Acta Geophys.* **68** 253–78

[12] Grimalsky V, Koshevaya S, Rapoport Y and Kotsarenko A 2016 Collapse of nonlinear electron plasma waves in a plasma layer *Phys. Scr.* **91** 105602

[13] Grimalsky V V, Rapoport Y G, Boardman A D and Koshevaya S V 2018 Nonlinear focusing of picosecond baseband pulses in paraelectric crystals *Opt. Quantum Electr.* **50** 102

[14] Rapoport Y G, Hayakawa M, Gotynyan O E, Ivchenko V N, Fedorenko A K and Selivanov Y A 2009 Stable and unstable plasma perturbations in the ionospheric F region, caused by spatial packet of atmospheric gravity waves *Phys. Chem. Earth.* **34** 508–15

[15] Yutsis V, Rapoport Y, Grimalsky V, Grytsai A, Ivchenko V, Petrishchevskii S, Fedorenko A and Krivodubskij V 2021 ULF activity in the earth environment: Penetration of electric field from the near-ground source to the ionosphere under different configurations of the geomagnetic field *Atmosphere* **12** 801

[16] Fedorenko A K, Kryuchkov E I, Cheremnykh O K, Voitsekhovska A D, Rapoport Yu G and Klymenko Yu O 2021 Analysis of acoustic-gravity waves in the mesosphere using VLF radio signal measurements *J. Atmos. Sol.-Terr. Phys.* **219** 105649

[17] Grimalsky V V, Rapoport Yu G, Koshevaya S V, Escobedo-Alatorre J and Tecpoyotl-Torres M 2021 Nonlinear focusing of picosecond baseband pulses in paraelectric crystals in a wide temperature range *Opt. Quantum Electr.* **53** 484

[18] Rapoport Y, Grimalsky V and Iorsh I *et al* 2013 Nonlinear reshaping of terahertz pulses with graphene metamaterials *JETP Lett.* **98** 503–6

[19] Grimalsky V V, Koshevaya S V and Rapoport Y G 2011 Superheterodyne amplification of electromagnetic waves of optical and terahertz bands in gallium nitride films *Radioelectron. Commun. Syst.* **54** 401

[20] Grimalsky V, Koshevaya S, Castrejon-Martinez C and Rapoport Y 2016 Amplification of optical phonons in semiconductor quantum wells at finite temperatures *2016 IEEE 36th Int. Conf. on Electronics and Nanotechnology (ELNANO)* 67–70

[21] Grimalsky S V, Koshevaya S, Rapoport Y, Tretiak N, Yanovsky F and Escobedo-Alatorre J 2019 Resonant properties of electron gas in n-InSb and graphene layers in magnetic fields for THz multilayered dielectric-plasma-like metamaterials *2019 IEEE 39th Int. Conf. on Electronics and Nanotechnology (ELNANO)* 164–8

[22] Yevtushenko F O, Dukhopelnykov S V, Zinenko T L and Rapoport Y G 2021 Electromagnetic characterization of tuneable graphene-strips-on-substrate metasurface over entire THz range: analytical regularization and natural-mode resonance interplay *IET Microw. Antennas Propag.* **15** 1225–39

Monographs/chapters in the monographs

[23] Boardman A D, Rapoport Y G, Aznakayeva D E, Aznakayev E G and Grimalsky V 2019 Graphene metamaterial electron optics: Excitation processes and electro-optical modulation *Handbook of Graphene* ed Mei Zhang vol 3 (Beverly, MA: Scrivener Publishing LLC) 263–96

[24] Boardman A D, Alberucci A, Assanto G, Rapoport Y G, Grimalsky V V, Ivchenko V M and Tkachenko E N 2017 Spatial solitonic and nonlinear plasmonic aspects of metamaterials *World Scientific Handbook of Metamaterials and Plasmonics* ed E Shamonina and S A Maier (Berlin: Springer) ch 10 pp 419–69

[25] Boardman A D, Tsakmakidis K L, Mitchell-Thomas R C, King N J, Rapoport Y G and Hess O 2015 From 'trapped rainbow' slow light to spatial solitons *Nonlinear, Tunable and Active Metamaterials Springer Series in Materials Science* ed I V Shadrivov, M Lapine and Y Kivshar vol 200 (Berlin: Springer) pp 161–91 https://ru.scribd.com/document/405050635/

[26] Boardman A D, King N and Rapoport Y 2010 Circuit model of gain in metamaterials *Nonlinearities in Periodic Structures and Metamaterials Springer Series in Optical Sciences* ed C Denz, S Flach and Y Kivshar vol 150 (Berlin: Springer) pp 259–71

[27] Rapoport Y G, Boardman A D, Grimalsky V V, Koshevaya S V, Zaspel C E and Ivanov B A 2008 Nonlinear vortex generation by forward volume magnetostatic waves *Electromagnetic, Magnetostatic and Exchange-Interaction Vortices in Confined Magnetic Structures* ed E O Kamenetskii (Kerala, India: Transworld Research Network) pp 29–44 (only in printed form)
[28] Grimalsky V V, Kremenetsky I A and Rapoport Y G 1999 Excitation of EMW in the lithosphere and propagation into magnetosphere *Atmospheric and Ionospheric Electromagnetic Phenomena Associated with Earthquakes* ed M Hayakawa (Tokyo: Terrapub) pp 777–87

Textbook

[29] Rapoport Yu G and Grytsai A V 2020 *Nonlinear Wave Processes in Plasmas* (Kyiv, Ukraine: Taras Shevchenko National University of Kyiv)

Yuriy Rapoport and Vladimir Grimalsky

Chapter 1

Introductory chapter

1.1 Metamaterials: the discovery of 2000th

In recent years, experimental and theoretical studies of nonlinear wave processes in layered and complex metamaterial (MM), gyrotropic, and plasma media have been intensively conducted. As MMs in the broadest sense, we mean artificial materials with a specially created structure, which provide a special response to the irradiation of the investigated media by the external fields (Tretyakov *et al* 2010).

This chapter presents the arguments that determine the relevance of the study of wave processes in MM, gyrotropic and plasma media.

The purpose of the book is to give a wide, tutorial-driven, presentation of the theory of wave processes occurring in layered nonlinear metamaterials, gyrotropic and plasma media; to determine the regularities of electromagnetic wave propagation and formation of wave structures; to investigate the new wave structures (solitons etc) and effects in MM, gyrotropic and plasma media with bulk, surface, resonant, moderate, strong nonlinearities; to study the effective methods of control of wave processes due to the use of inhomogeneous and non-stationary external fields.

Let us now dwell on the characteristic features of the considered structures and corresponding media and wave phenomena in them, which turned out to be important for achieving the set goals.

1.2 Characteristic features of the media and wave phenomena in metamaterials and the main approaches to their modeling

1.2.1 Typical properties of metamaterials

A typical example of a very important practical MM property of electromagnetic waves (EMWs) is the 'negative phase behavior' (or the negative phase of the waves,

doi:10.1088/978-0-7503-2336-9ch1

in contrast to the positive phase for 'ordinary' media), that is, the propagation of waves with oppositely directed group V_g and phase V_{ph} velocities in MMs with negative values of electric permittivity and magnetic permeability (Veselago 1968, Tretyakov *et al* 2010). This property follows from the consideration of a plane EMW with a frequency ω, a wavenumber k, component fields E_x, $H_y \sim \exp[i(\omega t - kz)]$ whose energy propagates in an infinite homogeneous MM medium in the direction $+z$ (in the presence of a source in the region $z = -\infty$), taking into account the principle of causality (Tretyakov *et al* 2010).

As follows from the Maxwell equations for a plane wave, the relations $k^2 = \varepsilon\mu k_0^2$, $P_z = (c/4\pi)(\varepsilon/k)|E_x|^2$ are satisfied, where $k_0 = \omega/c$, c is the speed of light, P_z is the value of the Poynting vector directed along $0Z$-axis. In accordance with the above relations, the condition of wave propagation (with $k^2 > 0$) and the energy transfer of the waves in the direction $+z$ (which coincides, in the absence or with a relatively weak dissipation, with the direction of the group velocity V_g) in the MM with negative ε, μ, is $k < 0$. Thus, we obtain that $V_g V_{ph} < 0$, that means a 'negative phase behavior' (in contrast to the 'positive phase behavior' that corresponds to the condition $V_g V_{ph} > 0$) in MMs with negative ε and μ. The negative phase behavior leads to a possible manifestation of a number of important consequences. Among them there are: the property of negative refraction, the construction of ideal lenses (Veselago); the realization of controllable ultra-high resolution lenses based on hyperbolic MMs (Jacob *et al* 2007); the construction, using appropriate combinations of materials with negative and positive phase behaviors, of miniature (with dimensions smaller than $\lambda/4$, where λ is a wavelength) resonators and filters; surfaces and devices with adjustable coefficients of transmission, reflection, and absorption (Eleftheriades 2005), which provide a set of exceptional mass and size characteristics, very important for space communication systems and scientific equipment for space research on microsatellites, including antennas, etc.

The properties of the negative refraction, the negative phase behavior, and the wave dispersion inherent in MMs are taken into account when setting problems and choosing the methods used in the book. Moreover, these properties are among ones that determine the type of nonlinear waves (wave packets) and structures formed in nonlinear layered media, and the coefficients (and their signs) in nonlinear evolution equations that characterize the propagation of nonlinear wave packets, and are derived using the general method delivered in the book.

These and other outstanding properties of MM and plasma-like media have led to their occupation of a worthy niche called metaphotonics. In this area of science and technology, for example, Epsilon-Near-Zero/zero index MMs, plasma-like linear (Davoyan and Engheta 2019, Ji *et al* 2021, 2020) and nonlinear (Boardman *et al* 2017a) media are used; layered plasma-like media of the 'dielectric-narrow-gap semiconductor-dielectric …' and hyperbolic MMs with highly efficient (Haugan *et al* 2021), control, including those based on topological transformation of the characteristics of the waves propagating in the corresponding MMs (Rapoport *et al* 2021, Grimalsky *et al* 2021). It is worth, in this context, also emphasizing the possibility of achieving a high density of states and a significant increase in the intensity of single-photon sources due to the Purcell effect (Makarova *et al* 2017); the other attractive possibility is generating Rogue waves and hotspots in the presence of an active component in the composition of MMs (Boardman *et al* 2017b); the possibility of constructing super-resolution lenses with overcoming the diffraction limit due to the hyperbolic nature of dispersion (Liu *et al* 2007). There are other intriguing effects with electromagnetic waves in nonlinear metamaterials (Ji *et al* 2021, Chu *et al* 2018), like metasurfaces (Kivshar and Quo 2021) of extremely high-Q resonances of various types, which may include Mie resonances (Khattak *et al* 2019), Kerker (electric–magnetic) resonance in the nanoparticles and the system of nanoparticles (Liu and Kivshar 2018), lattice resonances (Lucido *et al* 2021, Yevtushenko *et al* 2021), bound states in the continuum (BIC) resonances (Rybin *et al* 2017), Fano resonances (Kivshar and Quo 2021) etc) with a high concentration of electromagnetic fields. There also exists the possibility to strongly control nonlinear wave effects, including generation of higher harmonics and vortices (Kivshar and Quo 2021). It is worth mentioning the possibility of creating metasurfaces based on Huygens sources using chiral metaparticles and generation of harmonics by an incident wave in a reflectionless mode (Rapoport *et al* 2013a); full multiplexing using phase-singular vortex structures with finite optical orbital momentum from microwave to optical ranges (Willner 2021, Kai Pang *et al* 2018, Zhao *et al* 2021, Yao and Padgett 2011); hypersensitive sensor properties (Kivshar and Quo 2021), etc. The applications of nonlinear waves in layered MMs and plasma media (in particular, using structures such as 'Dielectric-Graphene-Dielectric …' and 'Dielectric-Narrow-Gap Semiconductor-Dielectric …') allows us to provide nonlinear switching, threshold in amplitude, between total reflection mode and the regime full passage of pulses (Rapoport *et al* 2013b), a threshold jump of the focusing point with the formation of a nonlinear self-localized resonator and generation of hot spots (Rapoport *et al* 2014b), generation of space-time solitons (bullets), including two-dimensional bullets in ferromagnetic (Buttner *et al* 2000) and MM waveguides with magneto-optical control (Rapoport *et al* 2014c); amplification with saturation and competition between pulses and other synergistic effects in hyperbolic nonlinear controllable active MMs (Grimalsky *et al* 2019). We also note the effects of non-reciprocity in non-stationary (Elnaggar and Milford 2018) and active (Kodera and Caloz 2018, Morgado and Silveirinha 2021) media without a magnetic field; controllable magnetostatic spin wave effects in MMs with a magnetic field (Sharaevsky *et al* 2018). Many other original and useful linear and nonlinear effects are also possible. Relevant applications include creating devices and systems with unique properties such as nonlinear communications and signal processing devices, supersensitive sensors, holographic devices, super-resolution imaging (Hart *et al* 2018), and, in perspective, on-chip single-photon sources with applications for high-speed quantum communications and quantum information processing including quantum cryptography (Shen *et al* 2020). Many other applications would be possible, located at and even behind the forefront of science and technology—'nanophotonics of the future.'

1.2.2 Bi-anisotropy (gyrotropy), non-reciprocity and controllability of environments of the metamaterial media

Bi-anisotropic (Tretyakov *et al* 2010, Serga 2006), or gyrotropic, in terms of Fedorov (1976) are media in which the macroscopic electric field contributes to the macroscopic value of magnetic induction, and in turn the macroscopic magnetic field contributes to the macroscopic value of electric induction. In other words, both magnetic $\vec{H}$ and electric $\vec{E}$ components of the incident field excite both electric $\vec{P}$ and magnetic $\vec{M}$ polarization:

$$\vec{D} = \widehat{\varepsilon}\vec{E} + \widehat{\xi}\vec{H}, \vec{B} = \widehat{\varsigma}\vec{E} + \widehat{\mu}\vec{H}, \tag{1.1}$$

where $\widehat{\varepsilon}$, $\widehat{\mu}$ are the tensors of electric permittivity and magnetic permeability, respectively, $\widehat{\xi}$, $\widehat{\varsigma}$ are the tensors of the magnetic-electric and electric-magnetic cross-coupling, respectively (Tretyakov *et al* 2010). The condition of the absence of dissipation (negative or positive) of the media is the hermiticity $\widehat{\varepsilon} = \widehat{\varepsilon}^{+}, \widehat{\mu} = \widehat{\mu}^{+}, \widehat{\xi} = \widehat{\varsigma}^{+}$. For reciprocal media (which do not include ferro-magnetics or plasma in an external magnetic field) we have (Tretyakov *et al* 2010): $\hat{\varepsilon} = \hat{\varepsilon}^{\mathrm{T}}$, $\hat{\mu} = \hat{\mu}^{\mathrm{T}}$, $\hat{\xi} = -\hat{\varsigma}^{\mathrm{T}}$, where the index 'T' denotes the transposition operation. The effects of non-reciprocity are measured by the antisymmetric parts of the tensors $\widehat{\varepsilon}$, $\widehat{\mu}$ and the non-reciprocal magnetoelectric coupling tensor (non-reciprocal cross-polarization tensor of Tellegin), denoted $\widehat{\chi}$ by the following equations (by Lindell–Sihvola (Lindell *et al* 1994, Tretyakov *et al* 2010)):

$\vec{D} = \widehat{\varepsilon}\vec{E} + (\widehat{\chi}^{\mathrm{T}} - i\widehat{\kappa}^{\mathrm{T}})\vec{H}$, $\vec{B} = (\widehat{\chi} + i\widehat{\kappa})\vec{E} + \widehat{\mu}\vec{H}$, where $\widehat{\chi} = 0.5(\widehat{\varsigma} + \widehat{\xi}^{\mathrm{T}})$, $\widehat{\kappa} = 0.5i(-\widehat{\varsigma} + \widehat{\xi}^{\mathrm{T}})$, and the time dependence of the electromagnetic field is selected as $\sim \exp(i\omega t)$. The tensor $\widehat{\kappa}$ is called the chirality tensor, according to (Tretyakov *et al* 2010) and is responsible for reciprocal magnetoelectric phenomena. Media for which in relations (1.1) $\widehat{\varepsilon}$, $\widehat{\mu}$, $\widehat{\xi}$, $\widehat{\varsigma}$ are reduced to a scalar form, are called bi-isotropic media (Lindell *et al* 1994). For such media, the coefficient χ is called the bi-isotropic parameter of non-reciprocity, or the Tellegin parameter (Tellegin proposed the idea of a non-reciprocal medium with $\chi \neq 0, \kappa = 0$ (Lindell *et al* 1994)). Chiral media ($\chi = 0, \kappa \neq 0$) have been known since their discovery by Louis Pasteur (1860), and are sometimes called isotropic optically active media (Tretyakov *et al* 2010).

The so-called 'intrinsic gyrotropy', i.e. gyrotropy inherent in the media in the absence of external stationary fields, corresponds to the material relation (1.1) with magnetoelectric coupling tensors in the form $\widehat{\xi} = -i\widehat{\kappa}^{\mathrm{T}}$, $\widehat{\varsigma} = i\widehat{\kappa}$. In this case there are $\widehat{\varepsilon} = \widehat{\varepsilon}^{\mathrm{T}}, \widehat{\mu} = \widehat{\mu}^{\mathrm{T}}$. In other words, in Fedorov (1976) the material relations for media with their own gyrotropy correspond to those for reciprocal bi-anisotropic media. In the case of induced gyrotropy (e.g., for non-reciprocal ferromagnets, plasma in a magnetic field, or magnetooptical medium), the tensors $\widehat{\varepsilon}, \widehat{\mu}$ have the form (Sugakov 1974, Fedorov 1976, Engheta *et al* 1992) $\varepsilon_{\alpha\beta} = \varepsilon^{(0)}_{\alpha\beta} + ie_{\alpha\beta\nu}g_{e\nu}$, $\mu_{\alpha\beta} = \mu^{(0)}_{\alpha\beta} + ie_{\alpha\beta\nu}g_{m\nu}$. Here the index '0' denotes the non-gyrotropic diagonal parts of the corresponding tensors, $e_{\alpha\beta\nu}$ is a single pseudo-tensor of the third rank, $\vec{g}$ is the gyration vector, $\alpha, \beta, \nu \to x, y, z$. For the case of gyrotropy, induced by external magnetic fields, the gyration vector is proportional to the external magnetic field (Sugakov 1974, Fedorov 1976, Engheta *et al* 1992). Regarding the qualitative physical definition of the phenomena of gyrotropy and chirality, the first of them is formulated as follows. The gyrotropy is related to the optical activity of the medium and the property of rotation of the plane of polarization. But, as noted in Fedorov (1976), another definition is more general. Namely, the gyrotropy is a property of material media, which is manifested in the fact that homogeneous plane waves propagating in the medium in almost all directions, in the absence of dissipation, have the elliptical polarization. The chirality (Engheta *et al* 1992) is associated with the presence of 'right' or 'left' properties in objects. Chiral objects cannot be combined with their mirror image by means of translation and rotation operations. The electromagnetic property of chirality is associated with optical activity, rotation of the plane of polarization of a linearly polarized wave, birefringence, and circular dichroism. In the book mutual bi-anisotropic and chiral electromagnetic active and nonlinear MMs (metasurfaces (Kildishev *et al* 2013)), including bi-anisotropic and chiral meta-particles (Engheta *et al* 1992) (Ω-particles and canonical chiral particles, respectively) are considered.

The concept of controllability is important for a number of disciplines, in addition to electromagnetism and optics, chemistry, biology, and particle physics (Engheta *et al* 1992). An excellent material to produce highly nonlinear and effectively controllable metamaterials is graphene. The chirality (spirality) $\hat{h}$ in graphene is the projection of the pseudo-spin $\widehat{\sigma}$ operator on the direction of the linear momentum direction, or wave vector $\vec{k}$ (Allain and Fuchs 2011), $\hat{h} = (\vec{k} \cdot \hat{\vec{\sigma}})/k$. The proper value of chirality is equal to $h = \pm 1$. This value coincides with the index of the valence band or the conduction band in graphene, $h = \alpha = \pm 1$ (Allain and Fuchs 2011), and for the kinetic energy we have

$$E_{\text{kin}} = E - V_0 = \alpha\hbar v_{\text{F}}\left(k_x^2 + k_y^2\right)^{1/2} \tag{1.2}$$

In (1.2), E, V_0, v_{F} are total energy, electric potential applied to the graphene, and Fermi velocity; $k_{x,\,y}$ are the values of the components of the wavenumber in the plane of graphene. All chiral and gyrotropic media considered in this section have, in particular, the following common feature: for them there is a splitting of energy. For example, when EMWs with a given value of the wavenumber propagate in an electron plasma along a constant magnetic field $\vec{H}_0$, the waves with right and left directions of rotation have frequencies that differ in magnitude $\Delta\omega \sim g \sim H_0$ (Faraday effect (Sugakov 1974)), and the corresponding difference in photon energies is equal to $\Delta E = \hbar\Delta\omega$. A similar situation occurs for EMW in gyrotropic magnetooptical MMs and ferromagnets. For quantum mechanical waves of electron states in graphene, as shown in (1.2), the energy difference for the states of the electron in the two energy zones adjacent to the Dirac point, associated with the presence of chirality (Allain and Fuchs 2011), is equal to $\Delta E = 2\hbar v_{\text{F}}(k_x^2 + k_y^2)^{1/2}$. Taking into account the relevant material relations and the above-mentioned splitting of energy states, the unity of approaches to wave processes in gyrotropic and chiral media of different physical nature is ensured. In the book, with the help of such a unified approach, the formation and control, through external fields, of nonlinear wave structures in layered gyrotropic ferrite, magnetooptical MM, and graphene media are investigated.

1.2.3 Graphene as a metamaterial chiral medium

Almost simultaneously with the MMs, the experimental stage of the development of the physics of graphene and (multi)layered dielectric-graphene media began (Allain and Fuchs 2011). Due to the relativistic chiral nature of the carriers, there is an inhibition of inverse scattering in the graphene p–n junction, known as Klein tunneling (Gómez *et al* 2012). For an electron beam transmitted through a single p–n junction, negative refraction occurs (Allain and Fuchs 2011). Indeed, let there be a transition 'from left to right' (from area 1 to area 2) with a potential step V_0 (potential in area 2), such that $V_0 > E$, where E is the total energy, the axes x, y are directed normally to the transition and along the interface between areas 1, 2, respectively, and the eigenvalues of chirality $\alpha = \pm 1$ in regions 1, 2. Then, taking into account formula (1.2), it is possible to obtain for normalized wave vectors of electron waves incident (in region 1) and passing (in region 2), respectively, that $\vec{k}_i = (k_{x1}, k_{y1}) = E(\cos\varphi_i, \sin\varphi_i)$ and $\vec{k}_t = (k_{x2}, k_{y2}) = -(E - V_0)(\cos\varphi_t, \sin\varphi_t)$, moreover $k_{y1} = k_{y2}$. Accordingly, $k_{x1} > 0$, $k_{x2} < 0$, and the refraction is negative (Allen and Fuchs 2011).

Based on this property, the Veselago lens for electrons based on the n–p–n junction is proposed (Cheianov *et al* 2007), which can be considered as the contact of two regions of graphene with opposite values of eigenvalues of chirality. Accordingly, the graphene can be considered as a chiral MM. Electron beam supercollimation based on the periodic potential and electron beam focusing based on a circular p–n junction have been proposed (Gómez *et al* 2012). Veselago's graphene lenses have been proposed as filter systems for spin-polarized electron beams.

The formation of two-dimensional resonators and diffraction gratings for quantum mechanical waves of electronic states, effectively controlled by external electric and magnetic fields, the exchange interaction due to graphene contact with ferromagnetic substrate and spin–orbit interaction (Bai *et al* 2010) had been presented, in particular, in Boardman *et al* (2019).

Layered dielectric-graphene structures are used as controlled electromagnetic MMs (Zhuang *et al* 2015), and with the help of an external electric potential the surface concentration of carriers and complex conductivity of graphene is effectively controlled and the dispersion of EMWs in the system is managed. Layered graphene-dielectric structures can be used as controlled hyperbolic MMs (Poddubny *et al* 2013), as well as controlled linear Fabry–Perot resonators, in particular in the THz range (Kaipa *et al* 2012). The modulation of electromagnetic pulses in such nonlinear resonators, associated with the electromagnetic (current) nonlinearity of graphene had been studied, first, in Rapoport *et al* (2013b). The resonant properties of electron gas in *n*-InSb and graphene layers in magnetic fields for THz multilayered dielectric-plasma-like MMs have been presented in Grimalsky *et al* (2019).

1.2.4 Nonlinearity in layered media. Wave beams and packets with a moderate spectrum width. Spatial and temporal solitons and magnetooptical control of solitons. Control of wave structures in inhomogeneous and non-stationary gyrotropic and metamaterial media

The study of controlled nonlinear wave processes occurring in layered media is very relevant, and this is due to the emergence of new research objects, media, and conditions for which research is conducted, as well as the synthesis of different levels of research—from micro to macroscopic. It turns out that in layered MMs (Boardman 2011, Plum *et al* 2008, Alù *et al* 2007, Singh *et al* 2011, Tretyakov 2003, Liu *et al* 2008, Engheta 2006, 2010, Dehbashi *et al* 2010, Kildishev 2011, Cortes *et al* 2012, Eleftheriades 2005, Capolino 2009), gyrotropic and plasma media, there are certain classes of nonlinear phenomena, such as solitons (Fernandes and Silveirinha 2014), strongly nonlinear waves (Kozyrev *et al* 2014), vortices (Boardman *et al* 2010a), parametric interactions in nonlinear magnetoactive (gyrotropic) space plasmas (Kryshtal *et al* 2014), which, despite all the features of certain media, can be described by similar analytical (Rabinovich and Trubetskov 2000), numerical-analytical, and numerical (Taflove and Hagness 2005) methods.

As will be shown, an interesting and productive idea is the control of soliton pulses in magnetooptical layered MMs due to certain laws of change in time of the magnetization field. Another possible way to control magnetooptical solitons in MMs is also to create a transversely inhomogeneous distribution of the stationary magnetization field (Boardman *et al* 2005f). Before the research included in this book, the question of the propagation of nonlinear waves in layered MM media in the presence of bulk and surface nonlinearities has not been studied in detail by other authors (Rowland 1999, Rabinovich and Trubetskov 2000, Leblond 2001, Agrawal 1996, Boardman *et al* 1994, Nayfeh 1984, Lukomsky and Gandzha 2009, Boyd 2003, Wang *et al* 2009, Paul *et al* 2011, Sipe *et al* 1980).

This applies to the relative role of bulk and surface nonlinearities in the evolution of nonlinear waves, the processes of formation of wave structures in active nonlinear layered media, nonlinear processes in layered bi-anisotropic MMs, and the relationship between the characteristics of these waves and the characteristics of bi-anisotropic nonlinear MMs, etc.

The simplest equation for the envelope amplitude, which describes the evolution of a two-dimensional video pulse with a narrow spectrum, is a nonlinear parabolic equation. We assume that the conditions, $\Delta k_{x,\,y} \ll k_0$, $\Delta\omega_{x,\,y} \ll \omega_0$, where x, y are the directions in which the pulse propagates and the direction transverse to the direction of propagation, respectively, $\Delta k_{x,\,y}$, $\Delta\omega$, k_0, ω_0 are the width of the spectra with respect to wavenumbers and frequency, and the values of the carrier wavenumber and pulse frequency, respectively. We assume that either the medium is isotropic or the direction x coincides with the anisotropy axis of the uniaxial medium. The dimensional and normalized versions of the nonlinear parabolic equation has the forms, respectively (Kadomtsev 1988)

$$i\frac{\partial E}{\partial t} + iv_g\frac{\partial E}{\partial x} + \frac{1}{2}g_1\frac{\partial^2 E}{\partial x^2} + g_2\frac{\partial^2 E}{\partial y^2} + \alpha|E|^2E = 0 \tag{1.3}$$

$$i\frac{\partial \overline{E}}{\partial \overline{t}} + i\frac{\partial \overline{E}}{\partial \overline{x}} + \frac{1}{2}\frac{\partial^2 \overline{E}}{\partial \overline{x}^2} + \frac{\partial^2 \overline{E}}{\partial \overline{y}^2} + |\overline{E}|^2\overline{E} = 0 \tag{1.4}$$

In equation (1.4), the values with a dash above are normalized. The first two terms in (1.3) correspond to the approximation of the geometric optics, the third, fourth and fifth terms correspond to the deviation from the approximation of the geometric optics and describe the effects of the linear dispersion, diffraction and cubic nonlinearity, respectively (Prokhorov 1988, Kalinin and Bayonets 1990, Boyd 2003, Vinogradova *et al* 1979). The coefficients $g_{1,2}$, α are called the coefficients of (linear) dispersion and diffraction, and nonlinearity, respectively (Boyle *et al* 1996). The coefficients $g_{1,2}$ are determined from the linear dispersion equation. If we use the 'running' coordinate frame $t' = t - x/v_g$ and consider one-dimensional motion, the normalized evolution equation takes the form (1.4) with $\bar{t} \to \bar{t}'$ (where $\bar{t}'$ is the normalized time in the running coordinate frame) and in the case $\partial/\partial\bar{y} = 0$ this equation reduced to the canonical nonlinear Schrödinger equation (Agrawal 1996).

Under the conditions $g_1\alpha > 0$, $g_2\alpha > 0$ (criteria of Lighthill and of self-focusing, respectively) there is a modulation instability (self-compression of the wave packet in the longitudinal direction) and self-focusing (localization in the transverse direction), respectively. In Boyle *et al* (1996), a mode of self-channeling of stationary beams of the backward volume magnetostatic waves (BVMSW) in gyrotropic layers (ferrite films) was found, under the conditions of compensation of diffraction and nonlinearity with relatively weak dissipation and formation of a nonlinear channel. In the case of nonlinear pulses, under the conditions of focusing both in the longitudinal and transverse directions and the presence of saturated nonlinearity, three-dimensional stable time-space solitons, so-called optical bullets (optical spheres) may form in an infinite isotropic medium (Skarka *et al* 1997). In Grimalsky *et al* (1997), Bauer *et al* (1997), and Buttner *et al* (2000) bullets based on the other principle than optical ones have been presented, for the first time. Namely, in these papers two-dimensional time collapse has been demostrated for stationary beams and pulses of BVMSW in gyrotropic layers (ferrite films) with a dispersion similar to those possessed by EMWs in MM waveguides. A possibility of the formation of BMSW bullets had been shown as well, in a gyrotropic layer, without saturation, but with a significant role of dispersion and dissipation (Grimalsky *et al* 1997, Bauer *et al* 1997, and Buttner *et al* 2000). In Boardman and Xie (2003) the possibility of magnetooptical (MO) control of spatial solitons in optical waveguides based on a positive phase medium in a transversely inhomogeneous magnetic field is shown. Note that the NSE equation (1.4) with nonzero coefficients (excluding the linear diffraction) does not and even cannot include any MM specificity, since all normalized equations are the same, regardless of the specific material.

The linear diffraction term adds another possibility of choosing the sign, but this is rather trivial and will not be specially discussed for equation (1.4) in the future. The entire specificity of the propagation of nonlinear wave packets in MMs manifests itself only when the higher linear and nonlinear effects are taken into account, as well as the effects of controlling the characteristics of EMW, such as MO interaction. This specificity is reflected by adding higher nonlinear terms and terms describing the aforementioned effects associated with the control of wave processes, in particular, to the left-hand sides of evolutionary equations like (1.3) or (1.4). Note that the first higher terms adding to the left-hand pat of the evolution equation, describe higher-order linear effects and nonlinear dispersion, diffraction, and the Raman interaction (Boyd 2003, Agrawal 1996, Boardman *et al* 2000b). An important approach from the point of view of the nonlinear wave control (including solitons) is one based on the manipulation of dispersion and diffraction characterizing by the coefficients ($g_{1,2}$). Appropriate common terms in nonlinear optics are the 'dispersion and diffraction management' (Ballav and Chowdhury 2006, Subha *et al* 2007).

In this book, the results of studying the MO control of spatial and temporal solitons in layered MMs (Boardman *et al* 2010b, 2010c), as well as the MO control, combined with the dispersion and diffraction management (Rapoport *et al* 2014c) are described. In the previous papers, it was shown that nonlinear dispersion affects the modulation instability and also leads to self-steepening of the leading edge of the pulse in nonlinear MMs with negative phase behavior (Scalora *et al* 2005). It is also known that the propagation of nonlinear pulses is influenced by the Raman scattering and the third-order linear dispersion (Agrawal 1996).

This book also presents the influence of higher-order nonlinear effects on the propagation of solitons in layered gyrotropic ferrite media.

The formation, collapse, amplification, and wave front reversion of two-dimensional nonlinear structures, including possible bullets, in layered nonlinear active gyrotropic ferromagnetic media under parametric pumping has been proposed in Grimalsky *et al* (2000a, 2000b, 2001), and then formation, amplification and stabilization of two-dimensional spatio-temporal solitons (bullets) in MM waveguides have been proposed in Rapoport *et al* (2014c). A possibility of controlling characteristics of linear and nonlinear magnetostatic waves (MSW) in ferrite films by means of spatially inhomogeneous bias magnetic field has been shown, respectively, in (Vashkovsky *et al* 1990) and (Grimalsky and Rapoport 1994, Grimalsky *et al* 1998). Nonlinear effects of higher order and magnetooptical control of bullets, and the possibility of stable propagation and amplification of bullets in an active periodic medium are described in this book.

At the same time, it is known that in a periodic medium without amplification (passive), the application of the periodicity of the parameters of the medium can lead to a significant increase in the distance at which the bullets can (quasi-) stably propagate without destruction. In a homogeneous medium, the bullets must collapse after propagation over a certain distance, since (2 + 1) time-space solitons are only quasi-stable (Malomed *et al* 2005). In the notation '(2 + 1)', standard for nonlinear optics, the first digit (2) means the number of spatial, and the second (1) the presence of time coordinates for the nonlinear wave structure (two-dimensional pulse) under consideration.

Active nonlinear metasurfaces (Tretyakov *et al* 2010, Minovich *et al* 2015, Yoo and Lim 2014) constitute a very important type of MMs. This applies in particular to multifunctional chiral nonlinear active metasurfaces. These studies are very relevant. Along with versatility, including the provision of unique controllable properties, such active surfaces are free from certain problems inherent in MM, at least in terms of energy losses due to the propagation of EMWs inside the material.

1.2.5 Active metamaterials. Problems of spatial amplification and convective instability

For assured use of and even for the study of nonlinear waves and, in particular, solitons, it is necessary to solve the fundamental problem of overcoming damping in MMs, where the most interesting modes (with negative phase, small values of dielectric permittivity and magnetic permeability, etc) occur near resonances of metaparticles that determine the properties of MMs. Such attenuation is immanent to MMs, due to the resonant properties of both the metaparticles and/or the structure of the MMs as a whole. To solve the problem of attenuation, a number of different options for giving the MMs the property of 'activity' are proposed, which implies the possibility of enhancing or compensating of the dissipation of EMW, see, for example, Noginov *et al* (2007) and Zheludev *et al* (2008).

But, as known from Fedorchenko and Kotsarenko (1981) and Lifshitz and Pitaevskii (1981), the spatial amplification and the absence of absolute instability (generation) is achieved for active media only under certain conditions that have not been previously studied for MMs. Here it is interesting to note the following paradoxical situation that arises for active MMs. In addition to other possible sources of the absolute instability in an active system (Fedorchenko and Kotsarenko 1981), there is one (Lifshitz and Pitaevskii 1981) that corresponds to $|\vec{k}| \rightarrow \infty$, where $|\vec{k}|$ is the absolute value of the wave vector characterizing a potentially dangerous perturbation in terms of the development of the absolute instability. But under such a condition, the MM approximation (Tretyakov *et al* 2010) is inapplicable. According to such approximation there should be $|\vec{k}|^{-1} \gg d$ where d is the maximum values from the following two: one of them is a distance between the metaparticles and the other one is a characteristic size of the metaparticles. Note that, at the same time, for the carrier wavenumbers of the propagating wave packets, the MM approximation is satisfied.

This means that for the study of active MMs, including the conditions of absence of the absolute instability, it was necessary to go beyond the MM approximation. Thus, it becomes necessary to consider wave processes at the scale levels of one metaparticle or one period of the MM, if it is a certain periodic structure. This was done for the first time in the papers discussed in the present book. Another case, when the development of the theory of MMs requires going beyond the formal limits of the MM approximation, is a fairly general approach to layered bi-anisotropic MMs. This approach should begin with the consideration of active nonlinear bi-anisotropic metaparticles, for which it is necessary to first go beyond the MM approximation.

Next, it is necessary to homogenize the medium, i.e. to move within the limits of the MM approximation, but taking into account the nonlinearity, activity, and resonant nature of the medium associated with the resonance of metaparticles. And the third step is to determine the characteristics of nonlinear wave processes in a layered MM medium in the presence, in general, of bulk and surface nonlinearities. Then it is possible to theoretically investigate nonlinear wave processes in a layered medium, starting from the characteristics of a single metaparticle, taking into account the influence of both the external field incident on the system and other metaparticles. The adequate method for studying the evolutionary dynamics of wave processes in a fairly general class of nonlinear layered media, including MMs, was not developed previously (Tretyakov *et al* 2010, Capolino 2009). The corresponding method is described in this book.

1.2.6 Field concentration, transformation optics, strong and resonant nonlinearities, nonlocality and spatial dispersion in layered media

One of the practically important opportunities provided by MMs is to obtain a significant field concentration that can be used to create highly sensitive bio- and chemical sensors (Tretyakov *et al* 2010), field or energy concentrators (Narimanov and Kildishev 2021, Sadeghi *et al* 2014), or devices for feeding nonlinear antennas. At a significant field concentration for layered MMs, the effects of strong nonlinearity are also manifested, when the approximation of amplitudes that change slowly does not work, and the effects of spatial dispersion (nonlocality) become important (De Ceglia *et al* 2013). The effects of the spatial dispersion can be important for a linear, in particular periodic medium, say hyperbolic MM, with a not very small value of the ratio of the period of the structure and the wavelength (Chebykin *et al* 2011).

The effects of the spatial dispersion and nonlinearity are very relevant for MM plasmonics (Pollard *et al* 2009, Jacob and Shalaev 2011, Gramotnev and Bozhevolnyi 2010, Wurtz *et al* 2011), which, in turn, is very relevant for the construction of future optical computers, a necessary element of which will be nonlinear active plasmonic transmission lines. There are some examples of the presence of nonlinearities in additional (so-called auxiliary) boundary conditions associated with the spatial dispersion (Pekar 1962, Gurevich and Melkov 1996). Since no universal methods have been developed to study an influence of the above indicated effects on wave propagation in layered nonlinear media, the problem of developing such methods is very relevant.

There are various ways to ensure a high field concentration, such as the use of nanoparticles or nanoantennas with edge effects that provide such a concentration (Zayats 2008, Zhao *et al* 2011) and high sensitivity sensors based on (nanoplasma) MMs constructed using such particles, or nanoantennas, respectively. One of the most promising approaches for the construction of nanomaterials with a high field concentration is the transformation optics (Leonhardt and Philbin 2009, Rahm *et al* 2008, Kildishev 2011, Yang *et al* 2009). Formally, the transformational optics is reduced to a coordinate transformation, which is in fact a special case of coordinate transformations adopted in the theory of relativity, which provides covariance of the form of the laws of electrodynamics (Sugakov 1974).

The transformation of coordinates ($x^{\alpha} \to x^{\alpha'}$) is determined by the Jacobian transformation matrix $\Lambda_{\alpha}^{\alpha'} = \partial x^{\alpha'}/\partial x^{\alpha}$, and the transformations of tensors of electric permittivity and magnetic permeability are determined by formulas $\varepsilon^{i'j'} = |\det(\Lambda_{\alpha}^{\alpha'})|^{-1}\Lambda_i^{i'}\Lambda_j^{j'}\varepsilon^{ij}$, $\mu^{i'j'} = |\det(\Lambda_{\alpha}^{\alpha'})|^{-1}\Lambda_i^{i'}\Lambda_j^{j'}\mu^{ij}$. Note that it is assumed here that the magnetoelectric interaction is absent. In the case of an isotropic medium, the above indicated transformations have the form $\varepsilon^{i'j'} = |\det(g^{i'j'})|^{-1} g^{i'j'}\varepsilon$, $\mu^{i'j'} = |\det(g^{i'j'})|^{-1} g^{i'j'}\mu$ where the metric tensor is $g^{i'j'} = \Lambda_k^{i'} \Lambda_l^{j'} \delta^{kl}$. As a result of the satisfaction of the above-mentioned relations, the spatially inhomogeneous tensors $\widehat{\varepsilon}$, $\widehat{\mu}$ provide the propagation of rays (energy) of EMWs in the desired directions (Rahm *et al* 2008, Schurig *et al* 2006). Based on this method of modeling and modern technologies, it is possible to ensure the implementation of devices, the achievement of which by other methods is questionable (Schurig *et al* 2006).

A convenient method of analysis of rays propagating under conditions of a created inhomogeneity is the method of the geometric optics (Schurig *et al* 2006).

On the other hand, in Narimanov and Kildishev (2021) and Rahm *et al* (2008) the equations of the geometric optics are reduced to equations of the Kepler type, which describe the motion of a particle in the field of an effective potential well. Note that these two methods—the transformational optics and the geometric optics, using the effective Hamiltonian, which describes the propagation of light, are very ideologically similar, as both are focused on ensuring the trajectory of light rays of a given type. In particular, in Narimanov and Kildishev (2021), to ensure the operation of a linear two-layer cylindrical (effectively two-dimensional) light concentrator, in the outer inhomogeneous region of such a concentrator a certain inhomogeneity of the dielectric constant is selected. Note that the dielectric permittivity is supposed to be dependent on the radius of the cylindrical region r, $\varepsilon = \varepsilon(r)$.

The effective Hamiltonian, which describes the propagation of light, has the form $H = p_r^2/[2\varepsilon(r)] + m^2/[2\varepsilon(r)r^2]$ where p and m are effective radial and angular components of the momentum. Such a Hamiltonian has the form of one for a mass in the central potential $V_{\text{eff}} = (1/2)(\omega/c)^2[\varepsilon_0 - \varepsilon(r)]$ (where ε_0 is the dielectric permittivity in the space surrounding the cylindrical concentrator). If the dependence of the dielectric permittivity in the outer region of the concentrator is chosen in the form $\varepsilon(r) \sim r^{-2}$, the proposed system is the optical equivalent of a trap system with a perfectly capturing potential (regardless of the angle of incidence of the particle).

In the model (Narimanov and Kildishev 2021), a linear material with strong dissipation was placed in the central homogeneous cylindrical region of the concentrator, which simulates some optical energy absorber, a linear electromagnetic 'black hole'. The question arises, what new and interesting wave phenomena can be expected in a nonlinear and active electromagnetic black hole when the central region includes a nonlinear focusing material? Moreover, since there is a pronounced effect even in a linear electromagnetic black hole, a very strong effect can be expected, including one corresponding to strong nonlinearity (when a slowly varying amplitude approximation is not suitable for describing wave processes). This problem is considered in the book.

The most significant and practically the most interesting features of graphene are unprecedented controllability by external fields and quite strong and broadband nonlinearity, at least from THz to optical frequencies. Since EMWs propagating along some layered structure with nonlinear layers of graphene are subject to quite strong attenuation, it is particularly interesting to use nonlinear effects when EMWs propagate in the direction transverse to a relatively multilayer nonlinear dielectric-graphene system. In Kaipa *et al* (2012) it was shown that the coefficients of transmission and reflection of THz EMWs through a multilayer linear structure are effectively controlled by an electric field (potential) applied to the graphene.

The surface impedance of periodic structures with graphene can be either inductive or capacitive, and switching between these modes can be achieved due to the electric potential applied to the graphene. Electromagnetic nonlinearity of current in graphene was demonstrated in Dong *et al* (2013). This can be seen, in particular, from the formula (Dong *et al* 2013) $\vec{J}_{\mathrm{p}} = (-e)v_{\mathrm{F}}[(p_x + eA)^2 + p_y^2]^{-1/2}(p_x + eA, p_y)$. In deriving this formula, a plane EMW with an electric field component was considered $E_x = -\partial A/\partial t$, where A is the vector-potential component. It was believed that this wave, propagating in a flat-layered system 'graphene-dielectric', is incident normally on the layer of graphene lying in the plane xy; v_{F}, e, $p_{x,y}$, $\vec{J}_{\mathrm{p}}$ are Fermi velocity in graphene, the electron charge, pulse components in the graphene layer, and the current component in the pulse space, respectively.

At the time of the beginning of this book, only linear effects in the propagation of electromagnetic pulses in a multilayer Bragg open resonator (a multilayer system 'dielectric-graphene') were investigated; nonlinear effects have not been investigated (Iorsh *et al* 2012, 2013). Note that this system, in the presence of current nonlinearity in graphene, due to almost (on the scale of the electromagnetic wavelength) infinitesimal layer thickness of graphene, belongs to those for which it is impossible to apply the method of slowly varying amplitudes, at least in space. In the book the corresponding nonlinear effects are investigated for the first time.

1.2.7 Hyperbolic metamaterials

Hyperbolic MMs are very popular and relatively easy to manufacture (Poddubny *et al* 2013, Guo *et al* 2020). A typical example is a medium with periodically alternating layers (for example, plasma layers of, say, metal or semiconductor, and dielectrics, with positive and negative values of the dielectric permittivity). Let us illustrate the important properties of hyperbolic MMs (Poddubny *et al* 2013, Jacob *et al* 2010, Jacob and Shalaev 2011) with a simple example. Consider the appropriate plane-layered medium and use a coordinate system where the axes x, y lie in the plane parallel to the layers and axis z is directed is along the normal to the layers.

For a hyperbolic MM, under certain conditions, including the performance of an MM approximation (layer thicknesses are much smaller than the wavelength), we obtain, in particular, $\varepsilon_{xx} > 0$, $\varepsilon_{zz} < 0$. Consider E (TM) (transverse magnetic) EMW with field components (E_x, H_y, E_z) propagating in a plane xz with a wave vector $\vec{k} = (k_x, 0, k_z)$ and a frequency ω. The dispersion relation has the form

$$\varepsilon_{xx}^{-1}k_z^2 + \varepsilon_{zz}^{-1}k_x^2 = k_0^2 \tag{1.5}$$

In (1.5), $k_0^2 = \omega^2/c^2$, c is the speed of light. For 'traditional' isotropic materials with a positive phase, for which $\varepsilon_{xx} = \varepsilon_{zz} = |\varepsilon|$, we obtain from (1.5) $k_z^2 + k_x^2 = |\varepsilon|k_0^2$. Therefore, the surface of constant energy in $\vec{k}$-space is a sphere. With respect to the hyperbolic MM with $\varepsilon_{xx} > 0$, $\varepsilon_{zz} < 0$, (1.5) can be re-written in the form $k_z^2 - |\varepsilon_{xx}/\varepsilon_{zz}|k_x^2 = |\varepsilon_{xx}|k_0^2$, so that the surface of the constant frequency is a hyperbola. Hence the possibility of a number of very important applications of hyperbolic MMs follows. One of such applications is the possibility of creating lenses with ultra-high (subwavelength) resolution, which follows from the following qualitative considerations (Poddubny *et al* 2013).

If the image is parallel to the plane, then for 'traditional' material the resolution is determined, by the order of value, by the size $\Delta x_0 \sim k_{x0}^{-1}$. Here $k_{x0} \approx |\varepsilon|^{1/2}k_0$, corresponds to the boundary between propagating and evanescent waves, which can no longer propagate along the axis z perpendicular to the source (which is considered flat) and the lens. Accordingly, the information transmitted by evanescent waves is lost. Nevertheless, in principle, such information can be stored using Veselago's ideal lenses (Eleftheriades 2005).

But in practice it is quite difficult to create such a lens. A much better solution is to use hyperbolic MMs. For them, according to equation (1.5), as shown above, evanescent waves, in the lossless material, do not exist. In practice, the subwavelength resolution is experimentally achieved in this way (Poddubny *et al* 2013). The replacement of a spherical or elliptical (for ordinary materials with a positive phase) constant frequency surface with a hyperbolic one (for hyperbolic MMs) is also associated with a whole class of important practical applications. The number of quantum states in the frequency range from ω to $\omega + \Delta\omega$ for a material with a positive phase and a surface of constant frequency in the form of a sphere (ellipse) is finite. In contrast, number of quantum states for hyperbolic MMs with a constant phase surface in the form of a hyperbola is, ideally, an infinite quantity.

Hence, in the implementation of various quantum devices based on the Purcell effect (Poddubny *et al* 2013), i.e. quantum resonance phenomena with a quantum source placed near a material with a high density of quantum states, hyperbolic MMs provide a significant, an order of magnitude or more, increase in the efficiency of the quantum yield. Thus, hyperbolic MMs are very promising for creating lenses, fluorescent light sources, collection and use of secondary heat energy, increasing the intensity of single-photon sources required for quantum cryptography (Shalaginov *et al* 2015) and other important applications. The use of nonlinear hyperbolic MMs with a combination of the principles of increasing the concentration of energy states and the concentration of the field is promising.

1.2.8 Wave processes in layered media of different physical nature in the presence of both volume and surface nonlinearities

Nonlinear layered gyrotropic (Cherkasskii and Kalinikos 2013, Serga 2006), plasma and/or MM media may include bulk and surface (Paul *et al* 2011, Wang *et al* 2009, Shramkova and Schuchinsky 2012a, 2012b) resonant (Lapine *et al* 2003, Zeng *et al* 2012) or nonresonant nonlinearities. Surface nonlinearities can be given by surface electric/magnetic polarizations/equivalent polarization currents, as well as nonlinear conductive currents, for example for plasma layered media. It is specific here that surface nonlinearities can be even stronger than volume ones, and in order to adequately take into account such nonlinearities, it is necessary to consider the specifics of the corresponding polarizations and derive and solve the corresponding material nonlinear equations. Moreover, since at the same time we are talking about nonlinear surface boundary conditions, it is necessary to develop a special method where it would be possible to very clearly and physically distinguish surface nonlinearity, in contrast to volume one, calculate its contribution to complete nonlinearity and embed it in nonlinear evolution equations. This task is very non-trivial.

Attempts to solve it using standard methods such as the reduction perturbation method (RPM) (Nayfeh 1984) or using the method of indeterminate coefficients (Lukomsky and Gandzha 2009, Lukomsky 1995, Lukomsky and Rapoport 1994) lead to very time-cumbersome calculations. Moreover, such an application of the pointed above methods remains very specific not only for certain physical media, but also for certain nonlinear structures. In practice, it should be expected that, for example, with the transition to a larger number of layers in the structures, the use of such a method becomes generally very unlikely. The use of another media and types of nonlinear polarization, nonlinearity of a higher order, and so on, itself it is in this context a new task, the complexity of which is difficult to predict.

It follows that such methods are not suitable for our universal approach. The analytical or analytical-numerical method suitable for us, in the presence of volume and/or surface nonlinearities, must be universal and suitable for different types of nonlinearities, material relations and media of different physical nature. It is intuitively clear that this is possible only if the new method is based on very general physical principles and characteristics of nonlinear wave structures. For such characteristics it is logical to choose the energy flux and (linear) dispersion relation, following the ideas (Rabinovich and Trubetskov 2000, Bonchkovsky 1954, Vainshtein 1966) and developing them for gyrotropic, plasma and bi-anisotropic nonlinear layered media, and checking on simple examples of isotropic magneto-electric dielectric (Tretyakov 2003) media.

In this book, a method is described that is very effective for deriving evolution equations, in particular for nonlinear electromagnetic and magnetostatic (Chen *et al* 1994, Lukomsky 1978, Slavin and Rojdestvenski 1994, Slavin *et al* 1994, Xia *et al* 1998, Zaspel *et al* 2001) waves in layered structures including bi-anisotropic and gyrotropic nonlinear layers, in the presence of both bulk and surface nonlinearities. It is also shown that this method is very general and suitable for the study of nonlinear waves of different physical nature propagating along layered structures.

In order to apply this method in a fairly general situation for waves in layered structures, we need to know only the following: (a) the linear dispersion law for these waves in a layered structure with a given geometry; (b) the nonlinearity must be known and it must be relatively weak or moderate, so that the method of slowly varying amplitudes is applicable; (c) this method works directly for nonlinear waves propagating in the form of narrow wave packets. Thus, it should be noted that taking into account the higher approximations of linear and nonlinear dispersion and diffraction, the condition of the narrowness of the wave packet becomes less rigorous.

A method for derivation of nonlinear evolution equations can be developed, based on those methods used to derive relations for excitation of linear waveguides and open resonators by given external sources (currents) (Bonchkovsky 1954, Vainshtein 1966) and for obtaining energy equations (the Poynting theorem) for nonlinear waves in layered structures (Grimalsky and Rapoport 1995, Rowland 1999). Namely, for the coefficients $C_{S,\,-S}$ that determine the amplitudes of the eigen sth (direct) and ($-s$)th (counter propagating) electromagnetic modes of a regular (longitudinally-homogeneous) waveguide with a cross-sectional area $S_{\perp}$ (Bonchkovsky 1954, Vainshtein 1966) are described by the following relations:

$$dC_s/dz = N_s^{-1}\int_{S_\perp}(\vec{j}^{\,e}\vec{E}_{-s} - \vec{j}^{\,m}\vec{H}_{-s})dS,\quad dC_{-s}/dz = N_s^{-1}\int_{S_\perp}(\vec{j}_e\vec{E}_s - \vec{j}_m\vec{H}_s)dS \quad (1.6)$$

In the relations (1/6), $N_s = (c/2\pi)\int_{S_\perp}[\vec{E}_s\vec{H}_s]\vec{l}\,dS$ is the norm, $\vec{E}_s$, $\vec{H}_s$, c, $\vec{l}$, $\vec{j}^{\,e,\,m}$ are the electric and magnetic fields of the sth mode, the speed of light, the unit vector in the direction coinciding with the normal to the cross section of the waveguide and the direction of propagation of the direct EMW (sth eigenmode), and the densities of effective electric and magnetic currents, respectively. These currents are considered as 'external sources', which are supposed to be given and are not described by a set of equations within the problem under consideration.

It is seen that the norm N_s, included in equation (1.6), is proportional to the energy flow through the cross section of the waveguide for the sth mode. To excite the eigenmodes in a bulk resonator, when the frequency of the external source ω is equal to the frequency ω_s of the eigenmode, the equations obtained in Bonchkovsky (1954) and Vainshtein (1966) are reduced to the form

$$dC_s/dt = (1/2)N_V^{-1}\int(\vec{j}_e\vec{E}_s - \vec{j}_m\vec{H}_s)dV \quad (1.7)$$

In (1.7), the norm $N_V = (1/4\pi)\int\varepsilon\vec{E}_s^{\,2}dV = -(1/4\pi)\int\mu\vec{H}_s^{\,2}dV$ is proportional to the energy concentrated in the resonator, ε, μ, V are the dielectric permittivity and the magnetic permeability of the medium filling the resonator, and the volume of the resonator, respectively. As shown in Grimalsky and Rapoport (1995) and Burlak *et al* (1990) when using the energy method, in particular, for the process of three-wave parametric magnetooptical interaction in a layered gyrotropic medium with the participation of two electromagnetic and one magnetostatic waves and taking into account the quadratic nonlinearity, the system of equations has the form

$$\partial W_i/\partial t + \partial P_i/\partial\xi = \alpha_i|\omega_i|I \quad (1.8)$$

In equation (1.8), P_i, W_i and ω_i are the energy flux, the energy density (per unit cross section of the wave beam), $\alpha_1 = -1$, $\alpha_{2,3} = 1$; the index '1' corresponds to the wave with the maximum frequency; it is assumed that the conditions of parametric synchronism, I is proportional to the corresponding overlap integral. As shown in Grimalsky and Rapoport (1995) and Burlak *et al* (1990), the equation of the form (1.8) for the three-wave parametric interaction of magnetostatic waves and electromagnetic ones with magnetostatic waves, respectively, in layered gyrotropic media can also be obtained using the method of Lagrange formalism. In the corresponding partial case, the above equations lead to Manley–Rowe relations (Rabinovich and Trubetskov 2000).

The advantages, in terms of the tasks of this book, of the methods (Bonchkovsky 1954, Vainshtein 1966), as well as (Grimalsky and Rapoport 1995) are their universality for a wide class of 'external sources', and the ability to take into account, using a very 'physically transparent' energy method, effects both volumetric and surface nonlinearities, respectively. In the absence of a three-wave interaction, the obtained equations are reduced simply to the law of conservation of energy for noninteracting waves, as shown in Rowland (1999). The methods (Bonchkovsky 1954, Vainshtein 1966) are linear, and the energy method (Grimalsky and Rapoport 1995, Rowland 1999) is applicable to the study of only the intensity of (nonlinear) waves, but not their phase.

With these methods it is impossible to study the evolution of the complex amplitude of the envelope of nonlinear waves. A method developed on the basis of generalization of ideas (Bonchkovsky 1954, Vainshtein 1966, Grimalsky and Rapoport 1995, Rowland 1999), see also equations (1.6), (1.8), which makes it possible to study the evolution of the envelope of a wave packet with a moderate spectrum width in a layered medium with bulk and surface nonlinearities, is presented in chapter 2 of the book. In fact, the role of external sources in the method of excitation of linear waveguides and resonators (Bonchkovsky 1954, Vainshtein 1966), see also equations (1.6), (1.7), is played by effective nonlinear sources. Moreover, the obtained relations will include, generally speaking, nonlinear boundary conditions, as well as both volume and surface nonlinearities.

The contribution to the surface nonlinearity from the effects connected to the spatial dispersion should also be included in the formulation of this method. The zero and second harmonics are also taken into account with the corresponding (nonlinear) boundary conditions (which is impossible using only the field averaging in the direction normal to the layered structure, as is done in Agrawal (1996) and Boardman *et al* (1994)), and this procedure in the developed method is significantly less cumbersome than in the method (Leblond 2001) in particular for thin gyrotropic layers. The method described in the book also includes the possibility of obtaining approximate expressions for higher nonlinear terms, in particular nonlinear dispersion, as well as the estimation of the influence of nonlinear diffraction in layered media.

1.2.9 Metamaterial approach

The MM approach (Smith *et al* 2004, Rawat *et al* 2016) has been developed and used to model seismogenic hydrodynamic–electromagnetic perturbations in the layered gyrotropic active system of lithosphere–atmosphere–ionosphere (LAI) (Lehtinen 2012, Zafiris and Sigalas 2015, Rapoport *et al* 1998). The approach to MM at the beginning of the development of the discipline involved the modeling of artificial components with a specific structure to ensure non-trivial electromagnetic properties, from ultralow frequencies to optical range. At the same time, special values of material parameters (ε, μ) and electromagnetic properties (for example, the activity of a media, the hyperbolic dispersion, high field concentration, etc) were provided. A well-known example of using a MM approach is the creation of lenses with subwavelength resolution.

Namely, it was possible to reproduce in images obtained with an ideal Pendry lens (Tretyakov 2003) or lenses based on hyperbolic MMs (Jacob and Shalaev 2011), complete information about the object to be imaged. This also applies to the information transmitted by reactive modes (those that do not propagate in the media, but are localized in the region of the excitation). As a result, the subwavelength resolution becomes achievable. In this case, the fields of reactive modes are effectively 'amplified' (actually, increase in the direction from the object to the image) in the MM of an ideal lens, which compensates for their reactive attenuation in the media with a positive phase velocity. In contrast to known approaches, the unified MM approach involves the use of both electromagnetic and acoustic wave phenomena in the media of different physical nature, including active ones, with frequencies and wavelengths that differ by many orders of magnitude.

A characteristic feature of the MM approach is that the problem-solving ideas which are characteristic for MMs are transferred from one medium and frequency range to another, despite the seemingly huge difference between the respective physical processes, as well as temporal and spatial scales. For example, the realization of the negative refraction and the creation of ideal lenses are possible using acoustic MMs, two-dimensional graphene, metasurfaces and three-dimensional solid-state MMs, in acoustic/ultrasonic, THz or IR, and optical ranges, respectively. To model the corresponding wave processes in such media, in particular, MM homogenization methods are used, as well as a full-wave approach taking into account the entire spectrum and the possibility of resonances taking into account reactive modes.

A striking example of an MM approach with the transfer of ideas proposed earlier to wave processes in the LAI system to MMs is the combined method of 'complex geometric optics—full-wave nonlinear solution of the Maxwell equations' for mesoscale MMs proposed and developed in the book. In one region of such media, the scale of inhomogeneity is much larger than the wavelength, which allows the use of complex geometric optics, and in another one, the value of the ratio between the scale of inhomogeneity and wavelength is such that it implies the need to solve Maxwell equations taking into account the full spectrum of waves. These ideas are a development on the nonlinear case of the approach proposed earlier for seismogenic magnetohydrodynamic (MHD) perturbations of the UHF range in the LAI system (Grimalsky *et al* 1999, Rapoport *et al* 2012b, 2012h, 2013c, 2014a, 2014b, Rapoport 2012e, 2013d, Sharma *et al* 2015). The proper details are described in the book.

The use of the metamaterial approach makes possible taking into account the entire spectrum of waves in media of various physical nature; to model quasi-two-dimensional vortices in the ionosphere as a gyrotropic nonlinear medium (Aburjania *et al* 2005, 2006; Aburjania and Chargazia 2011); use the NEELS method to qualitatively explain seismogenic disturbances in EMW characteristics (such as amplitude and phase) in the Earth–ionosphere waveguide (Molchanov and Hayakawa 1998, Rapoport *et al* 2006); to take into account the presence of reactive modes (Eleftheriades 2005) and the gravitational branch of atmospheric gravitational waves (AGWs) (Landau and Lifshitz 1986, p 55, Hines 1960, Rapoport *et al* 2004, 2009) in the atmosphere-ionosphere with a hyperbolic dispersion similar to EMW dispersion in hyperbolic MMs (Poddubny *et al* 2013, Jacob *et al* 2010, Jacob and Shalaev 2011). We note that the above-mentioned dispersion of gravity waves in the atmosphere-ionosphere is essential, in particular, for the characterization of the atmospheric surface gas antenna of hydrodynamic waves and atmospheric-ionospheric seismogenic lenses (Rapoport *et al* 2004, 2009).

1.3 Purpose, tasks, and structure of the book

Based on the review of the literature, the following conclusion is made. To implement the identified properties of layered active MMs, gyrotropic, and plasma media, in aspects of the development of the theory of nonlinear wave processes in such media, as well as new applications, it is necessary to solve the set of new problems. Such tasks include finding non-trivial wave structures in layered nonlinear and active media with bulk and surface nonlinearities, searching new wave effects in active and nonlinear layered media, developing methods for modeling and controlling the characteristics of wave processes in layered structures with non-stationary and nonlinear fields.

The structure of the book is the following. Chapter 1 is introductory. In chapter 2, MMs with active metaparticles (split-ring resonators and Ω-particles) and absolute and convective instability in the active MMs are considered. In chapter 3, the general method of the derivation of nonlinear evolution equations for layered structures (NEELS) with the volume and surface nonlinearities is delivered. The approaches presented in chapter 3, are used and developed farther in the subsequent chapters, in particular in chapters 4–7, 11 and 12. In chapter 4, the application of the method NEELS to the layered nonlinear passive gyrotropic and plasma-like structures with volume and surface nonlinearities is described. In chapter 5, controllable EMWs under the propagation and reflection in non-stationary layered gyrotropic and MM media are considered. In chapter 6, parametric interactions of the nonlinear waves in active layered MMs and gyrotropic structures are treated. In chapter 7, the formation, propagation, and control of bullets in MM waveguides with higher-order nonlinear effects and magnetooptic interaction are discussed. In chapter 8, the giant double-resonant second harmonic generation in the multilayered dielectric-graphene MMs is explored. In chapter 9, the method of complex geometrical optics-full wave solution of the nonlinear Maxwell equations for the mesoscopic structures is described. In chapter 10, wave processes in layered controllable and active MMs and plasma-like structures in the presence of resonances and strong nonlinearity are described. The effects of jumping focusing point, formation of self-localized nonlinear resonator in nonlinear MM field concentrator and formation of hot spots are presented. The approach is based on the methods described in chapter 9. In chapter 11, wave processes in nonlinear passive and active loss-compensating hyperbolic MMs are explored, accounting for the approaches developed, in particular, in chapters 3 and 10. In chapter 12, in fact, the synergetic approach to the wave processes in resonant active quantum hyperbolic MMs is considered. In chapter 13, Rogue waves in MM waveguides are presented. Chapter 14 is the conclusive one. In this chapter, the discussion and conclusions and the future scientific problems and perspective applications of the nonlinear wave processes in the MMs are presented.

List of abbreviations

AGW	Atmospheric gravitational waves
BIC	Bound states in the continuum
BVMSW	Backward volume magnetostatic waves
EMW	Electromagnetic wave
IR	Infrared
LAI	Lithosphere–atmosphere–ionosphere
MHD	MagnetoHydroDynamic
MM	Metamaterial
MO	Magnetooptical
NEELS	Nonlinear evolution equations for layered structures
NSE	Nonlinear Schrödinger equation
RPM	Reductive perturbation method
TM	Transverse magnetic

References

Aburjania G D, Alperovich L S, Khantadze A G and Kharshiladze O A 2006 A new model for the generation of large-scale ionospheric vortex electric field *Phys. Chem. Earth.* **31** 482–5

Aburjania G D, Chargazia K D and Jandieri G V *et al* 2005 Generation and propagation of the ULF planetary-scale electromagnetic wave structures in the ionosphere *Planet. Space Sci.* **53** 881–901

Aburjania G D and Chargazia K Z 2011 Self-organization of largescale ULF electromagnetic wave structures in their interaction with nonuniform zonal winds in the ionospheric E region *Plasma Phys. Rep.* **37** 177–90

Agrawal G 1996 *Nonlinear Fiber Optics* (Moscow: Mir) p 326

Allain P E and Fuchs J N 2011 Klein tunneling in graphene: Optics with massless electron *Eur. Phys. J.* **83** 301–17

Alù A, Silveirinha M G, Salandrino A and Engheta N 2007 Epsilon-near-zero (ENZ) metamaterials and electromagnetic sources: Tailoring the radiation phase pattern *Phys. Rev.* B **75** 155410–22

Bai C, Wang J and Jia S *et al* 2010 Spin–orbit interaction effects on magnetoresistance in graphene based ferromagnetic double junctions *Appl. Phys. Lett.* **96** 223102

Ballav M and Chowdhury A R 2006 On a study of diffraction and dispersion managed soliton in a cylindrical media *Progr. Electromagn. Res., PIER* **63** 33–50

Bauer M, Mathieu C, Demokritov S O, Hillebrands B, Kolodin P A, Sure S, Dotsch H, Grimalsky V V, Rapoport Y G and Slavin A N 1997 Direct observation of two-dimensional self-focusing of spin waves in magnetic films *Phys. Rev.* B **56** R8483–6

Boardman A 2011 Community perspective pioneers in metamaterials: John Pendry and Victor Veselago *J. Opt.* **13** 020401–6

Boardman A D and Xie M 2003 Magnetooptics—a critical review *Introduction to Complex Mediums for Optics and Electromagnetics* ed W S Weiglhofer and A Lakhtakia (Bellingham, WA: SPIE Press) pp 197–219

Boardman A D, King N and Velasco L 2005f Negative refraction in perspective *Electromagnetics (Special Issue on Exotic Electromagnetics)* **25** 365–89

Boardman A D, Wang Q, Nikitov S A, Chen J, Mills D and Bao J S 1994 Nonlinear magnetostatic surface waves in ferromagnetic films *IEEE Trans. Mag.* **30** 14–22

Boardman A D, Egan P, Mitchell-Thomas R C and Rapoport Y G 2010a Solitons, vortices and guided waves in plasmonic metamaterials *Proc. SPIE—Int. Soc. Opt. Eng.* **7757** 775711

Boardman A D, Hess O, Mitchell-Thomas R C, Rapoport Y G and Velasco L 2010b Temporal solitons in magnetooptic and metamaterial waveguides *Photon. Nanostruct. – Fundam. Appl.* **8** 228–43

Boardman A D, Mitchell-Thomas R C, King R and Rapoport Y G 2010c Bright spatial solitons in controlled negative phase metamaterials *Opt. Commun.* **283** 1585–97

Boardman A D, Marinov K, Pushkarov D I and Shivarova A 2000b Wave-beam coupling in quadratic nonlinear optical waveguides: Effects of nonlinearly induced diffraction *Phys. Rev.* E **62** 2871–7

Boardman A D, Alberucci A, Assanto G, Rapoport Y G, Grimalsky V V, Ivchenko V M and Tkachenko E N 2017a Spatial solitonic and nonlinear plasmonic aspects of metamaterials *Handbook of Metamaterials and Nanophotonics* vol 4 ed K Shamonina, S Guenneau, O Hess and J Aizpurua (Hackensack, NJ: World Scientific) ch 10 pp 419–70

Boardman A D, Alberucci A, Assanto G, Grimalsky V V, Kibler B, McNiff J, Nefedov I S, Rapoport Y G and Valagiannopoulos C A 2017b Waves in hyperbolic and double negative metamaterials including rogues and solitons *Nanotechnology* **28** 41

Boardman A D, Rapoport Y G, Aznakayeva D E, Aznakayev E G and Grimalsky V V 2019 Graphene metamaterial electron optics: Excitation processes and electro-optical modulation *Handbook of Graphene, Vol. 3: Graphene-like 2D Materials* ed M Zhang (Beverly, MA: Scrivener Publishing, Wiley) ch 9 pp 263–96

Bonchkovsky V F 1954 The slopes of the Earth's surface as one of the possible precursors of earthquakes *Tr. Geophys. Inst. USSR Acad. Sci.* **25** 134–53

Boyd R W 2003 *Nonlinear Optics* 2nd edn (New York: Academic) p 576

Boyle J W, Nikitov S A, Boardman A D, Booth J G and Booth K 1996 Nonlinear self-channeling and beam shaping of magnetostatic waves in ferromagnetic films *Phys. Rev.* B **53** 12173–81

Burlak G N, Kotsarenko N Y and Rapoport Y G 1990 Theory of three-wave magnetooptic interactions in layered materials *Sov. Phys. Solid State* **32** 1805–7

Buttner O, Bauer O, Demokritov S O, Hillebrands B, Kivshar Y S, Grimalsky V V, Rapoport Y and Slavin A N 2000 Linear and nonlinear diffraction of dipolar spin waves in yttrium iron garnet films observed by space-and time-resolved Brillouin light *Phys. Rev.* B **61** 11576–87

Capolino F 2009 *Theory and Phenomena of Metamaterials* ed F Capolino (Boca Raton, FL: CRC Press, Taylor and Francis Group) p 926

Chebykin A V, Orlov A A, Vozianova A V, Maslovski S I, Kivshar Y S and Belov P A 2011 Nonlocal effective medium model for multilayered metal-dielectric metamaterials *Phys. Rev.* B **84** 115438–46

Cheianov V M, Falko V and Altshuler B L 2007 The focusing of electron flow and a Veselago lens *Science* **315** 1252–5

Chen M, Tsankov M A, Nash J M and Patton C E 1994 Backward-volume-wave magnetic microwave-envelope solitons in yttrium iron garnet films *Phys. Rev.* B **49** 12773–90
Cherkasskii M A and Kalinikos B A 2013 Envelope solitons of electromagnetic spin waves in an artificial layered multiferroic *JETP Lett.* **97** 611–5
Chu H, Li Q, Liu B, Luo J, Sun S, Hang Z H, Zhou L and Lai Y 2018 A hybrid invisibility cloak based on integration of transparent metasurfaces and zero-index materials *Light: Sci. Appl.* **7** 50
Cortes C L, Newman W, Molesky S and Jacob Z 2012 Quantum nanophotonics using hyperbolic metamaterials *J. Opt.* **14** 129501–15
De Ceglia D, Campione S, Vincenti M A, Capolino F and Scalora M 2013 Low-damping epsilon-near-zeroslabs: Nonlinear and nonlocal optical properties *Phys. Rev.* B **87** 1551409–20
Davoyan A and Engheta N 2019 Nonreciprocal emission in magnetized epsilon-near-zero metamaterials *ACS Photonics* **6** 581–6
Dehbashi R, Fathi D, Mohajerzadeh S and Forouzandeh B 2010 Equivalent left-handed/right-handed metamaterial's circuit model for the massless dirac fermions with negative refraction *IEEE J. Select. Top. Quant. Electron.* **16** 394–400
Dong H and Conti C *et al* 2013 Terahertz relativistic spatial solitons in doped graphene metamaterials *J. Phys. B: At. Mol. Opt. Phys.* **46** 155401
Eleftheriades G I 2005 *Negative-Refraction Metamaterials: Fundamental Principles and Applications* ed K G Balmain (Piscataway, NJ: IEEE Press) p 418
Elnaggar S Y and Milford G N 2018 Controlling nonreciprocity using enhanced Brillouin scattering *IEEE Trans. Antennas Propag.* **66** 3500–11
Engheta E 2010 Taming light at the nanoscale *Phys. World* **23** 31–4
Engheta N, Jaggard L D and Kowarz M W 1992 Electromagnetic waves in faraday chiral media *IEEE Trans. Antennas Propag* **40** 367–74
Engheta N 2006 *Metamaterials Physics and Engineering Explorations* ed R W Ziolkowski (Piscataway, NJ: IEEE Press) p 414
Fedorchenko A M and Kotsarenko N Y 1981 *Absolute and Convective Instability in Plasma and Solids* (Moscow: Nauka) p 176
Fedorov F I 1976 *The Theory of Gyrotropy* (Minsk: Science and Technology) p 455
Fernandes D E and Silveirinha M G 2014 Bright and dark spatial solitons in metallic nanowire arrays *Photon. Nanostruct. – Fundam. Appl.* **12** 340–9
Gómez S, Burset P, Herrera W J and Levy A Y 2012 Selective focusing of electrons and holes in a graphene-based superconducting lens *Phys. Rev.* B **85** 115411
Gramotnev D K and Bozhevolnyi S I 2010 Plasmonics beyond the diffraction limit *Nat. Photonics* **4** 83–91
Grimal'skii V V and Rapoport Yu G 1994 Nonlinear magnetostatic surface waves in a transversely nonuniform magnetic field *Tech. Phys. Lett.* **20** 345–6
Grimalsky V V and Rapoport Y G 1995 Convolution of magnetostatic waves in ferrite films *J. Magn. Magn. Mater.* **140–144** 2195–6
Grimalsky V, Rapoport Y and Slavin A N 1997 Nonlinear diffraction of magnetostatic waves in ferrite films *J. Phys. IV France* **7C** 3931–4
Grimal'skii V V, Koshevaya S V and Resin A M 1998 Nonlinear surface magnetic polaritons in a nonuniform magnetic field *Tech. Phys.* B **43** 1091–3
Grimalsky V V, Kremenetsky I A and Rapoport Y G 1999 Excitation of electromagnetic waves in the lithosphere and their penetration into ionosphere and magnetosphere *J. Atmosph. Electr.* **19** 101–17

Grimalsky V V, Rapoport Y G, Slavin A N and Zaspel C E 2000a Parametric amplification and wave front conjugation of 2D pulse in ferromagnetic film *Proc. of Meeting of the APS Minneapolis (USA) (March 2000)* Z25–5 p 1035

Grimalsky V V, Mantha J H, Rapoport Y G, Slavin A N and Zaspel C E 2000b Numerical models of amplification and wave front reversal of two-dimensional spin wave packets in magnetic films *Materials Science Forum, 8th Europ. Magn. Mater. and Appl. Conf. (EMMA 2000) (Kyiv Ukraine)* 377–80

Grimalsky V V, Rapoport Y G, Zaspel C E and Slavin A N 2001 Parametric amplification of two-dimensional dipolar spin wave pulses in ferrite films *Proc. of the 8th Int. Conf. On Ferrites (ICF 8) (Kyoto and Tokyo (Japan)) (Kyoto: Japan Society of Powder and Powder Metallurgy)* 921–3

Grimalsky V, Koshevaya S, Rapoport Y, Tretiak N, Yanovsky F and Escobedo-Alatorre J 2019 Resonant properties of electron gas in *n*-InSb and graphene layers in magnetic fields for THz multilayered dielectric-plasma-like metamaterials *IEEE 39th Int. Conf. on Electronics and Nanotechnology (ELNANO) (Kyiv, Ukraine)* 164–8

Grimalsky V, Koshevaya S, Rapoport Y and Escobedo-Alatorre J 2021 Magnetoplasma terahertz solitons and collapsing pulses in narrow-gap semiconductors *Proc. 32nd Int. Conf. on Microelectronics (MIEL Serbia)* p 4

Guo Z, Jianga H and Chena H 2020 Hyperbolic metamaterials : From dispersion manipulation to applications *J. Appl. Phys.* **127** 071101

Gurevich A G and Melkov G A 1996 *Magnetization Oscillations and Waves* (New York: CRS Press) p 464

Hart W S, Bak A O and Phillips C C 2018 Ultra low-loss super-resolution with extremely anisotropic semiconductor metamaterials *AIP Adv.* **8** 025203

Haugan H J, Eyink K G, Urbas A M and Bas D A 2021 Engineering transient hyperbolic metamaterials using InAsSb-based semiconductor *AIP Adv.* **11** 075115

Hines C O 1960 Internal atmospheric gravity waves at ionospheric heights *Can. J. Phys.* **38** 1441–81

Iorsh I, Poddubny A, Orlov A, Belov P and Kivshar Y S 2012 Spontaneous emission enhancement in metal-dielectric metamaterials *Phys. Lett.* A **376** 185–7

Iorsh I V, Shadrivov I V, Belov P A and Kivshar Y S 2013 Tunable hybrid surface waves supported by a graphene layer *JETP Lett.* **97** 249–52

Jacob Z, Kim J-Y, Naik G V, Boltasseva A, Narimanov E E and Shalaev V M 2010 Engineering photonic density of states using metamaterials *Appl. Phys.* B **100** 215–8

Jacob Z and Shalaev V M 2011 Plasmonics goes quantum *Science* **334** 463–4

Jacob Z, Alekseyev L V and Narimanov E 2007 Semiclassical theory of the hyperlens *J. Opt. Soc. Am.* A **24** 52–9

Ji W, Zhou X, Chu H, Luo J and Lai Y 2020 Theory and experimental observation of hyperbolic media based on structural dispersions *Phys. Rev. Mater.* **4** 105202

Ji W, Luo J and Lai Y 2021 Hyperbolic media and zero-index media based on structural dispersions 2021 Abstract *15th Int. Congress on Artificial Materials for Novel Wave Phenomena (New York, USA, 20–25 September)* https://congress.metamorphose-vi.org/index.php/component/svisor/?task=showScheduleSvisor&Itemid=156

Kadomtsev B B 1988 *Collective Phenomena in Plasma* (Moscow: Nauka) p 304 (in Russian)

Kaipa C S R, Yakovlev A B, Hanson G W, Padooru Ya R, Medina F and Mesa F 2012 Enhanced transmission with a graphene-dielectric microstructure at low-terahertz frequencies *Phys. Rev.* B **85** 245407

Kalinin V A and Bayonets V V 1990 On the possibility of reversing the front of radio waves in an artificial nonlinear environment *Radiotech. Electron.* **35** 2275–81

Khattak H K, Bianucci P and Slepkov A D 2019 Linking plasma formation in grapes to microwave resonances of aqueous dimers *PNAS* **116** 4000–5

Kildishev A V 2011 Transformation optics and metamaterials *Phys.* **181** 59–70

Kildishev A V, Boltasseva A and Shalaev V M 2013 Planar photonics with metasurfaces *Science* **339** 1232009–14

Kivshar Y and Quo V 2021 Abstract metamaterials? *15th Int. Congress on Artificial Materials for Novel Wave Phenomena (New York, 20–25 September, 2021)* https://congress.metamorphose

Kodera T and Caloz C 2018 Unidirectional loop metamaterials (ULM) as magnetless artificial ferrimagnetic materials: principles and applications *IEEE Antennas Wirel. Propag. Lett.* **17** 1943–7

Kozyrev A B, Shadrivov I V and Kivshar Y S 2014 Soliton generation in active nonlinear metamaterials *Appl. Phys. Lett.* **104** 084105

Kryshtal A N, Gerasimenko S V and Voitsekhovska A D 2014 One type of three-wave interaction of low-frequency waves in magnetoactive plasma of the solar atmosphere *Kinemat. Phys. Celest. Bodies* **30** 147–54

Landau L D and Lifshits E M 1986 *Hydrodynamics* (Moscow: Nauka) p 736

Lapine M, Gorkunov M and Ringhofer K H 2003 Nonlinearity of a metamaterial a rising from diode insertions into resonant conductive elements *Phys. Rev.* E **67** 065601R–4R

Leblond H 2001 Rigorous derivation of the NLS in magnetic films *J. Phys. A: Math. General* **34** 9687–712

Lehtinen N G 2012 A waveguide model of the return stroke channel with a 'metamaterial' corona *Radio Sci.* **47** RS1003–12

Leonhardt U and Philbin T G 2009 Transformation optics and the geometry of light *Progress in Optics* vol 53 (Amsterdam: Elsevier) ch 2 pp 69–152

Lifshitz E M and Pitaevskii L P 1981 *Physical Kinetics* (Oxford: Elsevier) p 369

Lindell I V, Shihvola A H, Tretyakov S A and Vittanen A J 1994 *Electromagnetic Waves in Chiral and Bi-Isotropic Media* (Boston, MA: Artech House) p 335

Liu W and Kivshar Y 2018 Generalized Kerker effects in nanophotonics and meta-optics *Opt. Express* **26** 13085–105

Liu N, Kaiserand S and Giessen H 2008 Magnetoinductive and electroinductive coupling in plasmonic metamaterial molecules *Adv. Mater.* **20** 4521–5

Liu Z, Lee H, Xiong Y, Sun C and Zhang X 2007 Far-field optical hyperlens magnifying sub-diffraction-limited objects *Science* **315** 1686

Lucido M, Kobayashi K, Medina F, Nosich A and Vinogradova E 2021 Guest editorial: Method of analytical regularisation for new frontiers of applied electromagnetics *IET Microwaves Antennas Propag.* **15** 1127–32

Lukomsky V P and Rapoport Y G 1994 Method of undetermined coefficients for derivation of the nonlinear evolution equations with higher terms and its application to ion acoustic waves in plasma *Proc. of Int. Conf. of Plasma Phys., Brazil* Abstract Booklet p 410

Lukomsky V P 1995 Modulational instability of gravity waves on deep water accounting for nonlinear dispersion *J. Eksp. Theor. Phys.* **108** 567–76 (in Russian)

Lukomsky V P 1978 Nonlinear magnetostatic waves in ferromagnetic plates *Ukr. J. Phys.* **23** 134–9 (In Ukrainian)

Lukomsky V P and Gandzha I S 2009 Two-parameter method for describing the nonlinear evolution of narrow-band wave trains *Ukr. J. Phys.* **54** 207–15 (in Ukrainian)

Makarova O A, Shalaginov M Y, Bogdanov S, Kildishev A V, Boltasseva A and Shalaev V M 2017 Patterned multilayer metamaterial for fast and efficient photon collection from dipolar emitters *Opt. Lett.* **42** 3968–71

Malomed B A, Mihalache D, Wise F and Torner L 2005 Spatiotemporal optical solitons *J. Opt. B: Quantum Semiclass. Opt.* **7** R53–72

Melkov G A, Kobljanskyj Y V, Serga A A, Tiberkevich V S and Slavin A N 2001 Nonlinear amplification and compression of envelope solitons by localized nonstationary parametric pumping *J. Appl. Phys.* **89** 6689–91

Melkov G A and Serga A A 1998 Nonlinear parametric excitation of spin waves *Frontiers in Magnetism of Reduced Dimension Systems* ed V G Bar'yakhtar (Amsterdam: Kluwer Academic) pp 555–78

Melkov G A, Serga A A, Tiberkevich V S, Oliynyk A N, Bagada A V and Slavin A N 1999 Parametric interaction of spin wave pulse with localized non-stationary pumping: Amplification and phase conjugation *IEEE Trans. Magn.* **35** 3157–9

Minovich A E, Miroshnichenko A E, Bykov A Y, Murzina T V, Neshev D N and Kivshar Y S 2015 Functional and nonlinear optical metasurfaces *Laser Photon. Rev.* **9** 195–213

Molchanov O A and Hayakawa M 1998 Subionospheric VLF signal perturbations possibly related to earthquakes *J. Geophys. Res.* **103** 17489–504

Morgado T and Silveirinha M 2021 Drift-induced active and nonreciprocal plasmonics in graphene Abstract *15th Int. Congress on Artificial Materials for Novel Wave Phenomena (New York, USA, 20–25 September 2021)* https://congress.metamorphose-vi.org/index.php/component/svisor/?task=showScheduleSvisor&Itemid=156

Nayfeh A 1984 *Introduction to Methods of Perturbations* (Moscow: Mir) p 535

Narimanov E E and Kildishev A V 2009 Optical black hole: broadband omnidirectional light absorber *Appl. Phys. Lett.* **9** 041106–10

Noginov M A, Zhu G, Bahoura M, Adegoke J, Small C, Ritzo B A, Drachev V P and Shalaev V M 2007 The effect of gain and absorption on surface plasmons in metal nanoparticles *Appl. Phys.* B **86** 455–60

Paul T, Rockstuhl C and Lederer F 2011 Integrating cold plasma equations into the Fourier modal method to analyze second harmonic generation at metallic nanostructures *J. Mod. Opt.* **58** 438–48

Pang K *et al* 2018 400-Gbit/s QPSK free-space optical communication link based on four-fold multiplexing of Hermite–Gaussian or Laguerre–Gaussian modes by varying both modal indices *Opt. Lett.* **43** 3889–92

Pekar S I 1962 Supplementary light waves in crystals and exciton absorption *Sov. Phys. Usp.* **5** 515–21

Plum E, Fedotov V A and Zheludev N I 2008 Optical activity in extrinsically chiral metamaterial *Appl. Phys. Lett.* **93** 191911–3

Poddubny A, Iorsh I, Belov P and Kivshar Y 2013 Hyperbolic metamaterials *Nat. Photonics* **7** 958–67

Pollard R J, Murphy A, Hendren W R, Evans P R, Atkinson R, Wurtz G A and Zayats A V 2009 Optical nonlocalities and additionalwaves in epsilon-near-zero metamaterials *Phys. Rev. Lett.* **102** 127405–8

Prokhorov A M 1988 *Physical Encyclopedia, in 5 Volumes* vol 1 (Moscow: Sov. Encyclopedia) p 704 (in Russian)

Rabinovich M I and Trubetskov D I 2000 *Introduction to the Theory of Oscillations and Waves* (Moscow: Research Center 'Regular and Chaotic Dynamics') p 560

Rahm M, Schurig D, Roberts D A, Cummer S A, Smith D R and Pendry J B 2008 Design of electromagnetic cloaks and concentrators using form-invariant coordinate transformations of Maxwell's equations *Photonics Nanostruct.—Fundam. Appl.* **6** 87–90

Rapoport Yu G, Gotynyan O E, Ivchenko V M, Kozak L V and Parrot M 2004 Effect of acoustic-gravity wave of the lithospheric origin on the ionospheric F region before earthquakes *Phys. Chem. Earth* A/B/C **29** 607–16

Rapoport Yu G, Gotynyan O E, Ivchenko V N, Hayakawa M, Grimalsky V V, Koshevaya S V and Juarez-R. D 2006 Modeling electrostatic-photochemistry seismoionospheric coupling in the presence of external currents *Phys. Chem. Earth* A/B/C **31** 437–46

Rapoport Yu G, Hayakawa M, Gotynyan O E, Ivchenko V N, Fedorenko A K and Selivanov Yu A 2009 Stable and unstable plasma perturbations in the ionospheric F region, caused by spatial packet of atmospheric gravity waves *Phys. Chem. Earth* A/B/C **34** 508–15

Rapoport Y G, Boardman A D and Grimalsky V V *et al* 2012b Metamaterial based electromagnetic and acoustic field concentrators and new physical phenomena: nonlinear focusing switching *Proc. XXXII Int. Science Conf. on Electronics and Nanotechnology (ELNANO) (Kyiv Ukraine, 10–12 April)* 84–5

Rapoport Y, Boardman A, Grimalsky V, Selivanov Y and Kalinich N 2012h Metamaterials for space physics and the new method for modeling isotropic and hyperbolic nonlinear concentrators *Proc. of Int. Conf. on Mathematical Methods in Electromagnetic Theory (MMET) (Kharkiv Ukraine) (28–30 August)* pp 76–9

Rapoport Y G, Tretyakov S and Maslovski S 2013a Nonlinear active Huygens metasurfaces for reflectionless phase conjugation of electromagnetic waves in electrically thin layers *J. Electromagn. Waves Appl.* **27** 1309–28

Rapoport Y, Grimalsky V, Iorsh I, Kalinich N, Koshevaya S, Castrejon-Martinez C and Kivshar Y S 2013b Nonlinear reshaping of terahertz pulses with graphene metamaterials *JETP Lett.* **98** 503–6

Rapoport Y, Selivanov Y, Ivchenko V and Grimalsky V 2013c Modeling electromagnetic and hydromagnetic wave coupling in LAIM/MIAL system and possible application to space weather studying Second UK–Ukraine Meeting on Solar Physics and Space Science (Kyiv Ukraine, 16–20 September) 66–7 Abstracts

Rapoport Y G 2013d General method for modeling nonlinear waves in active metamaterial and gyrotropic layered structures Proc. of Int. Kharkov Symp. on Physics and Engineering of Microwaves, Millimeter and Submillimeter Waves, MSMW'13 (Kharkov Ukraine) June 23–28 pp 253–5

Rapoport Y G 2012e General method for modeling nonlinear waves in layered structures of diferent physical nature including bi-anisotropic and active metamaterials Progress in Electromagnetics Research Symp. (PIERS) (Moscow Russia) August 19–23 Abstracts pp 18–9

Rapoport Y G, Sirenko E K and Fedun V N 1998 Interaction of magnetosonic solitons with an ion beam in magnetized plasma *Ukr. J. Phys.* **43** 169–76 (in Ukrainian)

Rapoport Y G, Grimalsky V V, Koshevaya S V, Boardman A D and Malnev V N 2014a New method for modeling nonlinear hyperbolic concentrators Proc. of IEEE 34th Int. Scientific Conf. on Electronics and Nanotechnology (ELNANO) (Kyiv Ukraine) pp 35–8

Rapoport Y G, Boardman A D, Grimalsky V V, Ivchenko V M and Kalinich N 2014b Strong nonlinear focusing of light in nonlinearly controlled electromagnetic active metamaterial field concentrators *J. Opt.* (U. K.) **16** 0552029–38

Rapoport Y G, Grimalsky V V, Boardman A D and Malnev V N 2014c Controlling nonlinear wave structures in layered metamaterial, gyrotropic and active media Proc. of IEEE 34th Int. Scientific Conf. Electronics and Nanotechnology (ELNANO) (Kyiv Ukraine) April 15–18 pp 46–50

Rapoport Y, Grimalsky V, Vasnetsov M and Hensitskyi N 2021 Strongly nonlinear, singular and topological nanometaphotonic wave structure control Abstract 15th Int. Congress on Artificial Materials for Novel Wave Phenomena (New York, USA) *20–25 September https://congress.metamorphose-i.org/index.php/component/svisor/?task=showScheduleSvisor&Itemid=156*

Rawat V, Kitture R and Kumari D *et al* 2016 Hazardous material ssensing: An electrical metamaterial approach *J. Magn. Magn. Mater.* **415** 77–81

Rowland D R 1999 Conservation law for multimoded nonlinear optical waveguide interactions and its physical interpretation *Phys. Rev.* E **59** 7141–7

Rybin M V, Koshelev K L, Sadrieva Z F, Samusev K B, Bogdanov A A, Limonov M F and Kivshar Y S 2017 High-Q supercavity modes in subwavelength dielectric resonators *Phys. Rev. Lett.* **119** 243901

Sadeghi M M, Nadgaran H and Chen H 2014 Perfect field concentrator using zero index metamaterials and perfect electric conductors *Front. Phys.* **9** 90–3

Scalora M, Syrchin M S, Akozbek N, Poliakov E Y, D'Aguanno G, Mattiucci N, Bloemer M J and Zheltikov A M 2005 Generalized nonlinear Schrodinger equation for dispersive susceptibility and permeability: Application to negative index materials *Phys. Rev. Lett.* **95** 013902

Schurig D, Pendry J B and Smith D R 2006 Calculation of material properties and ray tracing in transformation media *Opt. Express* **14** 9794–804

Serga O O 2006 Parametric interaction of spin waves and oscillations with non-stationary local pumping *Abstract of the dissertation for the degree of Doctor of Physical and Mathematical Sciences. Science Kiev* p 36

Shalaginov M Y *et al* 2015 Enhancement of single-photon emission from nitrogen-vacancy centers with TiN/(Al,Sc)N hyperbolic metamaterial *Laser Photon. Rev.* **9** 120–7

Sharaevsky Y P, Sadovnikov A V, Beginin E N, Sharaevskaya A Y, Sheshukova S E and Nikitov S A 2018 Functional Magnetic Metamaterials for Spintronics *Functional Nanostructures and Metamaterials for Superconducting Spintronics. NanoScience and Technology* ed A Sidorenko (Berlin: Springer) p 270

Sharma R P, Kumari A and Yadav N 2015 Effect of background density fluctuations on the localized structures of inertial Alfvén wave and turbulent spectrum *Phys. Plasm.* **22** 112303

Shen L, Lin X, Shalaginov M Y, Low T, Zhang X, Zhang B and Chen H 2020 Broadband enhancement of on-chip single-photon extraction via tilted hyperbolic metamaterials *Appl. Phys. Rev.* **7** 021403

Shramkova O and Schuchinsky A 2012b Harmonic generation and wave mixing in nonlinear metamaterials and photonic crystals *Int. J. RF Microwave Comput-Aided Engineer* **22** 469–82

Shramkova O V and Schuchinsky A G 2012a Nonlinear scattering by anisotropic dielectric periodic structures *Adv. OptoElectr.* **2012** 154847

Singh R, Al-Naib I A I, Koch M and Zhang W 2011 Sharp Fano resonances in THz metamaterials *Opt. Express.* **19** 6312–9

Sipe J E, So V C Y, Fukui M and Stegeman G I 1980 Analysis of second-harfnonic generation at metal surfaces *Phys. Rev.* B **21** 4389–402
Skarka V, Berezhiani V I and Miklaszewski R 1997 Spatiotemporal soliton propagation in saturating nonlinear optical media *Phys. Rev.* E **56** 1080–7
Slavin A N and Rojdestvenski I V 1994 'Bright' and 'dark' spin wave envelope solitons in magnetic films *IEEE Trans. Magn.* **30** 37
Slavin A N, Kalinikos B A and Kovshikov N G 1994 Spin wave envelope solitons in magnetic films *Nonlinear Phenomena and Chaos in Magnetic Materials* ed P E Wigen (Singapore: World Scientific) pp 209–48
Smith D R, Pendry J B and Wiltshire M C K 2004 Metamaterials and negative refractive index *Science* **305** 788–92
Subha P A, Jisha C P and Kuriakose V C 2007 Stable diffraction managed spatial soliton in bulk cubic-quintic media *J. Mod. Opt.* **54** 1827–35
Sugakov V Y 1974 Theoretical physics *Electrodynamics* (Kyiv: Higher School) p 271 (in Ukrainian)
Taflove A and Hagness S C 2005 *Computationai Electrodynamics: The Finite-Difference Time-Domain Method* (Boston, MA: Artech House Antennas and Propagation Library) p 977
Tretyakov S 2003 *Analytical Modeling in Applied Electromagnetics* (Boston, MA: Artech House) p 275
Tretyakov S, Barois P, Scharf T, Kruglyak V, Beergmair I (Eds) and De Baas Anne (Ed. In Chief) 2010 Nanostructured metamaterials ed P Barois and T Scharf *et al* (Brussels: Office of the European Union) p 137
Vainshtein L A 1966 *Open Resonators and Open Waveguides* (Moscow: Sov. Radio) p 476 (in Russian)
Vashkovsky A V, Zubkov V I, Locke E G and Shcheglov V I 1990 Propagation of surface magnetostatic waves in an inhomogeneous constant magnetic field of the extended well type *Zh. Tech. Phys.* **60** 138–41
Veselago V G 1968 The electrodynamics of substances with simultaneously negative values of ε and μ *Sov. Phys. Usp.* **10** 509–14
Vinogradova M B, Rudenko O V and Sukhorukov A P 1979 *Wave Theory* (Moscow: Nauka) p 384 (in Russian)
Wang F X, Rodríguez F J and Albers W M 2009 Surface and bulk contributions to the second-order nonlinear optical response of a gold film *Phys. Rev.* B **80** 233402–25
Willner A 2021 High-capacity optical, THz, and millimeter-wave communications using multiple orbital-angular-momentum beams Abstract 15th Int. Congress on Artificial Materials for Novel Wave Phenomena (New York, USA) 20–25 September, 2021 https://congress.metamorphose-vi.org/index.php/component/svisor/?task=showScheduleSvisor&Itemid=156
Wurtz G A, Pollard R, Hendren W, Wiederrecht G P, Gosztola D J, Podolskiy V A and Zayats A V 2011 Designed ultrafast optical nonlinearity in a plasmonic nanorod metamaterial enhanced by nonlocality *Nat. Nanotechnol.* **6** 107–11
Xia H, Kabos P, Staudinger R A and Patton C E 1998 Velocity characteristics of microwave-magnetic-envelope solitons *Phys. Rev.* B **58** 2708–15
Yang J, Huang M, Yang C, Xiao Z and Peng J 2009 Metamaterial electromagnetic concentrators with arbitrary geometries *Opt. Exp.* **17** 19656–61
Yao A M and Padgett M J 2011 Orbital angular momentum: Origins, behavior and applications *Adv. Opt. Photon.* **3** 161–204
Yevtushenko F O, Dukhopelnykov S V, Zinenko T L and Rapoport Y G 2021 Electromagnetic characterization of tuneable graphene-strips-on-substrate metasurface over entire THz range:

Analytical regularization and natural-mode resonance interplay *IET Microw. Antennas Propag.* **15** 1225–39
Yoo M and Lim S 2014 Active metasurface for controlling reflection and absorption properties *Appl. Phys. Express* **7** 112204–7
Zafiris N A and Sigalas M M 2015 Large scale phononic metamaterials for seismic isolation *J. Appl. Phys.* **118** 064901–6
Zaspel C E, Mantha J H and Rapoport Y G 2001 Evolution of solitons in magnetic thin films *Phys. Rev.* B **64** 064416–20
Zayats A V 2008 Plasmonic metamaterials: From modelling to applications 12th Int. Conf. on Mathematical Methods in Electromagnetic Theory (MMET) (Odesa Ukraine) June 29–July 2 pp 114–6
Zeng Y, Dalvit D A R, O'Hara J and Trugman S A 2012 Modal analysis method to describe weak nonlinear effects in metamaterials *Phys. Rev.* B **85** 125107–15
Zhao Y, Engheta N and Alù A 2011 Effects of shape and loading of optical nanoantennas on their sensitivity and radiation properties *J. Opt. Soc. Am. B.* **28** 1266–74
Zheludev N I, Prosvirnin S L, Papasimakis N and Fedotov V A 2008 Lasing spaser *Nat. Photonics* **2** 351–64
Zhuang H-W, Kong F-M, Li K and Yue Q-Y 2015 A gating tunable planar lens based on grapheme *Opt. Quant. Electron.* **47** 1139–50
Zhao Z *et al* 2021 Modal coupling and crosstalk due to turbulence and divergence on free space THz links using multiple orbital angular momentum beams *Sci. Rep.* **11** 2110

Chapter 2

Metamaterials with active metaparticles. Absolute and convective instability in the active metamaterials

In this chapter, active metamaterials (MMs) based on metastable particles with active loads (diodes or others) are investigated. We include into consideration the active Ω-particles, which constitute bi-anisotropic MM. We propose a selection of characteristics of active loads (effective conductivity, etc), which provides the possibility of spatial amplification/convective instability of electromagnetic waves (EMWs), and, thus, the absence of the absolute instability (AI). We determine the characteristics of amplified waves in the MMs with a special gain. Then we will search the methods of the homogenization of the MM with nonlinear metaparticles.

2.1 Artificial molecules (AMs) and their individual polarizations

We consider a general approach to the characterization of nonlinear EMWs in artificial media with Ω-particles (Boardman *et al* 2009b, Cheng *et al* 2007, Lheurette *et al* 2008, Lheurette *et al* 2008, McCauley *et al* 2010, Ramaccia *et al* 2013, Rapoport *et al* 2006a, Rapoport and Boardman 2015b, Tretyakov and Mariotte 1995, Tretyakov 2003, Zheludev and Kivshar 2012), and then, as a special case, with split ring resonators (SRR). Let us start with the characterization of individual active/nonlinear artificial metaparticles. The details concerning the material of this section are given in appendix A on the basis of papers (Rapoport and Boardman 2015b, Rapoport *et al* 2004, 2006a, 2013, Rapoport 2012, 2013, 2014a, 2014b, 2015a, Boardman *et al* 2007a, 2007b, 2009a, 2009b, Zheng *et al* 2014,) and taking into account works (Eleftheriades *et al* 2003, Tretyakov *et al* 1994, 1996, Bosch and Thim 1974, Bosch and Engellmann 1975, Oleinik *et al* 2004, Auzanneau and Ziolkowski 1998, Vainshtein 1988, Fedorchenko and Kotsarenko 1981, Tretyakov

doi:10.1088/978-0-7503-2336-9ch2

2003, 2010, Eleftheriades and Balmain 2005, Scott 1970, Mariotte *et al* 1994, Ramaccia *et al* 2013, Katko *et al* 2010, Ionkin 1972, Migulin *et al* 1988, Kalinin and Shtikov 1990, Korn and Korn 1968). The details of the approach to linear and nonlinear homogenization (in the general case of bi-anisotropic) metaparticles (Auzanneau and Ziolkowski 1998, Boardman *et al* 2009b, Rapoport *et al* 2006a, Rapoport and Boardman 2015b), see also (Eleftheriades and Balmain 2005, Liu *et al* 2014, McCauley *et al* 2010, Zheludev and Kivshar 2012) are considered in appendices A.2 and A.3. A separate loaded particle and the elementary cubic lattice of MM with Ω-particles are depicted in figures 2.1(a) and (b), respectively; in figures 2.1(g) and (d) equivalent circuits are represented for Ω-particle, active/nonlinear element/diode, and split-ring resonators (SRRs), respectively; SRR and the corresponding MM will be treated as individual cases—particles and MMs on their basis, respectively. Note that the equivalent circuit in figure 2.1(a) in the linear case (corresponding to a linear load) actually reduces to the corresponding circuit in the form of a four-pole (Auzanneau and Ziolkowski 1998), which represents a loaded bi-anisotropic metaparticle. Accordingly, the expressions for individual polarization of metaparticles and the linear polarizability (linear loads) discussed below in section 2.1.1 and appendix A.2, namely (A.11) and (A.12), are reduced to the expressions obtained in Auzanneau and Ziolkowski (1998) in the relevant limiting cases. In the absence of loads, these expressions are reduced to the corresponding formulas (Tretyakov *et al* 1994).

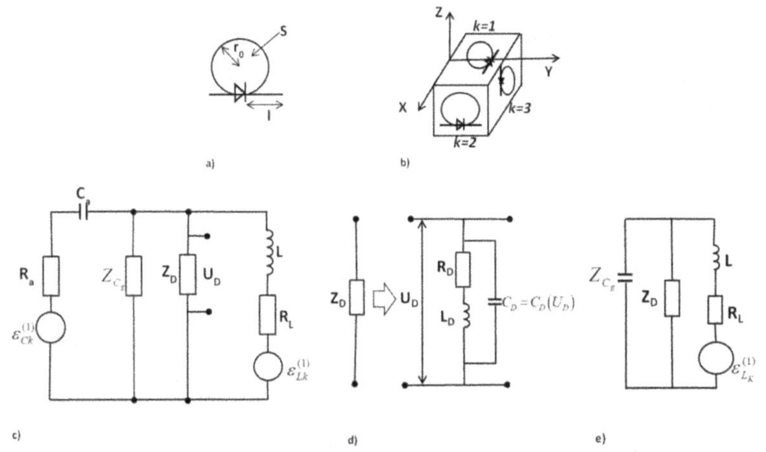

Figure 2.1. Part (a) is Ω-Particle; (b) is the elementary cubic cell of MM with Ω-particles, k = 1,2,3 are the corresponding faces of the cube; part (c) is the equivalent circuit of Ω-particle; (d) is the active element (diode)—loaded metaparticle; part (e) is SRR. The elements $Z_{C_g} \equiv (i\omega C_g)^{-1}$, Z_D and L, R_L describe the split in metaparticle without an active element, active/nonlinear element and magnetic loop, which is included both in Ω-particle, and in SRR, respectively; in (c) the elements C_a and R_a describe the electric dipole, which is included only in Ω-particle and in SRR; $\varepsilon^{(1)}_{ck;LK}$ are the electromotive forces induced in electric and magnetic dipoles, respectively; the values R_a, R_L describe dissipation in the corresponding elements; in (d) the elements R_D, L_D, C_D describe an active element/diode.

2.1.1 Individual polarizations of Ω-particles

Consider the equivalent circuit corresponding to the Ω-particle (Auzanneau and Ziolkowski 1998, Boardman *et al* 2007, 2009b, Rapoport 2013, 2014b, Rapoport and Boardman 2015b) in figure 2.1. Suppose, for the simplification (which does not change the qualitative picture 'as a whole'), that all the 'energy losses' refer only to magnetic and electric dipoles (elements $R_{a,\mathrm{L}}$, respectively). and we find the first harmonics $I_{C_a}^{(1)}$, $I_{\mathrm{L}}^{(1)}$ of (high-frequency) currents I_{C_a}, I_{L} passing through the electric and magnetic dipoles, where the electromotive forces ε_{ak}, $\varepsilon_{\mathrm{L}k}$ are applied accordingly. Electric and magnetic polarizations $\vec{P}^{\mathrm{incl}}$, $\vec{M}^{\mathrm{incl}}$, due to the presence of metaparticles or AMs, are also referred to as 'inclusions' in the literature (Auzanneau and Ziolkowski 1998), justifying the use of the index 'incl'. We are seeking here the first harmonics (proportional to $\exp(i\omega t)$) of these polarizations. They can be separated on the linear and nonlinear parts:

$$\vec{P}^{\mathrm{incl}(1)} = \vec{P}_{\mathrm{LIN}}^{\mathrm{incl}(1)} + P_{\mathrm{NL}}^{\mathrm{incl}(1)},\ \vec{M}^{\mathrm{incl}(1)} = \vec{M}_{\mathrm{LIN}}^{\mathrm{incl}(1)} + \vec{M}_{\mathrm{NL}}^{\mathrm{incl}(1)} \tag{2.1}$$

Polarizations are presented in general case in the form

$$\vec{P}^{\mathrm{incl}(1)} = (N/3)(\pi_x\vec{e}_x + \pi_y\vec{e}_y + \pi_z\vec{e}_z),\ M^{\mathrm{incl}(1)} = (N/3)(m_x\vec{e}_x + m_y\vec{e}_y + m_z\vec{e}_z), \tag{2.2}$$

In (2.1), $\vec{e}_{x,y,z}$ are unit vectors, N is concentration of the metaparticles,

$$m_{x,y,z} = (S/c)I_{L2,\,3,\,1}^{(1)},\ \pi_{x,y,z} = (l_{\mathrm{eff}}/i\omega)I_{C_a1,2,3}^{(1)} \tag{2.3}$$

For the EMW with the fields $\sim \exp[i(\omega t - kz)]$, where ω, k, z are the frequency, wavenumber, and a direction of the propagation, respectively. Similary to (2.2) and (2.3) the nonlinear part of the first harmonics of electric and magnetic polarizations have the form

$$\vec{P}_{\mathrm{NL}}^{\mathrm{incl}(1)} = (Nl_{\mathrm{eff}}/3i\omega)(\delta I_{C_a NL1}^{(1)}\vec{e}_x + \delta I_{\mathrm{LNL3}}^{(1)}\vec{e}_y + \delta I_{\mathrm{LNL1}}^{(1)}\vec{e}_z) \tag{2.4}$$

$$\vec{M}_{\mathrm{NL}}^{\mathrm{incl}(1)} = (NS/3c)(\delta I_{\mathrm{LNL2}}^{(1)}\vec{e}_x + \delta I_{\mathrm{LNL3}}^{(1)}\vec{e}_y + \delta I_{\mathrm{LNL1}}^{(1)}\vec{e}_z) \tag{2.5}$$

where $\delta I_{C_a\mathrm{NL1}}^{(1)}$, $\delta I_{C_a\mathrm{NL2}}^{(1)}$, $\delta I_{C_a\mathrm{NL3}}^{(1)}$ and $\delta I_{\mathrm{LNL1}}^{(1)}$, $\delta I_{\mathrm{LNL2}}^{(1)}$, $\delta I_{\mathrm{LNL3}}^{(1)}$ are nonlinear components of the currents on the faces 1, 2, 3 of the elementary cell. Note that the currents $\delta I_{C_a\mathrm{NL1}}^{(1)}$, $\delta I_{C_a\mathrm{NL2}}^{(1)}$, $\delta I_{C_a\mathrm{NL3}}^{(1)}$ and $\delta I_{\mathrm{LNL1}}^{(1)}$, $\delta I_{\mathrm{LNL2}}^{(1)}$, $\delta I_{\mathrm{LNL3}}^{(1)}$ are flowing through the electric and magnetic dipoles, respectively. On the other hand, the linear parts of the polarization can also be represented through the local fields in the form (Auzanneau and Ziolkowski 1998, Tretyakov 2003, Tretyakov and Mariotte 1995, Tretyakov *et al* 1994, Yatsenko *et al* 2003).

$$\begin{aligned}\vec{P}_{\mathrm{LIN}}^{\mathrm{incl}(1)} &= N(\langle a_{\mathrm{ee}}\rangle \vec{E}^{(1)}{}_{\mathrm{loc}} + \langle a_{\mathrm{em}}\rangle \vec{H}^{(1)}{}_{\mathrm{loc}}),\ \vec{M}_{\mathrm{LIN}}^{\mathrm{incl}(1)} \\ &= N(\langle a_{\mathrm{me}}\rangle \vec{E}^{(1)}{}_{\mathrm{loc}} + \langle a_{\mathrm{mm}}\rangle \vec{H}^{(1)}{}_{\mathrm{loc}})\end{aligned} \tag{2.6}$$

Diads of polarizabilities, $\langle a_{ee}\rangle$, $\langle a_{em}\rangle$, $\langle a_{me}\rangle$, $\langle a_{mm}\rangle$ (see the relations (A.11) and (A.12)) determine the individual linear polarizations of AMs. Expressions (A.11) and (A.12) coincide with the expressions obtained in Auzanneau and Ziolkowski (1998) in the corresponding boundary cases and correspond to Rapoport (2015a), Rapoport and Boardman (2015b) with an obvious correction for the location of all metaparticles for the geometry (Rapoport 2015a, Rapoport and Boardman 2015b) in one plane. In the absence of loads, these expressions are reduced to the corresponding formulas (Tretyakov *et al* 1994). After finding the currents passing through the elements (figure 2.1), respectively, the linear dyads were found by comparing (2.2) with (2.6). The relations are similar (2.1), (2.2) and (2.6), also for nonlinear polarizations. To find the linear and nonlinear parts of the first harmonics of currents, the Kirchhoff rules are used for contours of the circuit in figure 2.1(b). A system of nonlinear equations for currents and voltages is considered analytically or semi-analytically, after obtaining an analytical equation for voltage on a nonlinear element (figure 2.1). Then, for the solution of this nonlinear equation, an exact numerical method or an approximate analytic method of perturbations (or a small parameter) can be used. We will consider currents and voltages in a quasi-stationary form, the method of harmonic balance (Ionkin 1972, Kadomtsev 1988, Migulin *et al* 1988, Rabinovich and Trubetskov 1989), ignoring the slow dependence of their amplitudes on time, and assuming that for the first harmonic of all values, the dependence on time is concentrated only in the corresponding exponents. When using this method, we also take into account that for the considered cubic nonlinearities (see, for example, the relation (A.4)), the second and zero harmonics are insignificant/absent. The formulas (2.1)–(2.5) and (2.6), (A.11), (A.12) and (A.13)–(A.17), obtained in this approximation completely determine linear and nonlinear polarizations and dyads for Ω-particles by the perturbation method in an area not very close to the resonance of metaparticles (Rapoport *et al* 2006a, Rapoport 2014b, Rapoport and Boardman 2015b). Linear homogenization, that is, characterization for MMs as a continuous medium with metaparticles in the shape of Ω-particles, is considered in appendix A.2. The aforementioned approach to the characterization of (nonlinear) metaparticles is an integral part of the cross-sectional simulation method for determining the characteristics of wave processes in MMs (Boardman *et al* 2007b, 2009b, Rapoport *et al* 2006a, Rapoport 2013, 2014b). A separate case of nonlinear characterization of the MM beyond the perturbation method and the determination of the characteristics of nonlinear EMWs in the corresponding layered bi-anisotropic MM environment will be discussed in section 2.3.

2.1.2 Individual polarizations of the ring resonators with the gap. The necessary conditions for the change of the sign of the losses

This is one of the most important issues for MMs, fundamental for them, as for resonance environments since the signs of 'ε, μ' can change, and there is an 'interesting and unusual' dispersion, namely, the corresponding resonances. Let us consider the necessary conditions for compensating spatial losses, i.e. changes in the damping sign (Boardman and Marinov 2006) and start with MMs, which include metaparticles in the form of SRR. In the case of SRR, we have an equivalent circuit in figure 2.1(d), where only the elements shown to the right of the element $Z_{C_{\text{geff}}}$ in figure 2.1(c) are left. We take into account that in this case $l_{\text{eff}} = 0$ and we obtain, instead of (2.6) and (A.12), for linear polarizations:

$$\vec{M}_{\text{LIN}}^{\text{incl}(1)} = N\langle a_{\text{mm}}\rangle \vec{H}^{(1)}{}_{\text{loc}} \tag{2.7}$$

The electric polarization is the same as in the absence of metaparticles, in the absence of long electric rods, which are used to provide a 'negative' plasmon type of ε (Eleftheriades and Balmain 2005). Polarizabilities of such rods, according to Eleftheriades and Balmain (2005), are independent of those for rings, in the same approximation as one used in Eleftheriades and Balmain (2005) and other contemporary works. For the linear dyad, which is included in (2.7) and (A.11), we obtain (A.11) and (A.12):

$$a_{\text{mmzzLIN}} = a'_{\text{mmzzLIN}} + a''_{\text{mmzzLIN}} = -(i\omega S^2 Z_{00})/(c^2 Z_{L_1})[ZC_{\text{geff}}^{-1} + Z_D^{-1}] \tag{2.8}$$

where Z_{00}^{-1} are determined by the formula (A.5) with $Z_{a_1} \to \infty$. It is possible to present (2.8) in the form:

$$a_{\text{mmzzLIN}} = -(i\omega S^2/c^2)Z_{\text{eff}}^{-1} \equiv -(i\omega S^2/c^2)[Z_{L1} + (Z_D^{-1} + Z_{C_{\text{geff}}}^{-1})^{-1}]^{-1} \tag{2.9}$$

Present $Z_{C_{\text{geff}}}$ in the equivalent form, introducing the 'effective capacitance' C_{geff} with the help of the relation (see figures 2.1(c) and (e)) and corresponding figure captions:

$$Z_{C_{\text{geff}}}^{-1} \equiv i\omega C_{\text{geff}} = i\omega C_g + i\omega C_{\text{comp}} + (i\omega L_{\text{comp}})^{-1} \tag{2.10}$$

All other dyads in this case are equal to zero, i.e. metaparticles in the form of SRR give contributions only into the magnetic polarization. Then we obtain for the dyad a_{mmzzLIN}:

$$\begin{aligned} a_{\text{mmzzLIN}} &= -(i\omega S^2/c^2)(E + iF)^{-1}, \ E \equiv \text{Re}(Z_{\text{eff}}) \equiv Z'_{\text{eff}} \\ &= R_{\text{L}} + [1 + (\omega C_{\text{geff}} Z_{\text{D}})^2]^{-1} Z_{\text{D}} \end{aligned}$$

$$F \equiv \text{Im}(Z_{\text{eff}}) \equiv Z''_{\text{eff}} = \omega L - [1 + (\omega C_{\text{geff}} Z_{\text{D}})^2]^{-1} \omega C_{\text{geff}} Z_{\text{Deff}}^2 \tag{2.11}$$

Accounting for (A.5), it is possible to present (2.11) in the form:

$$E \equiv \text{Re}(Z_{\text{eff}}) \equiv Z'_{\text{eff}} = R_{\text{L}} + g_{0\text{eff}}{}^{-1}, \ F \equiv \text{Im}(Z_{\text{eff}}) \equiv \omega L - (\omega C_{\text{geff}}/g_0 g_{0\text{eff}}),$$

$$g_{\text{oeff}}^{-1} = [g_0^2 + (\omega C_{\text{geff}})^2]^{-1} g_0 \tag{2.12}$$

To provide the change of the sign of losses, in other words to replace the damping by amplification of EMW in the media, the sign of $\text{Im}(a_{\text{mmzzLIN}}) \sim (E \equiv \text{Re}(Z_{\text{eff}}))$ should change. The condition, when sign of E changes, takes the form:

$$E = \text{Re}(Z_{\text{eff}}) < 0, \quad \text{or } R_{\text{L}} + [g_0^2 + (\omega C_{\text{geff}})^2]^{-1} g_0 < 0 \tag{2.13}$$

At the same time, we are interested in the conditions, when the system is close to the resonance. In this case the effects of nonlinearity can be expressed much more strongly (Pendry *et al* 1999). Let us show that

$$\begin{gathered}\omega_{0\mathrm{res}}^2 \approx \omega_0^2 \equiv (LC_\mathrm{g})^{-1},\ \text{if}\ g_0^2 \ll (\omega C_\mathrm{geff})^2,\\ C_\mathrm{D} \ll C_\mathrm{g},\ L_\mathrm{comp} = 0,\ C_\mathrm{comp} = \infty,\end{gathered} \tag{2.14}$$

that is, under conditions (2.14), the resonance frequency $\omega_{0\mathrm{res}}$ is close to the resonance frequency ω_0 of the ideal/lossless oscillatory circuit. Indeed, if the loss (or gain) is not very significant (in more details this will be discussed in section 2.1.3.2), then the resonance, when performing (2.14) and the relation $C_\mathrm{geff} \approx C_\mathrm{g}$, and taking into account (2.12), is determined by the conditions

$$\begin{gathered}F = \mathrm{Im}(Z_\mathrm{eff}) = 0,\ \ F = \mathrm{Im}(Z_\mathrm{eff}) \approx (\omega / C_\mathrm{geff})(\omega_{0\mathrm{res}}^{-2} - \omega^{-2}),\ \omega_{0\mathrm{res}}^2\\ = (LC_\mathrm{geff})^{-1},\ \omega_{0\mathrm{res}}^2 \approx \omega_0^2\end{gathered} \tag{2.15}$$

The relation (2.15) proves the proximity of the resonant frequency of a single metaparticle to the corresponding frequency (2.14) in the 'ideal' oscillatory circuit. As will be shown below (see section 2.1.3.2), in the MM the resonance frequency shifts, as compared with (2.14), due to the presence of a set of metaparticles and the difference of the local field from the field in the media in the absence of artificial inclusions and scattering of EMWs on the metaparticles.

The relations (2.7) and the relations (A.17), obtained in appendix A.2, completely determine the linear and nonlinear (in the region of small nonlinearity) electrodynamic characteristics of the individual metaparticles in the shape of SRR, as well as macroscopic nonlinear polarizations; but the latter have been defined by the local values of the fields, which differ from the fields of the incident waves. The principles of linear and nonlinear homogenization based on the method of perturbation are presented in section 2.1.3 and appendix A.3. The approach to the nonlinear homogenization for a certain type of a bi-anisotropic MM 'out of the frames of the perturbation method' is described in section 2.3, and a concrete example of the use of the general method of modeling nonlinear waves in such MMs 'from the properties of a single metatparticle to determining the characteristics of nonlinear waves in a layered system' (Boardman *et al* 2007b, 2009b, Rapoport *et al* 2006a, Rapoport 2013, 2014b, Rapoport and Boardman 2015b) is discussed in section 6.2 of chapter 6.

2.1.3 Principles of linear and nonlinear homogenization based on the method of perturbations

Now let us show how local fields, polarizations, and dyads that characterize MMs with AMs in the shape of 'particles' and SRR are found, which, in general, is called the 'homogenization' (Capolino 2009, Eleftheriades and Balmain 2005, Tretyakov 2010). At the same time, the coefficients determining the material relations for the MM as an effective continuous medium (Boardman *et al* 2007, 2009a, 2009b,

Rapoport *et al* 2006a, Rapoport 2013, 2014b, Rapoport and Boardman 2015b) will be found.

2.1.3.1 General relations for the tensors of the linear polarizabilities and local susceptibilities and local fields for the homogenization of the metamaterials with Ω-particles by means of the perturbation method

In the previous sections, the concept of a local field was used as a field acting in a certain small area of an MM. Such a field includes also a field scattered by the metaparticles (Capolino 2009, Eleftheriades and Balmain 2005, Tretyakov 2010). We will now look at these fields in more detail, find them for certain structures, and obtain the effective material relations. We use firstly the known in literature approach and give the corresponding linear relations that determine the local fields for metaparticles of arbitrary shapes forming a cubic lattice in the Lorentz–Lorenz approximation (Boardman *et al* 2007a, 2009a, Rapoport *et al* 2006a, Rapoport 2013, 2014b, Rapoport and Boardman 2015b). The essence of this approach is that: (a) the electromotive forces included in the equivalent circuit of the corresponding metaparticle (see, for example, figures 2.1(a) and (e)), and then the polarization is expressed through the local fields; (b) the nonlinear polarization is sought by the method of perturbations (in particular, by the method of the harmonic balance) in the form of decompositions in a series of degrees of amplitudes of electromotive forces, which are reduced to the series in powers of components of local fields, which, finally, are expressed through the 'external' fields of incident EMWs (Rapoport *et al* 2006a). Later, for simplification, this method will be referred to as the method of local fields (MLF). In this section we describe these methods and corresponding results explicitly (see formulas (2.21)–(2.24) and (2.27)–(2.29)). We compare the results obtained by known methods with the results of our developed method of (nonlinear) homogenization for bi-anisotropic MMs in the nonlinear case and, as a particular case, in the linear approximation. In the nonlinear case we will consider the lowest order of perturbations (a small parameter), in which such a comparison is practically possible. Our proposed nonlinear homogenization method for bi-anisotropic media, in contrast to the known method, (a) includes the representation of polarizations in terms of voltages and currents (which arise with the participation of local fields) in the equivalent circuit, and (b) is not limited, in the general case, by the conditions of applicability of the perturbation method. This method, in the future for simplicity and to distinguish it from the MLF, will be called the method of local currents (MLC). The homogenization according to the standard MLF is as follows. As shown in Tretyakov *et al* (1996) and Tretyakov (2010), in the Lorentz–Lorenz approximation, the local fields are determined by the relations

$$\vec{E}_{\text{loc}} = \vec{E} + (4\pi/3\varepsilon)\vec{P}^{\text{incl}},\ \vec{H}_{\text{loc}} = \vec{H} + (4\pi/3\mu)\vec{M}^{\text{incl}} \tag{2.16}$$

Here ε, μ are the dielectric permittivity and magnetic permeability, respectively, of the host medium, which will be supposed for the simplification, to be isotropic. The 'host medium' means the medium-matrix, where the metaparticles are placed, in accordance with the applied technology (Tretyakov 2010). $\vec{P}_{\rm incl}$, $\vec{M}_{\rm incl}$ are electrical and magnetic polarizations, respectively, associated with available methods. $\vec{E}$, $\vec{H}$ are the average macroscopic fields that act in the MM. These fields are included into the 'homogenizing' material relations

$$\vec{D} = \widehat{\varepsilon}_{\rm eff}\vec{E} + \widehat{\alpha}_{\rm eff}\vec{H},\ \vec{B} = \widehat{\mu}_{\rm eff}\vec{H} + \widehat{\beta}_{\rm eff}\vec{E} \tag{2.17}$$

where $\vec{D}$, $\vec{B}$ are the vectors of the electric and magnetic induction, respectively. In accordance with Tretyakov and Mariotte (1995), in the linear case, the tensors, included in (2.17), may be presented in the form:

$$\widehat{\alpha}_{\rm eff} = 4\pi\widehat{T}_{\rm PH},\ \widehat{\beta}_{\rm eff} = 4\pi\widehat{T}_{\rm ME},\ \widehat{\mu}_{\rm eff} = \mu\widehat{I} + 4\pi\widehat{T}_{\rm MH},\ \widehat{\varepsilon}_{\rm eff} = \varepsilon\widehat{I} + 4\pi\widehat{T}_{\rm PE}, \tag{2.18}$$

$$\vec{D} = \varepsilon\vec{E} + 4\pi\vec{P}^{\rm incl},\ \vec{B} = \mu\vec{H} + 4\pi\vec{M}^{\rm incl} \tag{2.19}$$

Using (2.17)–(2.19), we obtain

$$\vec{P}^{\rm incl} = \widehat{T}_{\rm PE}\vec{E} + \widehat{T}_{\rm PH}\vec{H},\quad \vec{M}^{\rm incl} = \widehat{T}_{\rm ME}\vec{E} + \widehat{T}_{\rm MH}\vec{H}, \tag{2.20}$$

$$\widehat{T}_{\rm PE} = \widehat{K}^{-1}\left[N\langle\widehat{a}_{\rm ee}\rangle + N^2\langle\widehat{a}_{\rm em}\rangle\frac{4\pi}{3\mu}\left(\widehat{I} - N\langle\widehat{a}_{\rm mm}\rangle\frac{4\pi}{3\mu}\right)^{-1}\langle\widehat{a}_{\rm me}\rangle\right] \tag{2.21}$$

$$\widehat{T}_{\rm MH} = \widehat{K}_2^{-1}\left[N\langle\widehat{a}_{\rm mm}\rangle + N^2\langle\widehat{a}_{\rm em}\rangle\frac{4\pi}{3\varepsilon}\left(\widehat{I} - N\langle\widehat{a}_{\rm ee}\rangle\frac{4\pi}{3\varepsilon}\right)^{-1}\langle\widehat{a}_{\rm mm}\rangle\right] \tag{2.22}$$

$$\widehat{\alpha}_{\rm eff} = 4\pi N\widehat{K}_1^{-1}\langle\widehat{a}_{\rm em}\rangle\left[\widehat{I} + N\frac{4\pi}{3\mu}\left(\widehat{I} - N\langle\widehat{a}_{\rm mm}\rangle\frac{4\pi}{3\mu}\right)^{-1}\langle\widehat{a}_{\rm mm}\rangle\right] \tag{2.23}$$

$$\widehat{\beta}_{\rm eff} = 4\pi N\widehat{K}_2^{-1}\langle\widehat{a}_{\rm me}\rangle\left[\widehat{I} + N\frac{4\pi}{3\varepsilon}\left(\widehat{I} - N\langle\widehat{a}_{\rm ee}\rangle\frac{4\pi}{3\varepsilon}\right)^{-1}\langle\widehat{a}_{\rm ee}\rangle\right] \tag{2.24}$$

Here

$$\widehat{K}_1 = \left[\widehat{I} - N\langle\widehat{a}_{\rm ee}\rangle\frac{4\pi}{3\varepsilon} - N^2\langle\widehat{a}_{\rm em}\rangle\frac{4\pi}{3\mu}\left(\widehat{I} - N\langle\widehat{a}_{\rm mm}\rangle\frac{4\pi}{3\mu}\right)^{-1}\langle\widehat{a}_{\rm me}\rangle\frac{4\pi}{3\varepsilon}\right] \tag{2.25}$$

$$\widehat{K}_2 = \left[\widehat{I} - N\langle\widehat{a}_{\rm mm}\rangle\frac{4\pi}{3\mu} - N^2\langle\widehat{a}_{\rm me}\rangle\frac{4\pi}{3\varepsilon}\left(\widehat{I} - N\langle\widehat{a}_{\rm ee}\rangle\frac{4\pi}{3\varepsilon}\right)^{-1}\langle\widehat{a}_{\rm em}\rangle\frac{4\pi}{3\mu}\right] \tag{2.26}$$

As follows from (2.8), (2.11a) and (2.11b),

$$\widehat{\varepsilon}_{\rm eff} = \varepsilon\widehat{I} + 4\pi\widehat{K}_1^{-1}N\left[\langle\widehat{a}_{\rm ee}\rangle + N\langle\widehat{a}_{\rm em}\rangle\frac{4\pi}{3\mu}\left(\widehat{I} - N\langle\widehat{a}_{\rm mm}\rangle\frac{4\pi}{3\mu}\right)^{-1}\langle\widehat{a}_{\rm me}\rangle\right] \tag{2.27}$$

$$\widehat{\mu}_{\rm eff} = \mu\widehat{I} + 4\pi\widehat{K}_2^{-1}N\left[\langle\widehat{a}_{\rm mm}\rangle + N\langle\widehat{a}_{\rm me}\rangle\frac{4\pi}{3\varepsilon}\left(\widehat{I} - N\langle\widehat{a}_{\rm ee}\rangle\frac{4\pi}{3\varepsilon}\right)^{-1}\langle\widehat{a}_{\rm em}\rangle\right] \tag{2.28}$$

Using (2.20) and (2.16), we obtain the local fields:

$$\vec{E}_{\text{loc}} = \frac{1}{3}\left(2\widehat{I} + \frac{\widehat{\varepsilon}_{\text{eff}}}{\varepsilon}\right)\vec{E} + \frac{1}{3\varepsilon}\widehat{\alpha}_{\text{eff}}\vec{H},\ \vec{H}_{\text{loc}} = \frac{1}{3}\left(2\widehat{I} + \frac{\widehat{\mu}_{\text{eff}}}{\mu}\right)\vec{H} + \frac{1}{3\mu}\widehat{\beta}_{\text{eff}}\vec{E} \quad (2.29)$$

We emphasize again that formulas (2.18) correspond to the linear case. In the case of nonlinear electric and magnetic polarizations, they are found by means of equations (2.4) and (2.5), where the nonlinear components of the currents $\delta I^{(1)}_{\text{LNL}k}$, $\delta I^{(1)}_{C_a\text{NL}k}$ are determined by equations (A.9) and (A.10), (when the system is not very close to the resonant), by the method of successive approximations (Lakhtakia and Weiglhofer 1998) for MMs containing Ω-particles and SRR. The linear homogenization for MMs with metaparticles in the shape of particles, as well as the principles of the nonlinear homogenization by the perturbation method are considered in appendix A.3 (Boardman *et al* 2007b, Rapoport *et al* 2006a, Rapoport 2014b, 2015a, Rapoport and Boardman 2015b). The local fields and electromotive forces included in formulas (A.9) and (A.10) (appendix A.2) are determined by equations (2.29) and (A.27)–(A.32) (appendix A.3) respectively. As a result, in this section, taking into account appendices A.2 and A.3, the possibility of homogenizing the linear and nonlinear characteristics of a bi-anisotropic MM with Ω-particles is shown in the scope of the method of perturbations. It is valid in the case of weak nonlinearity and for frequencies that are not very close to the resonance frequency (Boardman and Marinov 2006, Boardman *et al* 2007b, Rapoport *et al* 2006a, Rapoport and Boardman 2015b). The original more effective method of the nonlinear homogenization for bi-anisotropic MMs 'beyond limits of the perturbation method' (MLC) is put forward in section 2.3. Note that in a series of works including (Lakhtakia and Weiglhofer 1998) (see also the reference in this paper to other articles involving the same authors) the perturbation method was used; in this way, the results of this section correspond to Lakhtakia and Weiglhofer (1998) qualitatively (by approach). But in the works of the series to which (Lakhtakia and Weiglhofer 1998) belongs, the results are not applied to a medium with metaparticles of a certain shape. In contrast, the nonlinear polarization (A.13) and (A.14), (appendix A.2) is obtained for a 'homogenized' MM with bi-anisotropic Ω-particles with nonlinear and active 'loads' that have a specific nonlinearity in the form (A.4) (appendix A.1).

2.1.3.2 Conditions of the change of sign of energy losses for the EMW in the approximation of the frequency-independent conductivity of active loads for the metamaterials with SRR

The question of the possible change in the sign of energy losses of EMW in an MM with metaparticles in the shape of SRR has already been discussed in part in section 2.1.2, but only 'from the standpoint of an individual particle.' We now consider this question in terms of an MM as a continuous medium in the Lorentz–Lorenz approximation (or Maxwell–Garnet (Tretyakov and Mariotte 1995, Tretyakov *et al* 1996, Tretyakov 2003, 2010)) used in Boardman and Marinov (2006). As a result, we obtain more precise conditions for amplification or partial compensation of energy losses in MMs with SRRs that interfere with 'active devices', such as Gunn diodes

(Bosch and Thim 1974, Bosch and Engellmann 1975) or resonance tunnel diodes (RTDs) (Kidner *et al* 1990, Ookawa *et al* 2006). We will not discuss, just now, the details of the gain's creation. Note only that placing biased Gunn, tunnel, or some other type of diodes (Andryieuski *et al* 2012, Bragin *et al* 1974, Bravo–Ortega and Glassor 1991, Bullough and Caudrey 1980) in the gap of SRR would be enough to provide the negative resistance and amplification, figure 2.2.

For an MM with metaparticles of the shapes of SRR, the metaparticles are only modified by the magnetic permeability, as can be seen from the relations of section 2.1.1, in particular (2.2) and (2.3) with $l_{\text{eff}} = 0$ taking into account (A.27). Using the relations (A.18), (2.8)–(2.12) and (2.6) we obtain for the linear part of the magnetic permeability (Boardman and Marinov 2006):

$$\mu''_{\text{eff0}} = 3\mu\varsigma_0\delta''_\mu/H = -3\mu\zeta_0\omega_1^2\omega_{0\text{res}}^2R_{\text{eff}}/(|Z_{\text{geff}}|\delta_\omega^4) \tag{2.30}$$

$$\mu''_{\text{eff0}} = \mu[1 + 3\varsigma_0(\delta'_\mu - \varsigma_0\delta_{\mu 2})/H] = \mu[1 - 3\zeta_0\omega_1^2(\Delta^2 + \zeta_0\omega_1^2)/\delta_\omega^4] \tag{2.31}$$

$$\Delta^2 = \omega^2 - \omega_{0\text{res}}^2,\ \varsigma_0 = 4\pi/9\mu,\ \delta_{\mu 2} = \delta'^2_\mu + \delta''^2_\mu,\ H \equiv 1 - 2\varsigma_0\delta'_\mu + \varsigma_0^2\delta_{\mu 2} \tag{2.32}$$

$$\delta_\omega^4 = (\Delta^2 + \zeta_0\omega_1^2)^2 + \omega_{0\text{res}}^4(R_{\text{eff}}/|Z_{\text{geff}}|)^2,\ \omega_1^2 = NS^2\omega_{0\text{res}}^2\omega/(|Z_{C\text{geff}}|c^2) \tag{2.33}$$

$$\begin{aligned}\delta'_\mu &= -(N\omega S^2/c^2)Z''_{\text{eff}}(Z'_{\text{eff}}2 + Z''_{\text{eff}}2)^{-1}\\ &= -[NS^2\omega_{0\text{res}}^2\omega/(|Z_{C_{\text{geff}}}|c^2)](\omega^2 - \omega_{0\text{res}}^2)X_0^{-1}\end{aligned} \tag{2.34}$$

$$\delta''_\mu = -(N\omega S^2/c^2)Z'_{\text{eff}}(Z'_{\text{eff}}2 + Z''_{\text{eff}}2)^{-1} = -[NS^2R_{\text{eff}}\omega_{0\text{res}}^4\omega/(c^2\,|Z_{C_{\text{geff}}}|^2)]X_0^{-1} \tag{2.35}$$

$$R_{\text{eff}} = g_0/(\omega C_{\text{geff}})^2 + R_L \approx Z'_{\text{eff}},\ X_0 \equiv (\omega^2 - \omega_{0\text{res}}^2)^2 + \omega_{0\text{res}}^4(R_{\text{eff}}/|Z_{C_{\text{eff}}}|)^2 \tag{2.36}$$

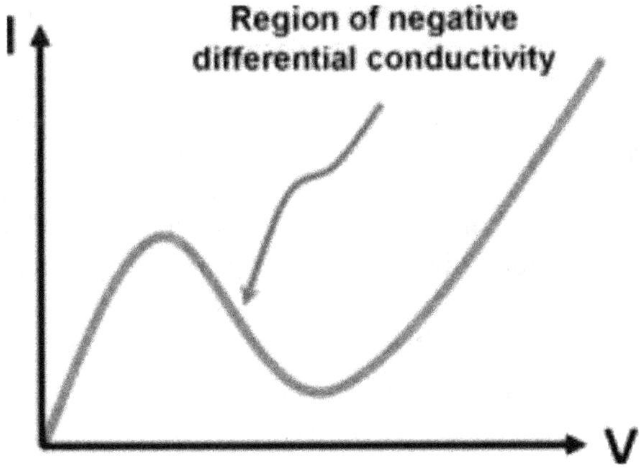

Figure 2.2. Example of current–voltage characteristic with negative differential conductivity in the case of tunnel diodes (Rabinovich and Trubetskov 1989, Buttner *et al* 2000a, Ookawa *et al* 2006, Chebykin *et al* 2011). Reprinted from figure 7 in Boardman *et al* (2011), copyright (2011) with permission from John Wiley & Sons.

$$\Delta^2 = \omega^2 - \omega_{0\mathrm{res}}^2,\ \delta_\omega^4 = (\Delta^2 + \zeta_0\omega_1^2)^2 + \omega_{0\mathrm{res}}^4(R_{\mathrm{eff}}/|Z_{\mathrm{geff}}|)^2,$$
$$\omega_1^2 = NS^2\omega_{0\mathrm{res}}^2\omega/(|Z_{C_{\mathrm{geff}}}|c^2) \tag{2.37}$$

and the quantities Z'_{eff}, Z''_{eff} are determined by the formulas (2.11). Note that the expressions (2.30) and (2.31) in the form (Boardman *et al* 2007a) were used in Perez-Molina *et al* (2009) to analyze laser modes based on the Fabry–Perot resonator with active MM. On the basis of this, a conclusion was drawn that the proposed approach is promising for further development in the field of MM lasers. The analysis of the expressions (2.30) and (2.31), taking into account (2.32)–(2.37), shows that the magnetic permeability has a pronounced resonant character. It is seen that the value describes the contribution of the collective effect of metaparticles to the magnetic permeability and, in particular, to the resonance, and is proportional to the concentration of metaparticles. To show the resonance conditions more clearly, consider the magnetic permeability in the absence of 'active losses', what means that we have a limiting case (taking into account (2.30) and (2.31))

$$R_{\mathrm{eff}}/|Z_{C\mathrm{geff}}| \ll 1,\ 3\zeta_0\omega_1^2 \geqslant (\Delta^2 + \varsigma_0\omega_1^2) > >\omega_{0\mathrm{res}}^2(R_{\mathrm{eff}}/|Z_{C_{\mathrm{geff}}}|) \tag{2.38}$$

$$\mu'_{\mathrm{eff0}} = \mu[1 - 3\zeta_0\omega_1^2(\Delta^2 + \zeta_0\omega_1^2)^{-1}],\ |\mu''_{\mathrm{eff0}}/\mu'_{\mathrm{eff0}}|$$
$$\sim \omega_{0\mathrm{res}}^2 R_{\mathrm{eff}}[|Z_{C_{\mathrm{geff}}}|(\Delta^2\zeta_0\omega_1^2)]^{-1} \ll 1 \tag{2.39}$$

Thus, as can be seen from (2.39), the presence of an 'average field' associated with the scattering ofni electromagnetic fields by a metaparticle leads to a decrease in the resonant frequency in the MM described as a continuous medium, in comparison with the resonance frequency of a single metatparticle in the shape of SRR. Moreover, the magnitude of the displacement (in the direction of decrease) of the resonant frequency of the MM in comparison with the frequency of a single metatparticle is proportional to the concentration of metaparticles in the MM in accordance with Capolino (2009), Eleftheriades and Balmain (2005) and Tretyakov (2010). Consider the case of finite dissipation, which may have a negative or positive sign in accordance with the sign of R_{eff}, see expressions (2.30) and (2.36). Determine the conditions in which the MM is a medium with negative values of dielectric permittivity and magnetic permeability, the EMWs are backward and amplifying, and at the same time, the 'negative dissipation' is relatively small, which is a prerequisite for the propagation of EMW in such an environment as waves with a slowly varying amplitude. Obviously, in order to complete the latter condition, it is necessary to satisfy the inequality

$$|\mu''_{\mathrm{eff0}}/\mu'_{\mathrm{eff0}}| \ll 1 \tag{2.40}$$

In appendix A.4, the more precise conditions are provided that the amplitude of the corresponding EMW changes slowly (see (A.35)). Let us denote

$$n_0 = (\Delta^2 + \zeta_0\omega_1^2)/\omega_{\mathrm{dis}}^2,\ \omega_{\mathrm{dis}}^2 = \omega_{0\mathrm{res}}^2(R_{\mathrm{eff}}/|Z_{C_{\mathrm{geff}}}|). \tag{2.41}$$

As follows from (2.41), the value of n_0 is actually a measure (at least in the case of the frequency close to the resonant one) of the relative frequency deviation from its resonant value in units of the value of the effective 'dissipative frequency' ω_{dis}. The square of ω_{dis} which is proportional to the square of the resonance frequency and the normalized value of the 'resistance of the losses/enhancement R_{eff}'. Taking into account (2.40) and (2.38), it can be shown that, to fulfill the condition

$$\mu''_{eff0} < 0 \tag{2.42}$$

which is necessary for obtaining a 'negative phase medium' (Eleftheriades and Balmain 2005), in turn, the following conditions are necessary:

$$\begin{gathered} 3\zeta_0\omega_1^2(\Delta^2 + \zeta_0\omega_1^2) \geqslant \delta_\omega^4,\ \omega_{0res}^2\left(\frac{R_{eff}}{|Z_{geff}|}\right) \ll (\Delta^2 + \zeta_0\omega_1^2),\ n_0 \gg 1, \\ 3\zeta_0\omega_1^2 \geqslant n_0\omega_{0res}^2\frac{R_{eff}}{|Z_{geff}|} \end{gathered} \tag{2.43}$$

Under the fulfillment of relations (2.43), the inequality (2.39) is valid and takes the form

$$|\mu''_{eff0}/\mu'_{eff0}| \sim n_0^{-1} \ll 1 \tag{2.44}$$

Numerical estimations for both μ', μ'', and 'characteristic amplification length' of the EMW in the active medium (Rapoport *et al* 2006a), are given in appendix A.4 (see (A.34)). Taking into account the obtained relations (A.34) and (A.35), it is possible to realize both the conditions specified in Rapoport *et al* (2006a), namely $L_{ampl}/\lambda \gg 1$ or $|\mu''_{eff0}/\mu'_{eff0}| \sim 1$; $L_{ampl}/\lambda \sim 1$ (the condition of 'supereffective gain'). Note that for MMs with SRR the quantities ε_{eff0} and μ_{eff0N} are independent, in the approximation used, for example, in Capolino (2009), Eleftheriades and Balmain (2005) and Tretyakov (2010) and in this chapter. Therefore, jointly with the negative μ_{eff0N}, it is possible to provide, using the standard method (for example, with application of metal rods (Capolino 2009, Eleftheriades and Balmain 2005, Tretyakov 2010)), also the negative value of ε_{eff0}. Thus, the possibility of constructing negative phase medium (NPM) is shown. The aspects of the stability of the NPM taking into account the frequency dependence of the 'conductivity of active loading' of AM (SRR) and the possibility of constructing an MM with spatial amplification of EMW are discussed in section 2.2. This problem is very nontrivial, since each separate active metaparticle is, in fact, an elementary generator, and the medium as a whole may be subject to the **absolute instability (AI) (generation)** (Lifshitz and Pitaevskii 1981, Fedorchenko and Kotsarenko 1981), while the spatial amplification becomes impossible.

Let us make a note on the absolute and convective instabilities (Boardman *et al* 2007a, 2008, 2011). As is well-known from communication systems, such as ordinary radio/TV transmission systems or radar systems, there are two main devices transforming energy of applied external electric field into electromagnetic oscillations or waves, namely generators and amplifiers. Two types of instabilities,

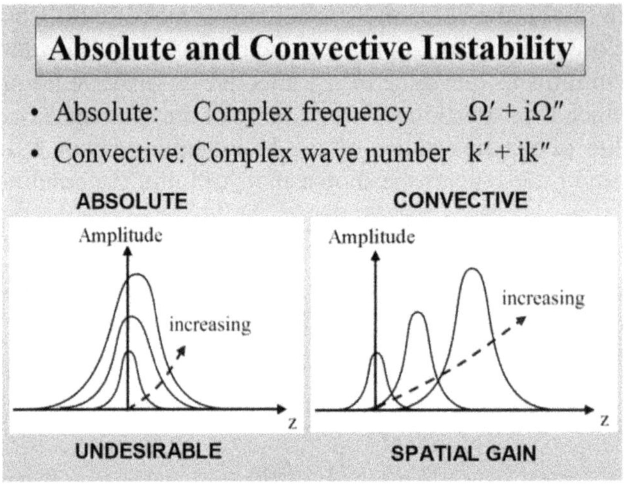

Figure 2.3. Interpretation of the absolute and convective instabilities using complex frequency ($\Omega' + i\Omega''$) and wavenumber ($k' + ik''$), respectively. Reprinted from figure 6 in Boardman *et al* (2011), copyright (2011) with permission from John Wiley & Sons.

namely, absolute and convective instabilities, correspond to the two above-mentioned regimes, namely, (linear) amplification and generation. In the first case, field grows with time in any point of space, while in the second case the pulse/beam amplitude increases in space along the direction of propagation (figure 2.3). These two types of instabilities can be associated with complex frequency or wavenumber, respectively (figure 2.3). Therefore to get spatial amplification, a system should possess a convective instability

Note that a concept of 'analogous nanoparticle transmission lines' becomes important now on optical frequencies (Boardman *et al* 2007a, 2007b). The analysis of EMWs moving through an unbounded lossy, or amplifying, MM leads naturally to a dispersion equation. The behavior of the system relates, in general, a complex frequency ω to a complex wave vector $\boldsymbol{k}$. Taking, for simplicity, a particular direction in an isotropic medium, some basic rules concerning AI can be identified (Boardman *et al* 2008, Popov and Shalaev 2006, Mikhailovskii 1974, Pendry *et al* 1999, Govyadinov *et al* 2007).

2.2 Possibility of the existence of active metamaterials with spatial amplification and negative phase behavior

2.2.1 Simplified consideration based on the approximation of the frequency-independent conductivity of the active loading

We start with a simplified analysis of the active MM with metaparticles (SRR), which include active elements whose characteristics do not depend on the operating frequency (Boardman *et al* 2007a) (see also appendix A.5), and then we proceed to a more detailed model taking into account the frequency dependence (Boardman *et al* 2008, 2011). The possibility of creating an active MM, which provides spatial amplification and does not have AI relative to the generation of EMWs, is demonstrated. Consider an environment with metaparticles in the shape of SRR (see figures 2.1(a), (b) and (d), where we consider that 'electric dipoles' in Ω-particles are absent, and Ω-particles are converted into SRRs). As shown in appendix A.5 (see also Boardman *et al* 2007a), the magnetic permeability for such a medium is equal to

$$\mu_{\text{eff0}} = \mu[1 + (8\pi/9\mu)\delta_\mu][1 - (4\pi/9\mu)\delta_\mu]^{-1} \tag{2.45}$$

where

$$\delta_\mu = -i(N\omega S^2/c^2 Z_{\text{eff}}),\ Z_{\text{eff}} \equiv Z'_{\text{eff}} + iZ''_{z\text{eff}} \approx R_{\text{D}}(1 + i\omega C_{\text{g}}R_{\text{D}})^{-1} + R_{\text{L}} + i\omega L$$

$$Z'_{\text{eff}} = R_{\text{L}} + [1 + (\omega C_{\text{g}}R_{\text{D}})^2]^{-1}R_{\text{D}},\ Z_{\text{eff}} \equiv Z'_{\text{eff}} + iZ''_{z\text{eff}}$$
$$\approx R_{\text{D}}(1 + i\omega C_{\text{g}}R_{\text{D}})^{-1} + R_{\text{L}} + i\omega L$$

$$Z'_{\text{eff}} = R_{\text{L}} + [1 + (\omega C_{\text{g}}R_{\text{D}})^2]^{-1}R_{\text{D}},\ Z''_{\text{eff}} = \omega L - [1 + (\omega C_{\text{g}}R_{\text{D}})^2]^{-1}\omega C_{\text{g}}R_{\text{D}}^2$$

$$R_{\text{D}} = 1/g',\ g = g' + ig'',\ C_{\text{g}} \gg C_{\text{D}} \tag{2.46}$$

In the last of the relations (2.46), R_{D}, g, g', g'', C_{D} are the equivalent impedance of the active element (diode), the conductivity of the diode, its real and imaginary parts, and the capacitance that can be included in the equivalent circuit of the diode, respectively (see also (2.14)). For NPM with relatively weak amplification or dissipative losses, in other words when $|\mu'| \gg |\mu''|$, $|\varepsilon'| \gg |\varepsilon''|$, we get (Boardman *et al* 2007a),

$$k_y^2 = k_0^2\varepsilon\mu,\ k_y = k'_y + ik''_y,\ k'_y \approx -\sqrt{\varepsilon'_{\text{eff}}\mu'_{\text{eff}}}\,k_0 < 0,\ k_0 \equiv \omega/c$$

$$k''_y \approx -(1/2)\sqrt{\varepsilon'_{\text{eff}}\mu'_{\text{eff}}}\,k_0[(\mu''_{\text{eff}}/\mu'_{\text{eff}}) + (\varepsilon''_{\text{eff}}/\varepsilon'_{\text{eff}})] \tag{2.47}$$

Assuming that the energy losses are related only to the magnetic permeability, using (2.47) we find that for the MM with negative values of electrical and magnetic permeability, with amplification performed

$$\varepsilon'_{\text{eff}} < 0,\ \mu'_{\text{eff}} < 0,\ k''_y > 0,\ \mu''_{\text{eff}} > 0 \tag{2.48}$$

The latter condition, given (see appendix A.5), is reduced to

$$Z'_{\text{eff}} < 0 \tag{2.49}$$

As can be seen from (2.46), to satisfy the condition (2.49) and the achievement of the spatial amplification of EMW in the MM under consideration, the negative conductivity of the diode, in particular, is necessary. When performing the necessary condition of spatial amplification (2.48) (or (2.49)), the absence of AI of EMW (Lifshitz and Pitaevskii 1981, Fedorchenko and Kotsarenko 1981, Scott 1970) in MM is a sufficient condition. There are several possible sources of AI. As shown in Boardman *et al* (2007, 2008, 2011), such sources that correspond to the poles of the dispersion relation for EMW (Lifshitz and Pitaevskii 1981, Fedorchenko and

Kotsarenko 1981, Scott 1970) cannot be exclusively considered only with the use of the model of a continuous medium. Therefore, they will be discussed in the following paragraphs, section 2.2.2. At the present time, the conditions for the absence of AI will be considered only with the roots of the dispersion relation, which we will analyze in the simplest form for plane waves with fields $\sim \exp[i(\omega t - kz)]$, and at special points $\omega'' > 0$ must be performed. Thus, as shown in Boardman *et al* (2007a, 2009a), in order to find the conditions for the absence of AI (and generation) and the presence of spatial gain determined by the points of bifurcation/the roots of the dispersion function (Lifshitz and Pitaevskii 1981, Fedorchenko and Kotsarenko 1981, Scott 1970), it is necessary to consider the relation

$$V_g k_y'' > 0,\ D(\omega, k_y) = k_y^2 - \varepsilon_{\text{eff}}\mu_{\text{eff}} k_0^2 = 0,\ D_{k_y} = 0 \tag{2.50}$$

Using (2.50) and taking into account that the roots of ε_{eff} in the 'plasmon form' (Eleftheriades and Balmain 2005) do not correspond to the AI, we obtain the conditions $k_y = 0$, $\varepsilon_{\text{eff}}\mu_{\text{eff}} = 0$, that are reduced to the equation for determining the points of the bifurcation in the form

$$\mu_{\text{eff}} = 0 \tag{2.51}$$

As described in appendix A.5, we obtain from (2.51):

$$\omega'_{\text{bifl},2} = \pm[1 + (R_L/R_D) - (1/4)(\omega_{01}C_g Z_0)^2]^{1/2}\omega_{01},\ \omega''_{\text{bifl},2} = (\omega_{01}/2)(\omega_{01}C_g Z_0) \tag{2.52}$$

where

$$Z_0 = R_L + R_D/(\omega_{01}C_g R_D)^2,\ \omega_{01}^2 = [1 - 2(\varsigma_0 NS^2/c^2L)]^{-1}\omega_0^2,\ \varsigma_0 = 4\pi/(9\mu),\ \omega_0^2 = (LC_g)^{-1} \tag{2.53}$$

Taking into account (2.50)–(2.53) and (2.48), we obtain the necessary conditions for the absence of AI and the presence of the spatial gain for NPM/MM with negative ε_{eff}, μ_{eff} at working frequency ω (Lifshitz and Pitaevskii 1981, Fedorchenko and Kotsarenko 1981, Boardman *et al* 2007a):

$$\omega''_{\text{bifl},2} > 0,\ Z_0 > 0,\ \varepsilon'_{\text{eff}} = 1 - \omega_p^2/\omega^2 < 0,\ \mu'_{\text{eff}} < 0,\ \mu''_{\text{eff}} > 0$$

$$Z'_{\text{eff}} = R_L + [1 + (\omega C_g R_D)^2]^{-1} R_D < 0 \tag{2.54}$$

Let us first analyze the possibility of performing the conditions (2.50)–(2.54) necessary for spatial amplification analytically, also taking into account that the presence of an active load/diode should not violate the normal operation of SRR as a resonant metaparticle providing a negative magnetic permeability (Boardman *et al* 2007a). The analysis shows that this is achieved with $(\omega C_g R_D)^2 \gg 1$. Taking into account (2.46) we can obtain that at the resonance frequency (corresponding to the resonance of $|Z''_{\text{eff}}|$ and $|\mu''_{\text{eff}}|$), the value of $|\mu''_{\text{eff}}|$ is relatively very small, is determined approximately as

$$\omega_{\text{res}}^2 \approx [1 + \varsigma_0 NS^2/(c^2L)]^{-1}\omega_0^2 \approx [1 - \varsigma_0 NS^2/(c^2L)]\omega_0^2 \tag{2.55}$$

while for the frequency ω_{01} we have

$$\omega_{01}^2 \approx [1 + 2\varsigma_0 NS^2/(c^2L)]\omega_0^2 \tag{2.56}$$

As seen from (2.46), (2.52) and (2.53) ω_{01} is an effective frequency that defines the values (and signs) of $\omega''_{\text{bifl},2}$, Z_0 at the bifurcation points $\omega_{\text{bifl},2}$. As seen from (2.46), the square of the frequency, at which $Z''_{\text{eff}} = 0$, is equal to $\omega_{00}^2 = [1 + (C_g\omega_0 R_D)^{-2}]\omega_0^2 \approx \omega_0^2$. Taking into account (2.55) and (2.56), the analysis shows that to satisfy the conditions (2.54), or the corresponding conditions (Boardman *et al* 2007, 2009) necessary to achieve the spatial amplification and the absence of AI, it is possible to choose the operating frequency ω so that:

$$(\omega_{01}C_g|R_D|)^{-2}(-R_D) < R_L < [1 + (\omega C_g|R_D|)^2]^{-1}(-R_D) \tag{2.57}$$

The condition necessary to satisfy (2.57)–(5.57) at $\omega \sim \omega_0 \sim \omega_{01}$ is

$$\omega < \omega_{01},\ 6C_g^2|R_D|^2\omega_1^2 > 1;\ \omega_1^2 \equiv [NS^2/(Lc^2)]\omega_{01}^2 \tag{2.58}$$

A very interesting result is that the ability to satisfy the necessary conditions for spatial amplification (2.54) and (2.55), (Boardman *et al* 2007a) according to (2.58) is related to the effect of interaction between particles, which also leads to an appropriate shift of the resonant frequency, proportional to the concentration the metaparticles (2.55) (Capolino 2009, Eleftheriades and Balmain 2005, Tretyakov 2003, 2010) the possibility of conditions (2.54), (2.57) and (2.58) is confirmed by the estimations for the parameters given in appendix A.5 and the corresponding calculations, the results of which are shown in figure A.1 in appendix A.5. Thus, for the MM in the microwave (GHz) range, with metamolecules, loaded with active diodes with a quite small value $|Z'_{\text{eff}}|$, where Z'_{eff} is the real part of effective conductivity, we can provide the spatial amplification of EMW. In other words, for not too powerful diodes, it is possible to obtain MMs with negative ε, μ, where the EMW propagate with slowly varying amplitudes, in the wavelength scale (see also appendix A.4, relation (A.34)). At the same time, in the approximation of independent conduction frequencies of active diodes in the mode of linear amplification of waves, the possibility of avoiding the AI (generation) is practically only due to the roots of the dispersion function (Boardman *et al* 2007a). In order to completely 'neutralize' the possible AI and obtain the spatial amplification of EMW in MMs, in the following sections (see sections 2.2.2, 2.2.3 and appendices A.5–A.7) the active MMs are considered with the negative phase behavior in the presence of active loads of metaparticles with a frequency dependence of a certain form (Boardman *et al* 2008, 2011) (see section 2.2.3).

Thus, the importance of our analysis (Boardman *et al* 2007a) of the absence of the AI for constructing and choosing the 'architecture' of a strong active magnetic MM, consisting of active cells, with the controlled magnetic permeability, is noted in Yuan *et al* (2009). The above-mentioned active MMs are important in applications with significant material losses, such as antennas, based on active MMs, and devices that amplify evanescent waves. Such MMs are promising for the expansion of the use of

MMs in communication systems at higher frequencies, where losses are more critical (Yuan *et al* 2009).

2.2.2 Qualitative estimations of the possibility of creating an active metamaterial with negative phase behavior on the basis of periodic infinite structures. Taking into account the frequency dependence of the active loads of metaparticles

2.2.2.1 Statement of the problem

Consider the NPM on the basis of the periodic transmission line, which can be described using the equivalent circuit depicted in figure 2.5(a). An elementary transmission line cell is shown in figure 2.5(a) in the form of an equivalent T-circuit. The goal is to create a medium with a spatial amplification, where the absence of the AI (and generation) is guaranteed. The preferences of using periodic transmission lines for creation and analysis of active metamaterials is due to the following reasons: (1) experimentally created media on the basis of transmission lines with negative refraction (Eleftheriades and Balmain 2005), and the transmission line model qualitatively correctly reflect the basic characteristics of the medium with circular resonators and wires, such as NPM; (2) only on the basis of consideration of the periodic medium, one can investigate all the main types of instabilities, including those related to very short-wave disturbances. Within the framework of the MM approximation, the real structure 'on fully legitimate grounds' is taken into account only through the averaged (homogenized) material parameters (ε, μ). At the same time, the corresponding wave perturbations correspond, on the scale of the wavelength $\lambda \sim k^{-1}$, to the limiting case $k \to \infty$. When we want to investigate the effect of a small (in a scale λ), but a finite period of the structure, the model of the periodic transmission line (LP) is the most adequate and simple. In this way, we would be able to investigate the manifestations of all the main potential sources of the AI of the material. Details are given below, as well as in appendices A.6 and A.7.

If we consider the circuit in figure 2.5(a) by analogy with a medium including CRP/metal pins, the correspondence between the parameters of an elementary cell and their 'distributed analogues' can be represented as

$$L_1 = L'd,\ C_1 = C'd \tag{2.59}$$

where L', C' are the equivalent distributed parameters, which in this case are linear inductance and capacitance. Thus, the circuit in figure 2.5(a). includes distributed parameters, when converting to equivalent concentrated parameters per cell length d, formulas (2.59) are used. Other elements shown in figure 2.5(a), can be considered as concentrated, loading the elementary cell of periodic TL. This model takes into account the finiteness of the width (effective) $\Delta\omega$ of the frequency band, where there is a negative equivalent conductivity of the active element (in particular, the diode). This approach is similar to that used for the effective bandwidth parameter, where there is a **negative (differential) impedance** in the equivalent circuit used in the simulation of a continuous laser medium in Scott (1970).

Consider the variable field is proportional to

$$\sim \exp[i(\omega t - kx)] \tag{2.60}$$

where ω, k are frequency and wavenumber of running EMW, respectively. The negative conductivity of the active element (diode) can be qualitatively represented as

$$g = g(\omega) = \{1 + i[(\omega^2 - \omega_0^2)/(\omega\Delta\omega)]\}^{-1} g_0;\ g_0 = \text{const},\ g_0 < 0 \tag{2.61}$$

In equation (2.61), ω_0 is frequency, where the absolute value of the equivalent negative conductivity of the active element reaches its maximum. Make the following note relative to the frequency relations (2.60) and (2.61). The simplest form, which provides one maximum and a finite bandwidth, is the Lorentz function (Scott 1970). We will show further that the complex function $g(\omega)$ (see relation (2.61)), taking into account the real and imaginary parts, provides, for a medium simulated with the application of the equivalent circuit shown in figure 2.2, absence of the AI. Moreover, as shown below, there is a certain class of complex functions of the conductivity of the active elements, the use of which this property of the artificial media becomes achievable. The function (2.61) is one that qualitatively corresponds to the properties of a certain type of active devices (Oleinik *et al* 2004). This approach is discussed in more detail in section 2.2.3. Considering the equivalent circuit in figure 2.2 in the approximation of equivalent lumped elements, which corresponds to the MM approximation (or with the approximation of the continuous medium), when the period of the structure d satisfies the inequality

$$d \ll \lambda \sim k^{-1} \tag{2.62}$$

one obtains the dispersion equation (DE) in the form:

$$\cos(kd) = 1 + YZ/2,\ \text{or}\ F(\omega, k) = \cos(kd) - 1 - YZ/2 = 0 \tag{2.63}$$

In relation (2.63), $Y(\omega)$ and $Z(\omega)/2$ are the admittance of the equivalent four-pole in the vertical part and the equivalent four-pole impedance in the horizontal part of the T-circuit, as shown in figure 2.2, respectively:

$$Z = i\omega L_1 + [i\omega C_r + (i\omega L_r)^{-1} + g(\omega) + R^{-1}]^{-1} \tag{2.64}$$

In the 'long-wave approximation' (2.62), we obtain from (2.63) DE in the form

$$(kd)^2 = -Y \cdot Z \tag{2.65}$$

The behavior of the system relates, in general, a complex frequency ω to a complex wave vector k. Taking, for simplicity, a particular direction in an isotropic medium, the relationship between complex k and complex ω, needs to be investigated, in accordance with basic rules concerning the AI presented in Lifshitz and Pitaevskii (1981), Fedorchenko and Kotsarenko (1981), Scott (1970), Boardman *et al* (2008), Popov and Shalaev (2006), Mikhailovskii (1974), Pendry *et al* (1999) and Govyadinov *et al* (2007). Given the definitions of complex wavenumber as $K = K' + iK'$ and the complex frequency as $\Omega = \Omega' + i\Omega'$, we assume that a plane wave is propagating along the z-axis, which is selected arbitrarily. Corresponding complex planes are shown in figure 2.4. Such a wave has a spatio-temporal form $\exp i(\omega t - kz) \equiv \exp i[\Omega t' - Kz']$, where t' and z' are also dimensionless.

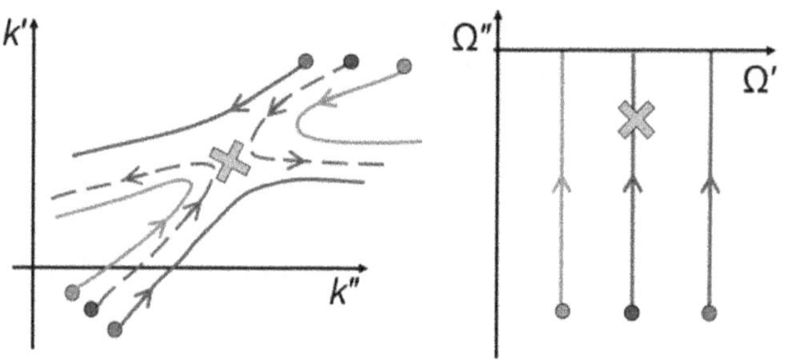

Figure 2.4. An example of outcomes of the mapping of the complex Ω-plane, where $\Omega = \Omega' + i\Omega''$, onto the complex *K*-plane, where $K = K' + iK''$. A saddle point is designated by the cross (Boardman *et al* 2011).

Figure 2.4 demonstrates what could happen to the complex K root locations traveling between the upper and lower halves of the complex wavenumber plane. Specific details of the system are not required at this stage. As Ω'' is varied, two roots approach each other, figure 2.4 (Boardman *et al* 2011). Respectively, saddle point forms in the complex (k', k'') plane. **For the existence of the AI, a saddle point must exist, and a double root is reached for which $\omega'' < 0$. Indeed, the roots of the dispersion equation approach the saddle point from different halves of the complex *k*-plane. If k_+ corresponds $z > 0$, and k_- to $z < 0$, then $k_+ = k_-$ reveals the existence of a resonance corresponding to a divergent growth without the necessity of a source.** To investigate mathematically the conditions of a possible AI and obtain stability conditions, we use the approach outlined in Lifshitz and Pitaevskii (1981), Fedorchenko and Kotsarenko (1981) and Scott (1970) and find the roots of the system of equations:

$$F(\omega, k) = 0,\ \partial F / \partial k = 0 \tag{2.66}$$

Given the frequency of the system being studied (figure 2.2), consider the value (kd) in the range

$$-\pi < kd < 0 \tag{2.67}$$

Strictly speaking, the physically meaningful roots of equation (2.61) should be within the interval (2.67) in the absence of dissipation (positive or negative); in other words when $R = \infty$, $g = 0$. If we have finite values, then we must look for the roots for which the value $k'd$, where $k' = \text{Re}(k)$, is in the region (2.67). The application of the condition (2.66) to equation (2.63) gives two roots for $k \times d$ and the corresponding equations for ω. The first root is determined by the relationships

$$kd = 0,\ Y(\omega) \times Z(\omega) = 0, \tag{2.68}$$

The second root is determined by the relationships

$$kd = -\pi,\ Y(\omega)Z(\omega) = -4 \tag{2.69}$$

The relations (2.68) are sufficient for studying possible sources of the instability associated with the roots of the systems (2.63) and (2.66) of formally finite values. The word 'formally' is used here to emphasize that the roots (2.68) that correspond to the ‘short-wave resonance’ of the periodic structure are taken into account, in contrast to the roots (2.68) that correspond to the MM long-wave approximation (2.62) and may be obtained in this approximation, also using DE in the form (2.65). Thus, when considering the AI, caused by short-wave disturbances (2.69), we already should go beyond the purely MM approximation. The reason is that short-wave disturbances can theoretically arise even for signals carrying the frequency and the wavelength of which, by themselves, fully fit into the framework of the MM approximation. Nevertheless, it is necessary to consider the third type of perturbation (Lifshitz and Pitaevskii 1981).

Namely, the perturbation, which corresponds to the poles of the dispersion function, may also be a source of AI. This is possible if these poles lie in the lower half-plane of the ‘complex ω’. The poles of the function formally mean that for them

$$|k| \to \infty \tag{2.70}$$

Physically, this condition corresponds to the limiting case for very short-wave perturbations for which the condition $|k| \gg 1/d$ is sastisfied. The smallest scale of this problem, associated with a possible AI, can be put equal to the period of a structure. Therefore (2.70) can be considered as an adequate limiting case and as an approximation, in which one can consider all perturbations for which the condition $|k| \gg 1/d$ is performed. Schematically, all major types of perturbations that can potentially lead to the AI are depicted in figure 2.2(b).

2.2.2.2 Conditions of the existence of a medium with spatial amplification/absence of the AI associated with the zeros and poles of the dispersion function

It can be shown that the poles and zeros (see equation (2.68)) of the dispersion function are determined, respectively, by the relations

$$Z(\omega) \to \infty,\ Y(\omega) \to \infty,\ Z(\omega) = 0,\ Y(\omega) = 0 \tag{2.71}$$

It can be shown that zeros and poles $Y(\omega)$ (Boardman *et al* 2008, 2011) do not lead to AI. Focus on the roots and poles $Z(\omega)$ of the function, in accordance with the first equations of the first and second pairs in (2.71). The details are discussed in appendix A.6. The corresponding conditions of the spatial amplification (absence of AI) were obtained for the first time in the form (Boardman *et al* 2008):

$$R_r^{-1} < (-g_0) < (1 + S)R_r^{-1},\ S = \min(S_1, S_2) \tag{2.72}$$

Here $S_{1,2}$ are the values that characterize the possibility of the AI due to the poles $|k| \to \infty$ (relation (2.70)) and the ‘long-wave roots’ (relations (2.68) and (2.71)), respectively. The expressions for $S_{1,2}$, namely (A.44) and (A.45), are given in appendix A.6. Let us note the following important results of the theory: (1) the frequency window, where there is a spatial amplification (the AI is absent) is inside a frequency window where the backward waves propagate. This is demonstrated by the numerical calculations given in section 2.2.2.4; (2) we also note the fundamental importance of the finiteness of the bandwidth $\Delta\omega$ with the negative real part of the differential conductivity $g(\omega)$. In fact, with the help of formulas (A.44) and (A.45) we can obtain $\Delta\omega/\omega \to \infty$ (and $g(\omega) = g_0$, in accordance with (2.61)) leads to $S = 0$, and, thus, conditions (2.72) cannot be satisfied. The physical sense of the effect of the finiteness of the bandwidth is the impossibility of overcoming the AI of the MM throughout all the frequency range, provided that each of the active metaparticles is, in effect, an effective 'elementary generator' that causes a generation (of the elementary absolutely unstable perturbations) in the whole infinitely wide frequency range The possibility of AI, caused by the finiteness of the period of the structure of the MM, is investigated in section 2.2.2.3.

2.2.2.3 *Conditions of the existence of a medium with spatial amplification associated with the finiteness of the period of the structure (in the absence of the AI)*

If the structure period is finite, then the roots of (2.69) of the system (2.66) must be analyzed along with the roots (2.68). The stability conditions corresponding to the roots (2.68) (or the second pair of equation (2.71)) and the poles (the first pair of equation (2.71)) of the dispersion function have already been obtained (conditions (2.72)). It is shown here that, under certain conditions, the stability problem associated with the roots (2.69), which correspond to the finiteness of the structure period and the short-wave resonance, with the wavelength of perturbation of the order of the period of the structure, can be solved by means of the perturbation method. First we find the roots of the equation (2.69) in a lossless approximation, in which (see also equation (2.61) and figure 2.5)

$$R_r \to \infty, g \to 0 \tag{2.73}$$

and then we obtain the imaginary parts of the corresponding frequencies, including their signs based on the perturbation method. Of course, we showed that the perturbation method is adequate, in particular, under the following conditions. The task is to analyze the roots of the polynomial of the sixth order, and such an analysis with the help of the approximate analytical method (perturbations) is given in appendix A.7. It is shown that under conditions (A.52), namely

$$\omega_0 \sim \omega_{rs} \sim \omega_r; \ \omega_0 L_r R_r^{-1} \sim \omega_0 L_r |g_0| \ll 1; \ \omega_0 L_r |g_0| \leqslant (\Delta\omega/\omega_0) \tag{2.74}$$

the special points of equation (2.69) do not lead to the AI. In equation (2.74), ω_0 and g_0 are the parameters characterizing the frequency dependence and the maximum value of the effective conductivity of the active element (see relation (2.61)), $\omega_{rs,r}$ is the characteristic frequencies for the circuit under consideration (figure 2.2(a)) (these frequencies are determined by relations (A.46) and (A.49)), L_r, R_r are the value of the corresponding elements of the equivalent circuit in figure 2.2. Thus, the condition of spatial amplification, when performing (A.52), is the correlation obtained for us for the first time (A.51). We now consider numerical results that describe qualitatively some characteristics of an MM with negative phase behavior and spatial amplification (corresponding to the absence of AI, as shown in Fedorchenko and Kotsarenko (1981)).

These calculations show that for $k''d > 0$ there is a frequency domain with a negative phase behavior where $V_g V_{ph} < 0$, and there is a spatial gain (the AI is absent).

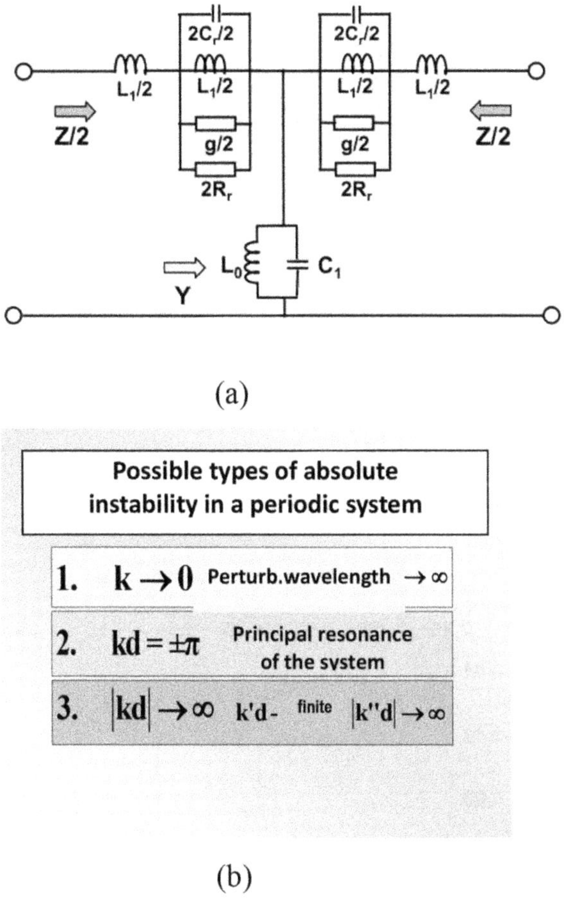

Figure 2.5. (a and b) (a) is the equivalent (symmetric) T-circuit, which presents an elementary cell of 1D MM/ transmitting line with a period *d*, loaded by an active element having complex frequency-dependent conductivity *g* (such as an active diode etc), C_r, L_r, R_r are equivalent components, describing SRR. L_1, C_1, L_0 describe a dielectric host medium and elements providing 'negative ε'/rods. Single cell 'works' in transmitting line, as a frequency-dependent impedance $Z(\omega)$ and admittance $Y(\omega)$. (b) is the main types of the AI of the periodic system with the period *d*. (b) Reprinted from figure 9 in Boardman *et al* (2011), copyright (2011) with permission from John Wiley & Sons.

2.2.2.4 Results of numerical calculations

In Eleftheriades *et al* (2003), for the passive MM on the basis of the transmission line, a model based on small wavenumbers was constructed. In our paper (Boardman *et al* 2008), in addition to constructing a model for active MMs, we have removed the limitations and the wavenumber (Eleftheriades *et al* 2003), using the exact DE (2.63). Figures 2.6(a) and (b) show that the use of the approximate and exact dispersion equations (2.65) and (2.63), respectively, gives very close results. The difference between them is more pronounced for the real parts of the wavenumbers in the 'short-wave' part of the dispersion dependence, but this difference is only quantitative.

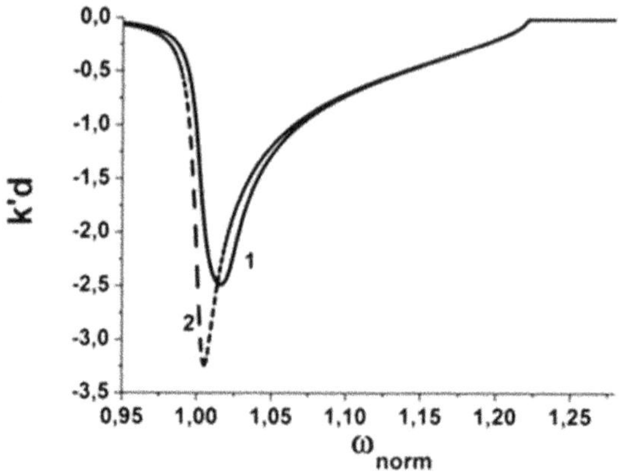

(a)

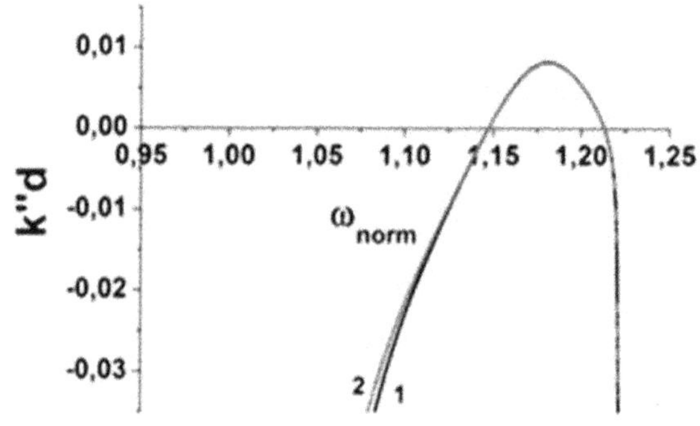

(b)

Figure 2.6. (a and b) Variations of real (a) and imaginary (b) parts of wavenumber $k'd$ in the range of normalized frequency $\omega_{\text{norm}} = \omega'/\omega_{\text{r}}$ with the 'negative phase behavior' for the periodic 'resonant' circuit in figure 2.2(a) (Boardman *et al* 2008); curve 1 is computations based on the exact dispersion relation (2.63); 2 is computations based on the approximate dispersion relation (2.65); $\bar{g}_0 = \omega_{\text{r}} L_{\text{r}} g_0 = 1.56 \times 10^{-2}$, $r = L_{\text{r}}\omega_{\text{r}}/R_{\text{r}} = 1.18 \times 10^{-2}$, $\Delta\omega/\omega_0 = 0.097$. Other input parameters correspond to Eleftheriades *et al* (2003). $k''d > 0$ corresponds to the amplification of EMWs in a MM medium. Reprinted from figures 3 and 4 in Boardman *et al* (2008), copyright (2008) with permission from Elsevier.

2.2.3 A general approach to the problem of stability of resonant negative metamaterials

It can be shown (Boardman *et al* 2011) that there is a certain class of functions that describe a possible conductivity complex of 'active devices', the loading of which by metaparticles ensures the spatial amplification (and the absence of AI (Fedorchenko and Kotsarenko 1981)). This conductivity has the form

$$g(\omega) \equiv g_0 e^{i\varphi} / \left\{ 1 + i[(\omega^2 - \omega_0^2)/(\omega\Delta\omega)] \right\};\ g_0 < 0,\ -\pi < \varphi < \pi \tag{2.75}$$

where the phase value $\varphi = -\pi$ is excluded to keep the negative sign of the actual conduction part in the frequency range of finite width. In Boardman *et al* (2011), the conditions of a possible stability/instability are investigated in detail. The denominator $g(\omega)$ is zero when

$$\omega = (1/2)\left[i\Delta\omega \pm \sqrt{4\omega_0^2 - \Delta\omega^2}\right] \tag{2.76}$$

where the actual part is positive. This ensures that there is no AI associated with such 'special points'. The value $\varphi = 0$ corresponds simultaneously to the change in the sign of the imaginary part of the conductivity and to the maximum (in absolute value) of its real part (Oleinik *et al* 2004). This case is considered in Boardman *et al* (2008). The value $\varphi = -\pi/2$ corresponds qualitatively to a resonant behavior of $g(\omega)$, which is described by the analytical model of the subcritical Gunn diode of Bosch and Thim (1974), Oleinik *et al* (2004). When $\varphi = -\pi/2$ one obtains:

$$\mathrm{Re}(g(\omega)) \equiv g(\omega)' = -g_0[(\omega^2 - \omega_0^2)/(\omega\Delta\omega)]/\left\{1 + [(\omega^2 - \omega_0^2)/(\omega\Delta\omega)]^2\right\}$$

$$\mathrm{Im}(g(\omega)) \equiv g(\omega)'' = -g_0/\{1^2\} + [(\omega^2 - \omega_0^2)/(\omega\Delta\omega)] \tag{2.77}$$

In this case, $|g(\omega)'|$ has a maximum equal to $|g_0|/2$ at the (real) frequency

$$\omega = \omega'_{\max} \approx \omega_0 - (1/2)\Delta\omega \tag{2.78}$$

if $\Delta\omega \ll \omega_0$. $\Delta\omega$ defines effective bandwidth, where $g(\omega)' < 0$, and amplification is possible.

Despite the fact that the aforementioned example is taken from the microwave theory and experiment, the very concept of finding a certain class of effective complex conductivity, which describes the active component of the MM and provides the spatial amplification of the EMW in the presence of the negative phase behavior, retains the value for MMs in the wide frequency range, at least from microwave to optical ranges.

Let us show the possibility of the spatial amplification for MMs based on metaparticles loaded with active elements having characteristics in the form (2.75)–(2.77). We will assume that the conditions for the suitability of the perturbation method (Boardman *et al* 2011)

$$(\omega_0 L/R_r) \sim \omega_0 L_r |g_0| \ll (\Delta\omega/\omega_r) \ll 1, \omega_r = (L_r C_r)^{-1/2} \tag{2.79}$$

The analysis of the roots (saddle points) (see relation (2.66)) and large complex wavenumbers ($|k| \to \infty$) (poles) leads to certain conditions, which must correspond to the conductivity of the active element (diode), to ensure the absence of AI and the presence of the spatial amplification, namely

$$1/R(1 + S_*) < (-g_0) < (1 + S)(1/R),\ S = \min(S_{\mathrm{saddle}}, S_{\mathrm{pole}}) \tag{2.80}$$

The values S_{pole}, S_{saddle} and S_* are obtained, but not presented here because the relevant calculations are similar to those conducted for output (A.51). The values S_{pole} and S_{saddle} correspond to the poles $k(\omega)$ and saddle points of the dispersion function $F(\omega, k)$, respectively. Note that the special points associated with the poles do not lead to the AI. This can be shown using (2.76) and (2.77). The right-hand side of (2.80) defines the conditions for the absence of an AI, while the left-hand side (including the value S_*) determines the possibility of amplification, at least at the closest neighborhood of the frequency $\omega'_{\max}$, where $|g(\omega)'|$ reaches its maximum value.

The preliminary analysis was related to certain models of active media that correspond to the active conductivity in the form (2.75) for certain values $\varphi(0)$ and $\varphi(-\pi/2)$. Consider now the problem in a more general formulation, investigating a whole class of active media with active loads (active elements) having the conductivity in the form (2.75), and the phase lies in the range $-\pi < \varphi < \pi$. Consider a discrete set that unites all special points, including poles and roots, namely $\omega_{1\mathrm{rt}}, \omega_{2\mathrm{rt}}, \ldots \omega_{n_{\mathrm{saddle}}}$ and $\omega_{1\mathrm{pole}}, \omega_{2\mathrm{pole}}, \ldots, \omega_{n_{\mathrm{pole}}}$, respectively, which are supposed to be close to the axis $\mathrm{Re}\,\omega$. This corresponds to a situation where the conditions of a type (2.79) are satisfied, and the gain/losses are sufficiently small. Here $n_{\mathrm{saddle}}, n_{\mathrm{pole}}$ is the full number of roots (nodes of points) and poles, respectively. After defining them, construct an ordered sequence of elements of the union of both sets. Here $N = n_{\mathrm{saddle}} + n_{\mathrm{pole}}$ (figure 2.7(a)). All special points corresponding to the poles must have positive imaginary parts (see, for example, equation (2.76), and these special points are not related to the AI. These special points are excluded from further consideration and from a set of points ω'_l, $l = 1, \ldots N$. Denote the frequency that corresponds to the maximum value of the (negative) real part of the conductivity of the active elements $g(\omega)'$, through $\omega'_{\max}$. The frequency $\omega'_{\max}$ can be located either to the left of the first frequency point of the set of singular points, $\omega'_{\max} < \omega'_1$, or to the right of the last point of this set $\omega'_{\max} > \omega'_N$, or between neighboring points ω'_k and ω'_{k+1} which belong to this set. Consider, for example, the case $\omega'_k < \omega'_{\max} < \omega'_{k+1}$ (the result is qualitatively the same for other possible placements of $\omega'_{\max}$). Even when the media is described by a more complicated equivalent circuit than those shown in figure 2.7(a), one can always connect some positive resistance R, for example, parallel to the active element, to control the 'effective conductivity' $g'(\omega') + R^{-1}$. The finite bandwidth $\Delta\omega$, conductivity g_0, and resistance R can be selected so that the value is negative only in the frequency region with the center at $\omega'_{\max}$ which is indicated by the arch in figure 2.7(a), and lies completely inside the interval $(\omega'_k, \omega'_{k+1})$.

Assume that the frequency $\omega'_{\max}$ and the frequency interval (shown by the arc in figure 2.7(a)), where the effective conductivity of the active element, taking into account the parallel resistance (the corresponding conductivity), is negative, is localized within the region with a negative phase behavior. This can be achieved by selecting the appropriate parameters, such as ω_0 and $\Delta\omega$. At the same time, it can be shown that in the framework of the suitability of the perturbation method, the imaginary part for the set of singular points is proportional to the corresponding value of effective conductivity and is positive, $\omega''_l \sim (g'(\omega'_l) + R_r^{-1}) > 0$, $l = 1, \ldots N$. These values are positive because, for the selected parameters, all special points are located near the axes of the actual and outside of the region where the effective conductivity is negative (indicated by an arch in figure 2.7(a)). The positivity of all ω''_l

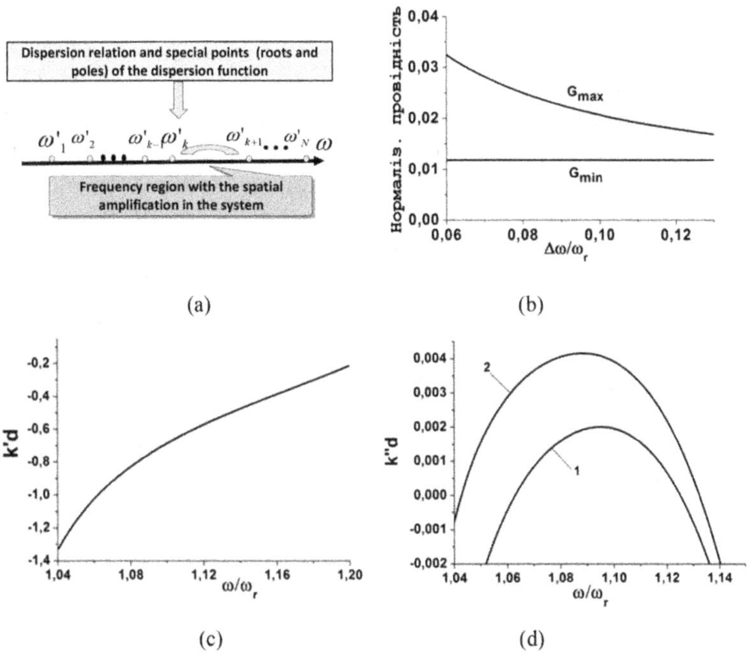

Figure 2.7. (a) Ordered set of points (zeros of the dispersion function, equation (2.63)–(5.63)) and poles of the wavenumber) and the range of frequencies (denoted by arc), where the system is active (the spatial amplification is present). The frequency ω'_{max} (corresponding to the maximum of the absolute value of conductivity) and the total frequency range (designated by the arch), where there is a spatial amplification, belonging to the region of the negative phase behavior, where $V_g V_{ph} < 0$. (b) is the area of the normalized conductivity $G = |g_0|\omega_r L_r$ corresponding to the spatial amplification, depending on the relative bandwidth $\Delta\omega/\omega_r$ for the case $\varphi = -\pi/2$. The values G_{min} and G_{max} correspond to the left and right sides of the inequalities (2.80), respectively. The dependence of the real (c) and imaginary (d) parts of the normalized wavenumber from the normalized frequency ($\omega_r = (C_r L_r)^{-1/2}$) for the case $\varphi = -\pi/2$; $\Delta\omega/\omega_r = 0.14$; the curve 1 in figure (d) is built for $g_0\omega_r L_r = 0.013$; curve 2 is for $g_0\omega_r L_r = 0.014$. Reprinted from Boardman et al (2011), copyright (2011) with permission from John Wiley & Sons

ensures the absence of AI and spatial amplification (Fedorchenko and Kotsarenko 1981). This is an important qualitative result. Note that when gain/loss is sufficiently small, the presence of gain/loss does not violate the 'negative phase behavior', at least in the narrow band of frequencies marked by an arch in figure 2.7(a). Thus, the medium also has the property of 'negative phase behavior'. In fact, this approach requires only the following very general properties of the 'enclosing medium and active elements,' namely: (a) the active element has no 'internal instability', in other words, the poles must have a positive imaginary part; (b) the imaginary part is negative only in the finite band of frequencies determined by some parameter; (c) the set of saddle and pole points is discrete; (d) the amplification/attenuation must be relatively small, so that the perturbation method is applicable; (e) the frequency domain with negative phase behavior is compatible with the area of spatial amplification for the parameters chosen accordingly. The proposed approach based on satisfying conditions (a)–(e) of spatial amplification is qualitatively suitable for a relatively broad class of a media with negative phase behavior. Thus, the active MM, which is described by the equivalent circuit shown in figure 2.5(a) and the ratio (2.75) is only one of the possible objects belonging to this class.

Such an approach can be applied not only for analysis, but even for the 'design' of new active MMs with the negative phase behavior.

The numerical results shown below are obtained for the MM media, which is described by the equivalent circuit shown in figure 2.7(a) with the conductivity of the active element corresponding to formula (2.75) with $\varphi = -\pi/2$. In figure 2.7(b), the limits of the normalized conductivity $G = |g_0|\omega_r L_r$ are shown. The lower and higher curves correspond to the left and right sides of the inequality (2.80), respectively. The lower curve determines the minimum normalized conductivity $G = |g_0|\omega_r L_r$ required at least to compensate the losses at the frequency point ω'_{max}, where the absolute value of the real part of the (negative) conductivity (see relation (2.77)) reaches its maximum value. The upper curve determines the maximum possible magnitude of the active conductivity $G = |g_0|\omega_r L_r$, which is still compatible with the absence of AI. Note that in this case (for selected parameters), for $\varphi = -\pi/2$, the maximum possible value G is determined by the roots corresponding to the condition $k'd = -\pi$, while in the case $\varphi = 0$, the corresponding boundary G is determined by the roots $k'd = 0$.

In figures 2.7(c) and (d) the frequency dependences of the real and imaginary parts of the normalized wavenumber for $\varphi = \pi/2$ are shown, and the frequency is normalized to $\omega_r = (C_r L_r)^{-1/2}$. The dispersion is calculated using the perturbation method, while the conditions (2.79) are valid and $|k'| \gg |k''|$.

Thus, it has been shown that spatial amplification (with $k''d > 0$) is achievable in the region of negative phase behavior (where $V_g V_{ph} < 0$), where V_g and V_{ph} are the group and phase velocities of the EMW in the MM, respectively, figures 2.7(c) and (d). Note that the concept of absolute and convective instability for active MMs can be expanded with the inclusion of the optical range.

2.3 Nonlinear homogenization out of the frames of the perturbation method and constitutional (material) nonlinear relationships

Recently, much attention has been paid to bi-anisotropic and chiral MMs, in particular on the basis of Ω particles (Auzanneau and Ziolkowski 1998, Cheng *et al* 2007, Lheurette *et al* 2007, McCauley *et al* 2010, Tretyakov and Mariotte 1995, Zheludev and Kivshar 2012) (figures 2.1(a) and (b)), which can provide original electrodynamic properties for a number of applications in the processing of signals, space communications, medicine and ecology. In the works (Boardman *et al* 2007b, 2009b, Rapoport *et al* 2006a, Rapoport 2013, 2014b) a new and integrated approach to modeling the characteristics of nonlinear waves in layered structures with nonlinear bi-anisotropic metaparticles was developed. In this case, we consider nonlinear Ω-particles, but the proposed approach, in general, can be developed and applied to the more general case of bi-anisotropic or chiral nonlinear particles of various forms or a continuous medium with nonlinear chiral characteristics. Such an approach involves the sequential modeling of a separate nonlinear metaparticle, the application of a new method for 'nonlinear homogenization' and the development of a new method for the derivation of nonlinear evolution equations for waves in layered bi-anosotropic structures in the presence of both bulk and surface non-linearities (nonlinear surface currents or polarizations). The idea of a method for deriving evolution equations for the wave amplitudes of the enveloping waves in layered bi-anisotropic MMs (figure 3.1) with bulk and surface nonlinearities is presented in section 3.1 of chapter 3. The corresponding nonlinear evolution equation derived within the framework of the NEELS method has the form (3.18). We consider weakly nonlinear waves in a waveguide structure with a bi-anisotropic medium, where quadratic and cubic nonlinearities take place. The quadratic nonlinearity associated with nonlinear elements (a capacitance, for example) included in the Ω-particle (metaparticle) gives (three-wave) parametric gain, while the cubic nonlinearity is associated with the same nonlinear elements or the surrounding medium causes self-interactuion/self-influence. The possibility of effective amplification of counter-propagating nonlinear slow waves (pulses) in bi-anisotropic waveguide is shown.

Let us consider first the linear dispersion in a rectangular metallic waveguide (Boardman *et al* 2009b). Dimensional linear dispersion relations for TM (E_x, E_y, H_z component) and TE (H_x, H_y, E_z) modes (where abbreviations 'TM' and 'TE' mean transverse magnetic and transverse electric modes, respectively) for the geometry shown in figure 2.8(c), have the forms, respectively

$$k_{\mathrm{TM}}^2 = k_0^2(\varepsilon\mu + \alpha^2) - \frac{\varepsilon}{\varepsilon_{\mathrm{h}}}\left(\frac{\pi}{L_x}\right)^2,\ k_{\mathrm{TE}}^2 = k_0^2(\varepsilon\mu + \alpha^2) - \mu\left(\frac{\pi}{L_x}\right)^2,\ k_0 = \frac{\omega}{c} \quad (2.81)$$

Here, α, ε, μ-scalar are components of the corresponding tensors that influence the field components of E_l and H_n (figures 2.8(a) and (b)), α describes the chirality. The dispersion determined by relations (2.81) is shown in figure 2.8(d), where the frequency is normalized to the effective resonance frequency of the Ω-particle. The line over letters in the symbols on the vertical and horizontal axes of the coordinates is used to denote normalized values of wavenumbers and frequencies, respectively.

Let us first make a remark regarding the 'linear homogenization'. The corresponding formulas obtained using MLC are given in Rapoport (2015a) and Rapoport and Boardman (2015b), as well as in appendix A.8 (see the formulas (A.53) and (A.54)). We emphasize that the linear polarization obtained with the help of MLC (Rapoport 2015a, Rapoport and Boardman 2015b) corresponds to the material parameters obtained with the help of the MLC (see relations (2.27) and (2.28)). Take into account that in this section, using MLC, we consider an MM with bi-anisotropic Ω-particles located in equivalent positions (all parallel to each other). The nonlinear polarization of the medium is determined, in accordance with the above-mentioned approach, in two stages (Boardman *et al* 2007b, 2009b, Rapoport *et al* 2006a, Rapoport 2012, 2013, 2014b, 2015a, Rapoport and Boardman 2015b). First, we consider the nonlinear characteristics of a single metaparticle (figures 2.1(a) and (b)). Then we investigate the nonlinear interaction of various metaparticles, and

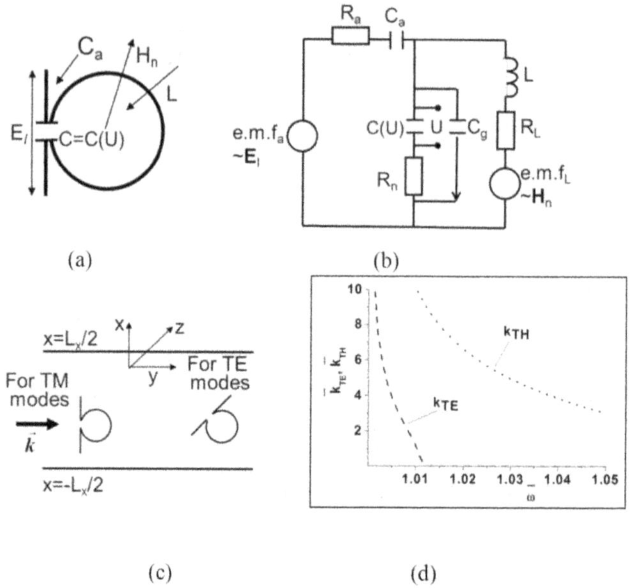

Figure 2.8. (a) is Ω-particle and its nonlinear equivalent circuit; E_l and H_n are electric and magnetic components of the electromagnetic field in the medium, parallel to the 'electric dipole' and normal to the 'inductive loop', respectively; $C = C(U)$ is the nonlinear capacity in the gap of Ω-particle (U is the voltage on the capacity); (b) is the elements L, C_a characterize the (linear) electric dipole and magnetic dipole/inductive loop in the Ω-particle; The left and right parts of the equivalent circuit characterize the electric dipole and the magnetic dipole/inductive loop, respectively. C_g and ($C(U)$, R_n) are elements that characterize the gap Ω and the nonlinear element on which it is loaded, respectively. R_L, R_a are active resistances for the relevant elements; e.m.f.$_{a,L}$ are electromotive forces, which are given on electric and magnetic dipoles, respectively; (c) is a bi-anisotropic rectangular waveguide. All Ω-particles are placed in the same position and in the same plane; (d) is dispersion for TE and TM modes in a bi-anisotropic rectangular metallized waveguide. Length of stem and radius of the loop of Ω-particle are denoted as l_0 and r_0, respectively.

derive an equation for nonlinear susceptibility, which, in the case of a bi-anasotropic (or chiral) medium, includes dyads (Tretyakov *et al* 1996, Tretyakov 2003, Lindell *et al* 1994, Kong 1972, Engheta *et al* 1992). The corresponding linear permeability has electric, magnetic, and magnetoelectric components $\widehat{\varepsilon}$, $\widehat{\alpha}$, $\widehat{\beta}$, $\widehat{\mu}$. The corresponding expressions for linear polarizations (A.53) are given in appendix A.8. We will note two original analytical results, obtained on the basis of MLC. (1) We emphasize that the linear permeabilities (A.53) obtained using MLC correspond to those obtained with the use of MLF (see relations (2.27) and (2.28)) (Lakhtakia and Weiglhofer 1998, Mariotte *et al* 1994, Tretyakov 2010), taking into account the location of particles, (A.53) and (A.54) directly point to the resonance frequency for the bi-anisotropic frequencies $\omega_{\text{res}}^2 = (L_{\text{eff}}C_{\text{eff}})^{-1}$. The resonance frequency of the medium with bi-anisotropic particles (Cheng *et al* 2007, Lheurette *et al* 2008, 2007, McCauley *et al* 2010, Ramaccia *et al* 2013, Tretyakov and Mariotte 1995, Zheludev and Kivshar 2012), as far as we know, was found for the first time from the first principles; namely, this frequency was found, based on the characteristics of both the individual particle and the interaction between the particles in Boardman *et al* (2009a),

Rapoport *et al* (2006a), Rapoport (2013, 2014b), Rapoport and Boardman (2015b) (see (A.53) and (A.54)). (2) In the used Lorentz–Lorenz approximation, (Tretyakov *et al* 1994, Tretyakov and Mariotte 1995, Tretyakov 2003, 2010), the whole effect of the interaction of bi-anisotropic metastable particles with an equivalent position is reduced only to the increase of the equivalent capacitance and inductance of the individual particle with the corresponding change in the components of the dyads of the effective polarizability (2.6), (A.53) and (A.54) and the decrease of the resonance frequency of the homogenized MM as a continuous medium. At the same time, the results on the resonance of the MM with Ω-particles obtained using the MLC, are well qualitatively consistent with the results obtained, in particular, in Mariotte *et al* (1994), Tretyakov *et al* (1994) and Tretyakov and Mariotte (1995) by means of numerical modeling and phenomenological model. Note that it was not explicitly indicated in Mariotte *et al* (1994) and Tretyakov and Mariotte (1995) that an influence of the interaction of metaparticles on the resonance frequency of the medium was accounted for in those papers.

The nonlinear material equations for bi-anisotropic medium, with nonlinear electric and magnetic polarizations $\vec{P}_{\rm NL}$, $\vec{M}_{\rm NL}$, have the form:

$$\vec{D} = \widehat{\varepsilon}\vec{E} + \widehat{\alpha}\vec{H} + 4\pi\vec{P}_{\rm NL},\ \vec{B} = \widehat{\beta}\,\vec{E} + \widehat{\mu}\vec{H} + 4\pi\vec{M}_{\rm NL} \tag{2.82}$$

The analysis of the nonlinear equivalent circuit shown in figure 2.8 gives

$$\vec{P}_{\rm NL} = \vec{P}_{\rm A} + \vec{P}_{\rm B},\ \vec{M}_{\rm NL} = \vec{M}_{\rm A} + \vec{M}_{\rm B} \tag{2.83}$$

In equation (2.83), the indices 'A' and 'B' correspond to the effects of the cubic polarization (self-influence) and quadratic polarization, respectively. Let us focus on these effects separately.

(a) Impact effects associated with the components of polarization $\vec{P}_{\rm A}$, $\vec{M}_{\rm A}$, the expressions for which they have the form (see (A.55)–(A.57) in appendix A.8):

$$P_{\rm Al} = \chi_{\rm ee}(|F|^2)E_l + \chi_{\rm em}(|F|^2)H_{\rm n},\ \ M_{\rm An} = \chi_{\rm me}(|F|^2)E_l + \chi_{\rm mm}(|F|^2)H_n \tag{2.84}$$

The 'l', 'n' indices correspond to the directions along the electric dipole and along the normal to the 'magnetic loop' (along the magnetic dipole) in the Ω-particle, respectively; $F = q_l E_l + q_n H_n$ is a linear combination of E_l and H_n (figure 2.8). The developed method (Rapoport 2015a, Rapoport and Boardman 2015b) allowed one to derive the equations for nonlinear polarizations as the functions $|F|^2$, $\chi_{\rm ee,em,me,mm}(|F|^2)$(see details in appendix A.8) given in this appendix as (A.53). We note that the proposed method of the 'nonlinear homogenization' is fundamentally different from the approach proposed in Lakhtakia and Weiglhofer (1998), where the local field is decomposed into a series by a small parameter ε characterizing weak nonlinearity, and then the 'ideology of weak nonlinearity' develops; the solution for susceptibility can be obtained only in the form of a series in powers of ε. In Boardman *et al* (2007b, 2008), Rapoport *et al* (2006a), Rapoport (2013, 2015a), Rapoport and Boardman (2015b) a more general method has been developed, that allows one to obtain equations for nonlinear susceptibilities (A.57) (appendix A.8), applicable in cases of considerable and/or resonant nonlinearity. Then, using such an equation (s), one can either solve it numerically (when nonlinearity is not weak) or satisfy a solution in the form of a series of powers ε (if the nonlinearity is weak and the frequency is not very close to the resonant one). Using the representation for the local fields, developed in Tretyakov and Mariotte (1995), Tretyakov (2003, 2010), and analyzing the nonlinear equivalent circuit, shown in figures 2.8(a) and (b), we get the cubic equation (A.57) relative to the square of the amplitude of the voltage U of the main (first) harmonic on the nonlinear element $C = C(U)$ shown in figures 2.8(a) and (b). Moreover, the free term of this equation is proportional to the magnitude included in the relations (2.84). After solving the equation obtained, we find all the voltages and currents that are included in the equivalent circuit in figures 2.8(a) and (b). As a result, we obtain both the electric and magnetic polarization of each metaparticle, and the corresponding (linear and nonlinear) polarization of the medium, as a whole, as recorded in relations (2.84). Thus, the proposed approach indeed gives the possibility to obtain all components of the nonlinear permeability for a bi-anisotropic medium, even with a not small or resonant nonlinearity. The Ampere–Volt characteristic of the nonlinear load (A.58, appendix A.8) differs from the corresponding characteristic (A.4). This difference lies in the presence of the second harmonic in the first case. As a result, for the active elements with the characteristics described by relation (A.58, appendix A.8), the possibility of parametric interaction and, respectively, of the spatial amplification (decreasing damping) of EMW is possible. Nevertheless, the corresponding equations (A.57) and (A.6) describing the nonlinear effect of the self-influence are of the same form. In accordance with the application of approaches based on MLF and MLC, in these cases the right-hand sides of the equations include the components of the incident EMW fields and the electromotive forces, which in turn, through the local fields, are expressed in terms of the fields of the incident EMWs (given the fourth relation with (A.6) and (A.27) and (A.28)).

(b) The polarization (see equation (2.83)) is related to the quadratic nonlinearity and parametric coupling (in the system with metaparticles shown in figure 2.8).

If the wave amplitude of the pumping does not depend on the coordinate Y, then the conditions of the parametric synchronism in the presence of the quadratic nonlinearity has the form

$$2\omega = \omega + \omega, \; 0 = \vec{k}_1 + \vec{k}_2 \tag{2.85}$$

The evolution equations and results of model calculations and the estimation of parametric amplification for parametrically interacting counter-propagating pulses in a bi-anisotropic nonlinear waveguide/layered medium are given in section 6.2.

2.4 Conclusions

1. It is shown for the first time that the frequency band where the corresponding elements are active must be finite to provide the spatial amplification of EMWs and negative phase behavior in MM with metaparticles loaded on active elements. The conditions of the spatial amplification are found jointly with negative phase behavior of MM with metaparticles loaded on active elements with frequency-dependent conductivity. A generalized theoretical scheme for constructing such an active MM that provides the spatial amplification of EMWs in a negatively phase MM medium is proposed. In the analysis of the AI, three main types of potentially dangerous sources of the AI in the infinite periodic system are considered, namely: (1) perturbations with a wavenumber k, which goes to zero; (2) the fundamental resonances of the system, $kd = \pm\pi$, where d is the period of the system; and (3) the perturbations that correspond to the condition $|kd| \to \infty$.
   - A model of a one-dimensional photonic crystal or a transmission line (TL) is used to determine the possibility of a new NPM with spatial amplification. This model corresponds to that used in Eleftheriades *et al* (2003) and Eleftheriades and Balmain (2005), but with an appropriately added active frequency-dependent element and dissipative resistance. As was shown experimentally in Eleftheriades *et al* (2003) and Eleftheriades and Balmain (2005), structures constructed on the basis of the corresponding TLs are indeed NPM that have the property of the negative refraction. The used model is a good model of an active NPM on the basis of SRRs and metal rods (with the same dispersion), and, on the other hand, it can be a model of a real active NPM based on TL or a periodic structure. It is shown that the AI is absent in such a material, while it is active, according to the chosen parameters and it is determined how the corresponding parameters should be chosen.
   - The new active medium with a negative phase has been proposed in Boardman and Marinov (2006, Boardman *et al* (2008, 2011) for the first time and is useful for many practical applications such as advanced lenses, special systems for linear and nonlinear signal processing, etc. Scaling the system and finding adequate active elements can lead to the application of the proposed principle of the new active medium in higher frequency bands.
2. A new nonlinear homogenization method for a bi-anizotropic medium is developed. It is applied to a particular case—particles with a nonlinear load, arranged parallel to each other. For a case of GHz frequencies, a single metaparticle is analyzed on the basis of the Kirchhoff rules for currents and voltages, taking into account nonlinear elements excited by the total action of local electromagnetic fields including the action of other metaparticles in the Lorentz–Lorenz approximation.

Problems to chapter 2

Problem 2.1

Make derivations of equation (A.17).

Consider the relations (A.5)–(A.14). Put $z_{a_1} \to \infty$. As a result, the relations (A.17) should be obtained.

Problem 2.2

Derive the inequality (A.43), as the condition of absence of AI in the SRR, basing on the consideration of the poles $z(\omega)$ (see equation (A.39)).

Consider the relation $z_2 = 0$ (see equations (A.39) and (A.40)). Consider the condition of absence of AI as of the one which needs displacement of the roots $z_2 = z_2(\omega)$ (see (A.39)) n the upper half of the 'complex' plane. Use Rauss–Hurwitz criteria (Korn and Korn 1968).

As a result, the relations (A.43) and (A.44) should be obtained.

Problem 2.3

Find the condition (s) of the absence of AI which may be connected with numerator $z(\omega)$(see (A.39)).

The numerator $z(\omega) = i\omega z_1$. Note that the root $\omega = 0$ does not correspond to the AI, because the gain in the scheme shown in figure 2.5(a) can exist only in a rather narrow frequency range. Accounting for Rauss–Hurwitz criteria (see also the condition of problem 2.2), the inequality (A.49) can be obtained, as the corresponding condition of the absence of AI.

List of abbreviations

AI	Absolute instability
AM	Artificial molecules
AP	To replace by AM—Artificial molecules
EMW	Electromagnetic wave
MP	Metaparticles
MM	Metamaterial
MLF	Method of local fields
MLC	Method of local currents beyond limits of the perturbation method
NDC	Negative differential conductivity
NEELS	Nonlinear evolution equations for layered structures
NPM	Negative phase medium
RTDs	Resonance tunnel diodes
SRR	Split ring resonator
TL	Transmission line
DE	Dispersion equation
TM	Transverse magnetic (mode)
TE	Transverse electric (mode)

Appendix A

A.1 Model of a new active bi-anisotropic metamaterial

We will consider an MM and EMW in it under conditions that allow us to apply the Lorentz–Lorenz, or Maxwell–Garnett, approximation/quasi-statics when considering the interaction between particles (figures 2.1(a)–(e)), (Eleftheriades *et al* 2003, Tretyakov *et al* 1996).

$$a/\lambda \ll 1,\ F \ll 1,\ F \sim a^3/a_0^3 \tag{A.1}$$

Here a, λ are the characteristic size of metaparticles or AM, and the wavelength of EMW, respectively, a_0 is the linear size of the elementary lattice (i.e. the actual characteristic distance between AM), therefore F is the effective coefficient of 'volume filling' of the lattice by AM. The loop and 'rods' included in the Ω-particle are represented as magnetic and electric dipoles, respectively, and Ω-particle is described as an equivalent circuit in figures 2.1(c) and (d). We assume that the variables in the linear approximation depend on the time in the form $\vec{E},\ \vec{H},\ U,\ I \sim \exp(i\omega t)$. Here $\vec{E},\ \vec{H}$ and $U,\ I$ are electric and magnetic fields and electric voltages and currents in the circuits in figures 2.1(a)–(d), respectively, ω is the frequency of EMW. The electromotive forces shown in the equivalent diagram in figures 2.1(a) and (d), are equal to

$$\varepsilon_{Lk}^{(1)} = -ik_0 S H_{k\mathrm{loc}}^{(1)},\ k_0 = \omega/c,\ \varepsilon_{ak}^{(1)} = E_{k\mathrm{loc}}^{(1)} l_{\mathrm{eff}} \tag{A.2}$$

$$E_{1\mathrm{loc}}^{(1)} = E_{x\mathrm{loc}}^{(1)},\ E_{2\mathrm{loc}}^{(1)} = E_{y\mathrm{loc}}^{(1)},\ E_{3\mathrm{loc}}^{(1)} = E_{z\mathrm{loc}}^{(1)},\ H_{1\mathrm{loc}}^{(1)} = H_{z\mathrm{loc}}^{(1)},\ H_{2\mathrm{loc}}^{(1)} = H_{x\mathrm{loc}}^{(1)},\ H_{3\mathrm{loc}}^{(1)} = H_{y\mathrm{loc}}^{(1)} \tag{A.3}$$

Here $k = 1, 2, 3$ is the face number of the unit cell shown in figure 2.1(c), $H_{k\mathrm{loc}}^{(1)}$ and $E_{k\mathrm{loc}}^{(1)}$ are the components of local magnetic and electric fields, respectively, acting on the face of the cell with the number k and inducing electromotive forces $\varepsilon_{Lk}^{(1)}$ and $\varepsilon_{ak}^{(1)}$ in the magnetic and electric dipoles located (within the corresponding metaparticle) on this face, c is the speed of light, S is the area of the magnetic loop included in Ω-particle. Examples of active elements with negative differential conductivity (NDC) are Gunn diodes (Bosch and Thim 1974, Bosch and Engellmann 1975, Oleinik *et al* 2004) and resonant tunnel diodes (RTD) (Kidner *et al* 1990, Ookawa *et al* 2006). The equivalent circuit of the active diode shown in figure 2.1 as a single element Z_D, shown separately in figure 2.1(c). We obtain for the 'Ampere–Volt' nonlinear characteristic of the active element with negative differential conductivity (Gunn diode or resonant tunnel diode (RTD) (Boardman *et al* 2007):

$$I_{0N} = Z_{0N}^{-1} U_{0N} + b U_{0N}^3;\ I^{(1)} = g_0 U^{(1)} + b_0 \left| U^{(1)} \right|^2 U^{(1)} \tag{A.4}$$

Here g_0 is the linear (negative) differential conductivity, b_0 is the coefficient of nonlinearity.

A.2 To the characterization of metamolecules of bi-anisotropic metamaterial

Considering the equivalent circuit in figure 2.1(a), using the Maxwell–Garnett approximation, investigating the fundamental harmonic of high-frequency current $I_{\mathrm{D}}^{(1)}$ and voltage on an equivalent nonlinear element Z_{D} (figure 2.1(a)), describing the active diode, taking into account (A.4), as a result of considering linear and nonlinear parts of currents and voltages for equivalent circuit for Ω-particles (figure 2.1(c)), we obtain the equation for the voltage $U_{\mathrm{D}}^{(1)}$ (Rapoport 2013, 2014b, 2015a, Rapoport and Boardman 2015b, Rapoport *et al* 2006a, Boardman *et al* 2007a):

$$U_D^{(1)}(Z_{00}{}^{-1} + q_0 b\,|U_D^{(1)}|^2) = Z_{L_1}{}^{-1}\varepsilon_L^{(1)} + Z_{a_1}{}^{-1}\varepsilon_a^{(1)};$$

$$Z_{00}{}^{-1} = Z_{a1}{}^{-1} + Z_{L1}{}^{-1} + Z_{C_{getf}}{}^{-1} + Z_D{}^{-1},\ Z_D = g_0^{-1} \quad (A.5)$$

where $q_0 = 1$, $Z_{a_1} = Z_{C_a} + R_a = (i\omega C_a)^{-1} + R_a$. Equation (A.5) can be written down also as:

$$U_D^{(1)}(A + B|U_D^{(1)}|) = \tilde{\varepsilon}_{\mathrm{eff}}^{(1)},\ A = Z_{00}{-}1,\ B = q_0 b,\ \tilde{\varepsilon}_{\mathrm{eff}}^{(1)} = Z_{L_1}{}^{-1}\varepsilon_{L_1}^{(1)} + Z_{a_1}{}^{-1}\varepsilon_{a_1}^{(1)} \quad (A.6)$$

Obtain the solution of (A.6) by the perturbation method:

$$U_{\mathrm{D}}^{(1)} = (\theta_0 + \theta_2\,|\tilde{\varepsilon}_{\mathrm{eff}}^{(1)}|^2 + \theta_4\,|\tilde{\varepsilon}_{\mathrm{eff}}^{(1)}|^4 + \ldots)\tilde{\varepsilon}_{\mathrm{eff}}^{(1)};\ \theta_0 = A^{-1},\ \theta_2 = -B(|A|^2A^2)^{-1}, \quad (A.7)$$

$$\theta_4 = -2\frac{B}{A^2}\left[\mathrm{Re}\left(\frac{1}{A}\right)\mathrm{Re}\left(\frac{B}{|A|^2A^2}\right) + \mathrm{Im}\left(\frac{1}{A}\right)\mathrm{Im}\left(\frac{B}{|A|^2A^2}\right)\right] + \frac{B^2}{|A|^4A^3} \quad (A.8)$$

For currents on the face of an elementary cell with a number k passing through the elements L, C_{a} (see figures 2.1(a) and (c)) we obtain, respectively:

$$\begin{aligned} I_{LK}^{(1)} &\equiv I_{L_k\mathrm{LIN}}^{(1)} + \delta I_{\mathrm{LNLK}}^{(1)} = Z_{L_1}{}^{-1}(\varepsilon_{LK}^{(1)} - U_{DK}^{(1)}) \approx \\ &Z_{L_1}{}^{-1}\left\{\varepsilon_{LK}^{(1)}[1 - (Z_{00}/Z_{L_1})] - (Z_{00}/Z_{a_1})\varepsilon_{a_1K}^{(1)}\right\} \\ &- Z_{L_1}{}^{-1}(\theta_2\,|\tilde{\varepsilon}_{\mathrm{eff}k}^{(1)}|^2 + \theta_4\,|\tilde{\varepsilon}_{\mathrm{eff}k}^{(1)}|^4)\tilde{\varepsilon}_{\mathrm{eff}k}^{(1)} \end{aligned} \quad (A.9)$$

$$\begin{aligned} I_{C_aK}^{(1)} &\equiv I_{C_a\mathrm{NLK}}^{(1)} + \delta I_{C_a\mathrm{NLK}}^{(1)} = Z_{a_1}{}^{-1}(\varepsilon_{aK}^{(1)} - U_{DK}^{(1)}) \approx \\ &Z_{a_1}{}^{-1}\left\{\varepsilon_{aK}^{(1)}[1 - (Z_{00_1}/Z_{a_1})] - (Z_{00}/Z_{L_1})\varepsilon_{a_1K}^{(1)}\right\} \\ &- Z_{a1}{}^{-1}(\theta_2\,|\tilde{\varepsilon}_{\mathrm{eff}k}^{(1)}|^2 + \theta_4\,|\tilde{\varepsilon}_{\mathrm{eff}k}^{(1)}|^4)\tilde{\varepsilon}_{\mathrm{eff}k}^{(1)} \end{aligned} \quad (A.10)$$

The first and second terms in (A.9) correspond to the linear and nonlinear components of the respective currents. The linear dyads of polarizations have the form:

$$\begin{aligned} \langle a_{\mathrm{ee}}\rangle &= (a_{\mathrm{eexxLIN}}/3)(\vec{e}_x\vec{e}_x + \vec{e}_y\vec{e}_y + \vec{e}_z\vec{e}_z),\ \langle a_{\mathrm{em}}\rangle \\ &= (a_{\mathrm{emxzLIN}}/3)(\vec{e}_x\vec{e}_z + \vec{e}_y\vec{e}_x + \vec{e}_z\vec{e}_y) \\ \langle a_{\mathrm{me}}\rangle &= (a_{\mathrm{mezxLIN}}/3)(\vec{e}_z\vec{e}_x + \vec{e}_x\vec{e}_y + \vec{e}_y\vec{e}_z),\ \langle a_{\mathrm{mm}}\rangle \\ &= (a_{\mathrm{mmzzLIN}}/3)(\vec{e}_x\vec{e}_x + \vec{e}_y\vec{e}_y + \vec{e}_z\vec{e}_z) \end{aligned} \quad (A.11)$$

where

$$a_{\rm eexxLIN} = a'_{\rm eexxLIN} + a''_{\rm eexxLIN} = (l_{\rm eff}^2 Z_{00}/i\omega Z_{a_1})[Z_{L_1}^{-1} + Z_{C_{\rm geff}}^{-1} + Z_D^{-1}],$$

$$\begin{aligned} a_{\rm emxzLIN} &= a'_{\rm emxzLIN} + a''_{\rm emxzLIN} = -a_{\rm mezxLIN} \\ &= -(a'_{\rm mezzxLIN} + a'''_{\rm mezxLIN}) = (l_{\rm eff} S Z_{00}/c Z_{a_1} Z_{L_1}), \end{aligned} \quad (A.12)$$

$$\begin{aligned} a_{\rm mmzzLIN} &= a'_{\rm mmzzLIN} + a''_{\rm mmzzLIN} \\ &= (l_{\rm eff}^2 Z_{00}/i\omega Z_{L_1})[Z_{L_1}^{-1} + Z_{C_{\rm geff}}^{-1} + Z_D^{-1}] \end{aligned}$$

In (A.12), the dash and two dashes denote the real and imaginary parts of the respective quantities. Expressions (A.11) and (A.12), coincide with the expressions obtained in Auzanneau and Ziolkowski (1998) in the corresponding limiting cases and correspond (Rapoport 2014b, 2015a, Rapoport and Boardman 2015b) with an obvious correction for the location of all metaparticles for geometry (Rapoport 2014b, 2015a, Rapoport and Boardman 2015b) in a single plane. In the absence of loads, these expressions are reduced to the corresponding formulas (Tretyakov *et al* 1994). After selection of nonlinear components of currents in (A.9) and (A.10), there are nonlinear components of polarizations

$$\vec{P}_{\rm NL}^{\rm incl(1)} = N\left\langle \left\langle \delta a_{\rm eeNL}\left(\vec{E}_{\rm loc}^{(1)}, \vec{H}_{\rm loc}^{(1)}\right)\right\rangle \vec{E}_{\rm loc}^{(1)} + \left\langle \delta a_{\rm emNL}\left(\vec{E}_{\rm loc}^{(1)}, \vec{H}_{\rm loc}^{(1)}\right)\right\rangle \vec{H}_{\rm loc}^{(1)} \right\rangle \quad (A.13)$$

$$\vec{M}_{\rm NL}^{\rm incl(1)} = N\left(\left\langle \delta a_{\rm meNL}(\vec{E}_{\rm loc}^{(1)}, \vec{H}_{\rm loc}^{(1)})\right\rangle \vec{E}_{\rm loc}^{(1)} + \left\langle \delta a_{\rm mmNL}(\vec{E}_{\rm loc}^{(1)}, \vec{H}_{\rm loc}^{(1)})\right\rangle \vec{H}_{\rm loc}^{(1)} \right) \quad (A.14)$$

with the nonlinear dyads of:

$$\left\langle \delta a_{\rm eeNL}(\vec{E}_{\rm loc}^{(1)}, \vec{H}_{\rm loc}^{(1)})\right\rangle = -iq_{\rm eeNL}\widehat{I}_{\rm ee}, \quad \left\langle \delta a_{\rm mmNL}(\vec{E}_{\rm loc}^{(1)}, \vec{H}_{\rm loc}^{(1)})\right\rangle = -iq_{\rm mmNL}\widehat{I}_{\rm mm}$$

$$\left\langle \delta a_{\rm meNL}(\vec{E}_{\rm loc}^{(1)}, \vec{H}_{\rm loc}^{(1)})\right\rangle = q_{\rm meNL}\widehat{I}_{\rm me}, \quad \left\langle \delta a_{\rm emNL}(\vec{E}_{\rm loc}^{(1)}, \vec{H}_{\rm loc}^{(1)})\right\rangle = q_{\rm emNL}\widehat{I}_{\rm em}$$

$$\begin{aligned} q_{\rm eeNL} &= (l_{\rm eff}^2/3\omega Z_{a_1}^2), \; q_{\rm mmNL} = (\omega S^2/3c^2 Z_{L_1}^2), \; q_{\rm meNL} = -q_{\rm emNL} \\ &= (q_{\rm eeNL} q_{\rm mmNL})^{1/2} \end{aligned} \quad (A.15)$$

where

$$\widehat{I}_{\rm ee} = \begin{bmatrix} Q_{\rm ee1} & 0 & 0 \\ 0 & Q_{\rm ee2} & 0 \\ 0 & 0 & Q_{\rm ee2} \end{bmatrix}, \; \widehat{I}_{\rm mm} = \begin{bmatrix} Q_{\rm ee2} & 0 & 0 \\ 0 & Q_{\rm ee3} & 0 \\ 0 & 0 & Q_{\rm ee1} \end{bmatrix}, \; \widehat{I}_{\rm me} = \begin{bmatrix} 0 & Q_{\rm ee2} & 0 \\ 0 & 0 & Q_{\rm ee3} \\ Q_{\rm ee1} & 0 & 0 \end{bmatrix}$$

$$\widehat{I}_{\rm em} = \begin{bmatrix} 0 & 0 & Q_{\rm ee1} \\ Q_{\rm ee2} & 0 & 0 \\ 0 & Q_{\rm ee3} & 0 \end{bmatrix}, \; Q_{\rm eek} = (\theta_2\,|\tilde{\varepsilon}_{L_{\rm eff}k}|^2 + \theta_4\,|\tilde{\varepsilon}_{l{\rm eff}k}|^4), \; k = 1, 2, 3 \quad (A.16)$$

The values $\varepsilon_{L_1}^{(1)}$, $\varepsilon_{a_1}^{(1)}$, which, according to (A.6), determine $\tilde{\varepsilon}_{L_{\mathrm{eff}k}}$ (see (A.16)), are expressed through the electric and magnetic fields of the EMW incident on the MM by the formulas, which will be given later in appendix A.3 (see (A.27)–(A.32)). With this mention, the formulas (2.2), (A.11), (A.12) and (A.13)–(A.16) completely determine the linear and nonlinear polarizations and dyads for 'Ω-particles' (Rapoport *et al* 2006a, 2014a, Rapoport 2015a, Rapoport and Boardman 2015b); the latter ones in this section are determined by the perturbation method and through local values of fields that differ from the fields of 'incident' waves.

The approach described in the present appendix A.2, leads to what corresponds for the nonlinear part of the magnetic polarization in the form:

$$\begin{aligned}\vec{M}_{\mathrm{NL}}^{\mathrm{incl}(1)} &= N\left\langle \delta a_{\mathrm{mmNL}}\left(\vec{E}_{\mathrm{loc}}^{(1)}, \vec{H}_{\mathrm{loc}}^{(1)}\right)\right\rangle \vec{H}_{\mathrm{loc}}^{(1)}, \left\langle \delta a_{\mathrm{mmNL}}\left(\vec{E}_{\mathrm{loc}}^{(1)}\vec{H}_{\mathrm{loc}}^{(1)}\right)\right\rangle \\ &= -iq_{\mathrm{mmNL}}\widehat{I}_{\mathrm{mm}}\end{aligned} \tag{A.17}$$

In equation (A.17), δa_{mmNL} are the components of the corresponding dyad associated with the single meta article in the form of SRR.

A.3 Homogenization for linear metamaterials with MC in the form of 'Ω-particles'. The principle of nonlinear homogenization of metamaterial by perturbation method

We will use the index 'LIN' for the linear parts of the tensors included in the 'material relations' (2.17). For a MM with an 'elementary cell' of cubic shape with Ω-particles (see figure 2.1(a–c)), we obtain using (2.17)–(2.29), (A.11) and (A.12):

$$\widehat{\varepsilon}_{\mathrm{effLIN}} = \varepsilon_{\mathrm{eff0}}\widehat{I}, \ \widehat{\mu}_{\mathrm{effLIN}} = \mu_{\mathrm{eff0}}\widehat{I}, \ \widehat{\beta}_{\mathrm{effLIN}} = \beta_{\mathrm{eff0}}\widehat{I_2}, \ \widehat{\alpha}_{\mathrm{effLIN}} = \widehat{\alpha}_{\mathrm{eff0}}\widehat{I_1} \tag{A.18}$$

$$\varepsilon_{\mathrm{eff0}} = \varepsilon + (4\pi/3K_{10})\varepsilon_{\varepsilon} = \varepsilon[1 - (4\pi\delta_{\varepsilon}/9\varepsilon)\delta_{\varepsilon}]^{-1}[1 + (8\pi\delta_{\varepsilon}/9\varepsilon)\delta_{\varepsilon}] \tag{A.19}$$

$$\mu_{\mathrm{eff0}} = \mu + (4\pi/3K_{20})\delta_{\varepsilon} = \mu[1 - (4\pi\delta_{\mu}/9\mu)\delta_{\mu}]^{-1}[1 + (8\pi\delta_{\mu}/9\mu)\delta_{\mu}] \tag{A.20}$$

$$\alpha_{\mathrm{eff0}} = (4\pi/3K_{10})[1 - (4\pi/9\mu)(Na_{\mathrm{mmzzLIN}})]^{-1}Na_{\mathrm{emxzLIN}}, \ \beta_{\mathrm{eff0}} = -\alpha_{\mathrm{eff0}} \tag{A.21}$$

$$K_{10} = 1 - (4\pi/9\varepsilon)\delta_{\varepsilon}, \ K_{20} = 1 - (4\pi/9\mu)\delta_{\mu} \tag{A.22}$$

$$\begin{aligned}\delta_{\varepsilon} &= Na_{\mathrm{eexxIN}} + (4\pi/9\mu)[1 - (4\pi/9\mu)(Na_{\mathrm{emzzLIN}})]^{-1}(Na_{\mathrm{emxzLIN}}) \\ &\quad (Na_{\mathrm{mezxLIN}}) \\ &= Na_{\mathrm{eexxIN}} - (4\pi/9\mu)[1 - (4\pi/9\mu)(Na_{\mathrm{emzzLIN}})]^{-1}(Na_{\mathrm{emxzLIN}})^{2}\end{aligned} \tag{A.23}$$

$$\delta_{\mu} = (Na_{\mathrm{mmzzLIN}}) - (4\pi/9\varepsilon)[1 - (4\pi/9\varepsilon)(Na_{\mathrm{eexxLIN}})]^{-1}(Na_{\mathrm{mezxLIN}})^{2} \tag{A.24}$$

$$\widehat{I_1} = \begin{bmatrix} 0 & 0 & 1 \\ 1 & 0 & 0 \\ 0 & 1 & 0 \end{bmatrix}, \ \widehat{I_2} = \begin{bmatrix} 0 & 1 & 0 \\ 0 & 0 & 1 \\ 1 & 0 & 0 \end{bmatrix}, \ \widehat{I} \text{ is the unit matrix} \tag{A.25}$$

$$\tilde{\alpha}_{\text{effLIN}} = -\widehat{\beta}^{T}{}_{\text{effLIN}};\ \beta_{\text{effLIN}mn} = -\alpha_{\text{effLIN}nm},\ m,\ n = x, y, z \tag{A.26}$$

The sign of 'effective losses' of EMW in the environment is affected, jointly with the signs of imaginary parts of components $\widehat{\varepsilon}$, $\widehat{\mu}$ (Vainshtein 1988), also by the signs of real parts of tensor components $\widehat{\alpha}$, $\widehat{\beta}$. A detailed study of the conditions of change of the attenuation sign and the possibilities of spatial amplification for MMs/NPM will be carried out for MMs with AM in the form of SRR, which are considered as a special case of 'Ω-particles' (see relations (A.18)–(A.26) and sections 2.1.2, 2.1.3). To determine the nonlinear polarizations (A.13)–(A.17) it is necessary to determine the values of the corresponding electromotive forces (A.2) and (A.3). In accordance with the values of local fields (2.29), the electromotive forces included in the circuits in figure 2.1(c), which correspond to the elements on the faces k = 1,2,3 (figure 2.1(b)), are equal to:

$$\varepsilon_{L1}^{(1)} = -ik_0SH_{z\text{loc}} = -(ik_0S/3)[(2 + \mu_{\text{eff0}}/\mu)H_z - (\alpha_{\text{eff0}}/\mu)E_y] \tag{A.27}$$

$$\varepsilon_{a1}^{(1)} = l_{\text{eff}}E_{x\text{loc}} = (l_{\text{eff}}/3)[(2 + \varepsilon_{\text{eff0}}/\varepsilon)E_x + (\alpha_{\text{eff0}}/\varepsilon)H_z] \tag{A.28}$$

$$\varepsilon_{L2}^{(1)} = -ik_0SH_{x\text{loc}} = -(ik_0S/3)[(2 + \mu_{\text{eff0}}/\mu)H_x - (\alpha_{\text{eff0}}/\mu)E_z] \tag{A.29}$$

$$\varepsilon_{a2}^{(1)} = l_{\text{eff}}E_{y\text{loc}} = (l_{\text{eff}}/3)[(2 + \varepsilon_{\text{eff0}}/\varepsilon)E_x + (\alpha_{\text{eff0}}/\varepsilon)H_x] \tag{A.30}$$

$$\varepsilon_{L3}^{(1)} = -ik_0SH_{y\text{loc}} = -(ik_0S/3)[(2 + \mu_{\text{eff0}}/\mu)H_y - (\alpha_{\text{eff0}}/\mu)E_x] \tag{A.31}$$

$$\varepsilon_{a3}^{(1)} = l_{\text{eff}}E_{z\text{loc}} = (l_{\text{eff}}/3)[(2 + \varepsilon_{\text{eff0}}/\varepsilon)E_z + (\alpha_{\text{eff0}}/\varepsilon)H_y] \tag{A.32}$$

As shown in Rapoport *et al* (2006a), Rapoport (2013, 2014b, 2015a), Rapoport and Boardman (2015b), Boardman *et al* (2007b), knowledge of linear tensors $\widehat{\varepsilon}_{\text{effLIN}}$, $\widehat{\mu}_{\text{effLIN}}$, $\widehat{\alpha}_{\text{effLIN}}$, $\widehat{\beta}_{\text{effLIN}}$ and nonlinear polarizations is sufficient to determine the nonlinear macroscopic characteristics of EMW in a (layered or homogeneous) medium with an MM. Thus, the possibility of obtaining the properties of (nonlinear) EMW in the active medium is substantiated, starting with the study of the microscopic properties of AM using the perturbation method.

A.4 Possibility of simultaneous amplification of EMW and negative signs of real parts of dielectric *ε* 'and magnetic *μ*' permeability of metamaterial based on split rings with active diodes. Estimation of the 'effective length' of spatial amplification of EMW in active metamaterial

To identify the condition of linear amplification of EMW, i.e. the creation of active MM, where active loads are included only in the SRR, consider equation (2.18) in the linear approximation. We introduce the 'effective length' or inverse spatial increment of EMW amplification (Boardman *et al* 2007a):

$$L_{\rm ampl} = 2k_y[\varepsilon k_0^2(-\mu''_{\rm eff0})]^{-1} \text{ or } L_{\rm ampl}/\lambda \sim \sqrt{\mu'_{\rm eff0}/\varepsilon}[(-\mu''_{\rm eff0})\pi]^{-1} \qquad \text{(A.33)}$$

where $\lambda = 2\pi/k_y$; the magnetic permeability μ is equal to $\mu \equiv \mu' + i\mu''$ where μ^1, μ'' are real and imaginary parts, respectively. Let us estimate for the following parameters of the MM and the 'active conductivity' ($g_0 \equiv 1/R_{\rm D} < 0$) of the Gunn diode/'active load' of SRR: $\omega = 2\pi \times 10^{10}$ s^{-1}, $C_{\rm geff} \sim 4.3 \times 10^{-2}$ pF, $|g_0| \sim 6.25 \times 10^{-7}$ Ohm^{-1}, $S_{\rm D} \sim 1\ \mu$m^2, $R_{\rm L} \sim 0.1$ Ohm, $R_{\rm eff} \gg R_{\rm L} + g_0(\omega C_{\rm g})^2 \sim$ $(-0.25) \times 10^{-2}$ Ohm, $N \sim 200$ cm^3, $\varepsilon \sim -2$, $n_0 = 200$, $3\varsigma_0\omega_1^2/(\Delta^2 + \varsigma_0\omega_1^2) \sim -2$. Here $S_{\rm D}$ is the effective cross-sectional area of the active crystal used in the diode (in particular, GaAs) (Bosch and Engellmann 1975, Oleinik *et al* 2004). In this case, we obtain from (2.30) and (2.31) taking into account (2.41), (2.44) and (A.33),

$$\mu'_{\rm eff0} \sim -1, \ |\mu''_{eff0}/\mu'_{\rm eff0}| \sim 10^{-2};\ L_{\rm ampl}/\lambda \sim 30 \qquad \text{(A.34)}$$

As seen from (A.34), in this case, and at rather small value $R_{\rm Deff}$, that is at small 'excess of gain over attenuation', we have the active environment where can extend with EMW amplification, and with amplitudes changing slowly. Under increasing the cross-sectional area of the active crystal up to $S_{\rm D} \sim 70\ \mu$m^2, we obtain the possibility of 'supereffective' amplification of EMW in the active MM with the 'negative phase behavior', namely

$$|\mu''_{\rm eff0}/\mu'_{\rm eff0}| \sim 0.3,\ L_{\rm ampl}/\lambda \sim 1 \qquad \text{(A.35)}$$

Given (A.34) and (A.35), it is possible to implement both conditions, namely $|\mu''_{\rm eff0}/\mu'_{\rm eff0}| \ll 1$; $L_{\rm ampl}/\lambda \gg 1$ or $|\mu''_{\rm eff0}/\mu'_{\rm eff0}| \sim 1$; $L_{\rm ampl}/\lambda \sim 1$, according to Rapoport and Boardman (2015b). The conditions for the possibility of the spatial amplification and, accordingly, the absence of the AI (Fedorchenko and Kotsarenko 1981) are investigated in section 2.2.

A.5 Simplified analysis of the possibility of creating an active metamaterial environment with spatial amplification: approximation of frequency-independent conductivities of 'active loads' of the metaparticles

MM with metaparticles in the form of SRR is characterized by the 'material relations'

$$\vec{B} = \widehat{\mu}_{\rm eff}\vec{H} = \mu_{\rm eff0}\vec{H} = \mu\vec{H} + 4\pi\vec{M}^{(\rm incl)},\ \vec{H}_{\rm loc} = \vec{H} + (4\pi/3\mu)\vec{M}^{(\rm incl)},$$

$$\vec{M}^{(\rm incl)} = N\widehat{a}_{\rm mm}\vec{H}_{\rm loc} = N\widehat{a}_{\rm mm}\Big[\vec{H} + (4\pi/3\mu)\vec{M}^{(\rm incl)}\Big],\ \widehat{a}_{\rm mm} = (a_{\rm mmLIN}/3)\widehat{I}$$

$$\vec{M}^{(\rm incl)} = [1 - (4\pi N/9\mu)a_{\rm mmLIN}](Na_{\rm mmLIN}/3)\vec{H} \qquad \text{(A.36)}$$

where μ, $\vec{H}$ are the magnetic permeability of the 'containing medium' and the 'averaged' magnetic field, $\vec{M}^{(\text{incl})}$ is the magnetic polarization associated with SRR, $\widehat{a}_{\text{mm}}$ is the magnetic susceptibility of SRR (Tretyakov 2003, 2010, Eleftheriades and Balmain 2005). Using (A.36), we obtain the effective magnetic permeability in the form (see (2.45) and (2.46))

$$\mu'_{\text{eff}} = \mu\left\{1 - 3\varsigma_0\frac{NS^2}{c^2L}\frac{\omega^2\left[\omega^2\left(1+\varsigma_0\dfrac{NS^2}{c^2L}\right) - \omega_0^2\left(\dfrac{(\omega C_g R_D)^2}{1+(\omega C_g R_D)^2}\right)\right]\Bigg\}}{\left(\dfrac{Z'_{\text{eff}}}{L}\omega\right)^2 + \left[\omega^2\left(1+\varsigma_0\dfrac{NS^2}{c^2L}\right) - \omega_0^2\left(\dfrac{(\omega C_g R_D)^2}{1+(\omega C_g R_D)^2}\right)\right]^2}\right]$$

$$\mu'_{\text{eff}} = -3\mu\varsigma_0\frac{NS^2}{c^2L}\left\{\frac{\omega^3\dfrac{Z'_{\text{eff}}}{L}}{\left(\dfrac{Z'_{\text{eff}}}{L}\omega\right)^2 + \left[\omega^2\left(1+\varsigma_0\dfrac{NS^2}{c^2L}\right) - \omega_0^2\left(\dfrac{(\omega C_g R_D)^2}{1+(\omega C_g R_D)^2}\right)\right]^2}\right] \tag{A.37}$$

Bifurcation points that determine the necessary conditions for the absence of AI (Lifshitz and Pitaevskii 1981, Scott 1970) in connection with the roots of the

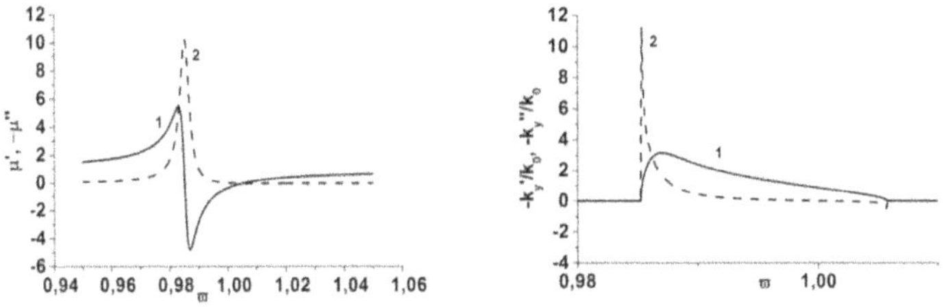

(a) (b)

Figure A.1. (a) is the dependence of the real (curve 1) and imaginary (curve 2) parts of the magnetic permeability on the normalized frequency $\varpi = \omega/\omega_0$ in the approximation of the frequency-independent conductivity of the active elements, $R_D{}^{-1}$; $R_L = 10$ Ohm, $R_D = -6.23 \times 10^2$ Ohm, $\varepsilon''_{\text{eff}} = 0$; at the resonant frequency $\varepsilon'_{\text{eff}} \approx -2.3$; (b) is the dependence of the real (curve 1, absolute value k'_y/k_{y0}) and imaginary (curve 2) parts of the normalized 'propagation constant' on the normalized frequency $\varpi = \omega/\omega_0$. This figure shows the value $|k'_y/k_{y0}| = -k'_y/k_{y0}$, where $k'_y/k_{y0} < 0$. In the frequency range, where $k_y'^2$ becomes negative, and EMW are non-propagating, it is put forward k'_y, $k''_y = 0$.We obtain using (A.38) the bifurcation frequencies (2.52) and the necessary conditions (2.54) for the spatial amplification (determined in this simplified consideration only) in connection with the roots of the dispersion relation while ignoring, at this stage, the poles (Lifshitz and Pitaevskii 1981, Fedorchenko and Kotsarenko 1981, Scott 1970). The necessary conditions for the spatial amplification in the form of (2.54) obtained under the above-mentioned restrictions are satisfied, in particular, by the parameters used in the calculations (Boardman *et al* 2007a), namely $\omega_0 = 2\pi f_0$, $f_0 = 10$ GHz, $N = 200$ cm^{-3}, $C_g = 0.2$nF, the radius of the CRC ring, the relative distance between frequencies close to $|\omega_{\text{res}} - \omega_{01}|/\omega_{01} \sim 3\varsigma_0(NS^2/c^2L) \sim 3 \times 10^{-2}3\varsigma_0(NS^2/c^2L) \sim 3 \times 10^{-2}$, $\varepsilon''_{\text{eff}} \approx -2.3$, $R_L \sim 0.1$ Ohm, $R_D = -6.23 \times 10^4$ Ohm. The possibility of the spatial amplification in the region of the 'negative phase' (i.e. the operation in the inverse wave mode) is illustrated in figure A.1.

dispersion function (see (2.50), (2.51), (A.37), (2.45) and (2.46)) are determined by the conditions $Z_{\text{eff}} - i(8\pi NS^2/9\mu c^2)\omega = 0$, $(1 + i\omega C_g R_D)R_D + R_L + i\omega L - i(8\pi NS^2/9\mu c^2)\omega = 0$. The last equation is reduced to a set of equations:

$$A\omega^2 + B\omega + C = 0,\ A = A' = C_g R_D,\ B = iB'';\ B'' = -[(\omega_{01}C_g R_L)(\omega_{01}C_g R_D) + 1],$$

$$C = C = -C_g(R_L + R_D)\omega_{01}^2;\ A'(\omega_{\text{bifl},2}^{\prime 2} - \omega_{\text{bifl},2}^{\prime\prime 2}) - B''\omega_{\text{bifl},2}^{\prime\prime} + C'' = 0$$

$$2A'\omega_{\text{bifl},2}^{\prime}2\omega_{\text{bifl},2}^{\prime\prime} + B''\omega_{\text{bifl},2}^{\prime} = 0 \tag{A.38}$$

A.6 The stability problem related to the roots and poles of the dispersion relation

As noted in section 2.2.2.2, the zeros and poles of the function do not lead to the AI, so we focus on the effects associated with the features of $Z(\omega)$ according to the first and third expressions from (2.71) for the roots and poles, respectively. The value Z included in (2.71) can be represented as:

$$Z \equiv Z(\omega) = i\omega\left[\sum_{i=0}^{4} Q_i\omega^i\right] \Big/ \left[\sum_{k=0}^{4} \beta_k\omega^k\right] = i\omega Z_1/Z_2, \tag{A.39}$$

where

$$Z_1 = \sum_{i=0}^{4} Q_i\omega^i,\ Z_2 = \sum_{k=0}^{4} \beta_k\omega^k \tag{A.40}$$

The coefficients in (A.39) and (A.40) are expressed by a system of relations: $Q_0 = iQ_{00} = -iL_1\omega_0^2$,

$$Q_1 = Q_{10} = L_1[\Delta\omega + \omega_0^2(L_r/R_r)] + L_r\Delta\omega$$

$$Q_2 = iQ_{20} = i\left\{L_1[1 + (\omega_0^2/\omega_r^2) + L_r\Delta\omega/R_r + g_0L_r\Delta\omega] + L_r\right\}$$

$$Q_3 = Q_{30} = -L_1[(\Delta\omega/\omega_r^2) + (L_r/R_r)],$$

$$Q_4 = iQ_{40} = -iL_1/\omega_r^2;$$

$$\beta_0 = i\beta_{00} = -i\omega_0^2,$$

$$\beta_1 = \beta_{10} = \Delta\omega + \omega_0^2(L_r/R_r),$$

$$\beta_2 = i\beta_{20} = i[1 + (\omega_0^2/\omega_r^2) + (L_r\Delta\omega/R_r) + g_0L_r\Delta\omega] \tag{A.41}$$

$$\beta_3 = \beta_{30} - [(\Delta\omega/\omega_r^2) + (L_r/R_r)],\ \beta_4 = i\beta_{40} = -i/\omega_r^2 \tag{A.42}$$

A.7 Conditions for the absence of AI associated with the effects of the finiteness of the structure period

Consider the stability of the respective roots semi-qualitatively, and then apply a more consistent approach based on the perturbation method. To do this, we first find the roots corresponding to equation (2.69) in the non-dissipative approximation (meaning the absence of dissipation of any sign), when

$$R_r \to \infty,\ g \to 0, \tag{A.47}$$

and then evaluate the dissipation and its sign by the perturbation method, considering the equivalent circuit in figure 2.2(a). Under the conditions (A.47), formula (2.64) is reduced to $Z(\omega) = i\omega L_l(\omega_r^2 - \omega^2)^{-1}(\omega_{rs}^2 - \omega^2)$, $Y(\omega) = i(C_l/\omega)(\omega^2 - \omega_{sh}^2)$. The second relation from (2.69) is reduced to

$$\omega^4 + A\omega^2 + B = 0,\ A = -(\omega_{sh}^2 + \omega_{rs}^2 + 4\omega_{shs}^2),\ B = 4\omega_{shs}^2\omega_r^2 + \omega_{rs}^2\omega_{sh}^2 \tag{A.48}$$

$$\omega_{sh}^2 + (C_l L_0)^{-1},\ \omega_{rs}^2 = (L_{rs}C_r)^{-1},\ L_{rs} = L_l L_r/(L_l + L_r),\ \omega_{shs}^2 = (C_l L_l)^{-1} \tag{A.49}$$

$$\begin{aligned}\omega_{10,\,20}^2 &= (1/2)[(\omega_{sh}^2 + \omega_{rs}^2 + 4\omega_{shs}^2) \pm \sqrt{D}],\ D \\ &= (\omega_{sh}^2 - \omega_{rs}^2) + 8\omega_{shs}^2(\omega_{rs}^2 + \omega_{sh}^2 - 2\omega_r^2) > 0\end{aligned} \tag{A.50}$$

If the values are R_r^{-1}, $|g_0|$ non-zero but relatively small, and the frequencies $\omega_{10,\,20}$ are not very close to ω_0 (the maximum frequency of the real part of $g(\omega)$, equation (2.61)), it can be expected that to provide stability/absence of the AI, it is necessary to satisfy the conditions

$$g'(\omega_{10,\,20}) + R_r^{-1} > 0 \tag{A.51}$$

Now consider the stability/instability criteria for the roots of the second of equation (2.69) more precisely, again based on the perturbation method, but without using the condition (A.47). At relatively small values g_0, R_r^{-1}, we obtain that the conditions of stability (Lifshitz and Pitaevskii 1981, Fedorchenko and Kotsarenko 1981) (or absence of the AI) is reduced to the form (A.51). For the other two roots of the second equation (2.69), under conditions

$$\omega_0 \sim \omega_{rs} \sim \omega_r,\ \omega_0 L_r R_r^{-1} \sim \omega_0 L_r|g_0| \ll 1,\ \omega_0 L_r|g_0| \leqslant (\Delta\omega/\omega_0) \tag{A.52}$$

it can be shown that they are not related to AI. Thus, when performing (A.51) and (A.52), AI is absent, and there is the spatial amplification of EMW in the MM (Lifshitz and Pitaevskii 1981, Fedorchenko and Kotsarenko 1981).

A.8 Homogenization and nonlinear constitutional equations for the bi-anisotropic metamaterial

The proposed approach (Boardman *et al* 2007b, 2009b, Rapoport *et al* 2006a, Rapoport 2012, 2013, 2014b, 2015a, Rapoport and Boardman 2015b) is fundamentally different from the perturbation method (Lakhtakia and Weiglhofer 1998), for homogenization of (nonlinear) bi-anisotropic medium (Auzanneau and Ziolkowski 1998, Mariotte *et al* 1994, Ramaccia *et al* 2013, Tretyakov *et al* 1994, Tretyakov 2003, 2010). Consider a bi-anisotropic MM, where all Ω-particles lie in one or parallel planes and are equally oriented/parallel to each other (see figures 2.5(a)–(c)). Linear M_n^{lin} and nonlinear M_n^{nl} components of magnetic polarization, directed parallel to the normal to the plane of the magnetic loop (for all metaparticles in the medium under consideration, this direction is the same, which we take as z and denote by the index 'n'). The linear P_l^{lin} and nonlinear P_l^{nl} components of electric polarization are directed along the 'tendrils' (all) of Ω-particles (take this direction x and denote the index 'l'). The components of linear polarizability are determined by the relations (Boardman *et al* 2007a, 2009b, Rapoport *et al* 2006a, Rapoport 2012, 2013, 2015a)

$$\chi_{\text{eett}}^{\text{lin}} = (Nl^2/i\omega)[(Z_{0\text{cg}}^{-1} + Z_{\text{LN}}^{-1})/(Z_{\text{an}}Z_{00N}^{-1})] = (\chi_{\text{ee0}}f/A)(F_{\text{me0}} + F_{\text{eel}}),$$

$$\chi_{\text{ee0}} = \varphi v_{\text{res}}^2(L_{\text{eff}}/L)^{-1}$$

$$\chi_{\text{mmnn}}^{\text{lin}} = -i(NS^2\omega/c^2)[(Z_{0\text{cg}}^{-1} + Z_{\text{aN}}^{-1})/(Z_{\text{Ln}}Z_{00N}^{-1})] = (\chi_{\text{mm0}}\varpi^2/A)(F_{\text{me0}} + F_{\text{mml}}),$$

$$\chi_{\text{mm0}} = \varphi(L_{\text{eff}}/L)^{-1},$$

$$\chi_{\text{ment}}^{\text{lin}} = -\chi_{\text{emtn}}^{\text{lin}} = (NlS/c)/(Z_{\text{aN}}Z_{\text{LN}}Z_{00N}^{-1}) = -(i\chi_{\text{me0}}f\varpi/A)F_{\text{me0}},$$

$$\chi_{\text{me0}} = \sqrt{\chi_{\text{ee0}}\chi_{\text{mm0}}} \tag{A.53}$$

The magnitudes presented in (A.53) are as follows

$$F_{\text{me0}} = 1 + i\psi\varpi,\ F_{\text{mml}} = -i\psi\varpi(C_{N0}/C_{\text{eff}}),\ F_{\text{eel}} = -(1 - f)\varpi^2 - i\psi\varpi^3(C_{\text{cg}}/C_{\text{eff}}),$$

$$\varphi = NS^2/c^2L_{\text{eff}},\ v_{\text{res}} = cl/S\omega_{\text{res}},\ \varpi = \omega/\omega_{\text{res}},\ \omega_{\text{res}}^2 = (L_{\text{eff}}C_{\text{eff}})^{-1},\ L_{\text{eff}} = (L + L_1),$$

$$C_{\text{eff}} = (C_{\text{cg}} + C_{N0} + C_{\text{aeff}})C_{\text{aeff}} = (1 - C_{\text{a}}C_{\text{p}}^{-1})^{-1}C_{\text{a}},\ L_1 = 4\pi S^2N/(3\mu_{\text{host}}c^2),$$

$$C_{\text{p}}^{-1} = 4\pi Nl^2/(3\varepsilon_{\text{host}}),\ A = 1 - \varpi^2 + iq\psi\varpi,\ \ q = [1 - \varpi^2(1 - C_{N0}/C_{\text{eff}})],$$

$$Z_{\text{aN}} = Zla + i(C_{\text{p}}\omega)^{-1} = (i\omega C_{\text{aeff}})^{-1},\ Z_{\text{a}} = (i\omega C_{\text{a}})^{-1},\ Z_{\text{LN}} = Z_{\text{L}} + i\omega L_1 = i\omega L_{\text{eff}},$$

$$Z_L = i\omega L,\ Z_{00N}^{-1} = Z_{0\text{Cg}}^{-1} + Z_{\text{aN}}^{-1} + Z_{\text{LN}}^{-1},\ Z_{0\text{Cg}}^{-1} = Z_0^{-1} + Z_{C_g}^{-1},\ Z_0^{-1} = (i\omega C_{0N})^{-1},$$

$$Z_{C_g}^{-1} = (i\omega C_{\text{g}})^{-1}, f = C_{\text{aeff}}/C_{\text{eff}} \tag{A.54}$$

In (A.54) A is the resonant denominator, the impedances Z_{L}, Z_0, Z_{Cg}, Z_{L} characterize the elements of the equivalent circuit of bi-anisotropic metaparticle (see figures 2.1(a) and (c), figures 2.5(a) and (b)); L, C_{g}, C_{N}, C_{a} are the inductance of the 'magnetic loop' and the capacitance of the gap of the Ω-particle, the linear part of the capacitance of the nonlinear element placed in the gap and the 'tendrils' of the electric dipole that is part of the particle; effective elements (inductance and inverse capacitance); L_1, C_p^{-1} include the 'collective effect' of the interaction between metaparticles in the Maxwell–Garnett approximation (Tretyakov *et al* 1994, 2010, Tretyakov 2003); ψ is the parameter of dissipative losses; N, c, l are the concentration of metaparticles, the speed of light in vacuum, and the length of metal 'tendrils' of Ω-particles, respectively; S and L_{eff} are the area and effective inductance of the magnetic loop of the bi-anisotropic particle, respectively, which includes the contribution of the effect of interaction between the particles; $\bar{\omega}$, ω, ω_{res} are the normalized frequency, the frequency of electromagnetic field, and the resonant frequency of AM, respectively.

For the nonlinear polarizations the formulas (Boardman *et al* 2007b, Rapoport *et al* 2006a, Rapoport 2012, 2013, 2014b, Rapoport and Boardman 2015b) (2.84) are obtained, which can be represented as:

$$P_{\text{NL}l} = \chi_{\text{ee}}(|F|^2)E_l + \chi_{\text{em}}(|F|^2)H_n,\ M_{\text{NL}n} = \chi_{\text{me}}(|F|^2)E_l + \chi_{\text{mm}}(|F|^2)H_n$$

The elements of nonlinear polarizations/susceptibilities are expressed as:

$$\chi_{\text{mm}}(|F|^2) = -\varphi(1 + i\psi\varpi)G(|F|^2),\ \chi_{\text{ee}}(|F|^2) = -\varphi v^2 f^2\varpi^2(1 + i\psi\varpi)G(|F|^2),$$

$$\chi_{\text{me}}(|F|^2) = -\chi_{\text{em}}(|F|^2) = -i\varphi v f\varpi(1 + i\psi\varpi)G(|F|^2) \qquad \text{(A.55)}$$

Using the notation $R \equiv |U_N^{(1)}|^2$, one can write down the expression for $G(|F|^2)$:

$$G(|F|^2) = [A(A + BR(|F|^2)]^{-1}BR(|F|^2),$$

where

$$F = q_l E_l + q_n H_n \equiv f\varpi^2\bar{\mu}_t + i\varpi\overline{H}_n,\ \bar{\mu}_t = lE_t/U_{0k},\ \overline{H}_n = v^{-1}(lH_n/U_{k0}) \qquad \text{(A.56)}$$

$\bar{U}_N^{(1)}$ and R are defined by equations

$$\bar{U}_N^{(1)}(A + B\,|\bar{U}_N^{(1)}|^2) = -F;\ |B|^2R^3 + 2\,\text{Re}(AB^*)R^2 + |A|^2R - f^2\ |F|^2 = 0 \quad \text{(A.57)}$$

$$B = -\bar{\omega}[\alpha_{N1}q_2 + (1/2)\alpha_{N2}],\ q_2 = \alpha_{N1}\bar{\omega}^2\left\{1 - 4\bar{\omega}^2 + 2i\bar{\omega}[1 - 4\bar{\omega}^2(1 - C_{N0}/C_{\text{eff}}]\right\}^{-1}$$

Here $q_{l,n}$ are the coefficients in the linear combination of electric and magnetic fields, which is included in F and determines the components of dyads of nonlinear susceptibility (A.55); these coefficients are easy to find from (A.56); (A.57) includes two equivalent forms of the equation for determination of $R \equiv |U_N^{(1)}|^2$, $U_N^{(1)}$ is the (complex) amplitude of the main (first) voltage harmonic on a nonlinear element (figures 2.1(a) and (c); figures 2.5(a) and (b)). To specify, the nonlinear/active load element of bi-anisotropic particles (figures 2.1(a) and (c); figures 2.5(a) and (b)) is considered capacitive (for instance, a diode), for the nonlinear current I_N, charge q_N, and voltage U_N at which we have

$$I_N = \text{d}q_N/\text{d}t = [C_{0N} + C_{1N}(U_N/U_{0k}) + C_{2N}(U_N/U_{0k})^2]\text{d}U_N/\text{d}t \qquad \text{(A.58)}$$

$\alpha_{N1} \equiv C_{1N}/C_{\text{eff}}$, $\alpha_{N2} \equiv C_{2N}/C_{\text{eff}}$; U_{0k} are the values of the typical voltage.

References

Andryieuski A, Sangwoo H, Sukhorukov A A, Kivshar Y S and Lavrinenko A V 2012 Bloch-mode analysis for retrieving effective parameter sofmetamaterials *Phys. Rev.* B **86** 035126

Auzanneau F and Ziolkowski R W 1998 Theoretical study of synthetic bianisotropic materials *J. Electromagn. Waves Appl.* **12** 353–70

Basu B 2002 On the linear theory of equatorial plasma instability comparison of different description *J. Geophys. Res.* **107** SIA 18-1–10

Boardman A D and Marinov K 2006 Electromagnetic energy in a dispersive metamaterial *Phys. Rev.* B **73** 16

Boardman A D, Rapoport Y G, King N and Malnev V N 2007a Creating stable gain in active metamaterials *J. Opt. Soc. Am. B.* **24** A53–61

Boardman A D, King N and Rapoport Y 2007b Metamaterials driven by gain and special configurations *SPIE Proc. Metamaterials II* **6581** 658108

Boardman A D, King N, Mitchell–Thomas R C, Malnev V N and Rapoport Y G 2008 Gain control and diffraction-managed solitons in metamaterials *Metamaterials* **2–3** 145–54

Boardman A D, King N and Rapoport Y 2009a Circuit model of gain in metamaterials *Nonlinearities in Periodic Structures and Metamaterials* (Berlin: Springer) pp 193–206

Boardman A D, Mitchell-Thomas R and Rapoport Y G 2009b Weakly nonlinear waves in layered bi–anisotropic *Proc. of Third Int. Congress on Adv. Electromagn. Materials in Microwaves and Optics: Metamaterials (London)* 495–7

Boardman A D, Grimalsky V V, Kivshar Y, Koshevaya S V, Lapine M, Litchinitser M, Malnev V N, Noginov M, Rapoport Y G and Shalaev V M 2011 Active and tunable metamaterials *Laser Photonics Rev.* **5** 287–307

Bosch R and Thim H W 1974 Computer simulation of transferred electron devices using the displaced maxwellian approach *IEEE Trans.* **ED-21** 16–25

Bosch B G and Engellmann R W H 1975 *Gunn-Effect Electronics* (London: Pitman) p 176

Bragin Y A, Tyutin A A and Kocheev *et al* 1974 Direct measurements of the vertical electric field of the atmosphere up to 80 km *Space Res.* **12** 279–81

Bravo–Ortega A and Glassor A H 1991 Theory and application of complex geometric optics in inhomogeneous magnetized plasmas *Phys. Fluids* B **3** 529–35

Bullough R K and Caudrey P J 1980 *Solitons* (Berlin: Springer) p 392

Buttner O, Bauer M, Demokritov S O, Hillebrands B, Kivshar Y S, Grimalsky V, Rapoport Y, Kostylev M P, Kalinikos B A and Slavin A N 2000a Spatial and spatiotemporal self-focusing of spin waves in garnet films observed by space–and time–resolved Brillouin light scattering *J. Appl. Phys.* **87** 5088–90

Capolino F 2009 *Theory and Phenomena of Metamaterials* (Boca Raton, FL: CRC Press, Taylor and Francis Group) p 926

Chebykin A V, Orlov A A, Vozianova A V, Maslovski S I, Kivshar Y S and Belov P A 2011 Nonlocal effective medium model for multilayered metal–dielectric metamaterials *Phys. Rev.* B **84** 115438–46

Cheng Q, Cui T J and Zhang C 2007 Waves in planar waveguide containing chiral nihility metamaterial *Opt. Commun.* **276** 317–21

Eleftheriades G V, Siddiqui O and Iyer A K 2003 Tramsmission line models for negative refractive index media and associated implementation without excess resonators *IEEE Microw. Wireless Compon. Lett.* **13** 51–3

Eleftheriades G I and Balmain K G 2005 *Negative-Refraction Metamaterials: Fundamental Principles and Applications* vol 5 (New York: IEEE Press, Wiley) p 418

Engheta N, Jaggard L D and Kowarz M W 1992 Electromagnetic waves in Faraday chiral media *IEEE Trans. Antennas Propag.* **40** 367–74

Fedorchenko A M and Kotsarenko N Y 1981 *Absolute and Convective Instability in Plasma and Solid State* (Moscow: Nauka) p 176 (in Russian)

Govyadinov A A, Podolskiy V A and Noginov M A 2007 Active metamaterials: sign of refractive index and gain-assisted dispersion management *Appl. Phys. Lett.* **91** 191103

Ionkin P A 1972 *Principles of Engineering Electrophysics* (Moscow: Visshaya Shkola) p 439 (in Russian)

Kadomtsev B B 1988 *Collective Phenomena in Plasma* (Moscow: Nauka) p 304 (in Russian)

Kalinin B A and Shtikov V V 1990 On the possibility of reversing the front of radio waves in the artificial nonlinear media *J. Commun. Technol. Electron.* **35** 2275–81 (in Russian)

Katko A R, Gu S and Barrett J P *et al* 2010 Phase conjugation and negative refraction using nonlinear active metamaterials *Phys. Rev. Lett.* **105** 123905

Kidner C, Mehdi I, East J R and Haddad G I 1990 Power and stability limitations of resonant tunneling diodes *IEEE Trans. Microwave Theory Tech.* **38** 864–72

Korn G A and Korn T M 1968 *Mathematical Handbook for Scientists and Engineers* (New York: McGraw-Hill) p 378

Kong J A 1972 Theorems of bi-anisotropic media *Proc. IEEE* **60** 1036–46

Lakhtakia A and Weiglhofer W 1998 Maxwell Garnett formalism for cubically nonlinear, gyrotropic, composite media *Int. J. Electronics* **84** 285–94

Lheurette E, Vanbesien O and Lippens D 2007 Double negative media using interconnected–type metallic particles *J. Microwave Opt. technol. Lett.* **49** 84–90

Lheurette É, Houzet G and Carbonell J *et al* 2008 Omega-type balanced composite negative refractive index materials *IEEE Trans. Anten. Propag.* **56** 3462–9

Lifshitz E M and Pitaevskii L P 1981 *Physical Kinetics* (Oxford: Elsevier) p 369

Lindell I V, Shihvola A H, Tretyakov S A and Vittanen A J 1994 *Electromagnetic Waves in Chiral and Bi-isotropic Media* (Boston, MA: Artech House) p 335

Liu M, Powell D A, Shadrivov I V, Lapine M and Kivshar Y S 2014 Spontaneous chiral symmetry breaking in metamaterials *Nat. Commun.* **5** 4441–9

Mariotte F, Tretyakov S and Sauviac B 1994 Isotropic chiral composite modeling: Comparison between analytical, numerical and experimental results *Microwave Opt. Technology lett.* **7** 861–3

McCauley A P, Zhao R, Homer and Reid M T 2010 Microstructure effects for Casimir forces in chiral metamaterials *Phys. Rev.* B **82** 165108–20

Migulin V V, Medvedev V I, Mustel E P and Parigin V N 1988 *Fundamentals of Oscilation Theory* (Moscow: Nauka) p 480 (in Russian)

Mikhailovskii A B 1974 Theory of plasma instabilities *Instabilities of a Homogeneous Plasma (Studies in Soviet Science)* vol 1 (Studies in Soviet Science) (London: Consultants Bureau)

Oleinik V F, Bulgach V L and Valaev V V *et al* 2004 Electron devices of milimiter and submilimeter range and basics of nanotechnology *State Univ. Telecommun. Kiev* p 389 (in Russian)

Ookawa Y, Kishimoto S, Maezawa K and Mizutani T 2006 Novel resonant tunneling diode oscillator capable of large output power operation *IEICE Trans. Electron.* **E89C** 999–1004

Pendry J B, Holden A J, Robbins D J and Stewart W I 1999 Magnetism from conductors and enhanced nonlinear phenomena *IEEE Trans. Microw. Theory* **47** 2075–84

Perez–Molina M, Carretero L and Blaya S 2009 Efficient computation of longitudinal lasing modes in arbitrary active cavities: the bidirectional time evolution method *J. Lightwave Technol.* **27** 3000–9

Popov A K and Shalaev V M 2006 Compensating losses in negative-index metamaterials by optical parametric amplification *Opt. Lett.* **31** 2169–71

Rabinovich M I and Trubetskov D I 1989 *Oscillation and Waves: In Linear and Nonlinear Systems (Mathematics and Its Applications)* (London: Spinger) p 598

Ramaccia D, Bilotti F and Toscano A 2013 Accurate analytical model of coupled omega particles for metamaterial design *Proc. of the IEEE 7th Int. Congress on Advanced Electromagnetic Materials in Microwaves and Optics–Metamaterial (Bordeaux)* 61–3

Rapoport Y, Grimalsky V, Hayakawa M, Ivchenko V, Juarez D, Koshevaya S and Gotynyan O 2004 Change of ionospheric plasma parameters under the influence of electric field which has lithospheric origin and due to radon emanation *Phys. Chem. Earth.* **29** 579–87

Rapoport Y G, Boardman A D, Kanevskiy V I, Malnev V N, King N J and Velasco L 2006a Modelling new active media based on metamaterials with artificial *IEEE Proc. of the 16th Int. Crimean Conf. 'Microwave and Communication Technology' CriMiCo'2006 (Sevastopol)* Catalog No. **06EX1376** 671–2

Rapoport Y G 2012 General method for modeling nonlinear waves in layered structures of diferent physical nature including bi–anisotropic and active metamaterials *Progress in Electromagnetics Research Symp. (PIERS) (Moscow)* 18–9 Abstracts

Rapoport Y G, Tretyakov S and Maslovski S 2013 Nonlinear active Huygens metasurfaces for reflectionless phase conjugation of electromagnetic waves in electrically thin layers *J. Electromagn. Waves Appl.* **27** 1309–28

Rapoport Y G 2013 General method for modeling nonlinear waves in active metamaterial and gyrotropic layered structures *Proc. of Int. Kharkov Symp. on Phys. and Engineering of Microwaves, Millimeter and Submillimeter Waves MSMW'13* (*Kharkov, Ukraine*) 253–5

Rapoport Y G, Grimalsky V V, Boardman A D and Malnev V N 2014a Controlling nonlinear wave structures in layered metamaterial, gyrotropic and active media *Proc. of IEEE 34th Int. Sci. Conf. Electronics and Nanotechnology ELNANO* (*Kyiv, Ukraine*) 46–50

Rapoport Y G 2014b General method for deriving the equations of evolution and modeling of nonlinear waves in layered active media with bulk and surface *Bull. Kyiv Nat. Taras Shevchenko Univ. Series: Phys.-Math. Sci.* **1** 281–88 (in Ukrainian)

Rapoport Y G 2015a Modeling 'From properties of a metaparticle (MP) to characteristics of nonlinear waves in layered active bi-anisotropic metamaterials *(BIAM) Ukr.–Germ. Symp. on Phys. and Chem. of Nanostructures and on Nanobiotechnology* (*Kyiv, Ukraine*) Abstract p 159

Rapoport Y G and Boardman A D 2015b Modeling 'from the characteristics of the metaparticle to the characteristics of nonlinear waves in layered active bi-anisotropic metamaterials with bulk and surface nonlinearities' *Bull. Kyiv Nat. Taras Shevchenko Univ. Series: Phys.* **3** 207–12 (in Ukrainian)

Scott A 1970 *Active and Nonlinear Wave Propagation in Electronics* (New York: Wiley) p 326

Tretyakov S A, Kharina T G, Simovski K R and Pavlov A A 1994 Frequency dispersion in chiral and omega media: An approximate theoretical model *Antennas Propagation Society Int. Symp. AP–S. Digest* **2** 722–5

Tretyakov S A and Mariotte F 1995 Maxwell Garnett modeling of uniaxial chiral composites with bianisotropic inclusions *J. Electromagn. Waves Appl.* **9** 1011–25

Tretyakov S A, Mariotte F, Simovski C R, Kharina T G and Heliot J-P 1996 Analytical antenna model for chiral scatterers: comparison with numerical and experimental data *IEEE Trans. Anten. Propagat.* **44** 1006–14

Tretyakov S 2003 *Analytical Modeling in Applied Electromagnetics* (Boston, MA: Artech House) p 275

Tretyakov S 2010 *Nanostructured Metamateriuals* (Brussels: : Office of the European Union Brussels) p 137

Vainshtein L A 1988 *Elektromagnetic Waves* (Moscow: Radio and Communication) p 440 (in Russian)

Yatsenko V V, Maslovski S I, Tretyakov S A, Prosvirnin S L and Zouhdi S 2003 Plane-wave reflection from double arrays of small magnetoelectric scatterers *Ttrans. Anten. Propag.* **51** 2–11

Yuan Y, Popa B I and Cummer S A 2009 Zero loss magnetic metamaterials using powered active unit cells *Opt. Express* **17** 16135–43

Zheludev N I and Kivshar Y S 2012 From metamaterials to metadevices *Nat. Mater.* **11** 917–24

Zheng Y, Meng Y and Liu Y 2014 Solitons in Gaussian potential with spatially modulated nonlinearity *Opt. Commun.* **315** 63–8

IOP Publishing

Waves in Nonlinear Layered Metamaterials, Gyrotropic and Plasma Media

Yuriy Rapoport and Vladimir Grimalsky

Chapter 3

General method of the derivation of nonlinear evolution equations for layered structures (NEELS) with the volume and surface nonlinearities

Plasmonic and metamaterial (MM) waveguides provide enhanced field localization and nonlinear interaction of optical waves with the material (Li *et al* 2018). The combination of MM and plasmonic approaches provides advantages that are effectively implemented in MM nanoplasmonics. Waveguides with layers of anisotropic MMs and plasma-like media possess properties unattainable in conventional dielectric waveguide, such as propagation of backward waves and modes propagating at frequencies below the cutoff frequencies of the conventional fundamental mode, zero group velocity (Bhardwaj *et al* 2020), working in Epsilon-Near-Zero (ENZ) regime (Li and Argyropoulos 2018), etc. A new approach to MM photonic waveguides and resonators can be based on toroidal systems using nonlinear plasmon resonators with Mie resonances and can be applied from microwave to optical frequencies (Yang *et al* 2021). A very interesting and important property of plasmonic waveguides is that the nonlinearity which determines their evolution is not volume, but surface nonlinearity on the plasma-dielectric surfaces of such waveguides, which was first shown in a series of works by the authors of this book, in particular (Grimalsky and Rapoport 1998a, Boardman *et al* 2017) to generate the second harmonic in the regime of only temporal resonance (ENZ) in the absence of spatial resonance. It is curious that this special property of the prevalence of surface nonlinearity over volume nonlinearity for plasmons in a layered dielectric-plasma waveguide was repeatedly 'rediscovered' after work (Grimalsky and Rapoport 1998a), albeit without reference to Grimalsky and Rapoport (1998a) and Boardman *et al* (2017), and once again this was done, in particular in the work Li *et al* (2018).

Using nonlinear MM and plasmonic waveguides and their exciting properties provides the excellent platform for designing low-threshold all-optical switches, nanolasers, nonlinear gap solitons, unidirectional coherent perfect absorbers (Li *et al* 2018), enhanced magnetic resonance imaging and electromagnetic shielding at radio frequencies, very interesting imaging applications and efficient coupling of the emitted radiation from small sources into waveguides at optical frequencies (Bhardwaj *et al* 2020); capabilities of multi-frequency plasmonic all-optical switches facilities allow their applications in miniaturized photonic circuits and in biosensors (Lotfi *et al* 2020). As shown in Peruch *et al* (2017), the optically-induced temperature distribution in the metal component of the plasmonic nanorod MM determines the geometrically-dependent dynamics of the optical response, which manifests itself in the properties of the coefficients of both reflection and transmission of electromagnetic waves in the MM layer; moreover, the dynamics of the optical response can be controlled via MM modal engineering. According to Yang *et al* (2021), systems with controlled toroidal moment open up new application possibilities, such as metaswitch, lasing spaser, toroidal circular dichroism, molecule detection and ultrasensitive sensors etc.

Chapter 3 is devoted to the method of derivation of the nonlinear evolution equation for layered media (NEELS) for the amplitudes of the envelope wave packets, in the presence of both bulk and surface nonlinearities and accounting for the effects of non-locality. The methods are applicable to studying wave processes in layered waveguiding media with weak and moderate nonlinearity of both artificial and natural origin in the wide frequency range, from ultra-low frequencies (ULF) no optics, where the corresponding wavelengths differ in many orders of magnitude. To the case of strong and resonant nonlinearities, chapter 10 is addressed.

3.1 A method for the derivation of the nonlinear evolution equations for the waves in layered structures with bi-anisotropic metamaterials

In this section, we present the main approaches for the derivation of the evolution equations for layered media with gyrotropy and bi-anisotropy.

3.1.1 Formulation of the problem for the media with weak nonlinearity and weak losses/gain. NEELS in a differential form

The technique of the derivation (Grimalsky *et al* 1996, 1997, Grimalsky and Rapoport 1998a, Rapoport *et al* 2006a, Rapoport and Grimalsky 2011, Boardman *et al* 2007, 2009, Rapoport 2014b) is used, similar to that used to obtain energy balance equations (like the Poynting type) in layered structures. At the same time, in contrast to the 'energy' method (Rowland 1999, Grimalsky and Rapoport 1995) and the Hamiltonian formalism (Slavin and Rojdestvenski 1994), this method implies the consideration of fields corresponding to two media that are 'slightly different' from one another. One of them is a hypothetical purely linear media in this layered structure (which differs from the real media only by the lack of a

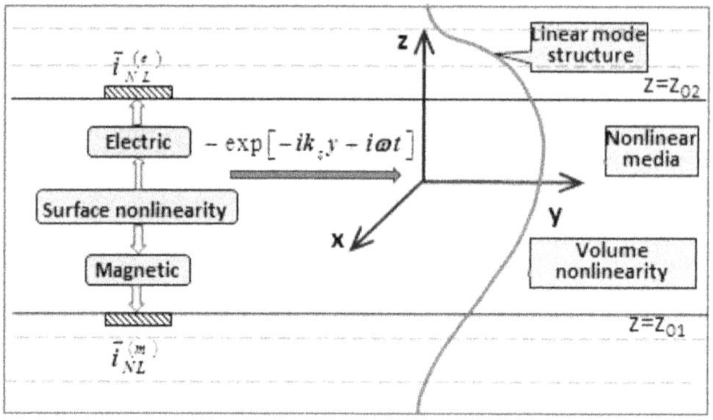

Figure 3.1. Wave packet in a nonlinear layered structure with volume and surface nonlinearities. Nonlinear MM waveguide structure. A similar structure is used when considering a nonlinear dielectric with spatial dispersion.

nonlinearity). The other one is a real medium in the same structure, where the nonlinearity is present. In the particular case, when there is no nonlinearity, the obtained relations are reduced to the Poynting relation (or to the set of such relations where many wave/multi-way interactions are considered). Consider the system shown in figure 3.1.

The layer $-L/2 < z < L/2$ is bi-anisotropic, half-space $z < -l/2$ and $Z > L/2$ may be, for example, isotropic or gyrotropic media, linear or nonlinear.

The surfaces of the section are parallel to the plane XY. In fact, there are no significant restrictions on the number of layers or types of environments. Suppose that:

(1) We have solved already the linear problem of the waveguide modes of a given system and obtained their dispersion equation.

$$D(\omega, \vec{k}) = 0 \tag{3.1}$$

This relation corresponds to the waves propagating in a plane XY in the corresponding structure without nonlinearity. We also consider the transverse distribution of fields corresponding to their own linear modes of the structure. Suppose that the components of the linear field (which is solutions of the corresponding linear problem) are denoted by the index 'l'; in Fourier presentation,

$$\begin{pmatrix} \vec{E}_l \\ \vec{H}_l \end{pmatrix} = A_0 \begin{pmatrix} \vec{f}_{\mathrm{e}} \\ \vec{f}_{\mathrm{h}} \end{pmatrix} \mathrm{e}^{j(\omega_0 t - \vec{k}_0 \vec{r})} \tag{3.2}$$

where ω_0, $\vec{k}_0$, and $\vec{r}$ are frequency, wave vector, and the coordinates in the plane XY, A_0 is the amplitude of a linear wave that can be used for normalization if necessary; in the following formulas we put $A_0 = 1$. The functions $\vec{f}_{\mathrm{e,\,h}}$, which we call 'polarization functions,' describe the distributions of the corresponding linear field components (indices 'e' and 'h' are used for electrical and magnetic components, respectively) in the direction Z, normal to the structure. These functions are obtained with taking into account the (linear) boundary conditions and the dispersion relation, so that

$$\vec{f}_{\mathrm{e.h}} = \vec{f}_{\mathrm{e.h}}(\omega, \vec{k}, z) \text{ and } |\vec{f}_{\mathrm{e,\,h}}| \to 0, \quad \text{when } z \to \pm\infty. \tag{3.3}$$

(2) We will also suppose that the field of a nonlinear spectrally narrow wave packet with a carrier frequency ω_0 and a wavenumber k_0 is represented as

$$\begin{pmatrix} \vec{E}_1 \\ \vec{H}_1 \end{pmatrix} = A_1(t, \vec{r}) \begin{pmatrix} \vec{E}_{1l} \\ \vec{H}_{1l} \end{pmatrix}, \quad \vec{r} = (x, y) \tag{3.4}$$

In equation (3.4), A_i is slowly varying amplitude (SVA) (slow amplitude (Anisimov 2003)); thus, we neglect, at the first approximation, variations in the field of the fundamental harmonic in the transverse direction, which is valid in the case of weak nonlinearity. The spectral narrowness of the wave packet is determined by the fact that

$$\Delta\omega \ll \omega_0, \quad \Delta k \ll k_0 \tag{3.5}$$

where $\Delta\omega$ and Δk are the frequency and wavenumber spectra of the wave packet. Suppose that the materials relations can present in the form (Lindell *et al* 1994, Fedorov 1976, Tretyakov 2003, 2010, Tretyakov *et al* 2007, Capolino 2009, Sugakov 1974, Nikolsky and Nikolskaya 1989, Kong 1972)

$$\begin{gathered} \vec{D} = \widehat{\varepsilon}_{\mathrm{eff}} \vec{E} + \widehat{\alpha}_{\mathrm{eff}} \vec{H}, \; \vec{B} = \widehat{\beta}_{\mathrm{eff}} \vec{E} + \widehat{\mu}_{\mathrm{eff}} \vec{H}, \; \widehat{\varepsilon}_{\mathrm{eff}} = \widehat{\varepsilon} + \widehat{\varepsilon}_{\mathrm{act}}, \; \widehat{\mu}_{\mathrm{eff}} = \widehat{\mu} + \widehat{\mu}_{\mathrm{act}}, \\ |\widehat{\varepsilon}| \gg |\widehat{\varepsilon}_{\mathrm{act}}|, \; |\widehat{\mu}| \gg |\widehat{\mu}_{\mathrm{act}}|, \; \widehat{\alpha}_{\mathrm{eff}} = \widehat{\alpha} + \widehat{\alpha}_{\mathrm{act}}, \; \widehat{\beta}_{\mathrm{eff}} = \widehat{\beta} + \widehat{\beta}_{\mathrm{act}}, \\ |\widehat{\alpha}| \gg |\alpha_{\mathrm{act}}|, \; |\widehat{\beta}| \gg |\widehat{\beta}_{\mathrm{act}}| \end{gathered} \tag{3.6}$$

where quantities with indices 'act' correspond to dissipative losses, or the presence of active elements in metamolecules that provide amplification of waves in the media. Therefore, we assume that quantities $\widehat{\varepsilon}$, $\widehat{\mu}$, $\widehat{\alpha}$, $\widehat{\beta}$ have the properties which correspond to the preserving energy, in other words (Tretyakov 2010, Fedorov 1976)

$$\widehat{\varepsilon}^{+} = \widehat{\varepsilon}, \widehat{\mu}^{+} = \widehat{\mu}, \widehat{\beta} = \widehat{\alpha}^{+}. \tag{3.7}$$

(3) Suppose that the nonlinear polarizations $\vec{P}_{\mathrm{NL}}$, $\vec{M}_{\mathrm{NL}}$ are known (or it is needed to consider nonlinear material equations), and their contribution is not included in (3.6). The goal is to derive an evolution equation for a nonlinear amplitude A_i (one can also obtain a set of equations for the corresponding amplitudes when cross-modulation or other interaction between several waves/modes is necessary).

3.1.2 Equations for the slowly varying envelope amplitude of a wave packet in unbounded media with a spatial dispersion

We write down the Maxwell equation for linear $\vec{E}_l$, $\vec{H}_l$ and nonlinear $\vec{E}$, $\vec{H}$ fields in relation (3.4). We use the approach (Grimalsky *et al* 1996, Grimalsky and Rapoport 1998a, Rapoport *et al* 2004b, Rapoport and Grimalsky 2011, Rapoport 2013, 2014b, 2015a, Rapoport and Boardman 2015b, Boardman *et al* 2007, 2009) to the Maxwell equations and material equations (Lindell *et al* 1994, Fedorov 1976, Tretyakov 2010, Sugakov 1974, Nikolsky and Nikolskaya 1989, Kong 1972, D'Aguanno 2008), while all linear currents are included in the respective inductions and magnetic and electrical nonlinearities are included in the corresponding polarizations

$$\vec{E}_l^* \cdot |\ \mathrm{rot}\vec{H} = \frac{1}{c}\frac{\partial \vec{D}}{\partial t} + \frac{4\pi}{c}\frac{\partial \vec{P}_{\mathrm{NL}}}{\partial t},\quad H_l^* \cdot |\ \mathrm{rot}\vec{E} = -\frac{1}{c}\frac{\partial \vec{B}}{\partial t} - \frac{4\pi}{c}\frac{\partial \vec{M}_{\mathrm{NL}}}{\partial t} \tag{3.8}$$

$$\vec{E} \cdot |\ \mathrm{rot}\vec{H}_l^* = \frac{1}{c}\frac{\partial \vec{D}_l^*}{\partial t} = -i\frac{\omega_0}{c}\vec{D}_l^*,\quad \vec{H} \cdot |\ \mathrm{rot}\vec{E}_l^* = -\frac{1}{c}\frac{\partial \vec{D}_l^*}{\partial t} = i\frac{\omega_0}{c}\vec{B}_l^* \tag{3.9}$$

Let us form bilinear combinations of fields $\vec{E}$, $\vec{H}$ similar to those used in the derivation of the energy conservation law, considering narrow wave packets for nonlinear waves with SVA and presenting wave fields as a series with small deviations from the carrier frequency and wavenumber. To do this, we multiply equations (3.8) and (3.9) by the corresponding fields, as is symbolically shown in the left-hand sides of these equations. For example, a symbolic record '$\vec{E}_l^*|$...' means that the first of equation (3.8), denoted here as '...', is multiplied by a scalar value. An asterisk for the field components in these equations denotes the complex conjugation of the corresponding fields. Applying the procedure described above, we obtain:

$$\begin{aligned}&\frac{c}{4\pi}\mathrm{div}[\vec{E}_l^* \times \vec{H}] + \frac{c}{4\pi}[\vec{E} \times \vec{H}_l^*] + \frac{1}{4\pi}\vec{H}_l^*\frac{\partial \vec{B}}{\partial t} + \frac{1}{4\pi}\vec{E}_l^*\frac{\partial \vec{D}}{\partial t}\\ &= -\left(\vec{H}_l^*\frac{\partial \vec{M}_{\mathrm{NL}}}{\partial t} + \vec{E}_l^*\frac{\partial \vec{P}_{\mathrm{NL}}}{\partial t}\right)\end{aligned} \tag{3.10}$$

We shall consider the propagation of a nonlinear wave packet using the Fourier method (Kadmontsev 1988). In this case, we will represent the values that depend on the frequency ω and wavenumbers (vectors) $\vec{k}$ in the form of an expansion in a series of deviations from the carrier frequency ω and wavenumber:

$$\omega = \omega_0 + \Delta\omega,\ \vec{k} = \vec{k}_0 + \Delta\vec{k}, \tag{3.11}$$

$$\begin{pmatrix}\widehat{\varepsilon}\\ \widehat{\mu}\\ \widehat{\alpha}\\ \widehat{\beta}\end{pmatrix} = \begin{pmatrix}\widehat{\varepsilon_0}\\ \widehat{\mu_0}\\ \widehat{\alpha_0}\\ \widehat{\beta_0}\end{pmatrix}_{\omega=\omega_0} + \frac{\partial}{\partial\omega}\begin{pmatrix}\widehat{\varepsilon_0}\\ \widehat{\mu_0}\\ \widehat{\alpha_0}\\ \widehat{\beta_0}\end{pmatrix}\Delta\omega + \frac{\partial}{\partial\vec{k}}\begin{pmatrix}\widehat{\varepsilon_0}\\ \widehat{\mu_0}\\ \widehat{\alpha_0}\\ \widehat{\beta_0}\end{pmatrix}\Delta\vec{k} + \dots, \tag{3.12}$$

$$\begin{pmatrix} \vec{f}_E \\ \vec{f}_H \end{pmatrix} = \begin{pmatrix} \vec{f}_{E0} \\ \vec{f}_{H0} \end{pmatrix}_{\omega=\omega_0} + \frac{\partial}{\partial \omega} \begin{pmatrix} \vec{f}_{E0} \\ \vec{f}_{H0} \end{pmatrix} \Delta\omega + \frac{\partial}{\partial \vec{k}} \begin{pmatrix} \vec{f}_{E0} \\ \vec{f}_{H0} \end{pmatrix} + \dots \quad (3.13)$$

We will consider conditions (3.6) and (3.7) and use, according to the Fourier operator method (Kadmontsev 1988), the substitution

$$i\Delta\omega \to \frac{\partial}{\partial t}, \; i\Delta\vec{k} \to -\frac{\partial}{\partial \vec{r}}. \quad (3.14)$$

As a result of these transformations, we obtain the following relation:

$$\begin{aligned} &\frac{c}{4\pi}\left[\vec{f}^*_H \frac{\partial}{\partial\omega}(\omega\widehat{\mu})\vec{f}_H + \vec{f}^*_E \frac{\partial}{\partial\omega}(\omega\widehat{\varepsilon})\vec{f}_E + \vec{f}^*_H \frac{\partial}{\partial\omega}(\omega\widehat{\alpha}^+)\vec{f}_E + \vec{f}^*_E \frac{\partial}{\partial\omega}(\omega\widehat{\alpha})\vec{f}_H\right] \\ &\times \frac{\partial A_1}{\partial t}\frac{c}{4\pi}\mathrm{div}\{[\vec{E}^*_{1l} \times \vec{H}_1] + [\vec{E}_1 \times \vec{H}^*_{1l}]\} - \frac{1}{4\pi}\left[\left(\vec{f}^*_H \frac{\partial\widehat{\mu}}{\partial\vec{k}}\vec{f}_H + \vec{f}^*_E \frac{\partial\widehat{\varepsilon}}{\partial\vec{k}}\vec{f}_E\right)\right] \\ &+ \left[\left(\vec{f}^*_H \frac{\partial\widehat{\alpha}}{\partial\vec{k}}\vec{f}_E + \vec{f}^*_E \frac{\partial\widehat{\alpha}}{\partial\vec{k}}\vec{f}_H\right)\right]\omega\frac{\partial A_1}{\partial\vec{r}} = \\ &-\frac{j}{4\pi}\omega\{(\vec{f}^*_H \widehat{\mu}_{\mathrm{act}}\vec{f}_H - \vec{f}_H \widehat{\mu}^*_{\mathrm{act}}\vec{f}^*_H) + (\vec{f}^*_E \widehat{\varepsilon}_{\mathrm{act}}\vec{f}_E - \vec{f}_E \widehat{\varepsilon}^*_{\mathrm{act}}\vec{f}^*_E) \\ &+ [(\vec{f}^*_E \widehat{\alpha}_{\mathrm{act}}\vec{f}_H - \vec{f}_H \widehat{\beta}^*_{\mathrm{act}}\vec{f}^*_E) + (\vec{f}^*_H \widehat{\beta}_{\mathrm{act}}\vec{f}_E - \vec{f}_E \widehat{\alpha}^*_{\mathrm{act}}\vec{f}^*_H)]\} \\ &- \left(\vec{H}^*_{1l}\frac{\partial\vec{M}_{\mathrm{NL}}}{\partial t} + \vec{E}^*_{1l}\frac{\partial\vec{P}_{\mathrm{NL}}}{\partial t}\right) \end{aligned} \quad (3.15)$$

Note that, according to the expansions (3.11)–(3.15), and as can be seen from (3.15), for the (present) case of an unbounded medium, the contributions to energy flows and densities of both temporal and spatial dispersion are taken into account. For this to become apparent, it is necessary to consider the case of linear waves when the first two terms on the left-hand side of equation (3.15) describe precisely the changes in density and the divergence of the energy flow (Kamenetskii 1996). The first term on the right-hand side of equation (3.15) corresponds to the 'dissipative loss density', the sign of which can be either positive or negative (for the passive or active medium, respectively), and the second term in the right-hand side (3.15) describes the volumetric nonlinearity. The developed version of NEELS equation (3.15), is suitable for structures with layers of nonlinear bi-anisotropic and gyrotropic structures and MMs with a negative phase behavior. The nonlinear equation (3.15) was first obtained in Boardman *et al* (2007) and Rapoport (2014b). A new method of nonlinear homogenization and obtaining the nonlinear electric and magnetic polarizations is proposed in Boardman *et al* (2007), Rapoport (2014b) and Rapoport and Boardman (2015b) and is described briefly in section 2.3 of chapter 2. In the particular case of the absence of nonlinearity, the energy conservation law (Lindell *et al* 1994, Kamenetskii 1996, Luan *et al* 2011) follows from equation (3.15). Note that in the absence of nonlinearity and dissipation (the zero right-hand side of (3.15)), and also in the absence of magnetoelectric coupling $\widehat{\alpha} = 0$, the relation (3.15) corresponds to the last (unnumbered) relation in Landau and Lifshits (1982, see paragraph 103, p 497).

For the generality of a consideration, in equation (3.15) the spatial dispersion is included in the form of the dependence of the tensors $\widehat{\varepsilon}$, $\widehat{\mu}$, $\widehat{\alpha}$ on $\vec{k}$. It is known that in the presence of such a dependence for layered media, 'additional light (electromagnetic) waves' (Pekar 1962) arise and the corresponding additional boundary conditions are required. But when integrating (3.15) for layered media in the direction normal to the layers, in the next paragraph we will consider the approximation of local fields (Lindell *et al* 1994, Tretyakov *et al* 1996, Tretyakov 2003). Mathematically, this is expressed in the absence of dependences of the values $\widehat{\varepsilon}$, $\widehat{\mu}$, $\widehat{\alpha}$ included into the left-hand side (3.15) on the wave vector $\vec{k}$. Nevertheless a way for extending the NEELS method to the case of layered media with the spatial dispersion, which is described not by a given dependence of material parameters on $\vec{k}$, but by the corresponding equations for polarization(s), is considered in sections 3.3.1–3.3.3 with examples of nonlinear wave processes in dielectric-ferromagnetic and dielectric-ferroelectric media. In these cases a presence of the nonlinearity in the corresponding auxiliary boundary conditions (Pekar 1962, Gurevich and Melkov 1996, Agranovich and Ginzburg 1965, Akhiezer *et al* 1967, Rapoport 2014b) is accounted for.

3.1.3 Equations for the envelope amplitudes NEELS in the integral form

Let us consider method NEELS for bi-anisotropic media in the approximation of local fields. At the same time, we consider a possible electromagnetic nonlinearity at the boundaries. Such a nonlinearity can be caused by artificial inclusions (meta-molecules), such as nonlinear loads with real or imaginary effective conductivity in the surface layer. Moreover, such loads in the form of nonlinear diodes can be tuned by an external bias electric field or optical (laser) radiation. Given that the boundary conditions for the tangential components of nonlinear fields (with the index '*tg*' at the surfaces $z = \pm L/2$) have a form

$$\vec{n} \times \Delta\vec{E}_1 = -\frac{4\pi}{c}\vec{i}_{tg}^{(m)}, \quad \vec{n} \times \Delta\vec{H}_1 = \frac{4\pi}{c}\vec{i}_{tg}^{(e)} \tag{3.16}$$

$$i_{x,\,y}^{(e)}|_{z=\pm L/2} = \int_{\pm L/2-\delta}^{\pm L/2} \frac{\partial P_{\mathrm{SNL}x,\,y}}{\partial t}\mathrm{d}z, \quad i_{x,\,y}^{(m)}|_{z=\pm L/2} = \int_{\pm L/2-\delta}^{\pm L/2} \frac{\partial M_{\mathrm{SNL}x,\,y}}{\partial t}\mathrm{d}z \tag{3.17}$$

In relations (3.16) and (3.17), $\vec{n}$, $\Delta\vec{E}_1$, $\Delta\vec{H}_1$ are the normal vector to the corresponding surfaces of the interface between the media and the 'surface jumps' of the respective electric and magnetic fields, δ is the thickness of the surface layer where a surface nonlinearity exists, and the value of δ is assumed to tend to zero when deriving the boundary conditions; $P_{SNLx,\,y}$, $M_{SNLx,\,y}$ are the corresponding components of surface nonlinear electric and magnetic polarizations, respectively, $i_{x,\,y}^{(e.m)}|_{z=\pm L/2}$ are the equivalent nonlinear electric and magnetic surface currents. Taking into account (3.16) and integrating (3.15), in the absence of the dependence of tensors $\widehat{\varepsilon}$, $\widehat{\mu}$, $\widehat{\alpha}$ on the wave vector $\vec{k}$, we obtain the relation of the NEELS method in the integral form:

$$\left\{\frac{\partial}{\partial t}+\left[V_g\frac{\partial}{\partial y}+\frac{i}{2}\frac{\partial^2\omega}{\partial k_y^2}\frac{\partial^2}{\partial y^2}+i\frac{\partial\omega}{\partial(k_x^2)}\frac{\partial^2}{\partial x^2}\right]+\frac{j\omega}{4\pi}\frac{Q}{W_0}\right\}A_1$$
$$=-\frac{1}{W_0}\int_{-\infty}^{\infty}\left(H_{1l}^*\frac{\partial\vec{M}_{\mathrm{NL}}}{\partial t}+\vec{E}_{1l}^*\frac{\partial\vec{P}_{\mathrm{NL}}}{\partial t}\right)\mathrm{d}z \tag{3.18}$$
$$-\sum_{n=1,2}\left[E_{1lx}^*i_x^{(e)}+E_{1ly}^*i_y^{(e)}+H_{1ly}^*i_y^{(m)}+H_{1lx}^*i_x^{(m)}\right]_{z=z_{0n}}$$

$$W_0=\int_{-\infty}^{\infty}[\vec{f}_H^*\frac{\partial}{\partial\omega}(\omega\widehat{\mu})\vec{f}_H+\vec{f}_E^*\frac{\partial}{\partial\omega}(\omega\widehat{\varepsilon})\vec{f}_E+\vec{f}_H^*\frac{\partial}{\partial\omega}(\omega\widehat{\alpha}^+)\vec{f}_E+\vec{f}_E^*\frac{\partial}{\partial\omega}(\omega\widehat{\alpha})\vec{f}_H]\mathrm{d}z \tag{3.19}$$

$$Q=\int_{-\infty}^{\infty}\{(\vec{f}_H^*\widehat{\mu}_{\mathrm{act}}\vec{f}_H-\vec{f}_H\mu_{\mathrm{act}}^*\vec{f}_H^*)+(\vec{f}_E^*\widehat{\varepsilon}_{\mathrm{act}}\vec{f}_E-\vec{f}_E\widehat{\varepsilon}_{\mathrm{act}}^*\vec{f}_E^*)$$
$$+[(\vec{f}_E^*\widehat{\alpha}_{\mathrm{act}}\vec{f}_H-\vec{f}_H\beta_{\mathrm{act}}^*\vec{f}_E^*)+(\vec{f}_H^*\widehat{\beta}_{\mathrm{act}}\vec{f}_E-\vec{f}_E\widehat{\alpha}_{\mathrm{act}}^*\vec{f}_H^*)]\mathrm{d}z\} \tag{3.20}$$

In (3.19), $V_g = \partial\omega/\partial k_y$ is the group velocity of the wave packet; $z = \pm L/2$ are two boundaries of the nonlinear layer shown in figure 3.1. It is assumed that the EMW propagates in the direction y and the medium is symmetric with respect to the directions $\pm x$, and the function that determines the linear dispersion, depends only on the even degrees of the wavenumber component k_x. If these conditions are not fulfilled, then the second, third and fourth terms in the left-hand part of equation (3.18) should be replaced by replaced by $[\vec{V}_g\partial/\partial\vec{r} + (i/2)\partial^2\omega/\partial k_i\partial k_j\partial^2/\partial x_i\partial x_j]A_1$, where $\vec{r}$ is the radius-vector in the plane (xy) of the wave propagation. It is assumed that the linear tangential electric field at the boundary of the MM layer is continuous. This means that an effective linear surface impedance is equal to zero. Such an impedance would characterize the difference of the thin surface layer at the boundary between the neighboring layers in a layered structures, in particular at $z = z_{01,2}$ in figure 3.1. Otherwise (when effective linear surface impedances are non-zero) a number of additional terms will appear on the right-hand side of (3.18). In such a case, corresponding additional nonlinear terms would include linear and nonlinear surface impedances. A detailed consideration of such effects is not included here. Note that the values W_0 and Q are proportional to the energy densities and specific losses (whose sign changes to the opposite in the active medium) in the linear medium. The first and second terms on the right-hand side of (3.15) determine the contributions of the volume and surface nonlinearities in the layered medium, respectively.

For testing the NEELS method, the corresponding results were compared with those obtained by calculations based on the traditional perturbation method for nonlinear fiber optics (Agraval 1996, Boardman and Xie 1997). In particular this has been done for the example of nonlinear waves in a magnetodielectric medium only with volumetric Kerr nonlinearities. Details relating to the material of this section are set out in appendix B on the basis of Grimalsky and Rapoport (1998a), Rapoport and Grimalsky (2011), Grimalsky *et al* (1996), Rapoport *et al* (2004a, 2004b, 2012a), Boardman *et al* (2008, 2010a), Slavin *et al* (2003), Buttner *et al* (2000b) and with reference to Akhiezer *et al* (1967), Kadomtsev (1988), Landau and Lifshits (1982), Paul *et al* (2011), Wang *et al* (2009), Kauranen and Zayats (2012), Chu and Sher (2008), Samson *et al* (2011), Zheludev and Emel'yanov (2004), Gromov *et al* (1999), Zvezdin and Popkov (1983), Damon and Eshbach (1961), Boardman *et al* (2000), Slavin and Rojdestvenski (1994), Leblond (2001). The exact coincidence with the corresponding results calculated by the perturbation method is obtained for (a) nonlinear frequency shift $\Delta\omega_{\rm NL}$ in a homogeneous infinite medium and (b) nonlinear shift of the wavenumber $\Delta k_{\rm NL}$ in a layered piecewise-continuous magnetoelectric medium, respectively (see appendix B.1). The NEELS method for these two cases was used in the forms of differential relations (3.15) and integral relations (3.18), respectively. The result obtained (see appendix B.1) for the nonlinear frequency shift using the differential relation of the NEELS method (3.15) also corresponds to the solution that follows in the corresponding approximation from equation (9) presented in Xiang *et al* (2012). For moderate-width spectral packets inequalities (3.5) become weaker, where sign '<<' would be replaced by '<'. In practice, this means that, for example, the spatial width of the packet is of the order of several wavelengths. In such a case nonlinearities and higher-order linear effects must be taken into account, beyond the parabolic approximation (Kadomtsev 1988), which was used to derive equation (3.18). In particular in the case of moderate spectrum width of a wave packet, nonlinear diffraction and dispersion and cubic linear dispersion should be included to the evolution equation. If necessary, the nonlinearity in the right-hand side of equation (3.18) can be considered much more accurately than in the cubic approximation, typical for the nonlinear Schrödinger equation (NSE) (Kalinin and Bayonets 1990). In the left-hand side of equation (3.18), it is easy to add the corresponding higher order linear terms, if necessary, which are omitted now for brevity.

It is worth noting that in the case of the cubic volume nonlinearity, the absence of the surface nonlinearity and absence of a dissipation, equation (3.18) for a narrow-spectrum wave packet reduces to equation (1.3), in other words to the nonlinear parabolic equation (Kadomtsev 1988, Boardman *et al* 1994, Zakharov and Shabat 1972, Agraval 1996.

Effective sources of the volume nonlinearity are included in the right-hand side of the evolution equation (3.23) (see the terms from the second to the fourth one) below in section 3.2. This equation appears as a result of the application of NEELS method to the surface plasmons in a layered plasma-dielectric medium. The volume sources in equation (3.23) correspond qualitatively to those given in equation (2.13) in Sipe *et al* (1980). The term describing the surface nonlinearity (the first in the right-hand side) in equation (3.23) and the corresponding very important effect of nonlinearity of the surface charge motion is new (see details in paragraph 3.2 below). The results of the paper (Pitilakis *et al* 2016), where nonlinear electromagnetic waves propagating along a dielectric-graphene waveguide have been considered, confirmed the adequacy of the NEELS method for problems of waveguide propagation of electromagnetic waves in the presence of surface nonlinearity, with the reference to our previous works (Rapoport and Grimalsky 2011, Rapoport 2014b). In contrast to the NEELS method described in this book, in the works (Grimalsky *et al* 1998c, Kolesik and Moloney 2004, Chen *et al* 2006, Afshar and Monro 2009), nonlinear envelope equations in a layered media have been derived based on a method similar to the derivation of the energy conservation law for electromagnetic waves; only volume, but not surface, nonlinearities have been included. Note that equation (3.18) has been written down without a direct inclusion of multimode interactions. The multimode case can be easily included in this theory.

3.2 Method NEELS for the giant resonance generation of the second harmonic of surface plasmons and the contribution of the surface and volume nonlinearities

We introduce the variations of the plasma concentration, velocity, electric and magnetic fields, respectively, $\tilde{n}_1$, $\vec{v}_1$, $\vec{E}_1$, $\vec{H}_1$ and $\tilde{n}_2^*$, $\vec{v}_2^*$, $\vec{E}_2^*$, $\vec{H}_2^*$. Here, the values with indices '1' correspond to nonlinear surface plasmons and are proportional to the amplitudes that change slowly, and the values with indices '2' describe the corresponding linear waves in 'almost the same' system, but in the absence of nonlinearity. The upper index '*' indicates a complex conjugation. The corresponding equations of motion for the plasma are as follows

$$\begin{aligned} &\frac{\partial \tilde{n}_1}{\partial t} + \mathrm{div}(n\vec{v}_1) = -\mathrm{div}(\tilde{n}\tilde{\vec{v}}), \quad \frac{\partial \vec{v}_1}{\partial t} + \frac{e}{m}\vec{E}_1 = -(\vec{v}_1\vec{\nabla})\vec{v}_1 - \frac{e}{m}[\vec{v}_{1^\times}\vec{H}_1] \\ &\mathrm{curl}\ \vec{H}_1 = \frac{1}{c}\frac{\partial E_1}{\partial t} + \frac{4\pi}{c}(-en_0\vec{v}_1 - e\tilde{n}\vec{v}), \quad \mathrm{curl}\ \vec{E}_1 = -\frac{1}{c}\frac{\partial \vec{H}_1}{\partial t} \end{aligned} \tag{3.21}$$

The equations similar to (3.21) (where nonlinear terms are included) must be written for linear components using indices '2'. The procedure proposed in Grimalsky *et al* (1996) (see also Rapoport 2014b, Rapoport and Grimalsky 2011), similar to the derivation of the energy conservation law (Akhiezer *et al* 1967, Kadomtsev 1988, Agranovich and Ginzburg 1965), is further applied for nonlinear surface plasmons. To account for surface nonlinearity, we consider the nonlinear motion of surface charges (with surface concentration n_S) at $z = 0$.

The differential and integral forms of NEELS for surface plasmons are obtained, on the basis of bilinear relations (obtained in the same way as described after relations (3.8) and (3.9)) in the forms, respectively:

$$\begin{aligned} &\frac{\partial}{\partial t}\left\{\frac{1}{4\pi}(\vec{E}_1\vec{E}_2^* + \vec{H}_1\vec{H}_2^*) + mn_0\vec{v}_1\vec{v}_2^*\right\} + \frac{c}{4\pi}\mathrm{div}\left\{[\vec{E}_1 \times \vec{H}_2^*] + [\vec{E}_2^* \times \vec{H}_1]\right\} = \\ &- mn_0\vec{v}_2^*(\vec{v}\vec{\nabla})\vec{v} - \frac{e}{c}n_0\vec{v}_2^*[\vec{v} \times \vec{H}] + eE_2^*(\tilde{n}\vec{v}) \end{aligned} \tag{3.22}$$

$$\begin{aligned} &\frac{\partial}{\partial t}\int_{-\infty}^{\infty}\left\{\frac{1}{4\pi}(\vec{E}_1\vec{E}_2^* + \vec{H}_1\vec{H}_2^*) + mn_0\vec{v}_1\vec{v}_2^*\right\}\mathrm{d}z + \frac{c}{4\pi}\frac{\partial}{\partial x} \\ &\int_{-\infty}^{\infty}\{[\vec{E}_1 \times \vec{H}_2^*]_x + [\vec{E}_2^* \times \vec{H}_1]_x\}\mathrm{d}z -= eE_{2x}^*\sigma v_x\,|_{z=+0} - m\int_0^{\infty} n_0\vec{v}_2^*(\vec{v}\vec{\nabla})\vec{v}\mathrm{d}z \\ &- \frac{e}{c}\int_0^{\infty} n_0\vec{v}_2^*[\vec{v} \times \vec{H}]\mathrm{d}z + e\int_0^{\infty} E_2^*(\tilde{n}\vec{v})\mathrm{d}z \end{aligned} \tag{3.23}$$

The first term on the right-hand side in equation (3.23) is due to the surface nonlinearity caused by the nonlinear motion of free surface charges. The following three parts of the volume nonlinearity are parts of the volume nonlinearity and correspond to the substitution nonlinearity, the nonlinear Lorentz force, and the concentration nonlinearity, respectively.

The details of the application of the NEELS method to the surface plasmons in the presence of the spatial (but in the absence of temporal) resonance with the second harmonic are given in chapter 4, section 4.1 and appendices B.2.1–B2.5, B.3 and B.4. Namely, in appendix B.2.1, linear motion of the charges on the plasma-dielectric surface is considered; in appendix B.2.2, the general method NEELS is formulated in the form, allowing one to include the nonlinearity accounting for the motion of surface charges; in appendix B.2.3, some details concerning the resonance of the second harmonic are presented; in appendix B.2.4, a set of equations for the coupled fundamental and resonant second harmonics is derived; in appendix B.2.5, the equation of the method NEELS in the integral form including the surface nonlinearity in the electrostatic approximation is written; in appendix B.3, quasi-solitons on surface plasmons with a second harmonic close to resonance are considered; in appendix B.4, estimation of conditions necessary for observation of nonlinear effects in layered plasmon structures from microwave to optical ranges are presented. Let us make an important methodological observation based on the results of this paragraph. The integral relation of the NEELS method applied to the surface plasmons (3.23) after the introduction of the slowly varying envelope wave amplitude can be represented as a separate case for the more general NEELS relation for a bi-anisotropic medium in the form (3.18) after multiplying this equation by W_0, and at $\widehat{\mu} = \widehat{I}$, $\widehat{\alpha} = 0$, $\vec{M}_{\rm NL} = 0$, $\vec{i}^{(m)} = 0$ (see appendix B.5).

3.3 Method NEELS for nonlinear electromagnetic and MSWs in the layered dielectric-ferromagnetic media with spatial dispersion and auxiliary boundary conditions

As an application of the NEELS method to nonlinear waves in gyrotropic media, let us consider an appropriate approach for the propagation of MSWs in gyrotropic layers (GL) (ferrite films).

3.3.1 BVMSWs in longitudinally magnetized ferrite films

Consider the pulse of MSWs (Gurevich and Melkov 1996) launched from the input antenna to the GL (ferrite film) (figure 3.2(a)). In a linear approximation, the magnetostatic potential φ_l, the components of the magnetic field $\vec{h}_l = -\vec{\nabla}\varphi_l$ and magnetization $\vec{m}$ are proportional to the common factor $\exp[i(\omega t - k_z z - k_y y)]$, where ω is frequency, $k_Z = k \cos\theta$, $k_Y = k \sin\theta$ are the components of the wave vector of MSW, the axis X is normal to the ferrite film, θ is the angle between the direction of the wave vector and the external magnetic field, which lies in the plane of the ferrite layer (film); the index 'l' will be used hereafter to denote a linear field. BVMSWs (Damon and Eshbach 1961, Danilov *et al* 1991, Gurevich and Melkov 1996) propagate in the GL along the direction of a bias magnetic field ($\theta = 0$). The method is based on the set of Maxwell and Landau–Lifshitz equations with the

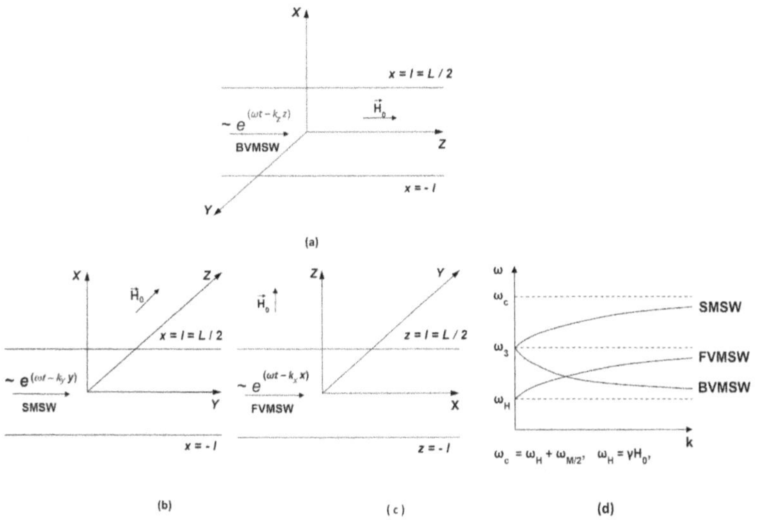

Figure 3.2. (a)–(c) Nonlinear gyrotropic layers (ferrite films) (a) in which magnetostatic waves (MSWs) of the three main types (Zvezdin and Popkov 1983, Gurevich and Melkov 1996) are propagating: backward volume magnetostatic waves (BVMSWs) (b), forward volume magnetostatic waves (FVMSWs) (c), and surface magnetostatic waves (SMSWs) (d), respectively. For all three main types of MSWs, it is assumed that the magnetic field is applied along the *Z*-axis. The qualitative picture of the linear dispersion for the three main types of MSW (d). For all three main types of MSWs, it is supposed that magnetic field $\overline{H_0}$ is applied along the axis *Z*; $\omega_H = \gamma H_0$, $\omega_M = 4\pi\gamma M_0$; H_0, M_0 and γ are the bias magnetic field within the film, saturation magnetization and gyromagnetic ratio, respectively.

inclusion of exchange interaction (Gurevich and Melkov 1996) (and, respectively, the spatial dispersion). An SVA is introduced for the wave pulse with a carrier corresponding to the basic BVMSW mode in the GL (assuming a weak nonlinearity for the ferrite film (Gurevich and Melkov 1996)). A procedure similar to that used for the derivation of the Poynting relation (energy conservation law) is applied, as was done in sections 3.1 and 3.2 for the electromagnetic waves. Higher, namely, second and zero harmonics with corresponding boundary conditions are taken into account (some details concerning these harmonics are given in appendices B.7.1 and B.7.2). The parametric interactions were considered (Rapoport *et al* 2004b, 2012a, Rapoport 2006b, 2014b, Grimalsky *et al* 2000a, 2000b, Gurevich and Melkov 1996) for two counter-propagating BVMSW pulses with identical carrier frequencies and absolute values of wavenumbers. In this case, the longitudinal magnetic field pumping at double frequency (2ω) is present. The parametric interaction is included in the formulation of the NEELS method. As a result, we obtain the following equation, which expresses the differential form of the NEELS method (Rapoport 2014b):

$$\frac{\partial}{\partial x_i} P_i + \frac{\partial}{\partial t} W_{1,\,1l} = F_{\mathrm{NL1}} + F_{\mathrm{PAR}} \qquad (3.24)$$

In equation (3.24), the effects of the pulse dispersion and diffraction are not accounted for; the summation by the repeated indices ($i = x,\, y,\, z$ in the first term) is used,

$$P_i = 1/(4\pi)div\left\{\left[\phi_{1l}{}^{*}\vec{b}_1\right] + \left[\phi_1\vec{b}_{1l}{}^{*}\right]\right\}_i + \alpha[(\partial m_{1x}/\partial t)(\partial m_{ilx}{}^{*}/\partial x_i)$$
$$+ (\partial m_{1lx}{}^{*}/\partial t)(\partial m_{1x}/\partial x_i) + (\partial m_{1y}/\partial t)(\partial m_{ily}{}^{*}/\partial x_i)$$
$$+ (\partial m_{1ly}{}^{*}/\partial t)(\partial m_{1y}/\partial x_i)]$$
$$W_{1,1l} = 1/(4\pi)(\varepsilon\vec{e}_1\vec{e}_{1l}{}^{*} + \vec{h}_1\vec{h}_{1l}{}^{*}) + (4\pi\omega_H/\omega_M)(m_{1x}m_{1lx}{}^{*} + m_{1y}m_{1ly}{}^{*}) + \tag{3.25}$$
$$\alpha[(\partial m_{1x}/\partial x_i)(\partial m_{ilx}{}^{*}/\partial x_i) + (\partial m_{1ly}{}^{*}/\partial x_i)(\partial m_{1y}/\partial x_i)] + \beta_S(m_{1x}m_{1lx}{}^{*})$$
$$F_{NL1} = -\,i\,\omega\left\{(4\pi\gamma/\omega_M)[m_{1ly}{}^{*}(\delta m_{x11} + \delta m_{x12}) + m_{1lx}{}^{*}(\delta m_{y11} + \delta m_{y12})\right.$$
$$\left. + h_{1lz}{}^{*}(m_{1z} + m_{2z})]\right\};\ F_{PAR} = \gamma\, h_0(h_{1ly}{}^{*}m_{2lx}{}^{*} - h_{1lx}{}^{*}m_{2ly}{}^{*})$$

where γ is the gyromagnetic ratio, $\omega_{H,\,M}$ are the cyclotron frequency, γH_0 and $4\pi\gamma M_0$, respectively, H_0 and M_0 are the bias magnetic field and saturation magnetization, respectively (for YIG, $4\pi M_0 = 1750$ Oe (Gurevich and Melkov 1996)). The incident signal and the idle (excited by parametric interaction) counter-propagating waves are indicated by the indices '1*l*' and '2*l*', respectively, $\vec{m}$ and $\vec{b}$ are the magnetization and magnetic induction of the corresponding waves. Equation (3.24) describes the evolution of the incident (signal) wave, and a similar equation can be written for the idle wave. The indices '*1l*' and '*2l*' denote the components of the electromagnetic fields and polarizations corresponding to the linear waves (where SVAs are not included). Index '*1*' denotes a nonlinear wave with the same carrier frequency and the distribution in the direction *X*, normal to the GL, a characteristic of linear mode. A wave with index '1' has an amplitude that varies slowly in time and in the spatial directions of y and z (lying in the plane of the gyrotropic layer). The term F_{PAR} describes the parametric interaction (Rapoport *et al* 2004b, 2012a, Rapoport 2006b, 2014b, Grimalsky *et al* 2000a, 2000b, Gurevich and Melkov 1996) of the signal wave and the idle wave; h_0 is the amplitude of the magnetic field of parametric pumping, which is directed along the axis *Z* (parallel to the external magnetic field). Pumping field is considered as almost a homogeneous one (in scale of MSW wavelength) in space, and has a frequency 2ω. In equation (3.25) α denotes the constant of exchange interaction for the gyrotropic ferrite medium (YIG), β_S is the constant characterizing the surface uniaxial anisotropy (Gurevich and Melkov 1996), whose axis is parallel to X direction, $\beta_S = \beta_{S+}\delta(x - L/2) + \beta_{S-}\delta(x + L/2)$, where $\beta_{S\pm}$ are the surface anisotropy constants. The values $x = \pm L/2$ and signs '±' on the right side of this equality correspond to the surfaces of the ferrite layer (film) while L is the thickness of the GL, $\delta(x \pm L/2)$ is the delta function. Being interested in the effects of surface nonlinearity, we will ignore the effects of the bulk anisotropy (they can be easily included in that considered, if necessary) . The value h_0 in the last of the relations included in (3.25) (for F_{PAR}) is the amplitude of the magnetic field (with frequency 2ω) used for parametic pumping. It was shown (Rapoport *et al* 2004b, 2012a, Rapoport 2006b, 2014b, Grimalsky *et al* 2000b) that the term F_{NL1} describing the nonlinearity 'felt' by wave 1 includes: (1) the self-action described by the terms of the form

$$\delta m_{x11} = \gamma(\tilde{h}_{y1}m_{z1} - \tilde{h}_{z1}m_{y1})^{(1)},\quad \delta m_{y11} = \gamma(\tilde{h}_{x1}m_{z1} - \tilde{h}_{z1}m_{x1})^{(1)}; \tag{3.26}$$

and (2) the cross-interaction with wave 2, which is described by terms of the form

$$\begin{aligned}\delta m_{x12} &= \gamma(\tilde{h}_{y2}m_{z1} - \tilde{h}_{z2}m_{y1})^{(1)} + \gamma(\tilde{h}_{y1}m_{z2} - \tilde{h}_{z1}m_{y2})^{(1)}\\ \delta m_{y12} &= \gamma(\tilde{h}_{x2}m_{z1} - \tilde{h}_{z2}m_{x1})^{(1)} + \gamma(\tilde{h}_{x1}m_{z2} - \tilde{h}_{z1}m_{x2})^{(1)}\end{aligned} \tag{3.27}$$

The notation $(\ldots)^{(1)}$ means that among all the nonlinear terms in the right-hand sides of relations (3.26) and (3.27), only terms proportional to the factor $\exp[i(\omega t - k_z z - k_y y)]$ are singled out, since the equation is now derived to study the effects of the self-action of nonlinear waves; $\tilde{h}_{1y,z} \equiv h_{1y,z} + \alpha\Delta m_{1y,z}$, $\tilde{h}_{1x} \equiv h_{1x} + \alpha\Delta m_{1x} + \beta_S m_{1x}$, '*' means the complex conjugation, Δ is the Laplace operator. To find the nonlinear term $F_{\mathrm{NL}1}$, the zeroth and second harmonics must be determined based on the solution of the corresponding nonlinear problems with the proper nonlinear boundary conditions. The equations for waves in the nonlinear and corresponding linear structures are used again and a procedure similar to that used in deriving the energy conservation law for the GL is applied. A special feature in this case is the consideration of additional boundary conditions that arise due to the presence of the spatial dispersion/exchange interaction in the ferrite film. In this case, we use auxiliary boundary conditions (Gurevich and Melkov 1996) for the magnetization, both linear and nonlinear, of waves 1 and $1l$ at $x = \pm L/2$ the surfaces of the ferrite layer (figure 3.2(b)), namely the Rado–Wirtman conditions (Gurevich and Melkov 1996). The nonlinear version of these conditions, in particular for the normal magnetization component, is of the form

$$(\pm\alpha\partial m_{x1}/\partial x - \beta_{S\pm}m_{x1} + M_0^{-1}R_{1\mathrm{NL}\pm})_{x=\pm L/2} = 0 \tag{3.28}$$

where $M_0^{-1}R_{1\mathrm{NL}\pm}$ describes the contribution of the nonlinearity into the boundary conditions. In the presence of the single (signal) wave we obtain the term which describes the surface nonlinearity in the form

$$R_{1\mathrm{NL}\pm} = -\beta_{S\pm}(m_{x1}m_{z1})^{(1)} \pm \alpha\left(-m_{x1}\frac{\partial m_{z1}}{\partial x} + m_{z1}\frac{\partial m_{x1}}{\partial x}\right)^{(1)} \tag{3.29}$$

In (3.29), the parametric interaction is not taken into account to simplify the expression, but it can easily be included in this relation if necessary. We assume that the nonlinearity is weak and the spatial dispersion in the ferrite film is moderate. Namely, we suppose that the relation $\Delta\omega_{\mathrm{NL}}/\omega_M \ll \alpha K^2 \ll 1$ holds. Here $\Delta\omega_{\mathrm{NL}}$ and K^2 are the characteristic nonlinear frequency shift and the square of the characteristic wavenumber for the selected spin wave mode, respectively. In this approximation, the linear structure of the spectrum of the dipole-exchange waves in the ferrite film is not distorted by nonlinearity and a single-mode approximation can be used at a correctly selected 'point' (ω, k) on the dispersion curve (Kalinikos *et al* 1988, Kalinikos and Slavin 1986). Let us integrate the relation (3.24), accounting for (3.25) describing the NEELS method in the differential form, taking into account relations (3.26) and (3.27) and linear and nonlinear boundary conditions for all the components $\vec{m}_{1l}$ (Gurevich and Melkov 1996), As a result we obtain the relation of NEELS method in the integral form, which reduces to the corresponding evolution equation for the MSW pulse in the layered ferrite structure. In a parabolic approximation, the evolution equation for SVA of

BVMSW pulse in the ferrite film (figures 3.1 and 3.2(a)) takes the form (Rapoport *et al* 2004b, 2012a, Rapoport 2014b)

$$\frac{\partial U_1}{\partial t} + (i/2)\left(\frac{\partial^2 \omega}{\partial k_z^2}\right)\left(\frac{\partial^2 U_1}{\partial z^2}\right) + (i/2)\left(\frac{\partial^2 \omega}{\partial k_y^2}\right)\left(\frac{\partial^2 U_1}{\partial y^2}\right) = F_{\mathrm{NLV}} + F_{\mathrm{NLS}} + F_{\mathrm{PARV}};$$
$$F_{\mathrm{NLV}} \equiv F_{\mathrm{NLV}x} + F_{\mathrm{NLV}y} + F_{\mathrm{NL}z};\ F_{\mathrm{NLV}} = F_{\mathrm{NL0V}}/W_{\mathrm{V1}l};\ F_{\mathrm{NL0}V} = \int_{-L/2}^{L/2} F_{\mathrm{NL1}}\mathrm{d}x \tag{3.30}$$

Here, the term F_{NL1} is determined by the third relation from (3.25) for the case of BVMSW, SVA U_1 for the magnetostatic potential φ_1 is defined as: $\varphi_1 = (1/2)U_1(t, z, y)f(x)\mathrm{e}^{i(\omega t - k_z Z)} + \mathrm{c.c.}$, where $f(x)$ is a function describing the distribution of magnetic potential for linear BVMSW along the direction x, normal to the GL (see figure 3.2(b)). In particular, for non-exchange BVMSW (figure 3.2(a)) (Gurevich and Melkov 1996), the transverse distribution has the form $f(x) = \sin \tau X$, $\tau = k_z/\sqrt{-\mu}$, where $\mu = (\omega^2 - \omega_\perp^2)/(\omega^2 - \omega_\mathrm{H}^2)$, $\omega_\perp^2 = \omega_\mathrm{H}(\omega_H + \omega_M)$. The four terms on the right-hand side of equation (3.30), which describes the volume nonlinearities and parametric interactions, take the following form:

$$F_{\mathrm{NLV}y} = -(1/W_{\mathrm{V1}l})(4\pi\gamma/\omega_M)i\omega\int_{-L/2}^{L/2} m_{1ly}^{*}(\delta m_{x11} + \delta m_{x12})\mathrm{d}x,$$
$$F_{\mathrm{NLV}z} = -(1/W_{\mathrm{V1}l})(4\pi\gamma/\omega_M)i\omega\int_{-L/2}^{L/2} [\tilde{h}_{1lz}^{*}(m_{1z} + m_{2z})]^{(1)}\mathrm{d}x,$$
$$F_{\mathrm{NL}x} = -(1/W_{\mathrm{V1}l})(4\pi\gamma/\omega_M)i\omega\int_{-L/2}^{L/2} m_{1lx}^{*}(\delta\tilde{m}_{y11} + \delta\tilde{m}_{y12})\mathrm{d}x$$
$$+ (1/W_{\mathrm{V1}l})(4\pi\gamma/\omega_M)i\omega\alpha\int_{-L/2}^{L/2} \frac{\partial m_{1x}}{\partial x}\frac{\partial}{\partial x}[(m_{1z} + m_{2z})m_{2lx}^{*}]\mathrm{d}x \tag{3.31}$$
$$F_{\mathrm{PAR}}\mathrm{V} = (1/W_{\mathrm{V1}l})\int_{-L/2}^{L/2} F_{\mathrm{PAR}}\mathrm{d}x.$$

Here $W_{\mathrm{Vi}l}$ can be obtained by substituting values with indices '$1l$', instead of values with index '1' in the expression for $W_{1,\,1l}$, given in the second of relations (3.25). Note that quantities $\delta\tilde{m}_{y11;y12}$ can be obtained from the values $\delta m_{y11;y12}$ defined in relations (3.26) and (3.27) by substituting $\tilde{h}_{x1,2} \to \tilde{\tilde{h}}_{x1,2}$ where $\tilde{\tilde{h}}_{1,2x} \equiv h_{1,2x} + \alpha(\partial^2/\partial y^2 + \partial^2/\partial z^2)m_{1,2x}$. The last term on the right-hand side of the first of relations (3.30) and (3.31) describes the parametric interaction and will be specified in section 6.1 of chapter 6. In the case of the parametric interaction of BVMSW with the slowly varying amplitude U_1 with other BVMSW with the slowly varying amplitude U_2 the equation similar to (3.30) should be written down for U_2, as done in chapter 6, see the system (6.2). The term describing surface nonlinearity for the case of BVMSW in the right-hand side of relation (3.30), *in the absence of parametric interaction*, is defined as

$$F_{NLS} = -\,(1/W_V)(4\pi\gamma/\omega_M)i\omega\,[(m_{1lx}{}^{*}R_{1NL\pm})_{x=+L/2} + (m_{1lx}{}^{*}R_{1NL\pm})_{x=-L/2}] \quad (3.32)$$

Using (3.26), (3.27) and (3.31), it can be shown that, for a single exchangeless BVMSW propagating in a thin ferrite film ($kL \ll 1$) without the parametric interaction ($F_{PAR} = 0$), the nonlinear term in the relations (3.25) and (3.30) is reduced, by the order of value, to (see some details in appendix B.7.2):

$$F_{NLV} \equiv F_{NLVx} + F_{NLVy} + F_{NLVz} \approx -iN_0|U_1|^2 U_1$$

$$N_0 = \frac{\omega\omega_M}{4\omega_H}\frac{(1-\mu)}{(4\pi M_0)^2}\frac{\int_0^{L/2}(\partial f_{BVMSW}/\partial x)^4 dx}{\int_0^{L/2}(\partial f_{BVMSW}/\partial x)^2 dx} \quad (3.33)$$

$$\approx [(4\pi M_0)^2 l^2]^{-1}(\omega\omega_M/4\omega_H)$$

In the second relation from equation (3.33), $f(x) = f_{BVMSW} \equiv f_{BVMSW}(x)$ is a function that describes the transverse distribution of the linear magnetostatic potential of BVMSW (Gurevich and Melkov 1996, Damon and Eshbach 1961), and the index '1' near the amplitude is omitted, since here we consider a single BVMSW pulse that 'feels' the nonlinear effect of self-action. Note that the relations (3.30), (3.31) are used for the modeling presented in figure 4.11 of section 4.3, chapter 4.

3.3.2 Developing NEELS method for the case of nonlinear pulses with the moderate spectrum in gyrotropic layered structures. Magnetized gyrotropic and dielectric structures with spatial dispersion and auxiliary nonlinear boundary conditions

Note the additional details of the development of the NEELS method associated with the inclusion in the consideration of higher nonlinear effects/corresponding coefficients of the evolution equation. Among the corresponding effects, there are nonlinear dispersion (Agrawal *et al* 1996, Gromov *et al* 1999) and diffraction (Boardman *et al* 2000, 2007, 2008), and other nonlinear terms, as well as higher linear effects (Gromov *et al* 1999). Consideration of the above mentioned effects is necessary to search the evolution of pulses with a relatively broader spectrum, is considered in Rapoport *et al* (2004b, 2012a). We obtain a (non-parabolic) normalized (dimensionless) nonlinear evolution equation of the following form:

$$\frac{\partial A}{\partial z_{\text{und}}} + \left(ig_1 + \delta\frac{\partial}{\partial\eta_{\text{und}}} + i\chi\frac{\partial^2}{\partial\eta^2_{\text{und}}}\right)\frac{\partial^2 A}{\partial\eta^2_{\text{und}}} + \left(ig_2 + g_3\frac{\partial}{\partial\eta_{\text{und}}} + ig_4\frac{\partial^2}{\partial\eta^2_{\text{und}}}\right.$$
$$\left. + i\rho\frac{\partial^2}{\partial y^2_{\text{und}}}\right)\frac{\partial^2 A}{\partial y^2_{\text{und}}} + \left(iN + b_{10}\frac{\partial}{\partial\eta_{\text{und}}}\right)(|A|^2 A) + ib_{20}A\frac{\partial^2}{\partial y^2_{\text{und}}}|A|^2 + \gamma U = 0 \quad (3.34)$$

The normalized amplitudes $A = U/U_0$ and coordinates $z_{\text{und}} = z/l_z$, $y_{\text{und}} = y/l_y$, $\eta_{\text{und}} = \eta t_{\text{NL}}$ are used here, where U_0 is the scale of the nonlinear wave amplitude (momentum envelope), $l_{z,\,y}$ and t_{NL} are the spatial scales for the z and y directions and the time scale, respectively. The details of the coefficients of the equation in (3.34) are given in appendix B.7.3. This approach, in special cases, is applied in section 5.4 of chapter 5. It should be noted that the use of the Fourier operator method (Kadomtsev 1988) and standard methods of reconstruction of evolution equations by the appearance of a nonlinear dispersion relation (Zvezdin and Popkov 1983) give only approximate formulas for nonlinear coefficients, while even the determination of the correct sign of the latter is not guaranteed. This especially concerns the nonlinear terms of higher order, in particular the nonlinear dispersion and diffraction. The new and more accurate 'method of equivalent nonlinear sources' gives a more accurate expression of the form (Rapoport *et al* 2012a) $b_{10}'|A|^2(\partial A/\partial t') + b_{10}''\partial(|A|^2A)/\partial t'$ with some coefficients b_{10}', b_{10}'', the second of which in order of magnitude and sign coincides with that included in equation (3.34). Thus, in equation (3.34), as well as in the standard works on nonlinear optics (see, for example, Agraval 1996) only one of the two terms describing the nonlinear dispersion is taken into account.

The comparison of the nonlinear coefficient for the BVMSW with those coefficients obtained by other authors based on other methods is presented in appendix B.7.4.

A modification of the NEELS method has been developed, which allows one to include electromagnetic effects (retardation) (Grimalsky *et al* 1996, 1998b, Rapoport 2014b) in the study of the evolution of nonlinear envelope waves in layered gyrotropic structures (Rapoport *et al* 2012a). We will start with the example of the MSW (Grimalsky *et al* 1997, Rapoport *et al* 2004b). In particular, the expressions for the values P_i, $W_{1,1l}$ included in the left part of equation (3.24) are modified as follows:

$$
\begin{aligned}
P_i = {} & c/(4\pi)div\left\{\left[\vec{e}_{il}{}^*\vec{h}_1\right] + \left[\vec{e}_1\vec{h}_{1l}{}^*\right]\right\}_i \\
& + \alpha[(\partial m_{1x}/\partial t)(\partial m_{ilx}{}^*/\partial x_i) + (\partial m_{1lx}{}^*/\partial t)(\partial m_{1x}/\partial x_i) \\
& + (\partial m_{1y}/\partial t)(\partial m_{ily}{}^*/\partial x_i) + (\partial m_{1ly}{}^*/\partial t)(\partial m_{1y}/\partial x_i)]\,; \\
W'_{1,1l} = {} & 1/(4\pi)(\varepsilon\vec{e}_1\vec{e}_{1l}{}^* + \vec{h}_1\vec{h}_{1l}{}^*) + (4\pi\omega_H/\omega_M)(m_{1x}m_{1lx}{}^* + m_{1y}m_{1ly}{}^*) + \\
& \alpha[(\partial m_{1x}/\partial x_k)(\partial m_{1lx}{}^*/\partial x_k) + (\partial m_{1ly}{}^*/\partial x_k)(\partial m_{1y}/\partial x_k)] + \beta_S(m_{1\varsigma}m_{1l\varsigma}{}^*)
\end{aligned}
\tag{3.35}
$$

In (3.35), c is the speed of light, $\vec{e}$, $\vec{h}$ (with the corresponding indices) are electric and magnetic fields, z is the direction along the magnetic field, ς denotes the direction along the normal to the ferrite film. In the absence of the parametric interaction, the nonlinear terms included in the right-hand side of equation (3.24) take the form

$$
\begin{aligned}
F_{\rm NL} &= -i\omega(4\pi\gamma/\omega_M)[m_{1ly}^*\delta m_{x11} + m_{1lx}^*\delta m_{y11} + h_{1lz}^*m_{1z}] + F_{\rm NLE} \\
F_{\rm NLE} &= -\frac{a_{\rm NL}}{4\pi}\vec{e}^*{}_{11}\frac{\partial}{\partial t}|\vec{e}_1|^2
\end{aligned}
\tag{3.36}
$$

The first three terms on the right-hand side of (3.36) characterize the magnetic nonlinearity, similar to relations (3.27) and (3.30), while the fourth term corresponds to a possible dielectric cubic nonlinearity, and $a_{\rm NL}$ is a proper Kerr constant.

The bilinear relations and the relations of the NEELS method in the integral form, as well as the evolution equations for solitons and nonlinear wave beams based on the NEELS method for nonlinear surface waves in plasma with retardation have been proposed in Grimalsky *et al* (1996), Grimalsky and Rapoport (1998a). In the form, which can be used for any of the three main directions of a homogeneous magnetic field and, accordingly, the three main types of MSW in layered dielectric-ferromagnetic structures (figure 3.2), in the absence of retardation, exchange interaction, and dielectric nonlinearity, the bilinear relations in the integral form and the evolution equations of the NEELS method for the envelope amplitudes of the type of (3.24) were obtained for the first time in Grimalsky *et al* (1997).

The equation in the differential form on the basis of bilinear relations for the electromagnetic waves (with retardation) in the structure of the 'ferromagnetic-dielectric' with a certain geometry corresponding to SMSW (figure 3.2(b)) with a transversely inhomogeneous magnetic field, as well as the corresponding nonlinear correction to the frequency had been proposed in Grimalsky *et al* (1998c). The bilinear relations deduced in Grimalsky *et al* (1997) for the SMSW and FVMSW in ferrite films with homogeneous magnetic fields are reduced to the corresponding relations given in Grimalsky *et al* (1998c) and Grimalsky (1998d) in the corresponding limiting cases. The results for the nonlinear frequency shift that follow from formulas (Grimalsky *et al* 1997) are reduced to the corresponding result, which can be obtained from equations (Grimalsky *et al* 1998c) in the limiting case of SMSW in a homogeneous magnetic field, as well as to the corresponding result given in Grimalsky (1998d) for FVMSW. The NEELS method was further developed, in particular, in Rapoport *et al* (2004b, 2012a), Boardman *et al* (2007) and Rapoport (2012b, 2014b), with a rather general consideration of the surface nonlinearity and the possibility of nonlinear effects with the spatial dispersion in the auxiliary boundary conditions.

The surface nonlinearity can be characterized by either the surface current $\vec{J}_{\rm SNL}$ or the corresponding nonlinear polarization $\vec{P}_{\rm NL}$. The corresponding wave effects may be important for surface waves at the interface between nonlinear media, including magneto-optical (Bertran *et al* 2001, Atkinson and Kubrakov 2001, 2002), MMs, bi-anisotropic (Rapoport *et al* 2006a, Boardman *et al* 2007), metallic or plasma (Fukui and Stegeman 1982), and other media. Our goal now is to demonstrate how the surface nonlinearity of a very general form can be included in the NEELS method. Thus, we assume that the surface nonlinear current $\vec{J}_{\rm SNL}$ or the surface nonlinear polarization $\vec{P}_{\rm SNL}$ are known (more precisely, we consider as known ones, their functional dependences on the field/field amplitude). As a result, we have a nonlinear boundary condition (Fukui and Stegeman 1982) in the form $h_z(x = l^+) - h_z(x = l^-) = -(4\pi/c)J_{{\rm SNL}y}$ where the values of coordinates $x = l^\pm$ are very close, but the corresponding points lie at different sides of the interface, $x = l$, while the surface current is equal to $J_{{\rm SNL}y} = \lim_{b\to 0}[(\partial P_{{\rm SNL}y}/\partial t)b]$. Here $b = |l^+ - l^-|$ is the thickness of a very thin surface layer, where the effective surface nonlinearity/surface nonlinear polarization is concentrated. For the sake of simplicity, we assume that the linear tangential electric field is continuous on the interface. Using the procedure described in section 3.1, we arrive at equations similar to (3.18), (3.30), with the additional term ${\rm NL}'_S$ in the right-hand side

$$\begin{aligned}&\partial A_1/\partial t + V_g\partial A_1/\partial\xi + (i/2)(\partial^2\omega/\partial k^2{}_\eta)(\partial^2 A_1/\partial\eta^2)\\&= NL'_V + NL'_S + NL'_{Vdiel}\end{aligned} \tag{3.38}$$

$${\rm NL}'_S = (1/W_{{\rm V}1l})e^*_{1ly}J^{(1)}_{{\rm SNL}y} \tag{3.39}$$

Here, the upper index (1) means, as in section 3.3.1, that the terms proportional to the $\exp(i\omega t)$ are revealed from the nonlinear surface current when equation (3.38) is used to investigate the effects of the nonlinear wave self-action. At the same time, a similar method can be used, for example, to simulate the generation of harmonics in layered structures (Bertran *et al* 2001, Atkinson and Kubrakov 2001, 2002) and in this case, in equation (3.39), instead of the nonlinearity at the fundamental frequency, one should reveal the terms responsible for generating harmonics. The first and third terms in the right hand side of equation (3.38) describe the possible volume magnetic and dielectric nonlinearities, respectively, while the second term describes the surface nonlinearity, in accordance with relation (3.39). The nonlinear term (3.39) can be regarded as a value proportional to the jump of the normal component of the Poynting vector due to the 'work' of a linear tangential electric field over the surface nonlinear current. It should be noted that the effective nonlinear surface current in MMs with metaparticles to which nonlinear elements (diodes or quantum dots or dye molecules) can be integrated can also be included in the model under the NEELS method (Dolgaleva *et al* 2009, 2011, Plum *et al* 2008, Boardman *et al* 2007, Oulton *et al* 2009).

The terms similar to (3.39) can also be obtained in the case of the surface magnetic nonlinearity, when the last is described in terms of the corresponding (effective) surface polarization.

3.3.3 Magnetized gyrotropic and dielectric structures with spatial dispersion and auxiliary nonlinear boundary conditions

We present two examples of the media with nonlinearities under auxiliary boundary conditions associated with the presence of spatial dispersion (Agranovich and Ginzburg 1965, Akhiezer *et al* 1967, Jardin *et al* 2002, Cottam *et al* 1984). In particular, the nonlinear layered ferromagnetics and ferroelectrics are considered (Jardin *et al* 2002, Cottam *et al* 1984). The general form of the evolution equation in the parabolic approximation has the form (Rapoport *et al* 2004b, 2012a), where, unlike in Kalinin and Bayonets (1990), the terms $F_{\mathrm{NLV,\,NVS}}$ corresponding to the volume and surface nonlinearities are revealed:

$$\begin{aligned}&\partial U_1/\partial t + V_{g1}\partial U_1/\partial z + (i/2)(\partial^2\omega/\partial k^2{}_z)(\partial^2 U_1/\partial z^2) +\\&(i/2)(\partial^2\omega/\partial k^2{}_y)(\partial^2 U_1/\partial y^2) = F_{NLV} + F_{NLS}\end{aligned} \tag{3.40}$$

At first, waves of the BVMSW type (Gurevich and Melkov 1996, Rapoport *et al* 2004b, 2012a) propagating along the direction of the magnetization and the axis Z lying in the plane of the GL are considered, whereas the axis X is directed along the normal to the ferrite film (figure 3.2(a)). In equation (3.40), U_1 is the amplitude of the magnetostatic potential (Rapoport *et al* 2004b, 2012a). In Akhiezer *et al* (1967), the following relation is obtained for the jump of the normal components of the Poynting flow at the boundary of the GL:

$$\vec{\Pi}_{-}\vec{n}_{-} - \vec{\Pi}_{+}\vec{n}_{+} = \left\{ n_k \left[\partial F / \partial\left(\frac{\partial \vec{M}}{\partial x_k}\right)\right]\left(\frac{\partial \vec{M}}{\partial t}\right)^*_{\mathrm{LIN}} \right\}_{\mathrm{exch}} \tag{3.41}$$

where $\vec{\Pi}_{\pm}$ are energy flows on both sides of the interface (of the ferrite and dielectric media), index 'LIN' means the corresponding value (the time derivative of magnetization) in the absence of nonlinearity. In the presence of nonlinearity, we can apply equation (3.41) where we put

$$\left[\partial F/\partial\left(\frac{\partial \vec{M}}{\partial x_k}\right)\right] = \left[\partial F/\left(\partial\frac{\partial \vec{M}}{\partial x_k}\right)\right]_{\mathrm{LIN}} + \left[\partial F/\left(\partial\frac{\partial \vec{M}}{\partial x_k}\right)\right]_{\mathrm{NL}} \tag{3.42}$$

The function F in equations (3.41) and (3.42) is the potential energy of the ferromagnetic (Akhiezer *et al* 1967). The partial derivative F is subdivided into linear and nonlinear parts, provided with indices 'LIN' and 'NL' in relation (3.42). Based on the nonlinear part of relation (3.42), the effect of self-action is considered and, respectively, the term, corresponding to the main harmonic, is revealed. The index '*exch*' in (3.41) is used to emphasize that in equation (3.40) and, accordingly, in relation (3.41), the contribution from the exchange interaction must be included in the energy. The energy has the form (Akhiezer *et al* 1967, formula (3.3.2)):

$$F\left(\vec{M}, \frac{\partial M_i}{\partial x_k}\right) = \frac{1}{2}\alpha_{iklm}\frac{\partial M_i}{\partial x_k}\frac{\partial M_l}{\partial x_m} + \gamma_{ik}(\vec{M})\frac{\partial M_i}{\partial x_k} + w_a(\vec{M}) + f(M^2) \tag{3.43}$$

Here $\hat{\alpha}$, $\hat{\gamma}$ are the corresponding tensors (Akhiezer *et al* 1967), w_{a} is the energy of (surface) anisotropy. In particular, the surface anisotropy energy can be represented as (Akhiezer *et al* 1967), $w_{\mathrm{a}} = -(1/2)\delta(X - X_0)\beta_S(\vec{M}\vec{n})^2$, where X_0 is the coordinate of the corresponding boundary (for the structure shown in figure 3.1) is the surface anisotropy constant. For a ferromagnetic with the inverse symmetry, we have in bulk $\hat{\gamma} = 0$, $i = l$, $\alpha_{iklm} \to \alpha_{km}$. In accordance with (3.41), (3.42), the linear exchange boundary conditions in the absence of surface anisotropy take the form (Akhiezer *et al* 1967)

$$\left\{n_k\left[\partial F/\partial\left(\frac{\partial \vec{M}}{\partial x_k}\right)\right]_{\mathrm{LIN}}\right\}_S = \left[n_k\alpha_{kl}\frac{\partial \vec{m}_l}{\partial x_l}\right]_S = 0 \tag{3.44}$$

where $\vec{M} = \vec{M}_0 + \vec{m}$, $\vec{m} = \vec{m}_l$; $\vec{M}_0$, $\vec{m}$, $\vec{m}_l$ are the saturation magnetization, the variable part of magnetization and the linearized variable part, respectively. In the presence of the surface anisotropy with the axis normal to the boundary, the corresponding linear auxiliary boundary conditions are reduced to the following form, for example, for the magnetization component normal to the boundary (Akhiezer *et al* 1967)

$$\pm\alpha\frac{\partial m_{lx}}{\partial x} - \beta_{S\pm}m_{lx} = 0 \tag{3.45}$$

The different signs in the first term in (3.45) correspond to different directions of the normal at the two boundaries of the GL shown in figure 3.2(a). Note that alternatively the same equation (3.45) can be obtained using equation (3.43) and the replacement $\gamma_{ik}(\vec{M}) \approx \gamma_{ikl}m_i$, $\gamma_{ikl} = (\partial\gamma_{ik}/\partial M_l)_{\vec{M}=\vec{M}_0}$ for the case of a small variable part of the magnetization (Akhiezer *et al* 1967). Next, it must be assumed that the inverse symmetry is broken and, accordingly, $\hat{\gamma} \neq 0$ only on the corresponding boundaries. As a result, one can obtain the linear boundary conditions in the form (3.45). The corresponding nonlinear (exchange) additional boundary conditions take the form (Gurevich and Melkov 1996), equations (3.28) and (3.29). Note that the last (nonlinear) term in (3.28) corresponds to the second (nonlinear) term on the right-hand side of relation (3.42). As a result, given (3.29), it becomes possible to estimate the nonlinear term on the right-hand side of relation (3.41), and the term to be added to the right-hand side of (3.30) and (3.40) is equal to (Rapoport *et al* 2004b, 2012a)

$$F_{\mathrm{NLS}} = -(1/W_{\mathrm{V}})(4\pi\gamma/\omega_M)i\omega[(m_{i1x}^{*}R_{1\,\mathrm{NL}\pm})_{x=+L/2} + (m_{i1x}^{*}R_{1\,\mathrm{NL}\pm})_{x=-L/2}] \tag{3.46}$$

Let us now demonstrate qualitatively how the approach described above can be extended to evaluate the contribution to the evolutionary equation of the surface nonlinearity related to the spatial dispersion in a layered dielectric medium. In contrast to wave processes in ferromagnetics, where the magnetic polarization is described by a first-order Landau–Lifshitz differential equation, the linear polaritons in a dielectric with the spatial dispersion, such as ferroelectrics (Jardin *et al* 2002, Cottam *et al* 1984), are described by second-order time equations

$$m\frac{\partial^2 P_i}{\partial t^2} = -m\omega_t^2 P_i + C\left[\left(\frac{\partial P_i}{\partial y}\right)^2 + \left(\frac{\partial P_i}{\partial z}\right)^2\right] + E_i - \gamma\frac{\partial P_i}{\partial t} \equiv -\frac{\delta F}{\delta P_i} - \gamma\frac{\partial P_i}{\partial t} \tag{3.47}$$

Here $\vec{P}$, m, ω_t, C, $\vec{E}$, γ, F are the polarization, the effective mass, the effective resonant frequency, electric field, the coefficient describing dissipative losses, and the free energy density, respectively. The density of the free energy corresponding to (3.47) in the volume of the nonlinear layer (figure 3.1) is of the form (Jardin *et al* 2002, Cottam *et al* 1984)

$$F = \frac{1}{2}\sum_i [m\omega_t^2 P_i^2 + C(\nabla P_i)^2] - \vec{E}\vec{P} \tag{3.48}$$

Note that the right-hand side of the identity included in (2.48) includes a variational derivative $\delta F/\delta P_k \equiv \partial F/\partial P_k - (\partial/\partial x_i)[\partial F/\partial(\partial P_k/\partial x_i)]$ of the function F. In thin surface layers with a thickness δ at the boundaries $z = \pm L/2$, the density of surface energy $(C_{\mathrm{R}}/2\delta)[P_i^2(z = -L/2) + P_i^2(z = +L/2)]$ is added to the right-hand side of equation (3.48). One can derive the following relation, which describes the law of the conservation of energy, which corresponds to equation (3.47):

$$\begin{gathered}\frac{\partial}{\partial t}(w_0 + F) + \frac{\partial}{\partial x_k}\Pi_{\mathrm{SD}k} = -\gamma\left(\frac{\partial \vec{P}}{\partial t}\right)^2, \\ w_0 = \frac{1}{2}m\left(\frac{\partial \vec{P}}{\partial t}\right)^2, \ \Pi_{\mathrm{SD}k} = -\frac{\partial P_i}{\partial t}\left[\partial F/\partial\left(\frac{\partial P_i}{\partial x_k}\right)\right]\end{gathered} \tag{3.49}$$

$$\vec{\Pi} = \vec{\Pi}_{\text{EM}} + \vec{\Pi}_{\text{SD}} = \frac{c}{4\pi}[\vec{E} \times \vec{H}] - \left[\partial F / \partial\left(\frac{\partial \vec{P}}{\partial \vec{r}}\right)\right] \frac{\partial \vec{P}}{\partial t} \tag{3.50}$$

In (3.49) and (3.50), w_0, $\vec{\Pi}_{\text{SD}}$ and $\vec{\Pi}$ are the effective kinetic energy, the flow associated with the spatial dispersion, and the total energy flow, respectively. To estimate the surface nonlinearity associated with the spatial dispersion, we summarize the above phenomenological calculations and apply the NEELS method. Applying this approach to nonlinear transversally-bounded polaritons in a spatially dispersive medium, one can obtain a result similar to relations (3.41) and (3.42). It is necessary to separate the nonlinear part in the relation similar to (3.41) (including also the relation similar to (3.28)), after the replacement $\vec{M} \rightarrow \vec{P}$ of the magnetic polarization with the electric one. One can then apply a linear approximation to β_{eff} in dependence on P_z. Further, considering (phenomenologically) known, according to each specific problem, the functional dependence of the nonlinear part NL_s in the expression for the nonlinear term $(\partial F / \partial(\partial \vec{P} / \partial x_k))_{\text{NL}}$ (see relation (3.42)), we obtain a nonlinear boundary condition similar to (3.28)

$$\left[\partial F / \partial\left(\frac{\partial \vec{P}}{\partial x_k}\right)\right] = \alpha_{\text{eff}} \frac{\partial P_x}{\partial t} + \beta_{\text{eff}} P_x + \text{NL}_S\left(P_i P_j, \frac{\partial P_i}{\partial x_k} \frac{\partial P_j}{\partial x_m}, \vec{P}^3, \ldots\right) = 0; \; \text{NL}_S \equiv \left[\partial F / \partial\left(\frac{\partial \vec{P}}{\partial x_k}\right)\right]_{\text{NL}} \tag{3.51}$$

In (3.51), α_{eff}, β_{eff} are the effective coefficients of the spatial dispersion and surface anisotropy, respectively. As a result, we obtain a relation similar to (3.41) in the form

$$[\vec{\Pi}_- \vec{n}_- - \vec{\Pi}_+ \vec{n}_+]_{\text{NL}} = \left\{ n_k \left[\partial F / \partial\left(\frac{\partial \vec{P}}{\partial x_k}\right)\right]_{\text{NL}} \left(\frac{\partial \vec{P}}{\partial t}\right)^*_{\text{LIN}} \right\} = \left[\text{NL}_{Si}\left(\frac{\partial P_i}{\partial t}\right)^*_{\text{LIN}}\right]_{z=L/2} \tag{3.52}$$

According to (3.52), the contribution to the complete nonlinearity in the dielectric-ferroelectric medium of the effect of surface nonlinearity, related to the spatial dispersion, included in the right-hand side of the evolution equation of the NEELS method similar to (3.30), is described by the term in the form $W_0^{-1}[\text{NL}_S(\partial \vec{P} / \partial t)^*_{\text{LIN}}]_{z=L/2}$. In the presence of the spatial dispersion, including the nonlinearity in the auxiliary boundary conditions, this extra nonlinear term must be added to the right-hand side of equation (3.18). According to the above conclusion, such a nonlinearity corresponds to surface $z = +L/2$. The resulting expression is the main result of this paragraph. Some details of the derivation, namely the equation of the NEELS method in the integral and differential forms, taking into account the nonlinearity under additional boundary conditions related to the spatial dispersion, are given in appendix B.6 (equations (B.14) and (B.15), respectively). The proposed method and the obtained relations can be used to estimate the contribution to the nonlinear wave evolution of the effects of the spatial dispersion, including the nonlinear additional boundary conditions, in ferroelectric/dielectric layered media.

3.3.4 Development of the NEELS method for space-time modeling solitons and nonlinear wave structures in controlled and active nonlinear gyrotropic and metamaterial waveguides: from microwave to the optical ranges

The following normalized higher-order evolution equation for electromagnetically controlled space-time wave structures can be obtained (Boardman *et al* 2010a, Rapoport *et al* 2014a, Rapoport 2014b, Rapoport and Boardman 2015b):

$$\begin{aligned} & i\frac{\partial\psi}{\partial Z} - \frac{1}{2}\mathrm{sgn}(\beta_2)\frac{\partial^2\psi}{\partial T^2} - i\delta_3\frac{\partial^3\psi}{\partial T^3} + \frac{1}{2}D\frac{\partial^2\psi}{\partial y^2} + K\frac{\partial^2}{\partial y^2}(|\psi|^2\psi) \\ & + \mathrm{sgn}(\overline{\chi}^{(3)})\left(|\psi|^2\psi + iS\frac{\partial}{\partial T}(|\psi|^2\psi) - \tau_{\mathrm{R}}\psi\frac{\partial}{\partial T}(|\psi|^2)\right) + i\gamma_0\psi - \nu\psi \\ & = 0 \end{aligned} \tag{3.53}$$

In (3.53), ψ is the normalized amplitude of the component of the field E_y of TM-mode in a layered MM structure, β_2, δ_3, $\overline{\chi}^{(3)}$, S, τ_{R} are the normalized coefficients of dispersion, the third-order linear dispersion, the cubic nonlinearity, the self-compression (nonlinear dispersion), and the time scale characterizing the Raman effect, respectively; D, γ_0, ν, K are the coefficients of the linear diffraction, the linear damping/ gain (depending on sign), the transverse magneto-optical effect (Voigt effect) and the nonlinear diffraction, respectively. Thus, in comparison with (3.34), equation (3.53) (derives similarly to equation (3.34)) describes additional higher nonlinear effects, namely, the Raman interaction, and at the same time, the magneto-optical interaction (terms proportional to R, τ and ν, respectively). Using equation (3.53), magneto-optical control of optical solitons in MMs has been demonstrated, first (Boardman *et al* 2010a). A number of new effects (Rapoport *et al* 2014a) for optically active (Boardman *et al* 2011) MM waveguide structures, including active ones, were suggested (Rapoport *et al* 2014a), including the stabilization of space-time (2 + 1) solitons (bullets) (Bauer *et al* 1998, Buttner *et al* 2000a, 2000b). These questions will be considered in more detail in sections 5.3 and 5.4 (chapter 5) and 7.2 and 7.3 (chapter 7). The derivation (Boardman *et al* 2010a, 2011) of the terms of the evolution equation (3.53) describing the above effects of higher nonlinearities and the active medium in layered MMs, is incorporated into the NEELS method.

3.4 Conclusions to chapter 3

The proposed NEELS method is formulated in a very general way. We may state that NEELS is one of the most effective modern methods for deriving evolution equations for waves in multilayered systems that include materials and nonlinearities of different physical nature, including bulk and surface nonlinearities. The NEELS method is developed, in particular, for layered gyrotropic, MMs, including bi-anisotropic, nano-structured and graphene MMs, plasma media, atmospheric-ionosphere systems. It is shown that NEELS can also be generalized to 2D wave propagation with the nonlinear dispersion and diffraction and the corresponding terms lead to a qualitative difference in the behavior of the propagation of nonlinear pulse in layered systems. Finally, the proposed method is also developed for systems including, in addition to higher-order nonlinear effects, the retardation, and the spatial dispersion (in particular, for the cases of ferroelectric-dielectric exciton structures and exchange interactions in ferromagnetics).

Problems to chapter 3

Problem 3.1

Obtain the solution (B.17) of equation (B.16) presented in appendix B.7.1.

To obtain the solution of (B.16), symmetrical respectively to x (see figure 3.2(a)), look for the corresponding solution in the form (B.17). The values $E_{0,2}$ can be obtained, accounting for that the first and the third terms in (B.17) presents the particular solution of the inhomogeneous differential equation (B.16). Look for the solution of the inhomomogeeous differential equation (B.16) for the second harmonic of the magnetostatic potential $\varphi_2^{(2)}$ in the form $\varphi_2^{(2)} = (1/2)F_{2+}\exp(i2\theta)\exp[-2k_z(|x| - l)] + \text{c.c.}$, $\theta = \omega t - k_z z$; $l \equiv L/2$ is a half-thickness of the GL/ferrite film. Then the values F_{2+} and E_2 can be found from the boundary conditions of the continuity of z-coordinate of the magnetic field (or potential $\varphi_2^{(2)}$) and normal component of the corresponding magnetic induction at the surfaces of the ferrite film. Finally, the solution for F_2 in the form (B.17), described in appendix B.7.1. should be obtained.

Problem 3.2

Estimate $N^{(0)}_{\text{BVMSW}}$ (see figure 3.2(a) for BVMSW), the result of which is given in appendix B.7.2 and equation (3.33).

Take a note of the following. The nonlinear term in the NEELS method for BVMSW includes the components $\text{NL}_{01,02,03,04}$ representing, respectively, the 1st, 2nd, 3rd and 4th integral terms included in the right-hand side of the relation (B.18). For BVMSW in a thin GL, for which, as shown above, the values $k_z l$, τl are small, we have $(k_z l)^2/(\tau l)^2 \sim (-\mu) \ll 1$, and we can get the 'degree of smallness' of the terms NL_{01}, NL_{031}, namely:

$$\text{NL}_{01} \sim \{\tau^3/[\sin^4(\tau l)]\}(k_z l)^2(\tau l), \quad \text{NL}_{03} \sim [\tau^3/\sin^4(\tau l)](\tau l); \; \text{NL}_{03} \gg \text{NL}_{01} \quad \text{(P.2.1)}$$

Finally, comparing all terms included in the nonlinear coefficient (equation (B.18)) for BVMSW, we obtain the following inequalities: $\text{NL}_{03} \gg \text{NL}_{01}, \text{NL}_{02}, \text{NL}_{04}$. Thus, the main nonlinear term is NL_{03}, and, thus, the relation is

$$\text{NL}_{0V} = \text{NL}_{\text{VBVMSW}} \approx -(i\omega/M_0)\int_{-\infty}^{\infty} m_{1lx}^{*}(h_{1x}m_{1z})^{(1)}\text{d}x. \quad \text{(P.2.2)}$$

Generally speaking, both the zeroth and the second harmonics contribute to equation (B.18) and (P.2.2). However, given that the 0th magnetization harmonic includes a term proportional h_{1x} (in this case, the direction X coincides with the normal to a thin GL), by the order of magnitude, relation (P.2.2) can be reduced to the form $\text{NL}_{0V} = \text{NL}_{\text{VBVMSW}} \approx -(i\omega/M_0)\int_{-\infty}^{\infty} m_{1lx}^{*}h_{1x}{}^{(1)}m_{1z}{}^{(0)}\text{d}x$. As a result, we obtain, for a nonlinear term in the evolution equation in the form (3.30), the relation:

$$F_{\rm NLV} = -iN^{(0)}_{\rm BVMSW}|U_1|^2 U_1,\ N^{(0)}_{\rm BVMSW} = -\partial\omega/\partial|U_1|^2 = -\Delta\omega_{\rm NL}/|U_1|^2$$
$$\approx (4\omega_H)^{-1}\omega\omega_M(1-\mu)/(4\pi M_0)^2$$
$$\times\left\{1 + [(\mu-1)^2 + \mu_a^2]^{-1}(\omega_M/\omega_H)\right\}^{-1}\left[\int_0^l(\partial f(x)/\partial x)^2 {\rm d}x\right]^{-1}\int_0^l(\partial f(x)/\partial x)^4 {\rm d}x\, U_0^2$$
$$> 0.$$

Here $\Delta\omega_{\rm NL}$, U_1, U_0 are the nonlinear frequency shift, the amplitude of the envelope (wave packet), which varies slowly, and the amplitude used for convenience, for normalization, respectively; $\mu_a \equiv \sigma_0 = -\omega_M\omega/(\omega^2 - \omega_H^2)$. By order of magnitude, we obtain for $N^{(0)}_{\rm BVMSW} \equiv N_0$ (Rapoport *et al* 2004a, 2012a), the expression (3.33).

Problem 3.3

Show that the term describing in the evolution equation (3.34) nonlinear diffraction and is conservative, is indeed in the form (see relations (B.22) and (B.23) in appendix B.7.3): $\sim b_{20}A\partial^2(|A|^2)/\partial y^2_{\rm und}$.

In order to get this result, you can act in the opposite way. We first show that the term with nonlinear diffraction in equation (3.34) has the form $\sim ib_{20}\partial^2[*(A|A|^2)]/\partial y^2_{\rm und}$; the meaning of the symbol '*' is explained below. For this purpose it is possible to use, as well as for derivation of equation (3.34), the 'operator method' (Kadomtsev 1988), based on an expansion in a series with a small deviation from the central wavenumber of the 'carrier wave': $\Delta k_z = (\partial k_z/\partial\omega)\Delta\omega + \ldots + (1/4)(\partial^4 k_z/\partial\omega^2\partial k_z^2)\Delta\omega^2\Delta k_y^2$, followed by replacement (B.19) in appendix B.7.3. The 'operator method' is based on the Fourier transform. Without a detailed consideration of the Fourier transform of a nonlinear term, we just note that the meaning of the operator '*' is not unambiguously defined. For example, it cannot be said that a nonlinear diffraction term in equation (3.34) includes only the second derivative acting on the value intensity $|A|^2$. Indeed, the nonlinear diffraction term may include, in principle, some other terms, except one included in equation (3.34) with the coefficient . The reason is that the corresponding differential operator describing the nonlinear diffraction acts on some nonlinear combinations of complex amplitudes; to accurately make the transformation from the representation in space of wavevectors and frequencies to the representation in space and time, it is necessary to apply the Fourier transform to the amplitude A and the whole combination ($|A|^2A$). The approach to the nonlinear diffraction and non-paraxiality is described in Boardman *et al* (2000, 2008) and Chi and Guo (1995).

Here we can restrict ourselves to a simpler, though not very strict, 'non-dissipative approximation' for nonlinear diffraction of magnetostatic waves. Namely, one can choose the term that describes the nonlinear diffraction in equation (3.34) in a form that corresponds to the condition of energy conservation for the pulse propagating in the GL. It is supposed that GL has finite width in the Y direction, and at the corresponding the conditions of zero magnetic field/magnetostatic potential are is satisfied, $A(y = \pm L_y) = 0$, $A(t' = 0;\ t' = T_m) = 0$, where T_m is some time that is several times longer than the pulse duration. To illustrate the satisfaction of the condition of the energy conservation for the pulse in the GL, consider only the first and last terms in equation (3.34). After proper integration, you can get a relation that expresses the law of the conservation of energy in the form

$$\mathrm{d}\left[\iint |A|^2 \mathrm{d}\eta_{\mathrm{und}} \mathrm{d}y_{\mathrm{und}}\right] \Big/ \mathrm{d}z_{\mathrm{und}} + 2\gamma \iint |A|^2 \mathrm{d}\eta_{\mathrm{und}} \mathrm{d}y_{\mathrm{und}} = 0.$$

Next, we obtain that the term $\sim\partial(A|A|^2)/\partial y^2$ is nonconservative, while the term describing the nonlinear diffraction in the form (B.22), namely: $\sim b_{20}A\partial^2(|A|^2)/\partial y^2$, is conservative. The corresponding term is used in equation (3.34) to describe the effect of the nonlinear diffraction in the form (B.22) with a coefficient may be used only to estimate the effect of nonlinear diffraction. Note that the term $\sim b_{10}\partial(A|A|^2)/\partial\eta_{\mathrm{und}}$ in equation (3.34) describing the nonlinear dispersion is also conservative.

Problem 3.4

Show that the term describing, in the evolution equation (3.34), the nonlinear diffraction, has the form (see relation (B.22)): $\sim b_{20}A\partial^2(|A|^2)/\partial y^2$, where the value b_{20} is expressed as follows:

$$b_{20} = \frac{1}{2}\frac{\partial^3 k_z}{\partial|U|^2 \partial k_y^2}\frac{l_z}{l_y^2}U_0^2 \approx \frac{1}{2}\frac{l_z}{l_y^2}U_0^2\frac{\alpha}{k_z^2 l}(2\mu_a^2 - 1);\ \alpha = \frac{1}{2}\frac{\omega_M\omega_H(\omega^2 + \omega^2)}{(\omega^2 - \omega_H^2)^3}$$

Taking into account (3.34) and (B.20), we obtain the estimation for the nonlinear diffraction coefficient (B.23), which solves this problem.

List of abbreviations

BVMSW	Backward volume magnetostatic waves
ENZ	Epsilon-near-zero
FVMSW	Forward volume magnetostatic waves
GL	Gyrotropic layer
NEELS	Nonlinear evolution equations for layered structures
NSE	Nonlinear Schrödinger equation
MSW	Magnetostatic waves
SMSW	Surface magnetostatic waves
SVA	Slowly varying amplitude

Appendix B to Chapter 3

Appendix B.1 Application of the method NEELS for the nonlinear waves in the homogeneous and layered magnetodielectric media. Shifts of frequency and wavenumbers due to a presence of volume nonlinearity

(1) Obtain (a) by the NEELS method using a differential relation of the form (3.15) and (b) by considering a nonlinear wave equation the same result, namely the nonlinear frequency shift (in terms of which a nonlinear effect can be expressed) is equal to $\Delta\omega_{NL} = -(\omega/4)[(\alpha_{NL}/\varepsilon_0)| E_0 |^2 + (\beta_{NL}/\mu_0)| H_0 |^2]$ where the linear frequency is ω and the wavenumber is K_y, $K_y^2 = \varepsilon_0\mu_0(\omega/c)^2$; E_0, H_0 are the amplitudes of electric and magnetic fields.

(2) By both methods, (a) based on the integral relation of the NEELS method (3.18) and (b) the standard perturbation method for nonlinear fiber optics (Zheludev and Kivshar 2012, Boardman *et al* 1994, 2009), obtain matching results for nonlinear wavenumber shift Δk_{NL} for nonlinear TM waves in a piecewise-continuous weakly waveguide layered structure 'the linear dielectric (layer 1)—the nonlinear magnetodielectric MM (layer 2)—nonlinear dielectric (layer 3)', with normal in the direction z: $\Delta k_{NL} \approx \mu_2(k_0^2/2k)(I_1/I_2)|A_1|^2$, where $k_0 = \omega/c$, $I_1 = \int_{z_{21}}^{z_{22}} \alpha_{NL}|f_{E_z}|^4 dz$, $I_2 = \int_{-\infty}^{\infty} |f_{E_z}|^2 dz$; ω, k, f_{E_z} are the frequency, the natural wavenumber, and the function of the transverse distribution of the transverse component of the electric field E_z of a linear electromagnetic wave propagating along a layered structure, respectively; $z_{21,\,22}$ are coordinates of the boundaries of the nonlinear layer 2; α_{NL} and μ_2 are the coefficient of the electrical Kerr nonlinearity and the value of the linear magnetic permeability in layer 2, respectively; A_1 is the value of the amplitude of the slowly changing transverse (zth) component of the electric field of the TE wave; it is assumed that there is only electrical nonlinearity and only in layer 2.

Appendix B.2 The method of derivation of nonlinear evolution equations for nonlinear plasmons in layered media by the method NEELS in the integral form

Appendix B.2.1 Linear motion of the charges

Consider nonlinear motion in an electron plasma in the simplest type of layered structure that supports surface plasmons, namely in the structure 'Plasma-linear dielectric'. Assume that the plasma occupies half-space $z \leqslant 0$, and the (linear) dielectric with dielectric permittivity ε occupies half-space $z > 0$, while the wave packet propagates along the axis x. When $\varepsilon = 1$ (the dielectric is a vacuum), the linear dispersion equation for surface plasmons takes the form

$$q/\varepsilon(\omega) + p = 0,\ p = (k^2 - k_0^2)^{1/2},\ k_0 = \omega/c,$$
$$q^2 = p^2\varepsilon^2(\omega),\ \varepsilon(\omega) = 1 - \omega^2/\omega_p^2 \tag{B.1}$$

Assume that the components of the field are proportional to $\sim e^{i(\omega t - kx)}$, where ω, k are the frequency and the wavenumber, respectively (the 'carrier' frequency and the wavenumber); ω_p and c are the plasma frequency and the speed of light, respectively. The corresponding linear boundary conditions (BC) are:

$$(1/4\pi)(\partial E_{2z}^*/\partial t\,|_{z=l+0} - \partial E_{2z}^*/\partial t\,|_{z=-l-0} = -e(\partial\sigma_2^*/\partial t) \tag{B.2}$$

Appendix B.2.2 The general method NEELS. The nonlinearity accounting for motion of surface charges

We assume that the nonlinearity in the system under consideration is moderate, so we can apply the approximation of slow varying amplitudes (SVA) (Anisimov 2003). We introduce variations in the plasma concentration, the velocity, the electric and magnetic fields, respectively $\tilde{n}_1$, $\vec{v}_1$, $\vec{E}_1$, $\vec{H}_1$, and $\tilde{n}_2^*$, $\vec{v}_2^*$, $\vec{E}_2^*$, $\vec{H}_2^*$ (see section 3.2). The equations of motion for a nonlinear plasma in this case will look like (3.21) (see section 3.2). Similar equations (but without nonlinear terms, and in the absence of stationary components of the magnetic field and particle velocity) should also be written for linear components with indices '2':

$$\partial\tilde{n}_2{}^*/\partial t + div(n\vec{v}_2)^* = 0,\ \ \partial\vec{v}_2{}^*/\partial t + (e/m)\vec{E}_2{}^* = 0,\ \ curl\ \vec{E}_2^{\,*} = -\ (1/c)\partial\vec{H}_2^{\,*}/\partial t \tag{B.3}$$

In equation (3.21), the electron concentration is considered equal to $n = n_0 + \tilde{n}$, where n_0 and $\tilde{n}$ are stationary and variable parts of the electron concentration, e and m are the electron charge and mass, respectively. We use the system (3.21) and the corresponding system of complex conjugate equations for a linear plasma, and a procedure similar to that used in deriving the law of energy conservation (Kadomtsev 1988, Landau and Lifshits 1982) (see section 3, paragraphs 3.1.1–3.1.3). The equations for nonlinear surface charge and boundary conditions on the surface ($z = 0$) 'plasma-dielectric', which takes into account the presence of free surface charges and the surface charge current j_{surf}, have the form, respectively:

$$\partial\sigma_1/\partial t + (nv_z)_1|_{z=+0} + \partial(\sigma v_x)_1/\partial x\,|_{z=+0} = 0$$
$$H_{1y}\big|_{z=+0} - H_{1y}\big|_{z=-0} = -(4\pi/c)j_{x\text{surf}},\ j_{x\text{surf}} = -e\sigma\vec{v}_x\big|_{z=+0} \tag{B.4}$$

where σ is the surface charge concentration.

Using (B.4), we obtain

$$[\vec{E}_1^{\,*} \times \vec{H}_2]_{z;z=+0} - [\vec{E}_2^{\,*} \times \vec{H}_1]_{z;z=-0} = (4\pi/c)E_{2x}^*e\delta v_x\,|_{z=+0} \tag{B.5}$$

In equation (B.5), the type designations $[...]_{z;z=\pm 0}$ denote the Zth component of the vector [...], taken at $Z = \pm 0$ (i.e. near the surface $Z = 0$ on both sides). As a result, we obtain a nonlinear evolution equation in the integral form (3.22). Note that the plasma concentration n_0 in the first term in equation (3.22) is non-zero only in the region $z \leqslant 0$. In expressions $E_2^*(\tilde{n}\vec{v})$ etc, the terms, which are proportional to ~ $\exp(i\omega t)$ are taken into account in the multiplier $(\tilde{n}\vec{v})$. The terms in integral relation (3.22) have a clear physical meaning. Nonlinear terms with index '1' are proportional to the amplitude A_1, which changes slowly, while the linear components are provided with index '1'. The first and second terms in the right-hand side of (3.22) are proportional to the temporal and spatial derivatives, respectively. The proportionality coefficients are the integral densities of the linear wave energy and energy flux, respectively. A more detailed analysis shows that a complete linear operator describing the evolution of a SVA in a parabolic approximation (Grimalsky and Rapoport 1998a, Rapoport and Grimalsky 2011) can be separated from the left-hand side (3.22), including the linear dispersion and diffraction. An evolution equation for slowly varying envelopes can be obtained from equation (3.22). To do this, take into account the presence of the zeroth and second harmonics in the nonlinearity included in equation (3.22), and these higher harmonics must be expressed in terms of the amplitude A_1 of the fundamental harmonic. The terms ~ $\exp(i\omega t)$ are separated from all combinations of quantities included in nonlinear terms in equation (3.22), whereas (complex conjugate) linear fields/quantities (with index '2') are proportional to $\exp(-i\omega t)$.

The rigorous hydrodynamic derivation of all nonlinear components of the zeroth and second harmonics is done. These harmonics are taken into account in ((3.22) and (3.23)), as well as in the set of related equations for the main and second resonant harmonics (see appendices B.2.3, B.2.4 and equation (B.8)). Two dimensionless small parameters are explicitly written down. One of them, namely $\chi_0 = [\bar{k}^{-2}(\omega/\omega_P)^{-2}]\bar{A}_1^2$ characterizes the smallness of the relative value of zeroth nonresonant harmonic concentration, whereas the second one $\chi_2 = 2\varepsilon(\omega)\bar{k}^{-2}\bar{A}_2$ is for the second resonant harmonic. Here $\bar{k} = k/k_0$, $k_0 = \omega/c$, k is the wavenumber of the surface wave, $\bar{A}_{1,\,2}$ are the amplitudes of the first and second harmonics, respectively, that are normalized to $E_0 = (m\omega_P^2/ek)$. The value of the linear perturbation of the concentration (its first harmonic) in this problem is identically equal to zero, because the divergence of the linear velocity of the surface plasma electromagnetic waves under consideration is equal to zero.

Note that the 'fast phase' is excluded from equation (3.22), and we obtain the evolution equation for the amplitude of the fundamental harmonic (signal), which varies slowly. To find the zeroth and second harmonics, they are considered as an 'effective forced oscillator' with an 'external force' determined by a nonlinear combination of terms with the amplitude A_1 of the fundamental harmonic. The corresponding nonlinear boundary conditions and the surface nonlinearity (due to the motion of the surface charge) are taken into account. To find the 'higher harmonics', the corresponding problem is solved, which includes the boundary conditions described above (see also appendices B.2.3, B.2.4).

Appendix B.2.3 Resonance of the second harmonic

Note that the plasma concentration n_0 in the first term in equation (3.22) is nonzero only in the region $z \leqslant 0$. And in the expressions like $E_2^*(\tilde{n}\vec{v})$, etc, the terms ~ $\exp(i\omega t)$ are separated in $(\tilde{n}\vec{v})$. For example, we demonstrate the result for complex amplitude of the second harmonic X-component of the electric field in the plasma region ($z \leqslant 0$):

$$E_x^{(2)} = E_0 e^{q(2\omega)z} - [8me^2q^2\varepsilon(2\omega)]^{-1} iek\omega_p^2\varepsilon(\omega)A_1^2 e^{2qz},\ z \geqslant 0$$

$$q(2\omega) = (k_2^2 - k_{02}^2\varepsilon(\omega_2))^{1/2}$$

$$q(2\omega) = (k_2^2 - k_{02}^2\varepsilon(\omega_2))^{1/2};\ k_2 = 2k,\ k_{02} = \frac{2\omega}{c},\ \omega_2 = 2\omega;$$

$$E_0 = \frac{iek\omega_p^2 A_1^2}{2m\omega^4 q}\,\frac{1 + [4c^2pq\varepsilon(2\omega)]^{-1}\omega^2\varepsilon(\omega)}{[2\varepsilon(2\omega)/q(2\omega)] + p^{-1}} \tag{B.6}$$

As can be seen from relation (B.6), under the condition of the 'temporal resonance', namely $\varepsilon(2\omega) \approx 0$, $\omega \approx \omega_p/2$, the amplitude of the second harmonic can become very large. Note that the condition presented above means well-known regime 'ENZ (epsilon near zero)' (Li and Agrypoulos 2018) for the second harmonic of the nonlinear surface plasmons. The term 'temporal resonance' is related to the fact that the 'spatial resonance' is absent because $k(2\omega) \neq 2k$. In the vicinity of the temporal resonance frequency (equation (B.6)), the amplitude of the second harmonic A_2 should be described by a separate evolution equation, and from the equations for the second harmonic we can obtain:

$$\begin{aligned} &E_x^{(2)} = (1/2)A_2 e^{i2(\omega t - kx)} f_2(z);\ f_2(z) = e^{q(2\omega)z} - e^{2qz}; \\ &\varepsilon(2\omega)A_2 = RA_1^2;\ R = (8mc^2q^2\omega^2)^{-1} iek\omega_p^2\varepsilon(\omega) \end{aligned} \tag{B.7a}$$

Appendix B.2.4 A set of equations for the coupled fundamental and resonant second harmonics

In the vicinity of the resonant frequency for the second harmonic, $\omega_2 = 2\omega \approx \omega_p$, it can be written down

$$\varepsilon(2\omega) \equiv \varepsilon(\omega_2) = (\omega_2 - \omega_{20} + \omega_{20} - \omega_{pe} + \omega_{pe}) = \omega(\omega_{pe}) + (\mathrm{d}\varepsilon/\mathrm{d}\omega)(\Delta\omega_0 + \Delta\omega)$$

$$\varepsilon(\omega_{pe}) = 0;\quad \Delta\omega_0/\omega_{pe} \ll 1;\quad \Delta\omega_0/\omega_{20} - \omega_{pe};\quad \Delta\omega \equiv \omega_2 - \omega_{20}$$

Using the Fourier transform and substituting $i\Delta\omega \to \partial/\partial t$, and using (B.7), for the amplitude of the second harmonic we can obtain the equation

$$\partial A_2/\partial t + i\Delta\omega A_2 = (\partial\varepsilon/\partial\omega)^{(-1)} RA_1^2 \tag{B.7b}$$

Finally, using equations (3.23) and (B.7b) results in the set of equation (B.8) for the fundamental and second (resonance) harmonics in the vicinity of the resonant region:

$$\partial A_1/\partial t + V_g \partial A_1/\partial x + (i\omega_{k_x k_x}/2)(\partial^2 A_1/\partial x^2) + ig_2(\partial^2 A_1/\partial y^2) + \alpha A_1^* A_2$$
$$+ i\alpha_0 \mid A_1 \mid^2 A_1 + \gamma A_1 = 0 \quad \text{(B.8)}$$
$$\partial A_2/\partial t + (i\Delta\omega_0 + \gamma_2)A_2 = \alpha_2 A_1^2$$

The nonlinear coefficients in the equations for the fundamental harmonic include contributions from the 'volume' and 'surface' nonlinearities (1st and 2nd of the following relations (B.9) (Grimalsky *et al* 1996, Grimalsky and Rapoport 1998a, Rapoport and Grimalsky 2011),

$$\alpha_0 = \alpha_{0\text{surf}} + \alpha_{0\text{volume}}, \alpha_1 = \alpha_{1\text{surf}} + \alpha_{1\text{volume}}; \alpha_{0\text{surf}}$$
$$\cdot \alpha_{0\text{volume}} < 0, \alpha_{1\text{surf}} \cdot \alpha_{1\text{volume}} < 0; \quad \text{(B.9)}$$
$$\mid \alpha_{0\text{surf}} \mid / \mid \alpha_{0\text{volume}} \mid \sim 3, \mid \alpha_{1\text{surf}} \mid / \mid \alpha_{1\text{volume}} \mid \sim 7$$

Note that for the first time the effect of the predominant role of surface nonlinearity, compared to bulk one, for surface plasmons, and in particular in the generation of the second harmonic, was proposed (based on the use of the NEELS method) in Grimalsky *et al* (1996) and Grimalsky and Rapoport (1998a). This effect was qualitatively confirmed in subsequent works, in particular in Paul *et al* (2011), Wang *et al* (2009) and Wang *et al* (2009) theoretically and experimentally, respectively. In this case, by the order of magnitude, the relations (B.9) for the coefficients describing the relative contribution to the total nonlinearity of the effects, caused by the volume and surface nonlinearities, correspond to the experimental results for the electron plasma in metals/metal films (Wang *et al* 2009).

The importance of the effect of surface nonlinearity compared to the volume for the plasma medium, which is described using a hydrodynamic model, was qualitatively confirmed in Kauranen and Zayats (2012) and Paul *et al* (2011). A very interesting result is that: (a) the main contribution to the nonlinear coefficient α_2 in the equation for the resonant second harmonic arises due to the volume nonlinearity; (b) in contrast, the main contribution to the nonlinearity coefficients in the equation for the fundamental harmonic is related to the 'surface effect' (5th and 6th of relations (B.9)). We emphasize that the contributions to the nonlinearity coefficients for the fundamental harmonic characterizing the volume and surface effects have different signs (3rd and 4th of the relations (B.9)).

As follows from relations (B.9), the inclusion of the surface nonlinearity can change the sign of the 'effective nonlinearity coefficient', and therefore we can expect qualitative changes in the behavior of the nonlinear system due to the effect of the surface nonlinearity. In particular, estimates show that for $\omega/\omega_{pe} > 0.65$, the time derivative in the second of the equation (3.24) can be neglected, $|\Delta\omega_0| > >|\partial/\partial t|$, $|\gamma_2|$, and the set of equation (3.24) is reduced to the standard nonlinear Schrödinger equation (NRS) with an effective nonlinear coefficient

$$\alpha_{\text{eff}} = \alpha - \alpha_1\alpha_2/\Delta\omega \quad \text{(B.10)}$$

Using (B.10), for an effective nonlinear coefficient we obtain

$$\omega_{k_x k_x} < 0, \ \alpha_{\text{eff}} > 0; \ \alpha_{\text{eff}}\omega_{k_x k} x < 0 \tag{B.11}$$

Thus, in this case, the 'formal' Lighthill criterion (Kadomtsev 1988) of the development of the modulation instability is not satisfied, and the formation of solitons is impossible. In contrast, if the surface nonlinearity were absent, the coefficient of the nonlinearity α_{eff} would change its sign and the soliton formation would be possible.

Appendix B2.5 Equation of the method NEELS in the integral form and surface nonlinearity in the electrostatic approximation

Consider how the NEELS method can be applied to plasmons in the electrostatic approximation in the plasma layer $|z| \leqslant l$ (Gramotnev and Bozhevolnyi 2010), and for the dielectric permittivity the relations $\varepsilon = 1$ at $|z| > l$ and $\varepsilon = \varepsilon(\omega)$(the dielectric permittivity for a plasma) at $|z| \leqslant l$, the distribution of the electrostatic potential is symmetric with respect to the directions $\pm z$, $\varphi = \varphi_0 f(z)$, φ_0 is the amplitude, $f(z) = \text{ch}(kz)/\text{ch}(kl)$, at $|x| \leqslant l; f(z) = \exp[-k(|x| - l)]$; at $| x | > l$. The corresponding dispersion equation has the form $\tanh(kl) = -\varepsilon(\omega)^{-1}$, where $k > 0$ is the wavenumber for plasmons propagating along the plasma layer in the direction x.

$$\frac{\partial}{\partial t}\left(\frac{1}{4\pi}\vec{E}_1\vec{E}_2^* + mn_0\vec{v}_1\vec{v}_2^*\right) + \text{div}\left[\varphi_1\left(\frac{1}{4\pi}\frac{\partial \vec{E}_2^*}{\partial t} - en_0\vec{v}_2^*\right) + \varphi_2^*\left(\frac{1}{4\pi}\frac{\partial \vec{E}_1}{\partial t} - en_0\vec{v}_1\right)\right]$$
$$= e\varphi_2\text{div}(\tilde{n}\vec{v}) - mn_0\vec{v}_2^*(\vec{v}\vec{\nabla})\vec{v}$$

The equation in the deferential and integral forms, obtained on the basis of bilinear relations and equation (3.25), is represented as

$$\begin{aligned}&\frac{\partial}{\partial t}\int_{-\infty}^{0}\left(\frac{1}{4\pi}\vec{E}_1\vec{E}_2^* + mn_0\vec{v}_1\vec{v}_2^*\right)\text{d}z + \frac{\partial}{\partial x}\int_{-\infty}^{0}\left[\varphi_1\left(\frac{1}{4\pi}\frac{\partial E_{2x}^*}{\partial t} - en_0\vec{v}_{2x}^*\right)\right]\\ &+ \left[\varphi_2^*\left(\frac{1}{4\pi}\frac{\partial E_{1x}}{\partial t} - en_0 v_1(x)\right)\right]\text{d}z = e\int_{-l}^{0}\varphi_2^*\text{div}(\tilde{n}\vec{v})\text{d}z - m\\ &\int_{-\infty}^{0} n_0 v_2^*(\vec{v}\vec{\nabla})\vec{v}\text{d}z + \varphi_2^*\,|_{z=-l}\left(\frac{\partial\sigma_1}{\partial t} + n_0 v_{z1}\right)_{z=-l+0}\end{aligned} \tag{B.12}$$

In the relation (B.12), σ_1 is the effective (two-dimensional) concentration of the surface charge. The first two and last terms in the right-hand side of (B.12) describe volume and surface nonlinearity, respectively.

Appendix B.3 Quasi-solitons on surface plasmons with a second harmonic close to resonance

The soliton of the NSE is formed when $\alpha_1 = 0$, $\alpha_2 = 0$, (see equation (3.23)). It is shown, when taking into account the dynamics of the second harmonic, that a quasi-soliton pulse can exist and propagate, at least for some time, starting from the input of the pulse into the system, for the case $\alpha_1 \neq 0$, $\alpha_2 \neq 0$. The 'formal' Lighthill criterion is satisfied when the amplitude and phase of the input pulse are chosen correctly. Namely, the input pulse has the form (Rapoport and Grimalsky 2011):

$$A_1(x=0) = A_{10}\,\mathrm{sech}[\{t - x/v_0\}/\tau]\exp(-i\Omega t),$$
$$\Omega = \alpha_{\mathrm{eff}}|A_{1\mathrm{inp}}|^2/2,\ \alpha_{\mathrm{eff}} = \alpha_0 - \alpha_1\alpha_2/\Delta\omega.$$

where A_{10}, v_0, τ are the pulse amplitude (soliton), the speed, and the characteristic duration of the input pulse, respectively. Thus we have an instability that is at least similar to the instability of the modulation type of the fundamental harmonic in the region of non-satisfaction of the Lighthill criterion, in the presence of a giant resonant second surface plasmon harmonic.

Appendix B.4 Estimation of conditions necessary for observation of nonlinear effects in layered plasmon structures from microwave to optical ranges

Let us estimate the typical parameters for which it will be possible to experimentally observe possible nonlinear effects in various layered plasmon structures.

(a) **Structure 'gas plasma-dielectric'.** The analysis of the nonlinear coefficient in equation (3.24) shows that, in order of magnitude, $|\alpha_0| \sim \omega_{pe}/8E_e^2$, $|\alpha_1| \sim \omega_{pe}/60E_e^2$, $|\alpha_2|\omega_{pe}/5E_e^2$, where $E_e = mc\omega_{pe}/e$. The typical parameters for the observation of nonlinear wave phenomena, including the giant generation of the second harmonic in the structure 'plasma-dielectric gas' are as follows. For a plasma with the electron concentration $n_0 \sim 10^{11}$ cm^{-3} and $\omega_{pe} \sim 1.6 \times 10^{10}$ s^{-1}, the amplitude of the input pulse should be of the order $A_1(x=0) \sim 0.1E_e \sim 100$ abs units, and the intensity of the pump wave/field of the fundamental harmonic at the input to the system $\sim(c/8\pi)A_1^2 \sim 1$ MW cm^{-2} at frequency $(\omega/2\pi) \sim 1.5$ GHz. The depth of localization, in this case, is of the order of 3 cm, and the duration of the pump pulse on the first harmonic must be not less than 100 ns.

(b) **Semiconductor-dielectric structure**. For a semiconductor with a narrow band gap (n-InSb), the nonlinearity of the Kane type is significant (Chu and Sher 2008). The corresponding dispersion law is analogous to the relativistic law, namely $E^2(p) = m^*(0)^2v_n^2 + p^2v_n^2$, where E, p are the energy in the conduction band and the quasi-momentum, respectively, $m^*(0)$ is the effective mass at the bottom of the conduction band, $v_n = [E_g/2m^*(0)]^{1/2} \sim 1.8 \times 10^8$ cm s^{-1}(for InSb) is the characteristic velocity, and E_g is the band gap (~ 0.2 eV for InSb); and 'effective quasi-relativistic mass' is equal to $m^* = m^*(0)(1 - v/v_n^2)^{-1/2}$. For InSb with $n_0 \sim (10^{15} - 10^{17})$ cm^{-3} and $\omega_{pe} \sim (10^{13} - 10^{14})$s^{-1}, $\omega \sim \omega_{pe}/\varepsilon_L{}^{1/2}$, ($\varepsilon_L \sim 10$ is the crystalline dielectric permittivity), the estimation gives for the amplitude of the fundamental harmonic a value of the order of 0.1 $0.1(m^*v_n\omega_{pe}/e) \sim 3 \cdot 10^1$ abs. units, the pulse length should be of the order of 100 ps, and the input intensity should be of the order of 100 kW cm^{-2}.

(c) **Metal-dielectric structure.** For a metal of Au type, with $n_0 \sim 10^{23}$ cm^{-3}, $\omega_{pe} \sim 2 \times 10^{16}$ s^{-1}, the characteristic nonlinear field is of the order of magnitude, $E_e \approx 1/(\chi^{(3)})^{1/2}$, where the coefficient of cubic nonlinearity $\chi^{(3)} \sim 10^{-9}$ abs. units, and to estimate for the first harmonic we have a value of the order of $\sim 0.1 E_c \sim 3 \times 10^3$ abs. units, the input pulse duration should be of the order of ~100 fs, and the input intensity of the first harmonic should be ~10 GW cm^{-2}. The estimations confirm the possibility of observing solitons in a 'metal-dielectric' structure (Samson *et al* 2011). Note that the possibility of generating a giant plasmonic second harmonic in the metal-dielectric structure is due to the presence of the quadratic nonlinearity. The best way to achieve this may be to use MM structures such as 'metal-dielectric', with a sequence of nonlinear nanoparticles with the appropriate asymmetry, which allows the presence of the quadratic nonlinearity. In this regard, we note that the excitation of the second harmonic in the optical range in the system, with an input wave intensity of the order of 1 GW cm^{-2} was considered in Zheludev and Emel'yanov (2004).

Appendix B.5 Relationships of the method NEELS for the surface plasmons as a particular case of bi-anisotropic metamaterials

Relation (3.18) for a lossless medium is presented as

$$\begin{aligned}
&\int_{-\infty}^{\infty}\left\{\frac{1}{4\pi}\left[\vec{f}_H^{\,*}\frac{\partial}{\partial\omega}(\omega\widehat{\mu})\vec{f}_H+\vec{f}_E^{\,*}\frac{\partial}{\partial\omega}(\omega\widehat{\varepsilon})\vec{f}_E+\vec{f}_H^{\,*}\frac{\partial}{\partial\omega}(\omega\widehat{\alpha}^{+})\vec{f}_E\right.\right.\\
&\quad\left.\left.+\vec{f}_E^{\,*}\frac{\partial}{\partial\omega}(\omega\widehat{\alpha})\vec{f}_H\right]\right\}\mathrm{d}z\frac{\partial A_1}{\partial t}\\
&+\frac{c}{4\pi}\frac{\partial}{\partial x}\int_{-\infty}^{\infty}\{[\vec{f}_E\times\vec{f}_H^{\,*}]_x+[\vec{f}_E^{\,*}\times\vec{f}_H 1]_x\}\mathrm{d}z\frac{\partial A_1}{\partial x}\\
&=-\int_{-\infty}^{\infty}\left(\vec{H}_{1l}^{*}\frac{\partial\vec{M}_{\mathrm{NL}}}{\partial t}+\vec{E}_{1l}^{*}\frac{\partial\vec{P}_{\mathrm{NL}}}{\partial t}\right)\mathrm{d}z\\
&\quad-\sum_{n=1,2}[E_{1lx}^{*}i_x^{(e)}+E_{1ly}^{*}i_y^{(e)}+H_{1ly}^{*}i_y^{(m)}+H_{1lx}^{*}i_x^{(m)}]_{z=z_{0n}}
\end{aligned}\tag{B.13}$$

Equation (B.13) when $\widehat{\mu}=\widehat{I}$, $\widehat{\alpha}=0$, $\vec{M}_{\mathrm{NL}}=0$, $\vec{i}^{(m)}=0$ is compared with equation (3.23) after use in the last SVA. It is shown that for the obtained transformations of the corresponding equations: (a) the first two terms in the left part correspond to each other in pairs, where the values proportional to the integral energy and energy flux of linear waves are included as coefficients at the corresponding slowly and temporally partial derivative of amplitudes; (b) the sum of the last three terms in the transformed equation (3.23) correspond to the first term (at $\vec{M}_{\mathrm{NL}}=0$) in the right part of the transformed equation (B.13), and characterize the contributions of the corresponding volume nonlinearities; (c) finally, the first term in the right part of the transformed equation (3.23) corresponds to the second term in the right part of equation (B.13) (for $\vec{i}^{(m)}=0$), characterizing the contributions of surface nonlinearities.

Thus, the relations (3.23) of the NEELS method for surface plasmons with bulk and surface nonlinearities can be considered as reducing, in the corresponding limiting case, to the integral relation of the NEELS method in a more general form for bi-anisotropic media (3.18).

Appendix B.6 Equations of the NEELS method in integral and differential form taking into account nonlinearity in auxiliary boundary conditions associated with spatial dispersion

For the sake of generality, in addition to nonlinearity in the additional boundary conditions associated with spatial dispersion (3.52), a nonlinear term of the form $\mathrm{NL}_{\mathrm{SDV}i} \equiv -\delta F/\delta p_i$ corresponding to the volume nonlinearity is added to the right-hand side equation (3.48). Using equation (3.48) and taking into account the boundary condition (3.52), it is possible to obtain bilinear relations similar to those used in deriving the law of conservation of energy and obtained similarly to that described in section 3.1. These relations in the differential and integral forms are obtained

$$\left\{\frac{1}{4\pi}(\vec{e}_1\vec{e}_{1l}^* + \vec{h}_{1l}^*\vec{h}_1) + \frac{1}{2}m\frac{\partial P_{1i}}{\partial t}\frac{\partial P_{1li}^*}{\partial t} + \frac{1}{2}m\omega_T^2 P_{1i}P_{1li}^*\right\}$$
$$+ \frac{\partial}{\partial x_k}\left[\frac{c}{4\pi}(\vec{e}_1\vec{e}_{1l}^* + \vec{h}_{1l}^*\vec{h})_k - C_R\left(\frac{\partial P_{1li}^*}{\partial t}\frac{\partial P_{1i}}{\partial x_k} + \frac{\partial P_{1i}}{\partial t}\frac{\partial P_{1li}^*}{\partial x_k}\right)\right] \tag{B.14}$$
$$- \mathrm{NL}_{\mathrm{SDV}i}\frac{\partial P_{1li}^*}{\partial t} = 0$$

$$\frac{\partial}{\partial t}\int_{-L/2}^{L/2}\left\{\frac{1}{4\pi}(\vec{e}_1\vec{e}_{1l}^* + \vec{h}_{1l}^*\vec{h}_1) + \frac{1}{2}m\frac{\partial P_{1i}}{\partial t}\frac{\partial P_{1li}^*}{\partial t} + \frac{1}{2}m\omega_T^2 P_{1i}P_{1li}^*\right\}\mathrm{d}z$$
$$+ \frac{\partial}{\partial y}\int_{-L/2}^{L/2}\left[\frac{c}{4\pi}(\vec{e}_1\vec{e}_{1l}^* + \vec{h}_{1l}^*\vec{h})_k - C_R\left(\frac{\partial P_{1li}^*}{\partial t}\frac{\partial P_{1i}}{\partial x_k} + \frac{\partial P_{1i}}{\partial t}\frac{\partial P_{1li}^*}{\partial x_k}\right)\right]\mathrm{d}z$$
$$+ \frac{C_R}{2\delta}\left(\frac{\partial P_{1li}^*}{\partial t}P_{1i} + \frac{\partial P_{1i}}{\partial t}P_{1li}^*\right) \tag{B.15}$$
$$- \int_{-L/2}^{L/2}\mathrm{NL}_{\mathrm{SDV}i}\frac{\partial P_{1li}^*}{\partial t}\mathrm{d}z$$
$$- \left[\left(\mathrm{NL}_{\mathrm{SDS}i}\frac{\partial P_{1li}^*}{\partial t}\right)_{z=-L/2} + \left(NL_{\mathrm{SDS}i}\frac{\partial P_{1li}^*}{\partial t}\right)_{z=+L/2}\right] = 0$$

Appendix B.7 Details of the derivation of the nonlinear evolution eauations for BVMSW by means of method NEELS

Appendix B.7.1 The contribution of terms with harmonics of second order in the nonlinear term for the BVMSW

We present only the final results for the second and zeroth harmonics of the components of potential, magnetization, and the magnetic field (only the results for the x-component of magnetization will be included for brevity). These results were obtained using the Landau–Lifshitz equations (Landau and Lifshits 1982) and div$\vec{b}$ = 0, written for the second and zeroth harmonics of the magnetic induction, as well as the nonlinear boundary conditions. The latter ones include the continuity of the normal components of the magnetic induction and the tangential components of the magnetic field. It is shown that in order to find the second harmonic of the magnetization, for BVMSW (figure 3.1), it is necessary to find the first and second ($\varphi^{(2)}$) harmonics of the magnetic potential and solve, taking into account the corresponding nonlinear surface conditions, the equation for the transverse field distribution in the form

$$\mu(2\omega)\frac{\partial^2 F_2}{\partial x^2} - 4k^2F_2 = [B_0 + B_2\cos(2\tau x)]A_0^2A_1^2. \tag{B.16}$$

Here $\phi^{(2)} = (1/2)F_2(x)\exp(2i\theta) + c.\ c.$, $\theta = \omega t - k_z z$, $\mu(2\omega)$ is the component of the magnetic permeability at the frequency 2ω

$$B_2 = -i\frac{k}{8\pi\mu_0}\frac{\tau^2}{\sin^2(\tau l)}\{[(\mu(2\omega) - 1)(\mu(\omega) - 1) + \sigma_0(2\omega)\sigma_0(\omega)]\}$$

$$+\frac{1}{2}\{[(\mu(\omega) - 1)^2 - \tau_0^2(\omega)]\}A_0^2A_1^2$$

the coefficients $B_{0,\,2}$ are as follows:

$$B_2 = -i\frac{k_z}{8\pi M_0}\frac{\tau^2}{\sin^2(\tau l)}\Big\{[(\mu(2\omega) - 1)(\mu(\omega) - 1) + \mu_a(\omega)\mu_a(2\omega)] +$$

$$\frac{1}{2}\Big[(\mu(\omega) - 1)^2 - \mu_a(\omega)^2\Big]\Big\}A_0^2A_1^2$$

$$B_0 = -i\frac{16_z}{8\pi M_0}\frac{\tau^2}{\sin^2(\tau l)}[(\mu(\omega) - 1)^2 - \mu_a(\omega)^2]A_0^2A_1^2$$

Here $l \equiv L/2$ is the semi-thickness of the GL/ferrite film, $\tau^2 = k_z^2/(-\mu)$. The second harmonics are determined by the following formulas:

$$\phi^{(2)} = (1/2)F_2(x)\exp(2i\theta) + \text{c.c.},\quad h_{x,z}^{(2)} = (1/2)H_{x,z}^{(2)}\exp(2i\theta) + \text{c.c.},\quad \theta = \omega t - k_z z.$$

Here $m_{x,\,y}^{(2)}$ and $h_{x,\,y}^{(2)}$ are the components of the second magnetization harmonic $m_{x,y,z}^{(2)} = (1/2)M_{x,y,z}^{(2)}\exp(2i\theta) + \text{c.c.}$ and magnetic field, respectively,

$$M_x^{(2)} = \frac{\mu(2\omega) - 1}{4\pi} H_x^{(2)} - \frac{\gamma}{2\omega_M} H_z^{(1)}[(\mu(2\omega) - 1)M_x^{(1)} + i\mu_a(2\omega)M_y^{(1)}]A_0^2 A_1^2$$

$$H_x^{(2)} = \frac{\partial F_2}{\partial x},\ H_x^{(1)} = \frac{\partial f}{\partial x},\ H_z^{(1)} = -ik_z f(x),\ M_x^{(1)} = \frac{\mu(\omega) - 1}{4\pi} H_x^{(1)},\ M_y^{(1)} = -\frac{i\mu_a(\omega)}{4\pi} H_x^{(1)}$$

In (B.16) A_0 is a constant used for normalization and A_1 is the SVA. The values $\mu(\omega;\ 2\omega)$, $\mu_a(\omega;\ 2\omega)$ are the components of the magnetic permeability tensor at the frequencies $\omega;\ 2\omega$. The solution of equation (B.16), using boundary conditions for the field component of the second harmonic, has the form:

$$F_2 = E_0 + E_2 \cos(2\tau x) + E_3 \cosh\left(\frac{2k_z x}{\sqrt{\mu(2\omega)}}\right) \tag{B.17}$$

In the relation (B.17),

$$E_0 = i\frac{1}{16\pi M_0 k_z}[(\mu(\omega) - 1)^2 - \mu_a^2(\omega)]\frac{\tau^2}{\sin^2(\tau l)} A_0^2 A_1^2,$$

$$E_2 = i\frac{k_z}{16\pi M_0 \mu(2\omega)\sin^2(\tau l)}\left\{[(\mu(2\omega) - 1)(\mu(\omega) - 1) + \mu_a(\omega)\mu_a(2\omega)] + \frac{1}{2}[(\mu(\omega) - 1)^2 - \mu_a^2(\omega)]\right\} A_0^2 A_1^2;\ E_3 = \frac{2S_1 + S_2/k_z}{2\cos(2\tau l) + \frac{2}{\sqrt{\mu(2\omega)}}\sinh(\frac{2k_z l}{\sqrt{\mu(2\omega)}})}$$

$$S_2 = i\frac{k_z}{8\pi M_0 \mu(2\omega)}[(\mu(2\omega) - 1)(\mu(\omega) - 1) + \mu_a(2\omega)\mu_a(\omega)]\frac{\tau\cos(\tau l)}{\sin(\tau l)} A_0^2 A_1^2 + E_2 2\tau\sin(2\tau l).$$

$$S_1 = -E_0 - E_2 \cos(2\tau l);\ \varphi^{(0)} = 0,\ m_z^{(0)} = -\frac{1}{64\pi^2 M_0}[(\mu(\omega) - 1)^2 + \mu_a^2(\omega)]A_0^2 A_1^2 \left(\frac{\partial f}{\partial x}\right)^2 \quad m_x^{(0)} = 0,$$

$$m_y^{(0)} = -\frac{\gamma}{4\omega_H}(h_z^{(1)} m_y^{(1)*} + \text{c.c.}) = -\frac{\mu_a(\omega)k_z}{64\pi H_0}|A_1|^2 A_0^2 \frac{\partial}{\partial x} f_1^2$$

After the calculations, all the nonlinear terms including ones in equation (3.30), accounting for relations (3.33) and conditions from both zero and second harmonics of the magnetic field and magnetization, are determined for the case of thin FFs, $kzl << 1$ (see also appendix B.7.2 and problem 3.2)

Appendix B.7.2 Evaluations of the contributions of the terms with higher harmonics in the nonlinear term for the BVMSW

The right-hand side of (3.24), in the absence of parametric interaction, after integration along the normal to the waveguide layered structure (figures 3.1 and 3.2(a)), is reduced to the form

$$
\begin{aligned}
Nl_{\mathrm{VBVMSW}} = \int_{-\infty}^{\infty} \{-i\omega h_{1lz} \times m_{1z} + (i\omega/M_0)\}[m_{1ly}^*(h_{1z}m_{1y})]^{(1)} - m_{1lx}^*(h_{1x}m_{1z})^{(1)} \\
+ \{[m_{1lx}^*(h_{1z}m_{1x})^{(1)}]\}\mathrm{d}x \equiv \mathrm{NL}_{01} + \mathrm{NL}_{02} + \mathrm{NL}_{03} + \mathrm{NL}_{04}
\end{aligned}
\tag{B.18}
$$

The dispersion relations for BVMSW in a thin ferrite film (Gurevich and Melkov 1996) is $\tan(\tau l) = \sqrt{-\mu(\omega)}$, where $\tau^2 = -k_z^2/\mu$; $\mu = (\omega^2 - \omega_0{}^2)/(\omega^2 - \omega_H{}^2)$, $\omega_0{}^2 = \omega_H(\omega_H + \omega_M)$; here z, k_z are the direction of the constant magnetic field H_0 and the wavenumber along this direction, respectively; $\omega_H = \gamma H_0$, $\omega_M = 4\pi\gamma M_0$; H_0 and M_0 are the magnetic bias field and the saturation magnetization, respectively; when $kzl << 1$, $\tau l << 1$; corresponding polarization equation is of the form $|h_{zl}/h_{xl}| \sim |k \sin(\tau x)/[\tau \cos(\tau x)]| \approx kx|<|kl|\ll 1$. Using (B.18), we obtain, by order of magnitude, for the nonlinear term in equation (3.30) (Rapoport *et al* 2004a, 2012a) the expressions (3.33); all terms $\mathrm{NL}_{01,02,03,04}$ are considered and, finally, the total nonlinear term F_{NLV} is determined (see also problem 3.2 and relations (3.30), (3.25)).

Appendix B.7.3 Coefficients in the evolution equations for the envelope amplitude with higher linear and nonlinear terms

We detail the further development of the NEELS method taking into account both the two-dimensionality of the pulses, the higher effects of diffraction and dispersion, and the higher linear terms in the evolution equation (Rapoport *et al* 2004b, 2012a). To derive the nonlinear evolution equation (3.29) for BVMSW in the parabolic approximation, the coordinates t,z,y were used. For the sake of convenience of studying higher nonlinear effects, including the nonlinear dispersion and diffraction (Boardman *et al* 2008, 2010a, 2010b, Rapoport *et al* 2004b, 2012a), which influence a single pulse, it is necessary to obtain the corresponding evolution equation in the coordinates z, y, $\eta = t - z/V_g$ ('running time'), where the group velocity is negative under the positive phase one for BVMSW (Damon and Eshbach 1961): $V_g = -|V_g|$. Note that this situation makes the study of nonlinear wave processes for BVMSW in layered gyrotropic media quite similar to those in MM waveguides (Boardman *et al* 2010a).

Accordingly, assume that there are functional dependences $k_z = k_z(\omega, k_y, |U|^2)$ or $U = U(z, \eta, y)$ and take into account that BVMSWs (which propagate along the direction (z) of the magnetization field), $\partial k_z/\partial k_y = 0$. Consider the case of a single (without the parametric interaction) pulse BVMSW, make a replacement $z \to -z$ (because BVMSW is a backward wave).

Let us use the Fourier transform and represent the operators on the left side of the nonlinear evolution equation as a series in small deviations of the wave vector and frequency from the corresponding carrier values characterizing a pulse with a narrow or moderate spectrum width (Kadomtsev 1988, Zvezdin and Popkov 1983, Slavin *et al* 2003, Buttner *et al* 2000b). This series also includes the lowest powers of the amplitudes of the pulse envelopes with a relatively weak nonlinearity. Let us make the inverse Fourier transform and apply the 'Fourier operator method' (Kadomtsev 1988, Zvezdin and Popkov 1983). Thus, in the series, we can apply the substitution

$$i\Delta\omega \to \partial/\partial t,\ i\Delta\vec{k} \to -i\partial/\partial\vec{r}, \tag{B.19}$$

As a result, we obtain the following (non-parabolic) normalized (dimensionless) nonlinear equation (3.34). In equation (3.34) the normalized amplitudes and coordinates are $A = U/U_0$, $z_{\text{und}} = z/l_z$, $y_{\text{und}} = y/l_y$, $\eta_{\text{und}} = \eta t_{\text{NL}}$, where U_0 is the scale of the amplitude of the nonlinear wave (pulse envelope); $l_{z,\,y}$ and t_{NL} are the spatial scales for z and y directions and the time scale, respectively. The normalized coefficients of equation (3.34) are equal to

$$\begin{gathered}
g_1 = \frac{1}{2}\frac{\partial^2 k_z}{\partial\omega^2}\frac{l_z}{t_{\text{NL}}^2};\ \delta = \frac{1}{6}\frac{\partial^3 k_z}{\partial\omega^3}\frac{l_z}{t_{\text{NL}}^3};\ \chi = -\frac{1}{24}\frac{\partial^4 k_z}{\partial\omega^4}\frac{l_z}{t_{\text{NL}}^4};\ g_2 = \frac{1}{2}\frac{\partial^2 k_z}{\partial k_y^2}\frac{l_z}{l_y^2} \\
g_3 = \frac{1}{2}\frac{\partial^3 k_z}{\partial\omega\partial k_y^2}\frac{l_z}{t_{\text{NL}} l_y^2};\ g_4 = -\frac{1}{4}\frac{\partial^4 k_z}{\partial\omega^2\partial k_y^2};\ \rho = -\frac{1}{24}\frac{\partial^4 k_z}{\partial k_y^4}\frac{l_z}{l_y^4};\ N = -\frac{\partial k_z}{\partial(|U|^2)}U_0^2 l_z \\
b_{10} = -\frac{\partial^2 k_z}{\partial\omega\partial(|U|^2)}\frac{l_z}{t_{\text{NL}}}U_0^2;\ b_{20} = \frac{1}{2}\frac{\partial^3 k_z}{\partial(|U|^2)\partial k_y^2}\frac{l_z}{l_y^2}U_0^2;\ \gamma = \frac{\omega_{\text{r}}}{|V_{\text{g}}|}l_z
\end{gathered} \tag{B.20}$$

Here ω_{r} is the relaxation frequency, which characterizes the dissipation of waves in ferrite films. The linear coefficients of equation (3.34), shown in relations (B.20), can be determined analytically or numerically using a linear dispersion relation for the fundamental BVMSW mode (Damon and Eshbach 1961). The nonlinear coefficients can be determined using the coefficient N_0 found on the basis of the NEELS method (see relations (3.31) and (3.33)) and the linear dispersion equation (Damon and Eshbach 1961). For example, it was found analytically that

$$\frac{\partial^2 k_z}{\partial\omega\partial((|U|^2)} = -\frac{1}{\left(\frac{\partial\omega}{\partial k_z}\right)^2_{U=const}}\left\{\frac{\partial^2\omega}{\partial k_z\partial(|U|^2)} - \frac{1}{\left(\frac{\partial\omega}{\partial k_z}\right)_{U=const}}\frac{\partial\omega}{\partial(|U|^2)}\frac{\partial^2\omega}{\partial k_z^2}\right\}$$

Given that in accordance with (B.20), (3.31) and (3.30), $b_{10} \sim \dfrac{\partial^2 k_z}{\partial\omega\partial((|U|^2)}$, $N_0 \sim \dfrac{\partial\omega}{\partial(|U|^2)}$, it is possible to express the coefficient b_{10} through N_0. Alternatively, a direct decomposition of the (nonlinear) dispersion equation can be used (similarly to what was done in Zvezdin and Popkov 1983). As a result, an estimate for the nonlinear dispersion coefficient is obtained:

$$b_{10} = b_{1\mathrm{DIM}} U_0^2 \frac{l_z}{t_{\mathrm{NL}}};\ b_{1\mathrm{DIM}} = -\frac{\partial}{\partial\omega}\frac{\partial k_z}{\partial|U|^2} = \frac{1}{l}\frac{\partial\kappa}{\partial\omega} \approx -\frac{1}{l}\frac{2\omega}{\omega_H\omega_M}\left[2 + \left(\frac{\omega}{\omega_H}\right)^2\right] \tag{B.21}$$

The form of a nonlinear diffraction term, proportional to the coefficient b_{20} in equation (3.34) is selected so as to guarantee the conservation of pulse energy when it is propagated in the GL. The term describing the nonlinear diffraction and being conservative has the form:

$$\sim b_{20} A\partial^2(|A|^2)/\partial y^2 \tag{B.22}$$

in relation (B.22),

$$b_{20} = \frac{1}{2}\frac{\partial^3 k_z}{\partial|U|^2\partial k_y^2}\frac{l_z}{l_y^2}U_0^2 \approx \frac{1}{2}\frac{l_z}{l_y^2}U_0^2\frac{\alpha}{k_z^2 l}(2\mu_a^2 - 1);\ \alpha = \frac{1}{2}\frac{\omega_M\omega_H(\omega^2 + \omega^2)}{(\omega^2 - \omega_H^2)^3} \tag{B.23}$$

Appendix B.7.4 Comparison of the nonlinear coefficient for the BVMSW with the coefficient obtained in Slavin and Rojdestvenski (1994) and Leblond (2001)

To compare the corresponding nonlinear coefficients, it is convenient to consider them for one-dimensional (1D) nonlinear equation (3.30) where only the time derivative and the nonlinear term corresponding to the volume nonlinearity are preserved:

$$\frac{\partial A_1}{\partial t} + \ldots + iN_{\mathrm{BVMSW}}|A_1|^2 A_1 = 0 \tag{B.24}$$

where, according to formulas (3.31) and (P.2.2) (see this equation in the comment to problem 3.2), as well as the solutions for second-order harmonics given in appendix B.7.1, we obtain

$$N_{\mathrm{BVMSW}} = \frac{\omega(1 - \mu(\omega))}{32\pi(4\pi M_0)^2} \times \frac{\left[\frac{3}{2}(\mu(\omega) - 1)^2 + \frac{1}{2}\mu_a^2(\omega)\right]\int_0^l \left(\frac{\partial f}{\partial x}\right)^4 \mathrm{d}x}{\frac{1}{8\pi}\int_0^\infty \left\{\left[1 + \frac{\omega_H}{\omega_M}((\mu(\omega) - 1)^2 + \mu_a^2(\omega))\right]\left(\frac{\partial f}{\partial x}\right)^2 + k_z^2 f^2\right\}\mathrm{d}x} \tag{B.25}$$

In the limiting case of a thin GL, $k_z l \ll 1$, relation (B.25) is reduced to the following form

$$N_{BVMSW} \approx \frac{\omega\omega_M}{4\omega_H} \frac{(1 - \mu(\omega))\left[\frac{3}{2}(\mu(\omega - 1)^2 + \frac{1}{2}\mu_a^2(\omega)\right]}{\left\{\frac{\omega_M}{\omega_H} + [(\mu(\omega) - 1)^2 + \mu_a^2(\omega)]\right\}} \frac{A_0^2}{(4\pi M_0)^2 l^2} \tag{B.26}$$

To compare the nonlinear coefficient (B.26) with the corresponding coefficient obtained in Slavin and Rojdestvenski (1994) and Leblond (2001), we take into account that in the first of these articles the nonlinear evolution equation is written with respect to the variable ψ_1, where

$$|\psi_1|^2 = \frac{|m_{x1}|^2 + |m_{y1}|^2}{2M_0^2} = \frac{1}{8} \frac{[(\mu(\omega) - 1)^2 + \mu_a^2(\omega)]}{(4\pi M_0)^2} A_0^2 |A_1|^2 \tag{B.27}$$

Using (B.27), it is possible to transform the coefficient (B.25) into a nonlinear coefficient of the corresponding equation, rewritten in terms of the variable. We use the nonlinear coefficient from Slavin and Rojdestvenski (1994), formula (Boardman *et al* 2009) and (Leblond 2001, formulas (78)), in the limiting case $kl \ll 1$ for the same parameters as in the above-mentioned works ($\omega_H/\omega_M = 1066/1750$, $L = 7.2$ μm). Note that the nonlinear coefficient obtained in Leblond (2001) in this limiting case is very close to that obtained in Slavin and Rojdestvenski (1994).

At the same time, the ratio of nonlinear coefficients obtained in our work and in Slavin and Rojdestvenski (1994) is ≈ 1.2. Therefore our method NEELS, accounting for all the relevant extensions, provides the possibilities of both the adequate calculations of the nonlinear coefficient(s) of the evolution equations for the amplitudes of envelopes of wave packets in layered nonlinear gyrotropic and plasma-like media (Slavin and Rojdestvenski 1994, Leblond 2001, Wang *et al* 2009, Pitikakis *et al* 2016) and the proper description of the experimental results (Buttner *et al* 2000a, Bauer *et al* 1998). The results of modelling nonlinear waves in layered nonlinear and active metamaterials will be presented in the next chapters.

References

Afshar V S and Monro T M 2009 A full vectorial model for pulse propagation in emerging waveguides with subwavelength structures part I: Kerr nonlinearity *Opt. Express* **17** 2298–318

Agraval G 1996 *Nonlinear Fiber Optics* (Moscow: Mir) p 326 (in Russian)

Agranovich V M and Ginzburg V L 1965 *Crystal Optics With Spatial Dispersion, Excitons* (Berlin: Springer) p 441

Akhiezer A I, Bar'yakhtar V G and Peletminsky S V 1967 *Spin Waves* (Moscow: Nauka) p 368 (in Russian)

Anisimov I O 2003 *Oscillations and Moments* (Kiev: Academic) p 280 (in Ukrainian)

Atkinson R and Kubrakov N F 2001 Boundary conditions in the simplest model of linear and second harmonic magneto-optical effects *Phys. Rev.* B **65** 014432

Atkinson R and Kubrakov N F 2002 Magneto-optical characterization of ferromagnetic ultrathin multilayers in terms of surface susceptibility *Phys. Rev.* B **33** 124414

Bauer M, Demokritov S O, Hillebrands B, Grimalsky V V, Rapoport Y G and Slavin A N 1998 Observation of spatiotemporal self-focusing of spin waves in magnetic films *Phys. Rev. Lett.* **81** 3769–72

Bertran P, Hermann C, Lampel G and Peretti J 2001 General analytical treatment of optics in layered structures: Application to magneto-optics *Phys. Rev.* B **64** 235421

Bhardwaj A, Pratap D, Semple M, Iyer A K, Jayannavar A M and Ramakrishna S A 2020 Properties of waveguides filled with anisotropic metamaterials *C. R. Phys., Tome* **21** 677–711

Boardman A D, Wang Q, Nikitov S A, Chen J, Mills D and Bao J S 1994 Nonlinear magnetostatic surface waves in ferromagnetic films *IEEE Trans. On Mag.* **30** 14–22

Boardman A D and Xie K 1997 Magneto-optic spatial solitons *J. Opt. Soc. Am. B.* **14** 3102–9

Boardman A D, Marinov K, Pushkarov D I and Shivarova A 2000 Influence of nonlinearly induced diffraction on spatial solitary waves *Opt. Quant. Electron.* **32** 49–62

Boardman A D, King N and Rapoport Y 2007 Metamaterials driven by gain and special configurations *SPIE. Proc. of Metamater. II* **6581** 658108

Boardman A D, King N, Mitchell-Thomas R C, Malnev V N and Rapoport Y G 2008 Gain control and diffraction-managed solitons in metamaterials *Metamaterials* **2–3** 145–54

Boardman A D, Mitchell-Thomas R and Rapoport Y G 2009 Weakly nonlinear waves in layered bi-anisotropic *Proc. of Third Int. Congress on Adv. Electromagn. Materials in Microwaves and Optics: Metamaterials* (London) pp 495–7

Boardman A D, Hess O, Mitchell-Thomas R C, Rapoport Y G and Velasco L 2010a Temporal solitons in magnetooptic and metamaterial waveguides *Photonics Nanostruct.—Fundam. Appl.* **8** 228–43

Boardman A D, Mitchell-Thomas R C, King N J and Rapoport Y G 2010b Bright spatial solitons in controlled negative phase metamaterials *Opt. Commun.* **283** 1585–97

Boardman A D, Grimalsky V V, Kivshar Y, Koshevaya S V, Lapine M, Litchinitser M, Malnev V N, Noginov M, Rapoport Yu G and Shalaev V M 2011 Active and tunable metamaterials *Laser Photon. Rev.* **5** 287–307

Boardman A D, Alberucci A, Assanto G, Rapoport Y G, Grimalsky V V, Ivchenko V M and Tkachenko E N 2017 From "World Scientific Handbook of Metamaterials and Plasmonics" *Spatial Solitonic and Nonlinear Plasmonic Aspects of Metamaterials* vol. 16 ed E Shamonina and S A Maier (Berlin: Springer) p 564 vol 1—pp 419–69

Buttner O, Bauer M, Demokritov S O, Hillebrands B, Kivshar Y S, Grimalsky V, Rapoport Y, Kostylev M P, Kalinikos B A and Slavin A N 2000a Spatial and spatiotemporal self-focusing of spin waves in garnet films observed by space- and time-resolved Brillouin light scattering *J. Appl. Phys.* **87** 5088–90

Buttner O, Bauer M, Demokritov S O, Hillebrands B, Kivshar Y S, Grimalsky V V, Rapoport Y and Slavin A N 2000b Linear and nonlinear diffraction of dipolar spin waves in yttrium iron garnet films observed by space- and time-resolved Brillouin light *Phys. Rev.* B **61** 11576–87

Capolino F 2009 *Theory and Phenomena of Metamaterials* (Boca Raton, FL: CRC Press, Taylor and Francis) p 926

Charney J G and Drazin P G 1961 Propagation of planetary-scale disturbances from the lower into the upper atmosphere *J. Geophys. Res.* **66** 83–109

Chen X, Panoiu N C and Jr Osgood R M 2006 Theory of Raman-mediated pulsed amplification in silicon-wire waveguides *IEEE J. Quant. Electron.* **42** 160–70

Chi S and Guo Q 1995 Vector theory of self-focusing of an optical beam in Kerr media *Opt. Lett.* **20** 1598–600

Chu J and Sher A 2008 *Physics and Properties of Narrow Gap Semiconductors* (Berlin: Springer) p 606

Cottam M G, Tilley D R and Zek B 1984 Theory of surface modes in ferroelectrics *J. Phys. C: Solid State Phys.* **17** 1793–823

D'Aguanno G 2008 Ultra slow light pulses in a nonlinear metamaterial *J. Opt. Soc. Am.* B **25** 1236–43

Damon R W and Eshbach J R 1961 Magnetostatic modes of a ferromagnetic slab *J. Phys. Chem. Solids* **19** 308–20

Danilov V V, Zavislyak I D and Balinsky M G 1991 *Spin-wave Electrodynamics* (Kiev: Libid)) p 212 (in Russian)

Dolgaleva K, Boyd R W and Milonn P W 2009 The effects of local fields on laser gain for layered and Maxwell–Garnett composite materials *J. Opt. A: Pure Appl. Opt.* **11** 024002

Dolgaleva K, Boyd R W and Milonn P W 2011 The effects of local fields on laser gain of composite optical materials *Int. Workshop on Theoretical Comput. Nanophotonic. AIP Conf. Proc.* vol 1398 pp 126–8

Fedorov F I 1976 *The Theory of Gyrotropy* (Minsk: Science and Technology) p 455 (in Russian)

Fukui M and Stegeman G I 1982 Non-linear optics of surface *Electromagnetic Surface Modes* ed A D Boardman (New York: Wiley) pp 725–72

Gramotnev D K and Bozhevolnyi S I 2010 Plasmonics beyond the diffraction limit *Nat. Photon.* **4** 83–91

Grimalsky V V and Rapoport Y G 1995 Convolution of magnetostatic waves in ferrite films *J. Magn. Magn. Mater.* **140–44** 2195–6

Grimalsky V V, Kotsarenko N Y and Rapoport Y G 1996 Nonlinear surface waves in electronic plasma *23rd European Phys. Soc. Conf. Controlled Fusion Plasma Phys.* **2** 396

Grimalsky V, Rapoport Y and Slavin A N 1997 Nonlinear diffraction of magnetostatic waves in ferrite films *J Phys. IV France* **7** C1-393-2

Grimalsky V V and Rapoport Y G 1998a Modulational instability of surface plasma waves in the second-harmonic resonance region *Plasma Phys. Rep.* **24** 980–2

Grimalsky V V, Kremenetsky I A and Rapoport Y G 1998b Excitation of EMW in the lithosphere and propagation into magnetosphere *Atmospheric and Ionospheric Electromagnetic Phenomena Associated with Earthquakes* ed M A Hayakawa and O A Molchanov (Tokyo: TERRAPUB) pp 777–87

Grimalsky V V, Koshevaya S V and Resin A M 1998c Nonlinear surface magnetic polaritons in an inhomogeneous magnetic field *Tech. Phys.* **68** 92–5 (in Russian)

Grimalsky V V 1998d Nonlinear magnetostatic waves in an inhomogeneous magnetic field *Radio Eng. Electron.* **43** 998–1001 (in Russian)

Grimalsky V V, Mantha J H, Rapoport Y G, Slavin A N and Zaspel C E 2000a Numerical models of amplification and wave front reversal of two-dimensional spin wave packets in magnetic films *Materials Science Forum, 8th Eur. Magn. Mater. Appl. Conf. (EMMA 2000) (Kyiv, Ukraine)* pp 377–80

Grimalsky V V, Rapoport Y G, Slavin A N and Zaspel C E 2000b Parametric amplification and wave front conjugation of 2D pulse in ferromagnetic film *Proc. Meeting APS Minneapolis (USA)* **Z25-5** 1035

Gromov E M, Piskunova L V and Tyutin V V 1999 Dynamics of wave packets in the frame of third-order nonlinear Schrödinger equation *Phys. Lett.* A **256** 153–58

Gurevich A G and Melkov G A 1996 *Magnetization Oscillations and Waves* (New York: CRC Press) p 464

Jardin J-P, Moch P and Dvorak V 2002 Polarization waves in dielectric films with spatial dispersion *J. Phys. Condens. Matter* **14** 1745–63

Kadomtsev B B 1988 *Collective Phenomena in Plasma* (Moscow: Nauka) p 304 (in Russian)

Kalinikos B A and Slavin A N 1986 Theory of dipole-exchange spin wave spectrum for ferromagnetic films with mixed exchange boundary conditions *J. Phys. C: Solid State Phys.* **19** 7013–33

Kalinikos B A, Kovshikov N G and Slavin A N 1988 Envelope solitons and modulation instability of dipole-exchange magnetization waves in yttrium iron garnet films *Zh. Eksp. Teor. Fiz.* **94** 159–76

Kalinin V A and Bayonets V V 1990 On the possibility of reversing the front of radio waves in an artificial nonlinear environment *Radiotech. Electron.* **35** 2275–81

Kamenetskii E I 1996 Energy balance equation for electromagnetic waves in bianisotropic media *Phys. Rev.* E **54** 4359–67

Kauranen M and Zayats A V 2012 Nonlinear plasmonics *Nat. Photon.* **6** 737–48

Kolesik M and Moloney J V 2004 Nonlinear optical pulse propagation simulation: from Maxwell's to unidirectional equations *Phys. Rev.* E **70** 036604

Kong J A 1972 Theorems of bi-anisotropic media *Proc. IEEE* **60** 1036–46

Landau L D and Lifshits E M 1982 *Electrodynamics of Continuous Media* (Moscow: Nauka) p 620 (in Russian)

Leblond H 2001 Rigorous derivation of the NLS in magnetic films *J. Phys. A: Math. General* **34** 9687–712

Li G, de Sterke M and Palomba S 2018 Fundamental limitations to the ultimate Kerr nonlinear performance of plasmonic waveguides *ACS Photon* **5** 1034–40

Li Y and Argyropoulos C 2018 Tunable nonlinear coherent perfect absorption with epsilon-near-zero plasmonic waveguides *Opt. Lett.* **43** 1806

Lindell I V, Shihvola A H, Tretyakov S A and Vittanen A J 1994 *Electromagnetic Waves in Chiral and Bi-Isotropic Media* (Boston, MA: Artech House) p 335

Lotfi F, Sang-Nourpour N and Kheradmand R 2020 Plasmonic all-optical switching based on metamaterial/metal waveguides with local nonlinearity *Nanotechnology* **31** 015201

Luan P-G, Wang Y-T, Zhang S and Zhang X 2011 Electromagnetic energy density in a single-resonance chiral metamaterial *Opt. Lett.* **36** 675–77

Nikolsky V V and Nikolskaya T I 1989 *Electrodynamics and Radio Wave Propagation* (Moscow: Nauka) p 543 (in Russian)

Oulton R T, Sorger V J, Zentgraf T, Ren-Min M, Gladden C, Dai L, Bartal G and Zhang X 2009 Plasmon lasers at deep subwavelenght scale *Nat. Lett.* **461** 629–32

Paul T, Rockstuhl C and Lederer F 2011 Integrating cold plasma equations into the Fourier modal method to analyze second harmonic generation at metallic nanostructures *J. Modern Opt.* **58** 438–48

Pekar S I 1962 Supplementary light waves in crystals and exciton absorption *Sov. Phys. Uspekhi* **5** 515–21

Peruch S, Neira A, Wurtz G A, Wells B, Podolskiy V A and Zayats A V 2017 Geometry defines ultrafast hot carrier dynamics and kerr nonlinearity in plasmonic metamaterial waveguides and cavities *Adv. Opt. Mater.* **5** 1700299

Plum E, Fedotov V A and Zheludev N I 2008 Optical activity in extrinsically chiral metamaterial *Appl. Phys. Lett.* **93** 191911–3

Pitilakis A, Chatzidimitriou D and Kriezis E E 2016 Theoretical and numerical modeling of linear and nonlinear propagation in graphene waveguides *Opt. Quant. Electron.* **43** 1–22

Rapoport Y G, Gotynyan O E, Ivchenko V N, Kozak L V and Parrot M 2004a Effect of acoustic-gravity wave of the lithospheric origin on the ionospheric F region before earthquakes *Phys. Chem. Earth.* **29** 607–16

Rapoport Y G, Zaspel C E, Grimalsky V V and Sanchez-Mondragon J 2004b Nonlinear Lorentz Lemma with the influence of exchange interaction and propagation of the magnetostatic waves with higher diffraction and dispersion *IEEE Proc. of the 14th Int. Crimean Conf. 'Microwave and Communication Technology' CriMiCo'2004 (Sevastopol, Crimea, Ukraine)* 361–63 Catalog No. 04EX843

Rapoport Y G, Boardman A D, Kanevskiy V I, Malnev V N, King N J and Velasco L 2006a Modelling new active media based on metamaterials with artificial *IEEE Proc. of the 16th Int. Crimean Conf. 'Microwave and Communication Technology' CriMiCo'2006 Sevastopol Crimea (Ukraine)* Catalog No. 06EX1376 671–72

Rapoport Y G 2006b Formation of structures under parametric coupling of two-dimensional nonlinear pulses of magnetostatic waves in magnetic films *IEEE Proc. of the 16th Int. Crimean Conf. 'Microwave and Communication Technology', CriMiCo'2006 (Sevastopol, Crimea, Ukraine)* 629–30 Catalog No. 06EX1376

Rapoport Y G and Grimalsky V V 2011 Nonlinear surface 2D plasmons and giant second harmonic generation *Proc. of the Int. Conf. Days on Diffraction (St. Petersburg, Russia)* pp 168–73

Rapoport Y, Grimalsky V and Zaspel C 2012a Method for the derivation of nonlinear evolution equations in layered structures (NEELS): an example of nonlinear waves in gyrotropic layers *Bull. Univ. Kyiv Phys.* **14/15** 72–6

Rapoport Y G 2012b General method for modeling nonlinear waves in layered structures of different physical nature including bi-anisotropic and active metamaterials *Progress in Electromagn. Research Symp. (PIERS) (Moscow, Russia)* Abstracts pp 18–9

Rapoport Y G 2013 General method for modeling nonlinear waves in active metamaterial and gyrotropic layered structures *Proc. of Int. Kharkov Symp. on Phys. and Engineering of Microwaves, Millimeter and Submillimeter Waves MSMW'13 (Kharkov, Ukraine)* pp 253–5

Rapoport Y G, Grimalsky V V, Boardman A D and Malnev V N 2014a Controlling nonlinear wave structures in layered metamaterial, gyrotropic and active media *IEEE 34th Int. Sci. Conf. Electronics and Nanotechnology (ELNANO)* pp 46–50

Rapoport Y G 2014b General method for deriving the equations of evolution and modeling of nonlinear waves in layered active media with bulk and surface *Bull. Kyiv Nat. Taras Shevchenko Univ. Series: Phys.-Math. Sci.* **1** 281–88 (in Ukrainian)

Rapoport Y G 2015a Modeling 'From properties of a metaparticle (MP) to characteristics of nonlinear waves in layered active bi-anisotropic metamaterials (BIAM)' *Ukr.–Germ. Symp. on Phys. and Chem. of Nanostructures and on Nanobiotechnology* Book of abstract p 159

Rapoport Y G and Boardman A D 2015b Modeling "From the characteristics of the metaparticle to the characteristics of nonlinear waves in layered active bi-anisotropic metamaterials with

bulk and surface nonlinearities" *Bull. of Kyiv Nat. Taras Shevchenko Univ. Series: Physics* **3** 207–12 (in Ukrainian)
Rowland D R 1999 Conservation law for multimoded nonlinear optical waveguide interactions and its physical interpretation *Phys. Rev.* E **59** 7141–7
Samson Z I, Horak P, MacDonald K F and Zheludev N I 2011 Femtosecond surface plasmon pulse propagation *Opt. Lett.* **36** 250–2
Sipe J E, So V C Y, Fukui M and Stegeman G I 1980 Analysis of second-harmonic generation at metal surfaces *Phys.Rev. B.* **21** 4389–402
Slavin A N and Rojdestvenski I V 1994 "Bright" and "dark" spin wave envelope solitons in magnetic films *IEEE Trans. Magn.* **30** 37
Slavin A N, Büttner O, Bauer M, Demokritov S O, Hillebrands B, Kalinikos B A, Grimalsky V V and Rapoport Y 2003 Collision properties of quasi-one-dimensional spin wave solitons and two-dimensional spin wave bullets *Chaos* **13** 693–701
Sugakov V Y 1974 *Theoretical Physics Electrodynamics* (Kyiv: Higher School) p 271 (in Ukrainian)
Tretyakov S A, Mariotte F, Simovski C R, Kharina T G and Heliot J-P 1996 Analytical antenna model for chiral scatterers: comparison with numerical and experimental data *IEEE Trans. Anten. Propagat.* **44** 1006–14
Tretyakov S 2003 *Analytical Modeling in Applied Electromagnetics* (Boston, MA: Artech House) p 275
Tretyakov S A, Simovski K R and Hudlicka M 2007 Bianisotropic route to the realization and matching of backward-wave metamaterial slabs *Phys. Rev.* B **75** 153104
Tretyakov S 2010 *Nanostructured Metamateriuals* (Brussels: Office of the European Union) p 137
Wang F X, Rodríguez F J and Albers W M *et al* 2009 Surface and bulk contributions to the second-order nonlinear optical response of a gold film *Phys. Rev.* B **80** 233402–25
Xiang Y, Dai X, Wen S and Guo J 2012 Nonlinear absorption due to linear loss and magnetic permeability in metamaterials *Phys. Rev.* E **85** 066604–10
Yang R, Xu J, Shen N-H, Zhang F, Fu Q, Li J, Li H and Fan Y 2021 Subwavelength optical localization with toroidal excitations in plasmonic and Mie metamaterials *InfoMat* Review Article **3** 577–97
Zakharov V E and Shabat A B 1972 Exact vtheory of two-dimensional self-focusing and one-dimensional self-modulation of waves in nonlinear media *Sov. Phys. JETP* **34** 62–9
Zheludev N I and Emel'yanov V I 2004 Phase matched second harmonic generation from nanostructured metallic surfaces *J. Opt. A: Pure Appl. Opt.* **6** 26–8
Zheludev N I and Kivshar Y S 2012 From metamaterials to metadevices *Nat. Mater.* **11** 917–24
Zvezdin A K and Popkov A F 1983 Contribution to the nonlinear theory of magnetostatic spin waves *Zh. Eksp. Teor. Fiz.* **84** 606–15

Yuriy Rapoport and Vladimir Grimalsky

Chapter 4

Application of the nonlinear evolution equations for layered structures (NEELS) method to the layered nonlinear passive gyrotropic and plasma-like structures with volume and surface nonlinearities

In the present chapter, the nonlinear evolution equations for layered structures (NEELS) method is applied for the giant resonance generation of the second harmonic of surface plasmons and contribution of the surface and volume nonlinearities. Then the same method is applied for the case of nonlinear pulses with the moderate spectrum in gyrotropic layered structures; for nonlinear magnetostatic waves (MSWs) in inhomogeneous ferrite films (FFs); for the formation of the nonlinear vortex magnetostatic structures in gyrotropic waveguides/FFs; formation and propagation of the bullets in the metamaterial (MM) waveguides accounting for higher-order nonlinearities; then instabilities of bullets in the MM waveguides with the influence of higher-order nonlinear effects are searched; the method NEELS, within the frames of the interdisciplinary metamaterial approach, is applied to the propagation of the waves in the linear waveguide Earth-Ionosphere. Finally, conclusions on the basis of this chapter are formulated.

4.1 Application of method NEELS for the giant resonance generation of the second harmonic of surface plasmons and contribution of the surface and volume nonlinearities

Details of the application of the NEELS method to surface plasmons in the presence of spatial (in the absence of temporal) resonance with the second harmonic are presented in appendices B.2.1–B.2.4. Based on the approach described in section 3.2

doi:10.1088/978-0-7503-2336-9ch4

of chapter 3 and appendices B.2.1–B.2.4, the system of equation (B.8), which is repeated here, for convenience is obtained (the coefficients are not shown here explicitly) for the envelopes $A_{1,2}$ of the coupled main and resonant second harmonic of the surface plasmon waves:

$$\begin{aligned}\partial A_1/\partial t + v_g\partial A_1/\partial x + (i\omega_{k_xk_x}/2)(\partial^2 A_1/\partial x^2) + ig_2(\partial^2 A_1/\partial y^2)\\ + \alpha_1 A_1^* A_2 + i\alpha_0 \mid A_1 \mid^2 A_1 + \gamma_1 A_1 = 0\end{aligned} \tag{4.1}$$

$$\partial A_2/\partial t + (i\Delta\omega_0 + \gamma_2)A_2 = \alpha_2 A_1^2 \tag{4.2}$$

To account for the surface nonlinearity, the nonlinear motion of surface charges (with the surface concentration n_S) at $z = 0$ (see figure 4.1) is taken into account (see figure 4.1).

In the first and second equation (4.1), v_g, ω_{kk}, g_2, $\Delta\omega_0$ and $\gamma_{1,2}$ are the group velocity, the coefficients of (linear) dispersion and diffraction, the deviation of the fundamental harmonic frequency from the resonant frequency, and coefficients of dissipative losses, respectively. Note that both quadratic (with coefficients $\alpha_{1,2}$) and cubic (with the coefficient α_0) nonlinearities are included in equation (4.1). The upper index '*' indicates a complex conjugation. The results of the analysis of the contribution of the volume and surface nonlinearities to the coefficients of the system of equations (4.1) and (4.2) are given in appendix B.2.4 in chapter 3, and the conclusion has been done on the prevailing role of the surface nonlinearity compared to the volume one at the resonant generation of the second harmonic. (Grimalsky *et al* 1996, Grimalsky and Rapoport 1998). This effect was qualitatively confirmed in subsequent works, in particular in Paul *et al* (2011), Wang *et al* (2009) and Wang *et al* (2009) theoretically and experimentally, respectively.

The numerical calculations based on equation (4.1) (Rapoport and Grimalsky 2011, Grimalsky and Rapoport 1998) are illustrated in figure 4.2. It is shown that the giant generation of the second harmonic leads to an amplitude whose value is greater than for the first harmonic; compare figures 4.2(b) of (a) and (d) of (c). In particular, it is shown that the contribution of the surface nonlinearity to the nonlinear coefficients that determine the generation of a giant resonant second harmonic is about three times greater than that of the volume nonlinearity. In the calculations and interpretation of

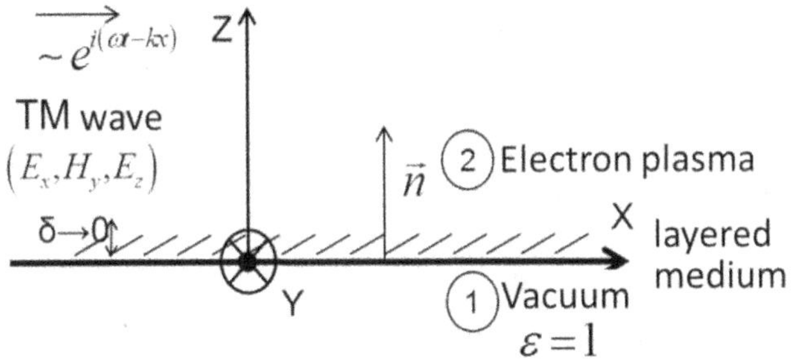

Figure 4.1. Surface plasmons in the system 'semi-infinite dielectric–semi-infinite plasma'. Geometry of the problem.

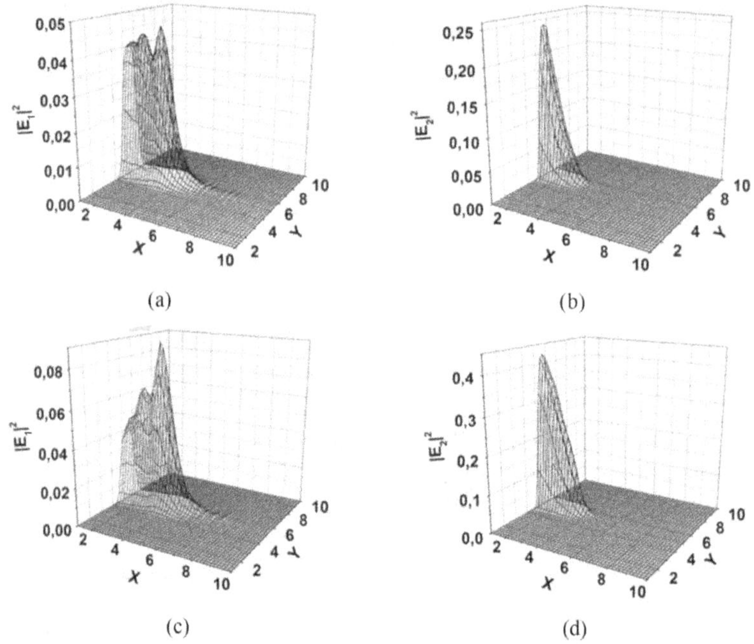

Figure 4.2. Dependence of normalized amplitude of the first (a and c) and second (b and d) harmonics on the normalized coordinates x,y for the value of the normalized time $T = 4.5$, which passed since the start of the action of the main pulse harmonic (pumping pulse) on the input of the system ($x = 0$). Both cubic and quadratic nonlinearities are present, and the signs of full nonlinearity coefficients are determined by the surface nonlinearity. The pumping field (the main harmonic at the input of the system), where $x = 0$, goes to the 'shelf' and then decreases so slowly over time that it can be approximately considered stationary, For (a)–(d) it is $E_1(x = 0) \approx \text{const} = 0.2$; for (a) and (b) the normalized frequency mismatch (from frequency $\omega_p/2$ for the main harmonic) and decrement of losses for the first and second harmonic are $\Delta\omega = \gamma_{1,2} = 0.1$; for (c) and (d) it is $(-\Delta\omega) = \gamma_{1,2} = 0.01$. Reproduced from Boardman *et al* (2017). Copyright (2017) World Scientific.

the results, the parameters $\chi_{0,2}$ characterizing the smallness of the relative values of the zero and second higher harmonics of the electron concentration are checked, respectively, and given in appendix B.2.2 in chapter 3, and correspond to the approximation of moderate nonlinearity. Section 10.1 of chapter 10 will demonstrate a tendency for the transition of surface plasmon generation to the mode of strong nonlinearity (Rapoport 2013, 2014, Rapoport and Grimalsky 2011). The possibility of formation of quasi-solitons on surface plasmons with the second harmonic close to the resonance is shown in appendix B.3 in chapter 3 Numerical estimations of the conditions required to observe nonlinear effects in layered plasmonic structures from microwave to optical range are done in appendix B.4 in chapter 3.

4.2 Vortex structures on the backward volume magnetostatic waves in ferrite films

A possibility of excitation of the vortex structures using the interaction of three nonlinear pulses of the forward volume magnetostatic waves (FVMSWs) in the normally magnetized FF is shown. Principles of the singular optics (Soskin and Vasnetsov 1998, Soskin *et al* 2016) had been applied to demonstrate the possibility of excitation of the stationary vortex structures on the MSW in microwave or millimeter wave ranges in FFs (Boardman *et al* 2005, Rapoport *et al* 2008). In the present paper, the excitation of the non-stationary vortex structures on the base of interaction of the three linear pulses of FVMSW in normally magnetized FF is searched. The interaction of three nonlinear pulses of FVMSW (Gurevich and Melkov 1996, Damon and Eshbach 1961) propagating in the normally magnetized FF is simulated. The pulses propagate under the angles 120° relatively to each other and pulses 1,2 and 2,3 are launched into the FF with the relative phase shifts of 120°. Their propagation is described by the set of three-dimensionless nonlinear Schrödinger equations for the amplitudes U_l of the magnetostatic potentials ϕ_l; $l = 1,2,3$ are the numbers of the pulses.

4.2.1 Excitation by a circular antenna of linear stationary FVMSW structures possessing phase defects

The excitation of the MSWs, needed to produce vortices, or phase defect structures, is shown using the example of stationary waves (Boardman *et al* 2005, Rapoport *et al* 2008). The principle of vortex excitation is the same for stationary and non-stationary vortices. The geometries for FVMSW propagation and the configuration for the excitation of the structures under consideration are shown in figure 4.3.

Figure 4.3 shows a linear FVMSW propagation in an FF along the Y axis. The magnetostatic potential and the field components are proportional to $\exp[i(\omega t - kY)]$, where ω and k are frequency and wavenumber, respectively. The field structure and dispersion are determined through the equation (Gurevich and Melkov 1996, Damon and Eshbach 1961)

$$\operatorname{div} \vec{b} = 0 \tag{4.3}$$

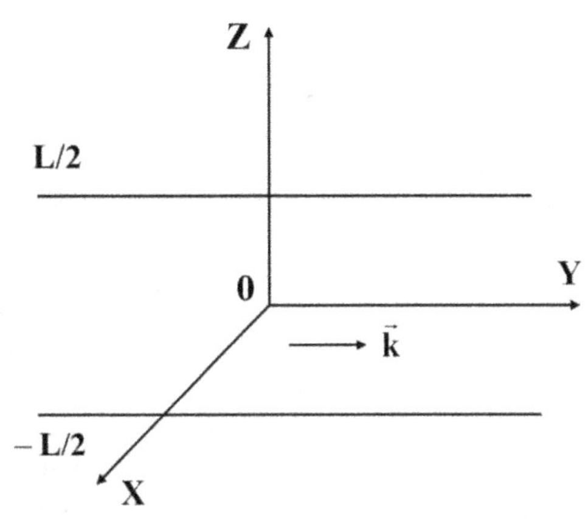

Figure 4.3. Geometry of the FF where the FVMSW propagates. Magnetic field, $\vec{H_0}$ is directed along axis Z, FVMSW propagates along axis Y, L is the thickness of FF. Reprinted from Kamenetskii (2008), copyright Transworld Research Network.

where

$$b_X = -(\mu\phi_X + i\mu_a\phi_Y),\ b_Y = -(\mu\phi_Y - i\mu_a\phi_X),\ b_Z = -\phi_Z,$$
$$\operatorname{div}\vec{b} = -\left[\mu(\phi_{XX} + \mu\phi_{YY}) + \phi_{ZZ}\right] \tag{4.4a}$$

inside the FF, and

$$\operatorname{div}\vec{b} = -(\phi_{XX} + \phi_{YY} + \phi_{ZZ}) \tag{4.4b}$$

outside the film. Here μ and μ_a are the components of the magnetic permeability tensor, and they are

$$\mu = \frac{\omega_\perp^2 - \omega^2}{\omega_H^2 - \omega^2},\ \mu_a = \frac{\omega\omega_M}{\omega_H^2 - \omega^2} \tag{4.4c}$$

where $\omega_H = |\gamma|(H_0 - 4\pi M_0)$, $\omega_M = |\gamma|4\pi M_0$, $\omega_\perp = (\omega_H(\omega_H + \omega_M))^{1/2}$. H_0 and M_0 are the magnitudes of the bias magnetic field (figure 4.3.) and the saturation magnetization of the FF, respectively, and γ is the gyromagnetic ratio. Jointly with the boundary conditions that impose the continuity upon the tangential components of magnetic field $\phi_{X,Y}$ and the normal component of magnetic induction b_Z at the FF boundaries, $Z = \pm L/2$ (figure 4.3.), equations (4.4a) and (4.4b) yield (Gurevich and Melkov 1996, Damon and Eshbach 1961) the spatial structure of the main mode of the magnetostatic potential as

$$\phi = \begin{Bmatrix} \cos(\sqrt{-\mu}|k|Z),\ |Z| \leqslant L/2 \\ e^{-|k|Z},\ |Z| > L/2 \end{Bmatrix} F(X, Y)e^{i\omega t} \equiv f_z(z)F(X, Y)e^{i\omega t} \tag{4.5a}$$

Here k is determined from the dispersion equation for the main mode of FVMSW that is

$$\tan\left(\sqrt{-\mu}kL/2\right) = \frac{1}{\sqrt{-\mu}} \tag{4.5b}$$

Equations (4.4a) and (4.4b), taking into account (4.5a), reduce to

$$\Delta_\perp F + k^2 F = 0,\quad \Delta_\perp = \frac{\partial^2}{\partial X^2} + \frac{\partial^2}{\partial Y^2} \tag{4.6}$$

For the propagation of FVMSW along the direction OY, $F(X, Y) = e^{-ikY}$(Gurevich and Melkov 1996). For the excitation of FVMSW by the circular antenna, shown in figure 4.4, equation (4.6) has a solution in terms of cylindrical functions. Only solutions with finite values of magnetostatic potential in the center of the circular antenna are considered however, figure 4.4, i.e.

$$\phi \text{ and } F(r,\theta) \text{ are finite, at } r = 0 \tag{4.7a}$$

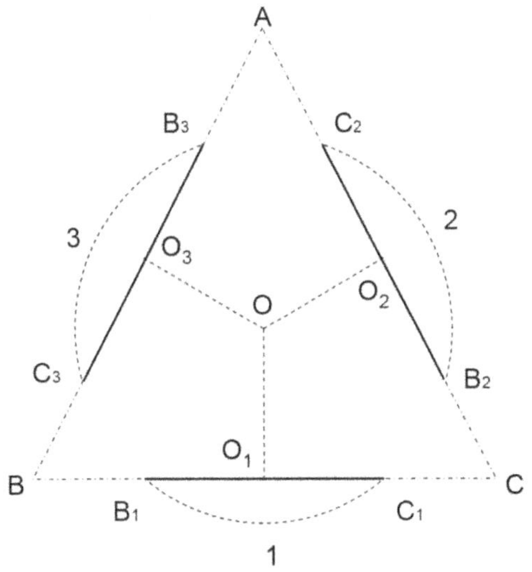

Figure 4.4. Excitation of three FVMSW by arc antennae. Reprinted from Kamenetskii (2008), copyright Transworld Research Network.

where r, θ are coordinates, in the cylindrical coordinate frame with the center coinciding with the center of circular microstrip antenna, figure 4.4. The Poynting vector in cylindrical coordinates is

$$P_r = -\frac{c}{8\pi}\,\mathrm{Re}\left[i\omega\phi\left(i\mu_a\frac{1}{r}\frac{\partial\phi^*}{\partial\theta} + \mu\frac{\partial\phi^*}{\partial r}\right)\right],\ P_\theta = -\frac{c}{8\pi}\,\mathrm{Re}\left[i\omega\phi\left(-i\mu_a\frac{\partial\phi^*}{\partial r} + \mu\frac{1}{r}\frac{\partial\phi^*}{\partial\theta}\right)\right], \tag{4.7b}$$

Consider now two examples of stationary linear vortex excitation (Boardman *et al* 2005).

A. Suppose that the magnetostatic potential at the microstrip antenna, of figure 4.4, with radius r_0, is determined by

$$F(r_0,\theta) = Ce^{i\theta} \tag{4.7c}$$

where C is a real constant. The solution of the equation (4.6), with boundary conditions (4.7a and 4.7b) is, then

$$F(r,\theta) = Ce^{i\theta}\frac{J_1(kr)}{J_1(kr_0)} \tag{4.8a}$$

Put $\tan(\Phi) = F_i/F_r$, where $F_{r,i}$ are real and imaginary parts of the magnetostatic potential, Φ is corresponding phase. To check the condition of vortex formation at the center of the structure (the condition of phase defect $F_r = 0$, $F_i = 0$, $\oint\Phi(\nabla\Phi)\mathrm{d}l = 2\pi n$, $n = 1,2,3\ldots$), see Nye and Berry (1974), Boardman *et al* (2005) consider, as a contour L, a circle with the radius $r = \rho$ around the phase defect point, $r = 0$. In accordance with (4.7c) and (4.8a) for the case under consideration,

$$F_r \equiv \text{Re}(F),\ F_i \equiv \text{Im}(F) = 0 \tag{4.8b}$$

and

$$\oint_L \mathrm{d}\Phi = \int_{0^+}^{2\pi} \mathrm{d}\theta = 2\pi \tag{4.8c}$$

therefore $n = 1$ and the topological charge is 1. Although the field considered here is determined through one scalar (magnetic potential) function, it is also interesting to consider the topological structure of the actual field components. For the points on the circle with a radius $r = \rho$, and using (4.8a), the result is

$$\begin{aligned}\vec{h}|_{Z=L/2} &= -\vec{\nabla}\phi|_{Z=L/2} = \vec{e}_Z h_Z + \vec{e}_r h_r + \vec{e}_\phi \vec{h}_\phi \\ &= \vec{e}_Z kCe^{i\theta}\frac{J_1(k\rho)}{J_1(kr_0)} + \vec{e}_r kCe^{i\theta}\frac{J_1'(k\rho)}{J_1(kr_0)} + \vec{e}_\theta \frac{1}{\rho}Ce^{i\theta}\frac{J_1(k\rho)}{J_1(kr_0)}\end{aligned} \tag{4.9}$$

From (4.9) it can be seen at the center of the circle the conditions of the formation of vortex structures, namely $F_r = 0$, $F_i = 0$, for h_Z and h_θ as well as for magnetostatic potential, ϕ. In contrast to this, the component h_r does not satisfy the conditions of the vortex formation at the center of the circle. The additional evidence for a vortex-like structure is derived from the Poynting vector that rotates about the center. There is only one non-zero component of Poynting vector in this case and that is

$$P_\theta = \frac{c}{8\pi}|C|^2 I_z \frac{J_1(kr)}{J_0^2(kr_0)}\left[\frac{\mu}{r}J_1(kr) - \mu_a k J_1'(kr)\right] \tag{4.10}$$

where $I_z = \int_{-\infty}^{\infty} f_z^2(z)\mathrm{d}z$, $f_z(z)$ is determined in (4.5a).

B. Suppose now that instead of (4.7c), there is a homogeneous distribution of the magnetic potential along the circular antenna,

$$F(r_0,\theta) = C \tag{4.11}$$

In this case

$$F(r,\theta) = C\frac{J_0(k\rho)}{J_0(kr_0)} \tag{4.12a}$$

$$h_Z = Ck\frac{J_0(kr)}{J_0(kr_0)},\ h_r = Ck\frac{J_0'(kr)}{J_0(kr_0)},\ h_\theta = 0 \tag{4.12b}$$

and the component h_r satisfies the conditions (4.8*b*) at the center of the circle, but $n = 0$. Neither the magnetostatic potential, ϕ nor the component h_Z satisfies the condition of the formation of vortex structures for $F_{i,r}$. Also the non-zero component of Poynting vector $\vec{P} = (c/8\pi)\mathrm{Re}(\varphi\partial\vec{b}^*/\partial t)$ is now

$$P_\theta = -\frac{c}{8\pi}|C|^2 I_z \frac{J_0(kr)}{J_0^2(kr_0)}\mu_a k J_0'(kr) \tag{4.13}$$

Finally, the field distributions along the circle antenna in the new forms have the structures with phase defects for the field component h_r with $n = 0$. A comparison of expressions (4.10) and (4.13) brings out the following points. In both the antenna potential distributions, a rotation of the Poynting vector around the center takes place ($P_\theta \neq 0$), but in the case of distribution (4.11), in distinction to (4.7b), this rotation is only due to the presence of a non-diagonal component of the magnetic permeability tensor, μ_a. Below, the 'plane wave' analog of distribution (4.7b), which provides a 'non-zero vortex charge', (4.8) will be investigated. From a practical point of view, however, it is easier to use a few short planes or arc antennae placed along the corresponding circle segments instead of using the entire circle antenna. In particular the case of three arc antennae is considered. To model the phase distribution, a phase shift $2\pi/3$ between the neighboring antennae will be used, but first of all, a simplified model using an interaction with three linear plane waves was adopted.

4.2.2 Formulation of the problem of non-stationary pulse interaction and main relations

Now, the interaction of three nonlinear pulses of FVMSW (Rapoport *et al* 2008) for a normally magnetized FF (figure 4.3) is considered. These pulses propagate at relative angles 120° to each other. As opposed to Boardman *et al* (2005), where coupling of stationary FVMSW has been considered, here a set of three coupled nonlinear non-stationary Schrödinger equations for (non-dimensional) amplitudes U_j of magnetostatic potentials ϕ_i, related to magnetic fields $\vec{h}_j$ through the relationships $\vec{h}_j = -\vec{\nabla}\phi_j$, where $j = 1,2,3$ is considered. These are

$$\frac{\partial U_j}{\partial t} + \frac{\partial U_j}{\partial y_j} + ig\frac{\partial^2 U_j}{dx_j^2} + ig_1\frac{\partial^2 U_j}{\partial y_j^2} + iN(|U_j|^2 + 2\sum_{l\neq j}|U_l|^2)U_j + \gamma U_j = 0 \tag{4.14}$$

Obviously $j,l = 1,2,3$, but g, g_1, N, and γ are dimensionless dispersion, diffraction, nonlinear coefficient, and the loss coefficient, respectively. y_j and x_j are the coordinate directions along and transverse to the directions of propagation of the corresponding pulses. The potential corresponding to the jth FVMSW is determined by the relation $\phi_j = (1/2)U_{dj}f_z(Z)\exp[i(\omega t - ky_j)] + \mathrm{c.c.}$, where $U_{dj} = U_j U_0$, U_{dj} are the true amplitudes, U_0 is an amplitude used for normalization, and $f_z(Z)$ is the transverse distribution function in the FF (see relations (4.5a) and (4.5b)), ω and k are the frequency and the wavenumber of the FVMSW, respectively. The parameters of diffraction, self-interaction, and loss are determined as $g = \frac{1}{2l_0 k}$, $N = N_d U^2{}_0 l_0$, $\gamma = \gamma_d l_0$, where $N_d = -\frac{1}{V_g}\left(\frac{\partial\omega}{\partial|U_d|^2}\right)_{k=\mathrm{const}}$ and γ_d are actual the dimensional nonlinear coefficient and the coefficient of losses, respectively, and l_0 is a length scale. The pulses 1,2 and 2,3 have at the input of the system a relative phase shift to each other, equal to 120°. The initial-boundary conditions have the form

$$U_l(y_l = 0, x_l, t) = U_{l0}(e^{i\frac{2\pi}{3}})^l \times \exp\left[-\left(\frac{t - t_1}{t_0}\right)^2\right]\exp\left[-\left(\frac{x_l - x_{l0}}{x_0}\right)^{2m}\right] \times F(x_l) \quad (4.15)$$

Here t_1, t_0 are the time shift, when FVMSW pulses come into the FF and the characteristic duration of the pulses, respectively, and x_0, x_{00} are, respectively, the width of the pulses and the shift of their centers in the direction, transverse to the direction of propagation. The function $F(x_l)$ describes the focusing of the pulses that are input into the system, when arc antennae are used, $m = 2$, and U_{l0} are the input amplitudes of the pulses. If $x_{00} = 0$, then three pulses propagate symmetrically along the bisectors of the equilateral triangle, in accordance with a symmetrical arrangement of the input antennae. The spatial distributions of the real part and phase of the magnetostatic potential and the Poynting vector for MSW are computed and presented in the next section. To show that, in particular, around the 'center of the interaction' point O of figure 4.4, a vortex structure with a 'phase defect' is formed, it is necessary to check, as well as in the stationary case, that: (a) the complex amplitude of the magnetic potential ϕ tends to zero; (b) the 'observation point' makes a whole revolution around the central point, then the change of phase of ϕ makes up a value equal to an integral number multiplied by 2π; and (c) the Poynting vector rotates around the center.

4.2.3 Numerical simulations

Typically, the following parameters are used for the computations: $l_0 = 1$ cm, $\gamma = 0.2$, $N = -1$, $L = 10$ μm, $|BC| = a = 1$ cm, $\omega \approx \omega_H \approx \omega_M \sim 3 \times 10^{10}$ s^{-1}, $k \approx 150$ cm^{-1}, and $l_0 = 1$ cm. The normalizing amplitude is such as $kU_0 \sim 2$ Oe, or $U_0 \sim 1.3 \times 10^{-2}$ Oe cm.

In figure 4.5, the real part of magnetostatic potential is shown at a number of times. It illustrates the resultant effect of 'purely linear' focusing, due to the arc antenna and also diffraction and nonlinear defocusing of FVMSW (Boardman *et al* 2005, Rapoport *et al* 2008).

In figure 4.6, the spatial distribution of the total magnetic potential in the presence of three pulses of FVMSW is shown for three consequent moments of time. A comparison of figure 4.6(c) (in the presence of nonlinearity) and 4.6(d) (where nonlinearity is absent) shows that the nonlinearity causes the broadening of the structures formed due to the interaction of the three pulses of FVMSW (as will be shown below), the structures are, in fact, vortex structures with a phase defect in the central point, namely O. If the centers of the input pulses are shifted from zero, i.e. ($x_{10} = x_{20} = x_{30} \neq 0$), then the formed structures have a shape shown in figure 4.7.

The phases of the nonlinear magnetostatic total potential of three focused, non-shifted, and shifted, pulses are shown in figures 4.8 and 4.9, respectively.

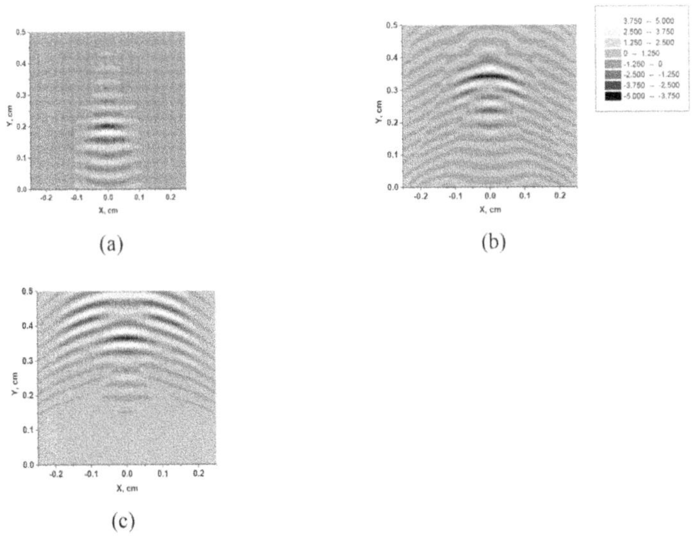

Figure 4.5. The real part of the nonlinear, non-stationary, magnetostatic field, Re(ϕ). $N = -1$; a single focused non-shifted beam, $x_{i0} = 0$, $i = 1,2,3$; time moments are (a) $t = 0.64$ (b) $t = 0.7886$ (c) $t = 0.9286$. Reprinted from Kamenetskii (2008), copyright Transworld Research Network.

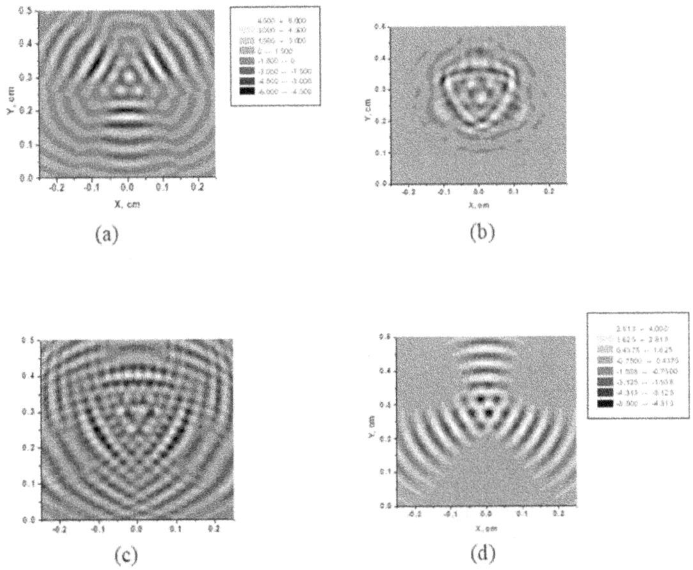

Figure 4.6. The real part of the nonlinear (a–c) and linear (d) non-stationary magnetostatic field, Re(ϕ). $N = -1$; 3 focused beams; no beam shift, $x_{i0} = 0$, $i = 1,2,3$. Time moments are equal to (a) $t = 0.64$ (b) $t = 0.7886$ (c and d) $t = 0.9286$. $t = 0.7886$ corresponds to a coincidence of maxima of the three pulses at the 'interaction center' point O. Reprinted from Kamenetskii (2008), copyright Transworld Research Network.

Finally, the spatial distribution of the Poynting vector in the central part of the FF, while figures 4.10(a) and (b) contours of constant phase in the close vicinity to the central point (point O) are shown.

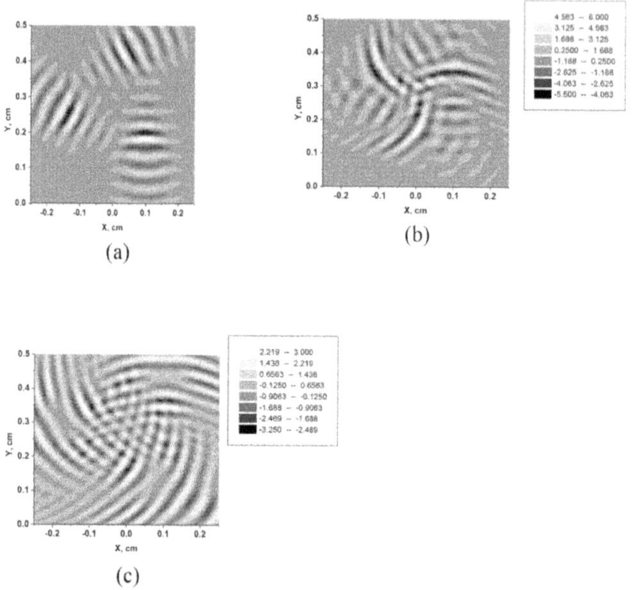

Figure 4.7. The real part of the nonlinear non-stationary magnetostatic field, Re(ϕ). $N = -1$; 3 focused beams; beams are shifted, $x_{i0} = 0.1, i = 1,2,3$. Time moments are (a) $t = 0.64$ (b) $t = 0.7886$ (c) $t = 0.9286$. $t = 0.7886$ corresponds to achieving a coincidence of maxima of the three pulses at the 'interaction center' point O. Reprinted from Kamenetskii (2008), copyright Transworld Research Network.

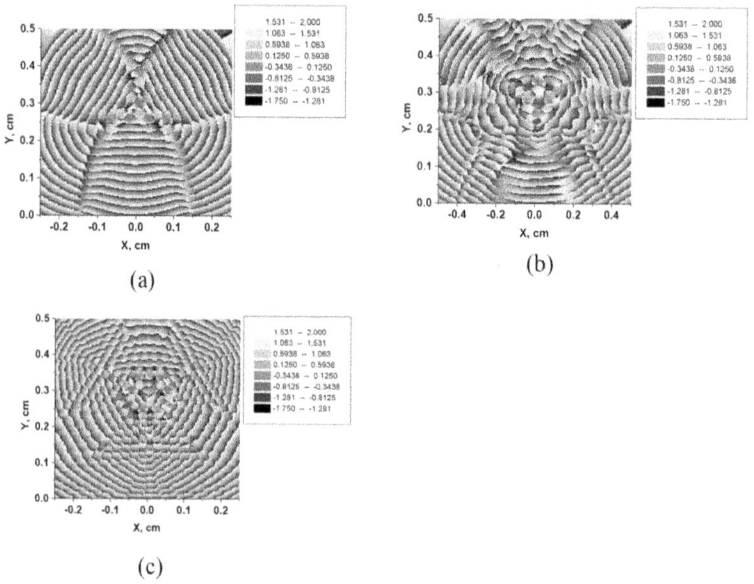

Figure 4.8. The phase of the nonlinear non-stationary magnetostatic field, Φ; $N = 1$; three focused, non-shifted beams; $x_{i0} = 0.$, $i = 1,2,3$. Time values: (a) $t = 0.64$ (b) $t = 0.7886$ (c) $t = 0.9286$. $t = 0.7886$ corresponds to achieving a coincidence of maxima of the three pulses at the 'interaction center' point O. Reprinted from Kamenetskii (2008), copyright Transworld Research Network.

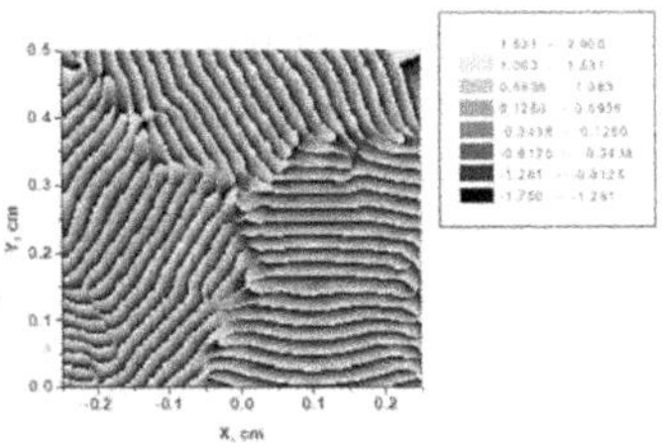

(a)

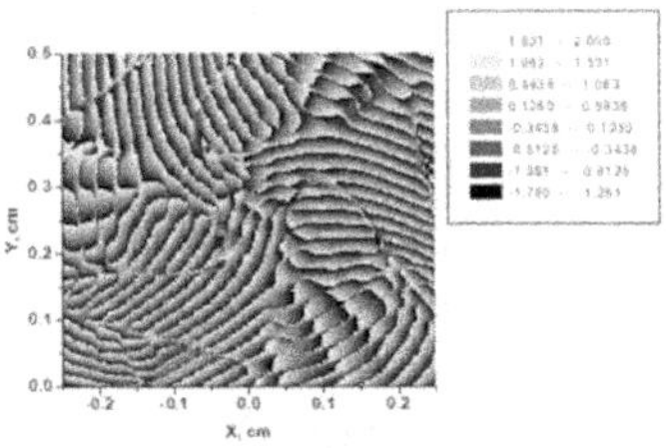

(b)

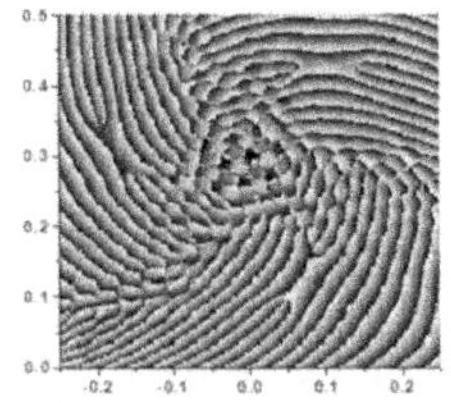

(c)

Figure 4.9. The phase of the nonlinear non-stationary magnetostatic field, Φ; $N = -1$; 3 focused shifted beams; $x_{i0} = 0.$, $i = 1,2,3$. Time values are (a) $t = 0.64$ (b) $t = 0.7886$, (c) $t = 0.9286$. $t = 0.7886$ corresponds to achieving a coincidence of maxima of the three pulses at the 'interaction center' point O. Reprinted from Kamenetskii (2008), copyright Transworld Research Network.

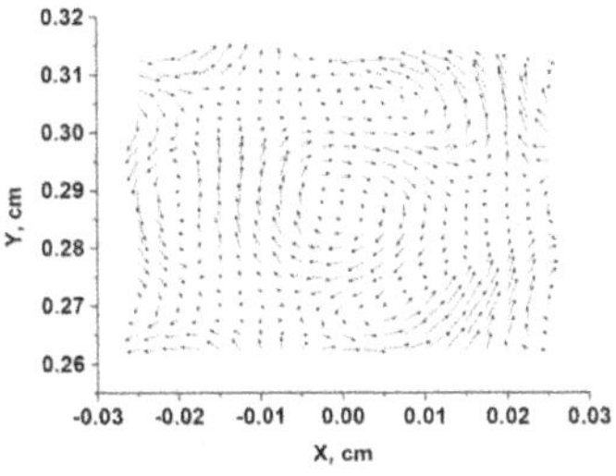

(a)

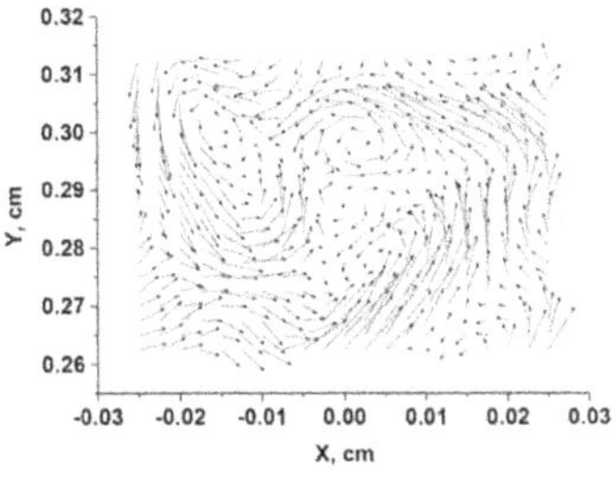

(b)

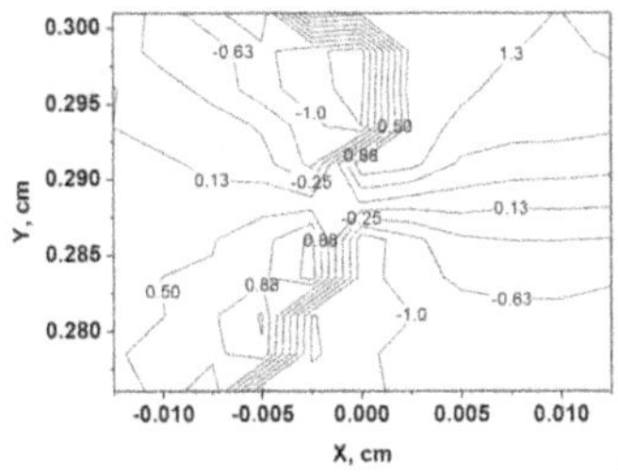

(c)

Figure 4.10. The Poynting vector structure in the region near the 'central point' for three focused nonlinear (a) non-shifted pulses (b) shifted pulses; (c) in the contours of constant phases in the close vicinity of the central point, O. All non-dimensional amplitudes are equal to 3. All figures (a–c) are built for $t = 0.7886$. This moment of time corresponds to achieving a coincidence of maxima of the three pulses at the 'interaction center' point O. Reprinted from Kamenetskii (2008), copyright Transworld Research Network.

As was shown in Boardman *et al* (2005) the analytical consideration of a phase integral along the contour surrounding the central point O, and in the close vicinity of this point, gives a value, equal to 2π in the linear case, for the stationary beams.

Numerical calculations (Boardman *et al* 2005) prove that, for nonlinear FVMSW beams, with equal amplitudes and a corresponding shift of the input phases, this integral is the same as in the linear case. The same is valid for three interacting nonlinear focused pulses of FVMSW, which is illustrated by figure 4.10(c). Therefore the 'vortex charge', both for stationary and non-stationary vortex structures, with the parameters assumed, is equal to 1. Numerical calculations also show that the amplitude of the magnetostatic potential, with proper numerical accuracy tends to zero at the central 'interaction point' O. This proves that, in particular, at point O, a 'phase defect' is created, and a vortex structure is formed around this point.

4.2.4 Conclusions to section 4.2

The interaction of three nonlinear pulses of FVMSW propagating at angles 120° relative to each other, and having the same relative phase shift is considered. A possibility of creating a vortex-like structure with a phase defect in the 'center of interaction' in ferromagnetic films is shown, accounting for 2D nonlinear diffraction and nonlinear interaction of FVMSW. For pulses with different amplitudes and a transverse shift of their centers, phase singularities could be also observed. Investigation of the MSW vector energy flow proves the presence of vortices, at least, at the interaction center, point O. It is shown that nonlinearity leads to broadening of the phase defect vortex structure.

4.3 Formation and propagation of the bullets in the gyrotropic waveguides accounting for higher-order nonlinearities

The possibility of forming finite-amplitude magnetic bullets that can collapse to reach very large values in finite time was demonstrated theoretically first in Grimalsky *et al* (1997). Figures 4.11(a) and (b) illustrate a highly focused bullet of a large but finite amplitude in a wide (with almost infinite width) gyrotropic layer/FF and show the possibility of reducing its amplitude by using the FF of a finite width, respectively. To get these results, equations (3.30), (3.31) of chapter 3 have been used. It is shown theoretically (Bauer *et al* 1998, Buttner *et al* 2000b) that two-dimensional nonlinear self-localized wave packets (magnetic bullets) can be formed in FFs due to the combined influence of nonlinearity, dispersion, diffraction and damping dissipation. Moreover, the damping time of the bullets is sufficient for their reliable observation, which was carried out for the first time by a group of co-authors of experimenters using the light scattering of Brillouin on magnetic perturbations in FFs in Bauer *et al* (1998) and Buttner *et al* (2000a). In Bauer *et al* (1998) and Buttner *et al* (2000b), for the first time, a comparison was between the corresponding theory, and, with the results of the above observations. The cross-sectional area of the bullets with typical parameters for BVMSW at their maximum focusing decreases approximately four times with increasing the power input from ~ 40 mW to 750 mW (Bauer *et al* 1998, Buttner *et al* 2000b).

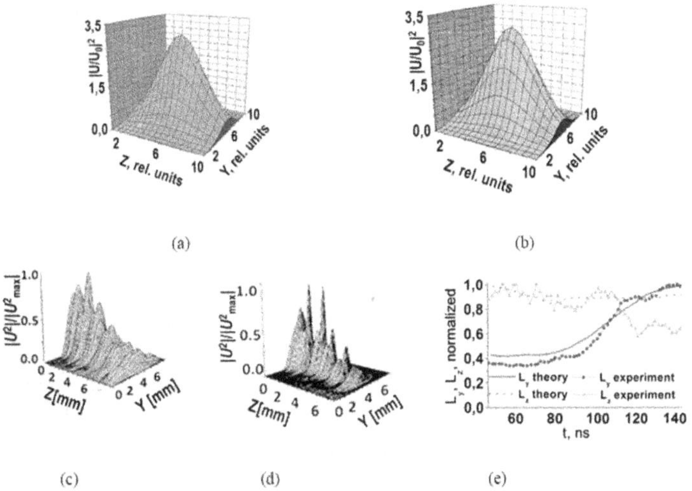

Figure 4.11. The formation of (2 + 1) of spatial-time solitons (magnetic bullets) of BVMSW in gyrotropic layers (FF) (Bauer *et al* 1998, Buttner *et al* 2000a, 2000b); (a) and (b) are the first theoretical demonstrations of the possibility of forming a magnetic bullet of a finite amplitude (Grimalsky *et al* 1997) in FFs with very large (almost infinite) (a) (Grimalsky *et al* 1997) and finite (b) (Grimalsky *et al* 1997) widths; (c) and (d) (reprinted with permission from Buttner *et al* 2000b, copyright (2000) by the American Physical Society) are theoretical calculations of the spatial distributions of the BVMSW intensity in the FF generated by a 29 ns pulse for five consecutive time points (30, 65, 90, 125, and 150 ns for successive peaks, starting from the second one); (c) linear (input power $P_{in} = 10$ mV; $|U_0/U_{thr}|^2 = 0.25$) and (d) nonlinear ($P_{in} = 460$ mV; $|U_0/U_{thr}|^2 = 6$) modes (reprinted with permission from Butner *et al* (2000b), copyright (2000) by the American Physical Society); the bullet is formed at a time of $T \sim 50$ ns and at a distance of ~2 mm from the input of the system; (e) is a comparison of theoretical calculations with the results of experimental observations (led by a team of co-authors (Bauer *et al* 1998, Buttner *et al* 2000a) using the Brillouin light scattering on magnetostatic perturbations) for the effective half-widths of the bullet in transverse and longitudinal directions $L_{y,z}$ respectively, determined by half of intensity.

The two-dimensional spatio-temporal soliton is not stable, but due to the combined effect of nonlinearity, dispersion, diffraction, and dissipation, magnetic bullets are quasi-stable and observed over time. The linear stage of propagation is shown in figure 4.11(c). Figure 4.11(d) shows the self-focusing, the formation of a two-dimensional pulse of BVMSW, propagation, and the onset of the decay of the nonlinear pulse. The results of modeling and observations agree well with each other (figure 4.11(e)). Then, for simulating the propagation of magnetic bullets in an FF, the Lagrangian formalism method was used (Buttner *et al* 2000b, Kivshar and Agraval 2003). Although in general the experiment, exact calculation and Lagrangian formalism are in agreement, the exact calculation using the description by the nonlinear Schrödinger equations (NSE) (Bauer *et al* 1998, Buttner *et al* 2000b) is more consistent with the results of direct observation of nonlinear wave structures in FFs by the Brillouin scattering, than the Lagrangian formalism. This concerns, in particular, the time in a range of 55–110 ns, when developed bullets exist (figure 4.11(e), see also Bauer *et al* 1998, Buttner *et al* 2000b). The calculations are less relevant to the experiment for relatively small values of time when the boundary effects associated with the input antenna are significant. The antenna also interferes with the observation of light scattering at the magnetic perturbations at the input to the system. For times greater than 120 ns, when the bullet collapses, the pulse expands and the pulse 'feels' the boundaries of the waveguide. Then the qualitative correspondence of theory and experiment is preserved, and the quantitative one decreases (see figure 4.11(e), as well as Bauer *et al* 1998, Buttner *et al* 2000b).

Consider the effect of nonlinear dispersion and diffraction on the propagation of bullets based on the solution of equation (3.34). As shown qualitatively in figure 4.12, for selected (phenomenologically) parameters when forming bullets in a 'model nonlinear medium' (e.g., in an FF or in an MM nonlinear waveguide), in particular before reaching maximum focusing when $Z = 6$, the effect of the nonlinear dispersion is most noticeable for the relative weak 'secondary' pulse that is extended in a direction transverse to the propagation direction. The 'secondary' pulse is ahead (figure 4.12(a)), or follows (figure 4.12(c)) a two-dimensional 'main' pulse, depending on whether the coefficient of the nonlinear dispersion b_{10} is negative (figure 4.12(a)) or positive (figure 4.12(c)), respectively. If $b_{10} = 0$(figure 4.12(b)), the secondary pulses are symmetrical along the propagation direction with respect to the main pulse. The following is an example where the secondary peak is larger compared to the main peak than for the case illustrated in figure 4.12. The width of the FF is 10 mm and the antenna length is 5 mm. The input pulse amplitude (the amplitude at the input to the FF, $z = 0$) $U_{\mathrm{inp}} \equiv U(z = 0)$ is characterized by a value $A \equiv A_{\mathrm{inp}} = U_{\mathrm{inp}}/U_0$. The scale of the amplitude U_0 is determined in such a way that $N = 1$. In this case $U_0^2 = U_{\mathrm{thr}}^2/\gamma$, where $U_{\mathrm{thr}}^2 \equiv N_0/\gamma$, and $A_{\mathrm{thr}} = \sqrt{k_{\mathrm{thr}}\gamma}$, $k_{\mathrm{thr}} \equiv |U_{\mathrm{inp}}/U_{\mathrm{thr}}|^2$ (Slavin *et al* 2003, Buttner *et al* 2000b). The calculation for the case with significant secondary pulses is shown in figure 4.13. The input pulse splits into three main pulses in the propagation direction (Rapoport *et al* 2012). If there are no nonlinear effects of higher orders ($b_{10,20} = 0$, see figure 4.13(a)), then the second and third (in time) pulses have comparable amplitudes, and for pulses influenced by higher order effects ($b_{10,20} \neq 0$, figure 4.13(b)) the amplitude of the main pulse is much larger than the amplitude of

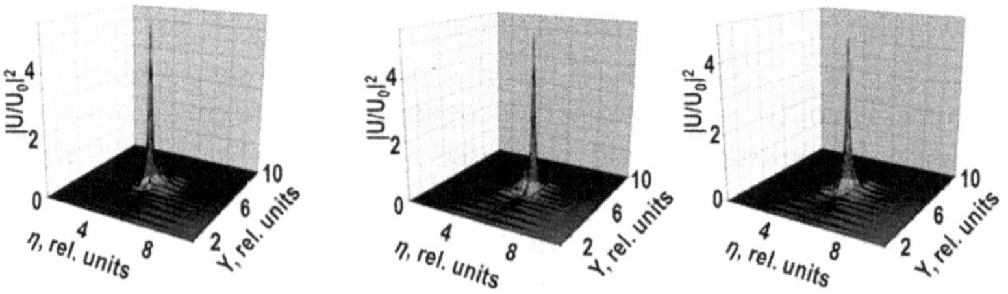

(a) (b) (c)

Figure 4.12. The formation of bullets in the 'model nonlinear medium' with the appearance of the other nonlinear terms; $Z = 6$ (Z is normalized distance from the input of the FF); normalized coefficient of nonlinear diffraction $b_{20} = 0.003$ and dispersion (a) is for $b_{10} = 0.133$; (b) is for $b_{10} = 0$; (c) is for $b_{20} = -0.133$, for a case of a model medium; the scale of the normalization is arbitrary and not connected with definite medium.

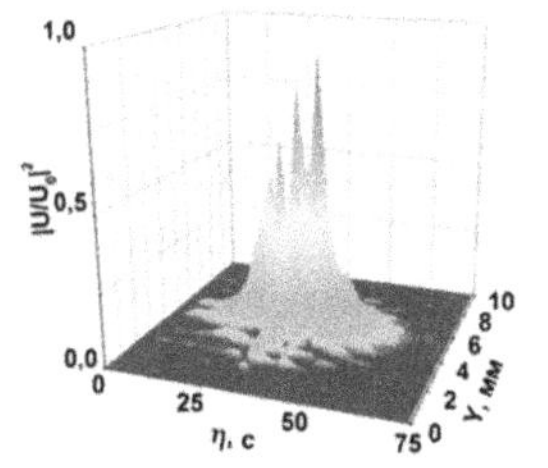

(a)

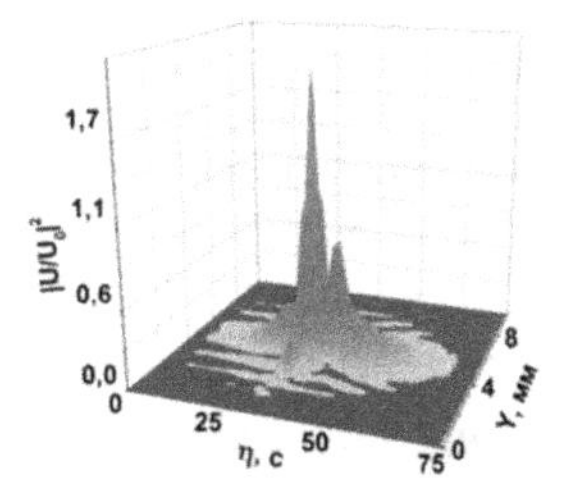

(b)

Figure 4.13. Evolution and the formation of the wave structures from the input pulse to the bell-like shape in the (YIG) FF (Rapoport *et al* 2012); input amplitude corresponds to $k_{\text{thr}} = 15$; the distance from input antenna = 3 mm; the dimensional values $\eta = t - z/|V_g|$, y are shown; (a) nonlinear effects of a higher order (nonlinear diffraction and dispersion) are absent, $b_{10,20} = 0$; (b) nonlinear diffraction and dispersion are present, $b_{10,20} \neq 0$.

the secondary pulses. Note that for $b_{10,20} = 0$ (figure 4.13(a)), the second pulse splits into the transverse direction (Y), unlike the situation when $b_{10,20} \neq 0$ (figure 4.13(b)).

Note also that in the case of the formation of a pronounced magnetic bullet (spatial-time soliton with relatively large amplitude and small cross-section) (Buttner *et al* 2000b) (figure is not shown here), it can become separated in a longitudinal direction with a monotonic (with distance) change with amplitude due to the presence of nonlinear dispersion. The influence of higher-order nonlinear effects on the instability in the propagation of bullets in MM waveguides and on the propagation of magnetooptic solitons in controlled metamaterials is discussed in chapter 7, section 7.2 with appendix E.1.

4.4 Application of the method NEELS for the propagation of the waves in the linear waveguide Earth-Ionosphere

Consider, in particular, the transverse magnetic (TM) mode with components H_y, E_z, E_x, where the electromagnetic wave propagates along the axis X, Z is directed vertically upwards, and consider a two-dimensional problem, $\partial/\partial y = 0$. Using the NEELS method, we arrive at the relation (Rapoport *et al* 2006):

$$\frac{\partial}{\partial t}\Phi = \int_{X_1}^{X_2} \Delta k_x(x)\mathrm{d}x$$

$$= -\frac{\omega}{16\pi V_g}\int_{X_1}^{X_2}\left[\frac{\displaystyle\int_{Z_1}^{Z_2}(\vec{F}_E^{*}\Delta\hat{\varepsilon}\vec{F}_E)\mathrm{d}z}{\displaystyle\int_0^{Z_{\max}}(\vec{F}_E^{*}\frac{\partial}{\partial\omega}(\hat{\varepsilon}\omega)\vec{F}_E + \vec{F}_H^{*}\vec{F}_H^{*})\mathrm{d}z}\right]\mathrm{d}x \tag{4.16}$$

In (4.16), $\omega = 2\pi f$; f, V_g are the frequency and group velocity of the electromagnetic wave, respectively; $\Delta\hat{\varepsilon}$, Δk_x,Φ are perturbation (due to seismogenic processes) of the dielectric permittivity, X-components of the wave vector of the electromagnetic wave, and integral change of the complex phase of the electromagnetic wave in the 'Earth-Ionosphere' waveguide, respectively; X_1, X_2, Z_1, Z_2 are the coordinates of the region where the perturbation of the environment parameters takes place, $\Delta X = X_2 - X_1$, $\Delta Z = Z_2 - Z_1$ are characteristic dimensions of the perturbation region in the directions X, Z, respectively; Z_{max} is the characteristic width of an equivalent waveguide for electromagnetic waves; $\vec{F}_{E,H}$ are functions describing the magnetic and electrical components of the electromagnetic wave field, respectively. The relations (4.16) can be used to estimate the seismogenic change of the complex phase of an electromagnetic wave in the Earth-Ionosphere waveguide (Rapoport *et al* 2006).

4.5 Conclusions to chapter 4

(1) The formation and behavior of a wave packet/space-time soliton/magnetic bullets in nonlinear gyrotropic layers/FF is investigated and explained for the first time in Buttner *et al* (2000b), Atkinson and Kubrakov (2002). In this case, the wave packet, which has a rather high initial amplitude at the input of the system, begins to converge, and its amplitude increases. The theory (Berge 1998) predicts that in a two-dimensional case a stable equilibrium between dispersion, diffraction, and nonlinearity is not possible, and the nonlinear self-focusing of a wave packet with a sufficiently high initial energy should lead to a wave collapse when all the energy of the packet is concentrated near one point. We have shown for the first time for a gyrotropic nonlinear layer that in a real medium with dissipation, the collapse is avoided because the wave packet loses energy. Thus, in some range of propagation distances, the nonlinear collapse is stabilized by dissipation and a quasi-stable highly localized two-dimensional wave packet, the so-called 'magnetic bullet', is formed. In contrast, the existence of stable two- and three-dimensional wave packets in focusing media, where the collapse is stabilized by saturation of nonlinearity, was predicted for optical wave packets in Silberberg (1990).

(2) It is first shown that in the case of simultaneous influence of nonlinear dispersion and diffraction on the propagation of pulses in a gyrotropic layered medium, the nonlinear dispersion has a complex effect on the formation of nonlinear structures in the transverse direction, and nonlinear diffraction, in turn, influences the formation of the structure's additional peaks in the longitudinal direction. Thus, the nonlinear dispersion and diffraction possess a complex effect on the stability and quasi-stable state of nonlinear structures.

(3) Nonlinear vortex structures can be excited in an FF on the base of the interaction of three beams with relative phase differences between adjacent beams which equals $2\pi/3$ (between the second and first beams and between the third and second ones). For beams with unbiased centers (with symmetric geometry), quasi-periodic wave structures with vortex singularities and a topological charge equal to 1 or -1 are formed. In the absence of nonlinearity, slightly different amplitudes of interacting beams do not, in

general, change the symmetry of the structures, but nonlinearity, jointly with a small difference in the amplitudes of the beams, can change the symmetry. Possibilities of formation of non-stationary nonlinear vortex structures at the interaction of three waves with corresponding mutual phase shifts in the GL are shown.

Problems to chapter 4

Problem 4.1

Derive relations (4.7b) for components of the Poynting vector $\boldsymbol{P}$ in cylindrical coordinates (Θ, r) magnetostatic waves (FMSW) with a magnetostatic potential φ in FF, namely:

$$P_r = -\frac{c}{8\pi}\mathrm{Re}\left[i\omega\varphi\left(i\mu_q\frac{1}{r}\frac{\partial\varphi^*}{\partial\Theta} + \mu\frac{\partial\varphi^*}{\partial r}\right)\right],\ P_\Theta = -\frac{c}{8\pi}\mathrm{Re}\left[i\omega\varphi\left(-i\mu_q\frac{\partial\varphi^*}{\partial r} + \mu\frac{1}{r}\frac{\partial\varphi^*}{\partial\Theta}\right)\right]$$

Use, to determine the pointing vector for FMSW, the following relation (Boardman *et al* 2005, Akhiezer *et al* 1967)

$\vec{P} = (c/8\pi)\mathrm{Re}(\varphi\partial\vec{b}^*/\partial t)$, where $\vec{b}$ is the magnetic induction which characterized transverse magnetooptic effect corresponding to FMSW.

Problem 4.2

Normalize the temporal modified NSE in the running coordinate system in the form

$$i\frac{\partial A}{\partial z} - \frac{\beta_2}{2}\frac{\partial^2 A}{\partial T^2} + \gamma\,|\,A\,|^2\,A - \frac{i\gamma}{\omega_0}\frac{\partial}{\partial T}(|\,A\,|^2\,A) - \frac{\omega}{C}\varepsilon_{xy}A = 0 \qquad \text{(P4.1)}$$

In equation (P4.1) A is an envelope amplitude of electromagnetic wave packet propagating in optical waveguide β_0, γ, γ/ω_0 are coefficients of dispersion, self-sleeping, and cubic nonlinearity, respectively; z and T are coordinate and time in the coordinate frame moving with the group velocity, respectively.

Equation (P5.1) corresponds to one presented in Agrawal (1996) with the edition of the last term describing the transverse magnetooptic (Voigt) effect (Boardman *et al* 2010a, 2010b). Use the normalization with the scales (Agrawal 1996)

$$A = \sqrt{P_0}u, z = L_0\xi,\ T = T_0 t, \qquad \text{(P4.2)}$$

where T_0 is characteristic time scale (for example, a pulse duration), $L_D = T_0^2/|\,\beta_2\,|$ is the diffraction length, P_0 is a value proportional to the field intensity.

Obtain the modified NSE, corresponding to equation (4.1), in the normalized form. Use the normalization based on the scales (P4.2). Present

$$i\frac{\sqrt{P}}{L_D}\frac{\partial U}{\partial\xi} + \frac{|\,\beta_2\,|}{2}\frac{\sqrt{P_0}}{T_0^2}\frac{\partial^2 U}{\partial t^2} + \gamma P_0\sqrt{P_0}\,|\,U\,|^2\,U + \frac{i\gamma_0}{\omega_0}\frac{P_0\sqrt{P_0}}{T_0}\frac{\partial}{\partial t}(|\,U\,|^2\,U) - \frac{\omega}{C}\varepsilon_{xy}\sqrt{P_0}\,U = 0 \qquad \text{(P4.3)}$$

Note that it is supposed that $\beta_2 < 0$.

Cancel $\sqrt{P_0}$ in equation (P4.3) and multiply by L_D:

$$i\frac{\partial U}{\partial \xi} + \frac{L_D}{2}\frac{\beta_2}{T_0^2}\frac{\partial^2 U}{\partial t^2} + t_0\gamma P_0 \mid U \mid^2 U + \frac{i\gamma}{\omega_0}\frac{P_0 L_D}{T_0}\frac{\partial}{\partial t}(\mid U \mid^2 U) - \frac{\omega}{C}\varepsilon_{xy} L_D U = 0 \quad \text{(P4.4)}$$

Take in to account that $L_D = T_0^2/\mid \beta_2 \mid$, and denote

$$L_{\mathrm{NL}} = \frac{1}{\gamma P_0} \quad \text{(P4.5)}$$

Accounting for (P4.5), reduce (P4.4) to the form:

$$i\frac{\partial U}{\partial \xi} + \frac{1}{2}\frac{\partial^2 U}{\partial t^2} + \frac{L_D}{L_{\mathrm{NL}}} \mid U \mid^2 U + \frac{i}{\omega_0 T_0}\frac{L_D}{L_{\mathrm{NL}}}\frac{\partial}{\partial t}(\mid U \mid^2 U) - \frac{\omega}{C}\varepsilon_{xy} L_D U = 0 \quad \text{(P4.6)}$$

Renormalize the envelope amplitude:

$$U = \frac{\sqrt{L_{\mathrm{NL}}}}{L_0}\psi \quad \text{(P4.7)}$$

Accounting for (P4.7) present equation (P4.6) in the following normalized form, finally:

$$i\frac{\partial \psi}{\partial \xi} + \frac{1}{2}\frac{\partial^2 \psi}{\partial t^2} + \mid \psi \mid^2 \psi + \frac{1}{\omega_0 T_0}\frac{\partial}{\partial t}(\mid \psi \mid^2 \psi) - \left(\frac{\omega}{C}\varepsilon_{xy} L_D\right)\psi = 0 \quad \text{(P4.8)}$$

In the absence of the last two terms, equation (P4.8) reduces to the canonical NSE (Agrawal 1996).

Problem 4.3

NSE in the system with running spatial coordinate $\xi = z - V_g t$, (where z and V_g are the spatial and time coordinates in laboratory system and group velocity, respectively) accounting the relations

$$\tau_0 = t,\ \frac{\partial}{\partial z} = \frac{\partial}{\partial \xi},\ \frac{\partial}{\partial t} = \frac{\partial}{\partial \tau_0} - V_g\frac{\partial}{\partial \xi} \quad \text{(P4.9)}$$

has the form:

$$i\frac{\partial E_0}{\partial \tau_0} + \frac{1}{2}V_g\frac{\partial^2 E_0}{\partial \xi^2} - \frac{\partial \omega}{\partial_0 \mid E_0 \mid^2} \mid E_0 \mid^2 E_0 = 0 \quad \text{(P4.10)}$$

In equation (P4.10) E_0 in envelope amplitude of the propagating nonlinear wave packet.

Equation (P4.10) can be rewritten in the form

$$i\frac{\partial E_0}{\partial \tau_0} + \frac{1}{2}V'_g\frac{\partial^2 E_0}{\partial \xi^2} - \alpha \mid E_0 \mid^2 E_0 = 0;\ V'_g \equiv \frac{\partial V_g}{\partial k} \quad \text{(P4.11)}$$

where $\alpha = \dfrac{\partial\omega}{\partial \mid E_0 \mid^2}$ is the coefficient of cubic nonlinearity. Note that during the derivation of equation (P4.10), the following expansion of the frequency of the wave packet has been used (Kadomtsev 1988) ($\mid \vec{\chi} \mid \ll \mid \vec{k}_0 \mid$):

$$\begin{aligned}&\omega_{\vec{k}_0+\vec{\chi}}\\&= \omega_{\vec{k}_0} + (\mid \vec{k}_0 + \vec{\chi} \mid - \vec{k}_0)V_g + \frac{1}{2}(\mid \vec{k}_0 + \vec{\chi} \mid - \vec{k}_0)^2 V'_g\\&+ \frac{\partial\omega}{\partial \mid E_0 \mid^2} \mid E_0 \mid^2 + \ldots = 0\end{aligned} \quad \text{(P4.12)}$$

Note that relation (P4.12) is written under for the isotropic media (Kadomtsev 1988).

In optics/for optical MMs, as a rule the other expansion is used, namely.

$$k_{\omega 0+\Delta\omega} = k_{\omega_0} + \Delta\omega\frac{\partial k}{\partial\omega} + \frac{1}{2}\Delta\omega^2\frac{\partial^2 k}{\partial\omega^2} + \frac{\partial k}{\partial \mid E_0 \mid^2} \mid E_0 \mid^2 + \cdots \quad \text{(P4.13)}$$

It is supposed, when relation (P4.13) is written, that $\Delta\omega \ll \omega_0$; ω_0 and k_{ω_0} are the carrier frequency and wavenumber of propagating in the MM waveguide with the relatively narrow spectrum width. Use the expansion (P4.13) and account for higher order nonlinear and transverse magnetooptic (Voigt) effects. The following evolution equation for the envelope (electromagnetic field) amplitude A_{Ex} of the TEM electromagnetic packet with moderate spectrum width in the MM waveguide can be derived (compare with equation (3.53) in chapter 3):

$$\begin{aligned}&\left\{i\frac{\partial}{\partial\xi} - \frac{1}{2}\frac{\partial^2 k}{\partial\omega^2}\frac{\partial^2}{\partial\tau^2} + \frac{1}{2k'}\frac{\partial^2}{\partial x^2}\right\}A_{E_x} + i\frac{(\omega/c)^2}{2k'}(\varepsilon'\mu'' + \varepsilon''\mu')A_{E_x}\\&+ \tilde{c}_s \mid A_{E_x} \mid^2 A_{E_x} + i\tilde{c}_s\frac{\partial}{\partial\tau}(\mid A_{E_x} \mid^2 A_{E_x})\\&+\frac{2\pi\chi_E}{k'\varepsilon'}\frac{\partial^2}{\partial x^2}(\mid A_{E_x} \mid^2 A_{E_x}) - \frac{\omega}{c\langle\varepsilon_{yx}\rangle}A_{E_x} = 0\end{aligned} \quad \text{(P4.14)}$$

Here $\varepsilon' = \mathrm{Re}(\varepsilon)$; $\varepsilon'' \equiv \mathrm{Im}(\varepsilon)$; $\mu' \equiv \mathrm{Re}(\mu)$; $\mu'' \equiv \mathrm{Im}(\mu)$; $k' \equiv \mathrm{Im}(k)$; μ, ε and k are magnetic permeability, electric permittivity and the carrier wavenumber of the wave packet, respectively ε_{xy} is the component of the electric permittivity tensor which characterized transverse magnetooptic effect; χ_E is the coefficient of electric cubic nonlinearity. In equation (P4.14) the second, third and fourth terms in the left-hand part describe the effects of the linear dispersion and diffraction and the linear losses, respectively; the fifth term is due to nonlinearity, the last ones describe the nonlinear dispersion, nonlinear diffraction, and magnetooptic effects. The brackets $\langle\rangle$ in the last term in the left hand side of equation (P4.14) mean the averaging of the corresponding term over the normal to the surfaces of (metamaterial) waveguide, where the electromagnetic wave packet propagates. The expression for $\widetilde{C}_s$ is $\widetilde{C}_s = C_{ne}\left(\frac{1}{\mu_c}\frac{\partial\mu}{\partial\omega}\bigg|_{\omega_c} + \frac{2}{\omega_c} - \frac{1}{V_{gc}k'_c}\right)$. Put that ($x$-component of the) nonlinear electric polarization has the form:

$$P_{\mathrm{NLX}} = \chi_E \mid A_{E_x} \mid^2 A_{E_x} \tag{P4.15}$$

Find the expression for the coefficient $\tilde{C}_{ne}$ in equation (P4.14)

Use the equation (P4.14), expression (P4.15) and take into account that

$$\tilde{c}_{ne} = \frac{\partial k}{\partial \mid A_{E_x} \mid^2} \tag{P4.16}$$

And dielectric permittivity ε can be presented as the sum of linear ($\varepsilon_{\mathrm{lin}}$) and nonlinear ($\varepsilon_{\mathrm{NL}}$) parts:

$$\varepsilon = \varepsilon_{\mathrm{lin}} + \varepsilon_{\mathrm{NL}} = \varepsilon_{\mathrm{lin}} + 4\pi\chi_E \mid A_{E_x} \mid^2 \tag{P4.17}$$

Note also that for the quasi-planar wave the following expression for the carrier wavenumber k is valid approximately:

$$k^2 \simeq \frac{\omega^2}{c^2}\varepsilon\mu \tag{P4.18}$$

Accounting for (P4.17) and (P4.18), one can write

$$2\Delta k_{\mathrm{NL}} k = \Delta_{\mathrm{NL}}(k^2) \simeq \frac{\omega^2}{c^2}\mu\varepsilon_{\mathrm{NL}} = \frac{\omega^2}{c^2}4\pi\chi_E \mid A_{E_x} \mid^2 \tag{P4.19}$$

In relation (P4.19) Δk_{NL} and $\Delta_{\mathrm{NL}}(k^2)$ are the nonlinear additions to the wavenumber and square of wavenumber, respectively, and, as follows from (P4.16) and (P4.19),

$$\tilde{c}_{ne} = \frac{\partial \Delta k_{\mathrm{NL}}}{\partial \mid A_{E_x} \mid^2} = \frac{2\pi}{kc}\frac{\omega^2}{c^2}\mu\chi_E \tag{P4.20}$$

Problem 4.4

Obtain, basing on equation (P4.14) (see problem 4.3), the corresponding normalized equation. To close, use the following normalization of the values included in equation (P4.14):

$$\begin{gathered} \xi = L_{\mathrm{Disp}}\bar{\xi},\ x = \bar{x}D_0,\ \tau = T_0\bar{\tau},\ L_{\mathrm{Disp}} = T_0^2/\left| \beta_2 \right|, \\ L_{\mathrm{Difr}} = \mid k'_{\mathrm{c}} \mid D_0^2,\ \xi = z,\ \tau = t - z/\left| V_{\mathrm{g}} \right| \end{gathered} \tag{P4.21}$$

In the relation (P4.21) L_{Disp}, L_{Difr}, D_0 and T_0 are corresponding spatial and temporal scales, (L_{Disp} and L_{Difr} are dispersion and diffraction lengths, respectively); $\xi = z,\ \tau = t - z/\left| V_{\mathrm{gc}} \right|$, t and τ are time in the system moving with group velocity V_{gc}, respectively, V_{gc} is the group velocity at the carrier frequency of the wave packet, $\beta_2 = \partial^2 k/\partial\omega^2 \mid_{\omega_{\mathrm{c}}}$ is dispersion coefficient. ω_{c} and k_{c} are the carrier frequency and wavenumber, respectively, $k'_{\mathrm{c}} \equiv \mathrm{Re}(k_{\mathrm{c}})$. Use also the following normalization for the envelope amplitude:

$$A = \psi A_0;\ L_{\text{Disp}} \mid \tilde{c}_{ne} \mid A_0^2 = 1 \tag{P4.22}$$

Using the normalization described by relations (P4.21) and (P4.22), present equation (P4.14) in the following normalized form:

$$\left[i\frac{\partial}{\partial \xi} - \frac{1}{2}\text{sgn}(\beta^2)\frac{\partial^2}{\partial \tau^2} + \frac{1}{2}\text{sgn}(k'_{\text{c}})\frac{L_{\text{Disp}}}{L_{\text{Difr}}}\frac{\partial^2}{\partial \overline{x}^2} + i\gamma_0\right]\psi + \text{sgn}(\tilde{c}_{ne})\mid \psi \mid^2 \psi$$
$$+ i\,\text{sgn}(\tilde{c}_{ne})b_\tau \frac{\partial}{\partial \overline{\tau}}(\mid \psi \mid^2 \psi) + K\frac{\partial^2}{\partial \overline{x}^2}(\mid \psi \mid^2 \psi) - \nu\psi = 0 \tag{P4.23}$$

In equation (P4.23),

$$K = \frac{2\pi\chi_E}{k'\varepsilon' D_0^2 \mid \tilde{c}_{ne} \mid} = \text{sgn}(k'_{\text{c}})\text{sgn}(\chi_E)\text{sgn}(\varepsilon')\frac{1}{D_0^2\left|\varepsilon'\mu'\left(\frac{\omega}{c}\right)^2\right|} \tag{P4.24}$$

$$K = L_{\text{Disp}}\frac{\omega}{c}\varepsilon_{xy} \tag{P4.25}$$

$$\tilde{c}_S = \left\{\frac{1}{\omega}\left[\left(2 + \frac{\omega}{\mu}\frac{\partial \mu}{\partial \omega} - \frac{\omega}{V_g k'}\right) + \frac{\omega}{x_E}\frac{\partial \dot{x}_E}{\partial \omega}\right]\right\}_{\omega=\omega_{\text{c}}} \tag{P4.26}$$

$$\text{sgn}\,(\tilde{c}_{ne}) = \text{sgn}\,(\chi_E)\,\text{sgn}\,(k')\,\text{sgn}\,(\mu')\,|_{\substack{\mu'<0\\ k'<0}} = \text{sgn}\,(\chi_E) \tag{P4.27}$$

Note that relation (P4.27) concerning MMs in the regime of the negative phase medium.

Problem 4.5

Obtain the necessary condition of the bullet formation, in particular for quasi-TEM electromagnetic wave packet in an MM waveguide in the regime of negative phase medium. Use equation (P4.23),

In the regime of negative phase medium we have

$$\text{sgn}(k'_{\text{c}}) = -1 < 0,\ \text{sgn}(\mu'_{\text{c}}) = -1 < 0,\ \text{sgn}(\varepsilon'_{\text{c}}) = -1 < 0 \tag{P4.28}$$

In accordance with Rabinovich and Trubetskov (1989), Kadomtsev (1988), Kivshar and Agraval (2003), and Buttner *et al* (2000b), the necessary conditions for a possibilities of bullets formation are as follows:

(i) the criterion of a possibility of self-focusing:

$$(-\text{sgn}(k'_{\text{c}}))(-\text{sgn}(\tilde{\tau}_{ne})) > 0 \tag{P4.29}$$

(ii) the criterion of a possibility of soliton formation in the direction of propagation:

$$[(-\text{sgn}(\tilde{\tau}_{ne}))(\text{sgn}(\beta_2))] > 0 \tag{P4.30}$$

Accounting for (P4.29), (P4.30) and (P4.28) the necessary conditions for a possibility of bullet's formation in MM waveguide (negative phase media) are:

$$\begin{aligned} &\mathrm{sgn}(\tilde{\tau}_{ne}) = \mathrm{sgn}(\chi_E) = -1 \\ &\mathrm{sgn}(\beta_2) > 0 \end{aligned} \quad \text{(P4.31)}$$

List of abbreviations

BVMSW	Backward volume MSW
FVMSWs	Forward volume magnetostatic waves
FF	Ferrite film
GL	Gyrotropic layer
MSW	Magnetostatic wave
NEELS	Nonlinear evolution equations in layered structure
TM	Transverse magnetic
YIG	Yttrium–iron garnet

References

Agrawal G 1996 *Nonlinear Fiber Optics* (Moscow: Mir) p 326

Akhiezer A I, Bar'yakhtar V G and Peletminsky S V 1967 *Spin Waves* (Moscow: Science) p 368 (in Russian)

Atkinson R and Kubrakov N F 2002 Magneto-optical characterization of ferromagnetic ultrathin multilayers in terms of surface susceptibility tensors *Phys. Rev.* B **33** 124414–5

Bauer M, Buttner O, Demokritov S O, Hillebrands B, Grimalsky V V, Rapoport Y G and Slavin A N 1998 Observation of spatiotemporal self-focusing of spin waves in magnetic films *Phys. Rev. Lett.* **81** 3769–72

Boardman A D, Grimalsky V V, Ivanov B, Koshevaya S V, Velasko L, Zaspel C and Rapoport Y G 2005 Excitation of vortices using linear and nonlinear magnetostatic waves *Phys. Rev.* E **71** 026614–24

Boardman A D, Hess O, Mitchell-Thomas R C, Rapoport Y G and Velasco L 2010a Temporal solitons in magnetooptic and metamaterial waveguides *Photonics Nanostruct.–Fundam. Appl.* **8** 228–43

Boardman A D, Mitchell-Thomas R C, King N J and Rapoport Y G 2010b Bright spatial solitons in controlled negative phase metamaterials *Opt. Commun.* **283** 1585–97

Boardman A D, Alberucci A, Assanto G, Rapoport Yu G, Grimalsky V V, Ivchenko V M and Tkachenko E N 2017 Spatial solitonic and nonlinear plasmonic aspects of metamaterials *World Scientific Handbook of Metamaterials and Plasmonics* vol 16 ed E Shamonina and S A Maier (Singapore: World Scientific) ch 10 419–69

Berge L 1998 Wave collapse in physics: principles and applications to light and plasma waves *Phys. Rep.* **303** 259–370

Buttner O, Bauer M, Demokritov S O, Hillebrands B, Kivshar Yu S, Grimalsky V, Rapoport Y, Kostylev M P, Kalinikos B A and Slavin A N 2000a Spatial and spatiotemporal self-focusing of spin waves in garnet films observed by space- and time-resolved Brillouin light scattering *J. Appl. Phys.* **87** 5088–90

Buttner O, Bauer M, Demokritov S O, Hillebrands B, Kivshar Yu S, Grimalsky V, Rapoport Y and Slavin A N 2000b Linear and nonlinear diffraction of dipolar spin waves in yttrium iron garnet films observed by space- and time-resolved Brillouin light scattering *Phys. Rev.* B **61** 11576–87

Damon R W and Eshbach J R 1961 Magnetostatic modes of a ferromagnetic slab *Phys. Chem. Solids* **19** 308–20

Grimalsky V V, Kotsarenko N Y and Rapoport Y G 1996 Nonlinear surface waves in electronic plasma *23rd European Phys. Soc. Conf. Controlled Fusion Plasma Phys. (Kyiv, Ukraine)* **vol 2** p 396

Grimalsky V, Rapoport Y G and Slavin A N 1997 Nonlinear diffraction of magnetostatic waves in ferrite films *J. Phys. IV France* **7C** 1–393

Grimalsky V V and Rapoport Y G 1998 Modulational instability of surface plasma waves in the second-harmonic resonance region *Plasma Phys. Rep.* **24** 980–2

Gurevich A G and Melkov G A 1996 *Magnetization oscillations and waves* (New York: CRS Press) p 464

Kadomtsev B B 1988 *Collective phenomena in plasma* (Moscow: Nauka) p 304

Kamenetskii E O 2008 Electromagnetic, Magnetostatic, and Exchange-interaction Vortices in Confined Magnetic Structures (Kerala: Transworld Research Network) pp 29–44

Kivshar Y S and Agraval G P 2003 *Optical Solitons* (Amsterdam: Elsevier Science Academic Press) p 527

Nye J F and Berry M V 1974 Dislocations in wave trains *Proc. R. Soc. London. Ser. A 336 Phys. Rev. Lett.* **76** 3955

Paul T, Rockstuhl C and Lederer F 2011 Integrating cold plasma equations into the Fourier modal method to analyze second harmonic generation at metallic nanostructures *J. Mod. Opt.* **58** 438–48

Rabinovich M I and Trubetskov D I 1989 *Oscillation and Waves: In Linear and Nonlinear Systems (Mathematics and Its Applications)* (London: Springer) p 598

Rapoport Y G, Gotynyan O E, Ivchenko V N, Hayakawa M, Grimalsky V V, Koshevaya S and Juarez D 2006 Modeling electrostatic–photochemistry seismoionospheric coupling in the presence of external currents *Phys. Chem. Earth* **31** 437–46

Rapoport Y G, Boardman A D, Grimalsky V V, Koshevaya S V, Zaspel C E and Ivanov B 2008 Nonlinear vortex generation by forward volume magnetostatic waves *Electromagnetic, Magnetostatic and Exchange—Interaction Vortices in Confined Magnetic Structure* ed E O Kamenetskii (Kerala, India: Transworld Research Network) pp 29–44

Rapoport Y G and Grimalsky V V 2011 Nonlinear surface 2D plasmons and giant second harmonic generation *Proc. of the Int. Conf. on Days on Diffraction (St. Petersburg, Russia)* pp 168–73

Rapoport Y, Grimalsky V and Zaspel C 2012 Method for the derivation of nonlinear evolution equations in layered structures (NEELS): An example of nonlinear waves in gyrotropic layers *Bull. Kyiv Nat. Taras Shevchenko Univ. Series: Phys.-Math. Sci.* **14/15** 72–6

Rapoport Y G 2013 General method for modeling nonlinear waves in active metamaterial and gyrotropic layered structures *Proc. of Int. Kharkov Symp. on Physics and Engineering of Microwaves, Millimeter and Submillimeter Waves, MSMW'13 (Kharkov, Ukraine)* 253–5

Rapoport Y G 2014 General method for deriving the equations of evolution and modeling of nonlinear waves in layered active media with bulk and surface *Bull. Kyiv Nat. Taras Shevchenko Univ. Series: Phys.-Math. Sci.* **1** 281–8 (in Ukrainian)

Silberberg Y 1990 Collapse of optical pulses *Opt. Lett.* **15** 1282–4

Slavin A N, Buttner O, Bauer M, Demokritov S O, Hillebrands B, Kostylev M P, Kalinikos B A, Grimalsky V V and Rapoport Yu G 2003 Collision properties of quasi-one-dimensional spin wave solitons and two-dimensional spin wave bullets *Chaos* **13** 693–701

Soskin M S and Vasnetsov M V 1998 Nonlinear singular optics *Pure Appl. Opt.* **7** 301

Soskin M, Boriskina S V, Chag V, Dennis M and Desyatnikov A 2016 Singular optics and topological photonics *J. Opt.* **19** 010401

Wang F X, Rodríguez F J and Albers W M *et al* 2009 Surface and bulk contributions to the second-order nonlinear optical response of a gold film *Phys. Rev.* B **80** 233402

Chapter 5

Controllable propagation and reflection of electromangetic waves in layered gyrotropic metamaterial media

5.1 The problems under consideration

Layered gyrotropic metamaterial media have a very attractive property of controllability, in particular, with the help of external electric or magnetic fields. Depending on the geometry, waves propagating in a layered system can also have the property of nonreciprocity. Alternatively, similar properties and controllability can be achieved in the media with artificial gyrotropy and nonreciprocity (Davoyan and Engheta 2019, Kodera *et al* 2018). An excellent method of magneto-optical control of nonlinear optical waves, in particular vortices, in a layered magneto-optical metamaterial medium is the use of a magnetic field with a specially selected inhomogeneity or time dependence. Controlling stationary magneto-optical solitons using an inhomogeneous field and time solitons in a magneto-optical waveguide using a nonstationary magnetic field was proposed, respectively, in Boardman and Velasco (2006b), Boardman *et al* (2010a) and Boardman *et al* (2010b). The control of spatial dissipative vortex solitons in an active magneto-optical medium, described using the complex Ginzburg-Landau (CGLE) equation, with spatially inhomogeneous magnetic field is considered in Boardman and Velasco (2006b) and Kochetov and Tuz (2018). The control of vortex solitons with a targeted change in angular momentum and vortex charge using a special periodic potential was proposed in Kochetov *et al* (2019). In Guixin (2017), the possibility of generation is shown to achieve a significant amplification of the electromagnetic field near meta-atoms using the excitation of localized surface plasmon resonances (which are sensitive to geometry). Thus, it is possible to provide the required nonlinear optical response of plasmonic metamaterials by choosing the geometric properties of meta-atoms.

doi:10.1088/978-0-7503-2336-9ch5

In this section the principles of electromagnetic wave control for propagation and reflection in non-stationary and non-uniform layered gyrotropic and metamaterial (MM) media are considered.

The role of the gyrotropic medium in the physical characteristics of surface waves and reflection from systems including structures with dielectric, ferromagnetic, and MM layers having the properties of 'negative phase medium (NPM)' are shown. This property is controlled by an external magnetic field. The Voigt configuration was considered (Boardman and Xie 2003, Boardman *et al* 2005a, 2006b, 2009a, 2010a, 2010b, 2010c, 2010d), in which the surface waves move in a direction transverse to the magnetic field lying in the plane of the interface between the layers. The surface waves and their dispersion are investigated. The reflection from the system 'isotropic dielectric prism-gyrotropic material-NPM' is investigated. The results are related to the resonances associated with surface excitations. The magnitude of the Goos–Hänchen shift (Mazur and Djafari 1984) has been shown to be associated with resonances in the reflection of pulses. For the first time, the possibility of obtaining both positive and negative Goos–Hänchen shifts by resonant excitation of oppositely directed SWs at the NPM boundary and gyrotropic medium having the same wavenumber values, belonging to the same dispersion branch and able to exist simultaneously is shown. Thus, the non-reciprocal dispersion of surface waves at the NPM-gyrotropic medium boundary is important for reflection. The fundamentally new mode types of electromagnetic waves in the layered system 'semi-infinite ferromagnetic negative phase metamaterial with negative' are investigated. Particular attention is given to the new phenomenon in the physics of surface magnetic polaritons, due to the presence in the layered structure of MM with negative values of electric permittivities and magnetic permeabilities. Namely, there is the existence in one frequency range of surface waves propagating in opposite directions, unlike the known system 'dielectric-ferromagnetic', where at a certain frequency there can be only one magnetic surface polariton. Correspondingly, the resonant 'Goos–Hänchen shift' was studied, for the first time, for the reflection of pulses from the 'ferromagnetic-metamaterial' structure, when the frequency of incident pulses corresponds to forward and backward surface magnetic polaritons, lying on the same dispersion curve. It turns out that, for two resonances corresponding to the same resonance curve, their shifts have different signs. For the first time the evolution equations for electromagnetic waves in MMs are obtained in the presence of Raman (Agrawal 1989) and magnetooptic (MO) (Boardman and Xie 2003, Boardman *et al* 2006b, 2009a, 2010a, 2010b, 2010c, 2010d) interactions under the action of a non-stationary magnetic field, nonlinear dispersion taking into account the dispersion of the nonlinearity coefficient, the third-order linear dispersion, and nonlinear diffraction. We have shown for the first time the ability to control MO solitons due to a non-stationary magnetic field, in particular the compensation due to the MO effect of soliton delay caused by the Raman interaction, the nonlinear dispersion, and the third-order linear dispersion. The possibility of controlling MO solitons in MMs by means of the diffraction control was demonstrated, first (Boardman *et al* 2008, 2010a, 2010c, 2010d). It will demonstrate the ability to effectively control the temporal position of the soliton maximum with the help of the proposed magnetooptical Lagrange function. The fundamental difference between negative-phase MMs and conventional positive-phase materials will be proven in terms of the space-time (2 + 1) solitons, or bullets. Namely, for the formation of bullets, the corresponding materials should have combinations of nonlinearity, dispersion, and diffraction coefficients with different signs.

5.2 The magnetooptic control of spatial and spatio-temporal solitons in metamaterial waveguides

Controlled spatial, temporal, and temporal-space solitons in MMs are a very relevant area of research and applications (Zhang *et al* 2010, 2014, Banerjee and Nehmetallah 2007, Zheng *et al* 2014, Argyropoulos *et al* 2013, Mihalache 2012, Torner and Kartashov 2010, Li *et al* 2010). A practically important approach to have control over the solitons in integrated optics and signal processing is the application of MO interactions (Boardman *et al* 2006b, 2009a, 2009b). Fundamentally new possibilities have opened up with the purposeful use of MO control of nonlinear waves and space-time structures, jointly with higher-order effects (linear and/or nonlinear) in MM layered structures (waveguides). For this purpose it is necessary to develop and apply methods of analytical and numerical solutions of the corresponding problems for wave processes in controlled nonlinear and active MM media. Such methods, in particular for nonlinear waves in layered MO MM media, are described in the present chapter. This includes details concerning the evolutionary equation for space-time solitons in the presence of higher-order effects. We will consider in particular stationary higher order equations, taking into account the nonlinear dispersion and the presence of both electrical and magnetic nonlinearities in MMs, obtained, firstly, in Boardman *et al* (2010a). MO control in a non-stationary medium is considered (Boardman *et al* 2010b), using the corresponding Lagrange function. Using the generalized Schrödinger equation (Scalora *et al* 2005), the modulation instability for short pulses was studied (Wen *et al* 2006a, 2006b, 2007, Xiang *et al* 2007, Wang and She 2005). Note that the presence of this instability is the condition for the formation of spatial and temporal solitons. Relevant solitons, including slow ones, were studied in D'Aguanno *et al* (2008). A consideration is given to the possibility of controlling solitons using the linear diffraction, diffraction management, and nonlinear diffraction (Kockaert *et al* 2006, Boardman *et al* 2000a, 2000b, 2005a).

5.3 Stationary equations and spatial solitons in the presence of the higher-order effect: nonlinear diffraction

The geometry of the problem with respect to stationary MO solitons under the influence of higher-order nonlinear effect (nonlinear diffraction) and with MO control is shown in figures 5.1(a) and (b). The direction of propagation of nonlinear waves is Z. The waveguides are uniform along X axis, whereas the propagating waves are generally non-uniform in the X direction. The core of waveguides is a

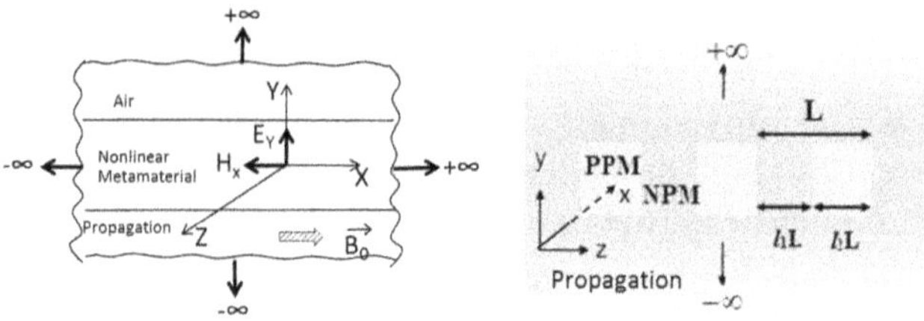

a b

Figure 5.1. Geometry of the problem. Part (a) is transverse magnetic (TM) (quasi-transverse electric magnetic (TEM)) polarized space beam in a planar MM waveguide infinite in the X direction; MO control based on the Voigt effect is used. Part (b) is a planar waveguide structure consisting of alternating layers of positive phase medium (PPM) and NPM for a realization of the diffraction management. Z is the direction of the wave propagation; the diffraction takes place in the X direction. The periodic medium has an elementary cell with a length L, much smaller than the wavelength; the frequency dependence of the nonlinear diffraction coefficient $\kappa = [4\pi^2 m^2(1-1/\Omega^2)(1-0.6/\Omega^2)]^{-1}$ is determined by the parameters: $(\omega_{\rm pm}/\omega_{\rm pe})^2 = 0.6$, the Drude model is adopted for both dielectric permittivity and magnetic permeability, $\kappa = [4\pi^2 m^2(1-1/\Omega^2)(1-0.6/\Omega^2)]^{-1}$, where $\Omega = \omega/\omega_{\rm pe}$, $w = m\lambda$ is the beam width at the input of the system. Typical Drude models are adopted that are of the form $\varepsilon(\varpi) = \varepsilon_{\rm D} - 1/\varpi^2$, $\mu(\varpi) = 1 - (\omega_{\rm m}^2/\omega_{\rm e}^2\varpi^{-2})$, where $\omega_{\rm e}$ and $\omega_{\rm m}$ are the plasma frequencies associated with the permittivity and the permeability, respectively, and ϖ is the dimensionless frequency, $\varpi = \omega/\omega_{\rm e}$. To allow for a first-order MO effect, the system is slightly asymmetric respectively to axis Y. (b) Reprinted from Boardman *et al* (2010a), copyright (2010), with permission from Elsevier.

nonlinear MM that possesses both the linear electric permittivity $\varepsilon(\omega)$ and the magnetic permeability $\mu(\omega)$. The Drude model is used for $\varepsilon(\omega)$, $\mu(\omega)$:

$$\varepsilon(\omega) = \varepsilon_{\rm D} - \frac{\omega_{\rm pe}^2}{\omega^2}; \quad \mu(\omega) = 1 - \frac{\omega_{\rm pm}^2}{\omega^2}, \tag{5.1}$$

where $\omega_{\rm pe}$, $\omega_{\rm pm}$ are corresponding plasma frequencies, $\varepsilon_{\rm D}$ is the non-plasma part of the permittivity; the dissipation terms are omitted here. The cubic nonlinearity in an MM includes both the electric nonlinear polarization $P_{\rm NL}$(E) and the nonlinear magnetization $M_{\rm NL}$(H). Here only the self-action nonlinear effects are considered, because the higher harmonics are assumed non-resonant. Generally for these nonlinear terms separate equations should be written down. But in the case of a moderate nonlinearity for quasi-monochromatic waves the simple expressions for $P_{\rm NL}$ and $M_{\rm NL}$ are used:

$$P_{\rm NL} = \chi_{\rm E}|E|^2 E; \quad M_{\rm NL} = \chi_{\rm M}|H|^2 H. \tag{5.2}$$

Here E, H, $P_{\rm NL}$, $M_{\rm NL}$ are slowly varying amplitudes for some carrier frequency of a wave beam.

The substrate is a linear dielectric that possesses the MO properties. Its dielectric permittivity is

$$\varepsilon_m = \begin{pmatrix} (n_m^2) & 0 & 0 \\ 0 & (n_m^2) & (-iQn_m^2) \\ 0 & (iQn_m^2) & (n_m^2) \end{pmatrix} = \begin{pmatrix} \varepsilon & 0 & 0 \\ 0 & \varepsilon & \varepsilon_{yz} \\ 0 & -\varepsilon_{yz} & \varepsilon \end{pmatrix} \tag{5.3}$$

The magnitude of the MO interaction Q is included in the MO tensor in the form (Boardman *et al* 2010b). The MO effect is considered as a small adding term to the basic equation for the wave beam propagation.

In (5.3), n_m is the refractive index of the substrate, $Q \sim 10^{-4}$, and, as a commonly accepted practical approximation (Mizumoto and Naito 1982, Boardman *et al* 2010a, 2010b), all diagonal terms in (5.3) are assumed to be the same. In the case of volumetric interaction in isotropic material, the Voigt effect is very weak and does not allow control because it is characterized by the value $\sim Q^2$. An interesting possibility to use this effect is to utilize a specially designed asymmetric (relatively to the reflection transform $x \to -x$) waveguide. In this case, the Voigt effect will be of the order Q, and it can be used for MO control of TM-polarized spatial solitons.

Two cases are under investigation. The first one is the nonlinear propagation in a waveguide uniform in Z direction, figure 5.1(a). The second case is related to the waveguide with the core that possesses the periodic structure along the direction of wave propagation Z figure 5.1(b). The period of this structure is much less than the wavelength. This makes it possible to consider the parameters averaged over the period of the structure. Each cell includes the layers with the NPM and PPM, so new interesting effects like the diffraction management occur in such a waveguide.

The following is a situation, which is quite characteristic of nonlinear integrated optics (Boardman and Xie 2003, Boardman *et al* 2006b, 2009a, 2009b), when the longitudinal component of the field is sufficiently small, namely for TM (E_x, E_z, H_y)—waves, $|E_z| \ll |E|_x$, and for TE (H_x, H_z, E_y) waves, $|H_z| \ll |H|_x$. Consider, in particular, TM mode. Note that to account for nonlinear dispersion, we will not use the approximation $\mathrm{div}\vec{H} = 0$, commonly applied in the nonlinear optics (Boardman *et al* 2000b). Instead, we will use a more accurate equation taking into account the nonlinear (in this case magnetic) polarization. The following set of equations is used:

$$\begin{aligned} \mathrm{rot}\vec{H} &= (1/c)\partial\vec{D}/\partial t + (4\pi/c)\partial\vec{P}_{\mathrm{NL}}/\partial t \\ \mathrm{rot}\vec{E} &= -(1/c)\partial\vec{B}/\partial t - (4\pi/c)\partial\vec{M}_{\mathrm{NL}}/\partial t \\ \mathrm{div}\vec{B} &= \mu\mathrm{div}\vec{H} + 4\pi\mathrm{div}\vec{M}_{\mathrm{NL}} = 0 \end{aligned} \tag{5.4}$$

In equations (5.4), $\vec{P}_{\mathrm{NL}}$, $\vec{M}_{\mathrm{NL}}$ are the electric and magnetic nonlinear polarizations, respectively. We assume that the component of nonlinear polarization $P_{\mathrm{NL}z}$ can be neglected, and then we obtain $(\mathrm{rot})_y(\partial\vec{P}_{\mathrm{NL}}/\partial t) = (\partial/\partial z)(\partial P_{\mathrm{NL}}/\partial t)$. Let us denote $k^2(\omega) = (\omega^2/c^2)\varepsilon(\omega)\mu(\omega)$.

We use the Fourier transform $\vec{E}(\vec{r}, \omega) = \int_{-\infty}^{\infty} \vec{E}(\vec{r}, t)\exp(i\omega t)\mathrm{d}t$, $\vec{H}(\vec{r}, \omega) = \int_{-\infty}^{\infty} \vec{H}(\vec{r}, t)\exp(i\omega t)\mathrm{d}t$, $\tilde{\vec{P}}_{\mathrm{NL}}(\vec{r}, \omega) = \int_{-\infty}^{\infty} \vec{P}_{\mathrm{NL}}(\vec{r}, t)\exp(i\omega t)\mathrm{d}t$, $\tilde{\vec{M}}_{\mathrm{NL}}(\vec{r}, \omega) = \int_{-\infty}^{\infty} \vec{M}_{\mathrm{NL}}(\vec{r}, t)\exp(i\omega t)\mathrm{d}t$. Consider, in particular, TM mode. Suppose that $E_y = A_{\mathrm{E}}\exp(k_c z - \omega_c t) + \mathrm{cc}$, $H_x = A_{\mathrm{H}}\exp(k_c z - \omega_c t) + cc$, $P_{\mathrm{NL}x} = P_{\mathrm{NL}}\exp(k_c z - \omega_c t) + cc$, $M_{\mathrm{NL}y} = M_{\mathrm{NL}}\exp(k_c z - \omega_c t) + cc$, where (ω_c, k_c), $k_c = k(\omega_c)$ are the carrier frequency and the wavenumber of the wave packet. We will consider the stationary case and assume that $A_E = A_E(z, y)$, $A_H = A_H(z, y)$. In this case we get

$$[(\partial^2/\partial z^2) + 2ik_c(\partial/\partial z) - k_c^2]A_H + \partial^2 A_H/\partial x^2 + k_c^2 A_H + (4\pi\omega_c^2/c^2)\varepsilon_c M_{NL} + (4\pi/\mu_c)\partial^2 M_{NL}/\partial x^2 - (4\pi i\omega_c/c)\partial P_{NL}/\partial z = 0; \ \varepsilon_c \equiv \varepsilon(\omega_c) \quad (5.5)$$

Considering that, for a structure with a weak nonlinearity, the polarization relations, as well as the transverse mode structure, differ a little from the linear case (Rapoport *et al* 2012, Rapoport 2013, 2014), we obtain from (5.5), in a paraxial (parabolic) approximation

$$i\partial A_H/\partial z + (1/2k_c)\partial^2 A_H/\partial x^2 + (2\pi\omega_c^2/c^2 k_c)\left(\varepsilon_c\chi_H + \left(\mu_c^2/\varepsilon_c\right)\chi_E\right)|A_H|^2 A_H + \left(2\pi\chi_H/\mu_c k_c\right)\partial^2(|A_H|^2 A_H)/\partial x^2 = 0 \quad (5.6)$$

In equation (5.6) χ_H is the coefficient of magnetic cubic nonlinearity. It can be shown that, to account for the Voigt MO effect, a term $-(\omega/c)\varepsilon_{xy}A_H$ must be added in the left-hand side of (5.6). Equation (5.5) is taken into account for the first time (Boardman *et al* 2010a, 2010b) for a layered MM simultaneously two types of the enhanced cubic nonlinearity, namely electrical and magnetic ones, the third term on the left side of equation (5.6), and the nonlinear dispersion, the last term. For a double nonlinear medium there are $\varepsilon_c < 0$, $\mu_c < 0$, $k_c = -|k_c|$. We choose the nonlinearity, the cubic nonlinear coefficients, as $\left(|\varepsilon_c|\,\chi_H + (|\mu_c|^2/|\varepsilon|_c)\chi_E\right) < 0$, where χ_E is the coefficient of the electric cubic nonlinearity. The last condition is necessary to provide the effect of self-compression and the existence of (stationary) spatial solitons (Agrawal 1989, Remoissenet *et al* 1984). We use the coordinate transformation/normalization, $z = z_0 Z$, $x = wX$, $z_0 = |k_c|w^2$, $A_H = A_{H0}\psi$, $A_{H0}^2 = [2\pi\omega_c^2 w^2/c^2|N_{TM}|]^{-1}$. The use of MMs has reached a visible optical range (Simovski and Tretyakov 2009, Wang *et al* 2007). Let us explore the possibilities of controlling solitons in the appropriate range. We introduce the coefficient of the nonlinear diffraction, MO interaction (Boardman *et al* 2010a, 2010b), and the value of the diffraction coefficient D, which describes the diffraction management. The diffraction management makes possible to choose the diffraction coefficient of a desired value different from its value in the uniform waveguide. Analytical calculations describing the coefficients of nonlinear evolution equations for transverse electric (TE) and TM equations are described in Boardman *et al* (2010b).

In the linear approximation we obtain, using the tensor (5.3), the MO perturbation to the NSE in the form

$$i\partial H_x/\partial z = -(\omega/c)n_m^2\overline{Q}H_x \quad (5.7)$$

We derive the normalized evolution equation with MO interaction, taking into account the dependence of the corresponding normalized coefficient ν on the transverse coordinate x and the nonlinear diffraction

$$i\partial\psi/\partial Z - (1/2)\partial^2\psi/\partial X^2 - |\psi|^2\psi + \kappa\partial^2(|\psi|^2\psi)/\partial X^2 = -\upsilon(X)\psi \quad (5.8)$$

In the quasi-TEM approximation there are $k_c^2 \approx |\varepsilon_c||\mu_c|(\omega_c^2/c^2)$, $\kappa = \mathrm{sgn}(\chi_H)\{k_c^2 w^2[1 + (\mu_c^2\chi_E/\varepsilon_c^2\chi_H)]\}^{-1}$. In doing so, we consider that for the selected MM $\chi_e = -|\chi_e|$, $\kappa < 0$. When $\chi_E = 0$, κ it reduces to $\kappa = -(k_c w)^{-2}$ (Boardman *et al* 2010b). It is important to note that if $\nu \sim Q$ is a constant, then the transverse MO effect will cause nothing but an additional phase shift to the soliton solution (5.8) without the right-hand side. This can be taken into account by redefining ψ to $\psi e^{i\nu Z}$ as can be seen from equation (5.8). Thus, the magnitude ν must be some function of transverse coordinate X. For example, $v(X) = v_{\max}\,\mathrm{sech}(X/X_0)$, where $v_{\max}$, X_0 are the maximum magnitude and half-width of the corresponding spatial distribution, respectively. In addition to MO control, the spatial solitons in the MM waveguide can be controlled by the diffraction management (Boardman *et al* 2010a), figure 5.1(b). We apply the principle of the diffraction management (figure 5.1(b)), based on the averaging over the unit cell of the structure parameters along the wave propagation direction Z with an appropriate selection of the coefficients $l_{1,2}$ and the parameters of the segments of the positive phase medium (PPM) and NPM (Boardman *et al* 2010b). The diffraction management makes it possible to get the desired value of the diffraction coefficient within a wide interval of values.

As shown in detail in Boardman *et al* (2010b), averaging the parameters over a unit cell (figure 5.1(b)), which includes the parts with the positive and negative phases and has a size much smaller than the wavelength, yields, instead of (5.8), the following equation:

$$i\partial\psi/\partial Z - (D/2)\partial^2\psi/\partial X^2 - |\psi|^2\psi + \kappa\partial^2(|\psi|^2\psi)/\partial X^2 = -v(x)\psi \tag{5.9}$$

Below, some numerical results are presented, that qualitatively describe the effect of the diffraction management and the width of the input beam on MO spatial solitons. Figure 5.2 illustrates the role of the nonlinear diffraction in the absence of the diffraction management, with the classical single-soliton profile selected at the input. It is seen that in single-soliton mode, the effect of the nonlinear dispersion is reduced, even for a sufficiently narrow pulse, of about of a wavelength order, to the trivial effect of a rather moderate broadening and a decrease in the amplitude of the pulse at the output of the system.

Figures 5.2(d) and (e) demonstrate that the nonlinear diffraction in the case of the higher-order breather soliton leads to a qualitatively new effect in the absence of the diffraction management (figure 5.2(d)). In the absence of the nonlinear diffraction, there is a periodic structure consisting of three first-order solitons that propagate together, having the zero binding energy. In the presence of the same control based on the nonlinear diffraction, such a soliton rapidly perturbs (figure 5.2(e)). As a result, the third-order soliton is transformed into the first-order soliton and some low energy beams that are emitted.

The position on the Z axis, where the beam begins to split, can be controlled in the MMs through the control of the carrier frequency, due to the presence of the frequency dispersion of waves. The cumulative effect of the diffraction management, nonlinear diffraction, and non-reciprocal MO interaction is shown in figure 5.2(f)–(h). In this case, in the absence of MO interaction, both the diffraction management (when the linear dispersion decreases by an order of magnitude) and the nonlinear dispersion are relatively important in this case. As a result of the reduction of the linear

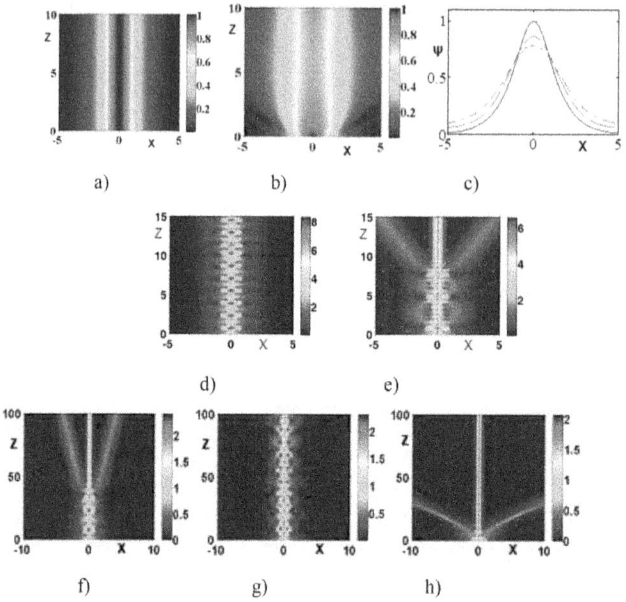

Figure 5.2. (Boardman *et al* 2010b); (a)–(c) are the input pulses: $\mathrm{sech}(X)$, $m = 1$, $D = 100\%$; (a) $\kappa = 0$ (b) $\kappa = 0.17$(c) Beam profiles after passing a distance from the input to the system equal to 10 Rayleigh lengths: the solid blue line ($\kappa = 0$), the dot-dashed green line ($\kappa = 0.1$), the dotted red line ($\kappa = 0.17$); (d) and (e) are the input pulses: $3\,\mathrm{sech}(X)$, $m = 3$, $D = 100\%$; (d) $\kappa = 0$ and (e) $\kappa = 0.0028$; (f), (g), (h) D = 10%, $\kappa = 0.00168$, $m = 3$, $v(X) = v_{\max}\,\mathrm{sech}(X/X_0)$; input: $\psi = \ \mathrm{sech}(X)$ (f) $v(X) = 0$, (g) $v(X) > 0$, and (h) $v(X) < 0$. Reprinted from Boardman *et al* (2010a), copyright (2010), with permission from Elsevier.

dispersion, the energy, which in the case of a 'normal' linear dispersion would correspond to a standard first-order soliton, is too large for it. A third-order breather soliton is formed, which splits into a first-order soliton and two radiating beams (figure 5.2(f)). With a correctly selected (positive) sign of the magnetic field (magnetization), the combined effect of the reduced linear dispersion, nonlinear dispersion, and MO interaction leads to the formation of an effective potential well, in which the breather soliton is fully trapped (figure 5.2(g)). With another sign of MO interaction (magnetization/magnetization field), the 'trapping' potential is not formed, but instead the breather soliton is quickly split and a single narrow beam is formed (figure 5.2(h)). The considered above theory demonstrates that a magnetic field can provide a very important control of spatial solitons in specially structured nonlinear MO MMs.

5.4 Non-stationary equations and spatial-temporal solitons in the presence of higher-order effects: nonlinear diffraction and dispersion, Raman interaction, and linear third-order dispersion. Generalization of NEELS method

5.4.1 Evolution equations for temporary solitons with higher-order effects

In Wen *et al* (2006a, 2006b, 2007), Xiang *et al* (2007), and Scalora *et al* (2005), there were obtained evolutionary equations in an unbounded environment and they investigated the modulation instability, including the influence of the nonlinear diffraction/self-action. In Boardman *et al* (2010b) (see some details in appendix C.4.1), a normalized evolution equation was derived for temporal solitons in layered MMs, taking into account the effects of the third-order linear dispersion, Raman interaction, nonlinear dispersion and MO effect. In the absence of an MO effect, but in the presence of higher linear and nonlinear dispersion and Raman interaction, this equation is derived in appendix C.1 (equation (C.10)) and, for convenience, repeated below. Namely, this equation has the form (see also equation (3.53) in chapter 3, the special case of which is given below equation (5.10))

$$i\frac{\partial\psi}{\partial Z} - \frac{1}{2}\,\mathrm{sgn}(\beta_2)\frac{\partial^2\psi}{\partial T^2} - i\delta_3\frac{\partial^3\psi}{\partial T^3} + \mathrm{sgn}(\chi^{(3)})\,\mathrm{sgn}\,(k')\,\mathrm{sgn}\,(\nu')\left(|\psi|^2\psi + iS\frac{\partial}{\partial T}(|\psi|^2\psi) - \tau_R\psi\frac{\partial}{\partial T}(|\psi|^2)\right) = 0 \tag{5.10}$$

Note that for EMW propagation there are $k' > 0$, ' > 0 for PPM, whereas $k' < 0$, ' < 0 for NPM (Boardman *et al* 2010b). In equation (5.10), T_0, $\beta_{1,2,3}$, δ_3 are the characteristic pulse duration, the inverse group velocity, the dimensional coefficients of the second- and third-order linear dispersion and the normalized (dimensionless) third-order linear dispersion coefficient, respectively, $\chi^{(3)}$ is the electrical cubic nonlinearity coefficient, S, μ, and k' are the non-dimensional coefficient of the nonlinear dispersion (self-steepening), the magnetic permeability, and the wave number of the wave in the direction of propagation, respectively, τ_R is a characteristic response time for the Raman interaction,

$$\beta_2 = \frac{\mathrm{d}}{\mathrm{d}\omega}\left(\frac{1}{V_g}\right),\ \delta_3 = \frac{\beta_3}{6\,|\,\beta_2\,|\,T_0},\ S = \frac{1}{\omega_0 T_0}\left[2 - \frac{\beta_1\omega_0}{\beta_0} + \left(\frac{\omega}{\mu}\frac{\partial\mu}{\partial\omega}\right)_{\omega_0} + \left(\frac{\omega}{\chi^{(3)}}\frac{\partial\chi^{(3)}}{\partial\omega}\right)_{\omega_0}\right],$$

$\tau_R = \dfrac{\int_{-\infty}^{\infty} tR(t)\mathrm{d}t}{T_0}$. The last term of equation (5.10), proportional to τ_R, arises as a result of a delayed nonlinear response and describes the effect of frequency self-shift (forced combination self-scattering (Agrawal 1989)). In derivation of (5.10), the dispersion is considered, and a response function $R(t)$ is introduced, which determines τ_R:

$$\vec{D}(\vec{r}, t) = \int_{-\infty}^{\infty}\varepsilon(t-\tau)\vec{E}(\vec{r}, \tau)\mathrm{d}\tau,\ \vec{B}(\vec{r}, t) = \int_{-\infty}^{\infty}\mu(t-\tau)\vec{H}(\vec{r}, \tau)\mathrm{d}\tau,$$

$$\vec{P}_{NL}(\vec{r}, t) = 4\pi\chi^{(3)}\vec{E}(\vec{r}, t)\int_{-\infty}^{\infty}R(t-\tau)|\vec{E}(\vec{r}, \tau)|^2\mathrm{d}\tau \tag{5.11}$$

where ε is the dielectric permittivity, $\vec{P}_{NL}$ is the nonlinear electric polarization. The normalization method used to obtain equation (5.10) is described in appendix C.1. In equation (5.10), the dispersion of nonlinearity is taken into account (see also appendix C.1), in other words the dispersion of the nonlinear coefficient $\chi^{(3)}$ and the Fourier transform for $\chi^{(3)}$ are represented, in the first approximation, in the form (Boyd 2003) $\chi^{(3)}(\omega) = \chi_0^{(3)} + (\omega - \omega_0)(\partial\chi^{(3)}/\partial\omega)$, where ω_0, $\chi_0^{(3)}$ are the carrier pulse frequency and the value of the cubic nonlinear coefficient at this frequency, respectively.

The question arises whether the MM provides a significant possibility of the frequency dispersion of the self-action coefficient (the nonlinear dispersion). The answer is illustrated in figure 5.3, where the contrast between the NPM and the PPM is emphasized. For a qualitative description of the MM, Drude models $\varepsilon(\bar{\omega}) = \varepsilon_D - \bar{\omega}^{-2}$, $\mu(\bar{\omega}) = 1 - (\omega_m^2/\omega_e^2)\bar{\omega}^{-2}$, are adopted where ω_e and ω_m are the plasma frequencies associated with the dielectric permittivity and the magnetic permeability, respectively, $\bar{\omega}$ is the dimensionless frequency, $\bar{\omega} = \omega/\omega_e$. For each application, we choose $\bar{\omega} = \bar{\omega}_0$ where $\bar{\omega}_0$ is the operating frequency, which is selected so that it lies in the window shown in figure 5.3. The corresponding coefficients are shown as functions of $\bar{\omega}$. In terms of the dimensionless values, $\bar{\beta}_2 \equiv \bar{\beta}_2(\bar{\omega})$ and $\bar{\beta}_3 \equiv \bar{\beta}_3(\bar{\omega})$, the coefficients $\bar{\delta}_3$ and $\bar{S}$ are represented as $\bar{\delta}_3 = (6\omega_e T_0)^{-1}(\bar{\beta}_3/|\bar{\beta}_2|)$, $\bar{S} = [\bar{\omega}(\omega_e T)]^{-1}[2 + (\bar{\beta}_1\bar{\omega}/|\bar{\beta}_0|) + (\bar{\omega}/\mu)\partial\mu/\partial\bar{\omega}]$ and are shown in figure 5.3(a) and (b).

A typical contribution of the Raman self-action effect is also shown in figure 5.3(a). Naturally, for a given pulse duration, it does not depend on frequency, so this magnitude is constant in the operating window of frequencies shown in figure 5.3(a). It is very interesting, however, that the use of MMs provides a strong frequency dependence for the self-action coefficient. The non-trivial consequence of the use of MMs is not only that they can provide a significant effect on the characteristic behaviors of solitons, but also the ability to compensate for the Raman scattering, as well as the effectiveness of MO control. It is important that in figure 5.3(a) the marked portions of the 'frequency window' include a very narrow frequency neighborhood (a special frequency region), where $\beta_2 \to 0$, and the nonlinear Schrödinger equation, presented in its actual normalization, becomes invalid. Figure 5.3 was made for a pulse of 21 fs duration. For pulses of such short or shorter duration, it is necessary to consider the third-order linear **dispersion** whose frequency dependence is shown in figure 5.3(b).

5.4.2 The principle of magnetooptical control of time solitons in a non-stationary metamaterial medium

To provide the MO control using the transverse MO effect (the Voigt effect), it is proposed to use the asymmetric structure shown in figure 5.4(a), and the MO effect

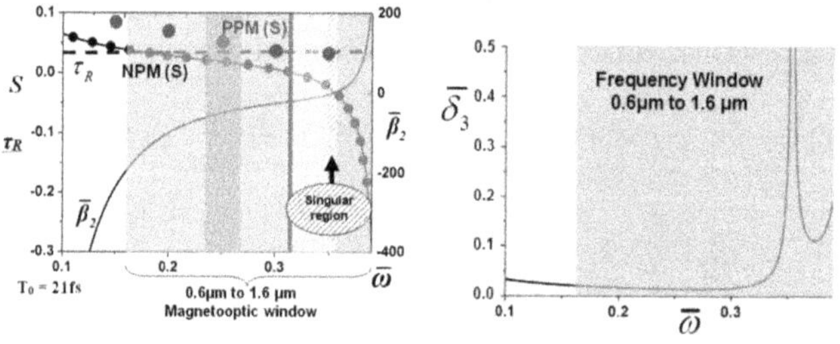

a) b)

Figure 5.3. (a) and (b) (modified figures from Boardman *et al* 2010a). (a) is the dispersion of the self-steepening coefficient (the nonlinear dispersion) s' for the positive-phase medium (PPM) and the negative-phase medium (NPM) for the normalized frequency $\bar{\omega}$, $\bar{\beta}_2$ is the dispersion coefficient (group velocity dispersion), and τ_R is the Raman interaction coefficient; $\omega_m/\omega_e = 0.4$; $\omega_e = 7.08 \times 10^{15}$ s^{-1}; $\varepsilon_D = 4$; b is the dependence of the normalized coefficient of the third-order linear dispersion $\bar{\delta}_3$ on the normalized frequency, $\bar{\omega}$. A sharp peak occurs when the coefficient of dispersion $\bar{\beta}_2$ tends to zero. Reprinted from Boardman *et al* (2010b), copyright (2010), with permission from Elsevier.

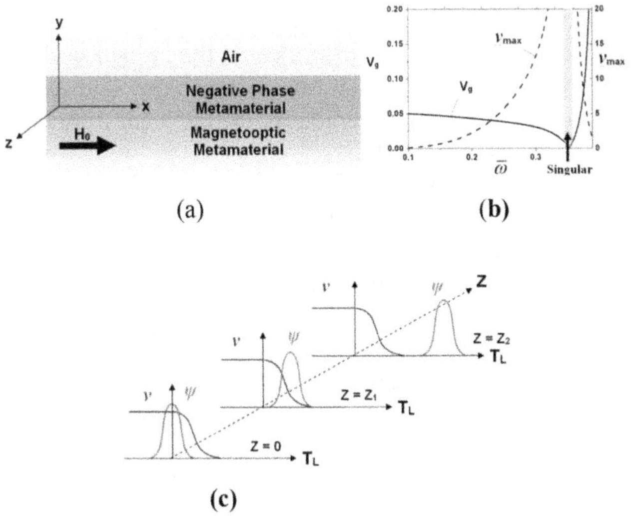

Figure 5.4. (a)–(c) (Boardman *et al* 2010a). MO control of temporal solitons in a layered MM environment. (a) is an asymmetric planar waveguide where a transverse MO effect (Voigt effect) is used to control solitons. The pulse propagates in the Z direction; the magnetic field H_0 is directed along X; (b) is the variations of the normalized group velocity and the maximum value of the MO coefficient for a typical MM in the range corresponding to figure 5.4(a). T_0 = 21 fs, n_{m}. The normalized value in the singular region highlighted in figure 5.4(b); (b) is the principle of MO control in an MM medium with a non-stationary magnetic field. Pulse and MO distributions, in the laboratory frame, for three typical positions along the propagation direction. T_{L} is the normalized time, Z is the normalized length. The pulse 'senses' the MO interaction as it passes an area with a variable parameter, and then always bears the mark of this interaction, in particular in the form of a change in speed/delay time. (c) Pulse and magnetic distributions, in the laboratory frame. These distributions are presented for the positions of three specific points along the propagation axis. T_{L} and Z are the dimensionless time and length respectively. Reprinted from Boardman *et al* (2010b), copyright (2010), with permission from Elsevier.

is described by the dielectric tensor of the form as in equation (5.3). Note that all diagonal elements in equation (5.3) are considered to be the same in this approximation. This approximation is generally accepted in practice and is based on experimental applications. For polarized beams, using the tensor included in equation (5.3), one can obtain MO perturbations for the Schrödinger equation. In fact, for the two main polarizations in the planar waveguide shown in figure 5.4(a), the results for the analysis of the relevant perturbations are as follows:

$$\text{TE polarizations: } i\frac{\partial A}{\partial z} = \text{previous terms} + 0;$$

$$\text{for TM polarization: } i\frac{\partial A}{\partial z} = \text{previous terms} + \frac{\omega}{c}\left\langle \varepsilon_{yz} \right\rangle A \tag{5.12}$$

where $\left\langle \varepsilon_{yz} \right\rangle$ is the effective averaged in the direction normal to the layers of the waveguide structure, the MO coefficient (Zvezdin and Kotov 1997, Brabec and Krausz 1997, Boardman and Xie 2003, Zharov and Kurin 2007, McGahan *et al* 1991), A is a slowly varying amplitude for the TE or TM mode electric field components. The normalization in equation (5.12) is made taking into account the corresponding modal field structure. At this stage it is possible to include the influence of the MO substrate shown in figure 5.4 and on the evolution of a soliton. The corresponding one-dimensional equation with the MO interaction, in the linear approximation, is

$$i\frac{\partial\psi}{\partial Z} - \frac{1}{2}\,\mathrm{sgn}(\beta_2)\frac{\partial^2\psi}{\partial T^2} - i\delta_3\frac{\partial^3\psi}{\partial T^3} + \nu\psi = i\gamma_0\psi \tag{5.13}$$

In equation (5.13) the values Z, T, y, β_2, and ∂^3 have the same physical meaning as the ones in equation (5.10); coefficients ν and γ_0 describe the effects of magnetooptic interaction and linear losses, respectively (Boardman *et al* 2010b). Note that the general evolutionary equation (5.10) for MM waveguides, taking into account the diffraction, MO control, and higher-order linear and nonlinear effects, is given in section 3.3.4. Since the MO effect is considered relatively small, obtaining the corresponding term for equation (5.13) also holds for more general equation (5.10). It is easy to check that if ν is a constant, then the MO effect results only in a trivial phase shift of the envelope soliton. Thus, it is necessary to make this parameter ν time dependent. Note that the corresponding function depends on the time given in the laboratory coordinate frame and equation (5.13) is given in the moving coordinate frame. Given this fact, the corresponding function is substituted into equation (5.13) in the form:

$$\nu(T_{\mathrm{L}}) = \nu(T + Z/V_{\mathrm{g}}) = (\omega/c)L_{\mathrm{D}}n_{\mathrm{m}}^2Q_{\mathrm{sat}}f(T + Z/V_{\mathrm{g}}) = \nu_{\mathrm{max}}f(T + Z/V_{\mathrm{g}}) \tag{5.14}$$

Here T_{L}, given in the laboratory coordinate frame, the normalization is determined by the characteristic time scale T_0 (for instance, the pulse duration), and f is determined by the functional dependence of the magnetization, the maximum value of the MO coefficient. As shown in figure 5.4(c), as Z increases, the pulse enters and then exits the region where the effect of the MO interaction is significantly decomposed (see also the caption to figure 5.4). Two forms of the functional dependence on time are considered:

$$\nu(T + Z/V_{\mathrm{g}}) = \nu_{\mathrm{max}}\mathrm{sech}[(T + Z/V_{\mathrm{g}} - T_{\nu})/\Delta T_{\nu}] \tag{5.15a}$$

$$\nu(T + Z/V_{\mathrm{g}}) = (\nu_{\mathrm{max}}/2)\{1 - \tanh[(T + Z/V_{\mathrm{g}} - T_{\nu})/\Delta T_{\nu}]\} \tag{5.15b}$$

Here T_{ν} is the time delay for the magnetization/MO coefficient, ΔT_{ν} is a characteristic time of magnetic field reduction/demagnetization. The relation (5.15(a)) corresponds to the presence of a maximum at a certain moment of time, and the relation (5.15(b)) corresponds to the transition from one 'shelf' to another. The normalized frequency dependences for ν, V_{g} are shown in figure 5.4(b). It can be shown that the normalized magnetization/MO coefficient has a singularity at $V_{\mathrm{g}} \to 0$. We will assume that the frequency domain, including the near vicinity of the corresponding 'singularity point', should be omitted at work. At least in this frequency neighborhood, the approximations used to derive the above equations (e.g. (5.13)) ceases to be true. Note also that $\beta_2 = (d/d\omega)V_{\mathrm{g}}$ and this function is included in ν_{max} because of the definition of the dispersion length.

5.4.3 Lagrangian dynamics and magnetooptical control of time solitons with higher-order effects

The application of Lagrangian formalism leads to a description of the soliton dynamics in the form of a set of (connected) first-order differential equations. This is done to obtain a set of equations describing the motion of the maximum (peak) position with regard to the self-steepening (nonlinear dispersion) and under the control of the MO interaction, in particular for the time dependence of the MO coefficient in the form (5.15(a)). The Lagrangian analysis is developed on the basis of use of test function in the form

$$\psi = \eta \operatorname{sech}(\eta(T - T_0)) \exp[i(\xi/2)(T - T_0) + i\phi] \tag{5.16}$$

where η is the amplitude, T_0 is the time corresponding to the maximum (center) of the pulse, ξ is the effective frequency shift of the pulse, ϕ is a phase. The nonlinear Schrödinger equation of the type of (5.10) with the 'linear and nonlinear higher order effects' for a one-dimensional pulse can be rewritten in the form where the 'relatively small effects' are included in the right-hand side using the Lagrangian formalism (Boardman and Marinov 2006a, Boardman and Velasco 2006b, Kivshar and Krolikowski 1995, Whitham 1974):

$$i\frac{\partial \psi}{\partial Z} - \operatorname{sgn}\frac{1}{2}(\beta_2)\frac{\partial^2 \psi}{\partial T^2} + \operatorname{sgn}(\chi^{(3)}) \mid \psi \mid^2 \psi + \nu\psi = R \tag{5.17}$$

where $R = -i \operatorname{sgn}(\chi^{(3)}) S\partial(\mid \psi \mid^2 \psi)/\partial T + i\delta_3 \partial^3 \psi/\partial T^3 + \tau_R \psi \partial(\mid \psi \mid^2)/\partial T$. Some details regarding the Lagrangian approach to the study of the soliton/soliton pulse (5.16) in an MM medium with MO control are presented in appendix C.2. As a result of applying this method, we obtain a set of equations ((C.16) and (C.17)) for the location of the center and the effective shift of the frequency of the soliton-like pulse (5.16), which, for convenience, are repeated below, respectively:

$$\frac{\partial T_0}{\partial Z} = -\frac{1}{2} \operatorname{sgn}(\beta_2)\xi + \eta^2 S \operatorname{sgn}(\chi^{(3)}) + \delta_3\left(\eta^2 + \frac{3}{4}\xi^2\right) \tag{5.18a}$$

$$\frac{\partial \xi}{\partial Z} = \eta \int_{-\infty}^{\infty} \nu(T) \frac{\partial}{\partial T_0} \operatorname{sech}^2[\eta(T - T_0)] \mathrm{d}T + \frac{16}{15}\tau_R \eta^4 \tag{5.18b}$$

Equations (5.18a) and (5.18b) show how the peak of the pulse, in a moving coordinate frame, moves with increasing Z, and how it depends on ξ, leading to a propagation rate in the laboratory frame that is greater or less than V_g, ξ, in turn, depends on the function $\nu(T)$ that describes the non-stationary MO interaction.

Returning to a more accurate evolution equation in partial derivatives with higher order effects equation (5.17), both the nonlinear dispersion and third-order linear one, and the MO addition of the form as in equation (5.14), we present the corresponding simulation results. In Boardman *et al* (2010a), first of all, the calculations were performed for frequency domains where the second-order linear dispersion, which is characterized by the coefficient β_2, is more important than the third-order dispersion characterizing by β_3. In the following, we consider in more detail the situation where the Raman scattering characterized by the parameter τ_R is important; the self-steepening (the nonlinear dispersion), characterizing by the parameter S; and also a third-order linear dispersion characterizing by a normalized parameter $\delta_3 = (\beta_3/6 \mid \beta_2 \mid T)$. The corresponding parameters are different functions of the normalized frequency ϖ (see figure 5.3(b) and 5.4(b)). β_2 is included in the normalization for other values. The frequency dependence of β_2 is such that this magnitude goes through zero value with increasing ϖ. To demonstrate the role of third-order linear dispersion, the frequency is chosen equal to $\varpi = 0.3442$. As can be obtained using the frequency dependences in figures 5.3(b) and 5.4(b), β_3 which is expressed through β_3 by the *dispersion relation*, gives a quite significant effect compared to the effect of the second-order dispersion. Thus it is necessary to take into account higher linear dispersion effects. The effects of the same type are important for a particular frequency range and for femtosecond pulses. It should be noted that Raman scattering leads to an effect on the pulse delay, which is the opposite of one arising from the negative self-steepening. In order to determine whether an external magnetic field can be used in this case to control the behavior of temporal solitons, three possible cases are illustrated in figure 5.5(a)–(d). In the first case (figure 5.5(a)), we neglect the third-order linear dispersion and include only the effects of the nonlinear dispersion and the Raman scattering. In the second case (figure 5.5(b)), we also include the third-order linear dispersion effect. Accordingly, the motion of the soliton is slowed down and significant radiation is observed, which leads to a significant decrease in its amplitude. The sign of the coefficient of the nonlinear dispersion is negative. Figure 5.5(c) illustrates a situation different from that, which corresponds to figure 5.5(b), by the presence of the MO effect. Comparing figure 5.5(a)–(c) shows that the inclusion of the magnetic field and the MO effect results in partial compensation of the soliton deceleration caused by the use of the cubic linear dispersion and Raman scattering.

Figure 5.5(d) summarizes the results of all cases (a–c). Figure 5.5(e) shows the results of comparing exact calculations with those based on the Lagrangian formalism. The coincidence of the results on motion of the pulse maxima obtained by means of Lagrangian formalism and the solution of the evolution equation is very good.

This section shows results based on standard Drude models for the magnetic permeability and the electric permittivity. The influence of higher order linear dispersion, nonlinear dispersion, and Raman scattering on the propagation of the pulse is included, and the possibility of effective MO control is demonstrated. Namely, the MO effect substantially compensates for the slowdown of the pulse caused by the presence of the Raman interaction and the linear third-order dispersion. Note that the analysis of (2 + 1) (space-time) solitons/quasi-solitons in MM waveguides (Buttner *et al* 2000, Boardman *et al* 2010a, Kadomtsev 1988) shows that bullets are possible, in the case of NPM with the following set of signs of coefficients in the nonlinear evolution equation (3.53), as a necessary condition: $D < 0$, $\beta_2 < 0$, $\chi^{(3)} < 0$. Note that in the case of NPM there are $m' < 0$, $k' < 0$, where m', k' are the real parts of the magnetic permeability and the wave number of EMW, respectively.

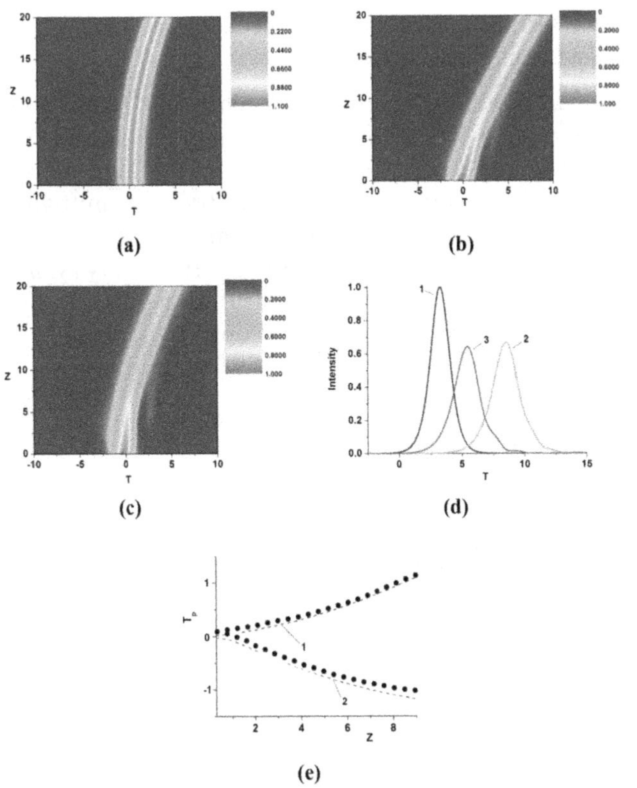

Figure 5.5. (a)–(e) (Boardman *et al* 2010b). The influence of the third-order dispersion on the propagation pulse in an MO waveguide; (a)–(c) are for $V_g = 0.0063$; (a) $\overline{S} = -0.023$, $\overline{\tau}_R = 0.033$, $\overline{\delta}_3 = 0$, $\nu_{max} = 0$; (b) $\overline{S} = -0.023$, $\overline{\tau}_R = 0.033$, $\overline{\delta}_3 = 0.2$, $\nu_{max} = 0$; (c) $\overline{S} = -0.023$, $\overline{\tau}_R = 0.033$, $\overline{\delta}_3 = 0.2$, $\nu_{max} = 97.91$; (d) the amplitude at the output of the system ($Z = 20$), curves 1, 2 and 3 correspond to the values specified in paragraphs. (a), (b) and (c), respectively; (e) is the position of the pulse maximum. Comparison of the results of using the Lagrangian formalism (dashed curves) with the results of solving the Schrödinger evolution equation ('dotted' line); curves 1 are without applied magnetic field; curves 2 are with the applied magnetic field; curves 1 are for $S = 0.0149$, $\delta_3 = 0.02779$, $\tau_R = 0.033$, $\nu_{max} = 0$; curves 2 are for $S = 0.0149$, $\delta_3 = 0.02779$, $\tau_R = 0.033$, $\nu_{max} = 7.435$. Reprinted from Boardman *et al* (2010a), copyright (2010), with permission from Elsevier.

5.5 New types of surface magnetic polaritons and reflection of electromagnetic waves in metamaterial–dielectric systems

A very effective way of controlling surface polaritons is to use gyrotropy and a magnetic field. In particular, in the case of the system 'semi-infinite dielectric–semi-infinite ferromagnet' there are non-reciprocal surface waves (Remer *et al* 1984, Bonod *et al* 2004, Hermann *et al* 2001, Sugano and Kojima 2000). This approach is also used to investigate for the first time new types of magnetic polaritons (Boardman *et al* 2005c) in the system of 'semi-infinite ferromagnetic–semi-infinite metamaterial'.

5.5.1 Equations for surface polaritons at the boundary of semi-infinite layered gyrotropic and metamaterial media

Consider the configuration shown in figure 5.6, where two semi-infinite materials are separated by a surface at $Y = 0$. Consider two situations where (1) the medium 1 is gyromagnetic, or (2) gyroelectric with tensor electric $\widehat{\varepsilon}$ and scalar magnetic μ or tensor magnetic $\widehat{\mu}$ permeabilities and scalar electric ε permittivities, respectively, with the external field applied in Z direction. The medium 2 is an isotropic MM with the scalar electrical permittivity and the magnetic permeability ε_2, μ_2. Let us take the notation: κ is an exponential decay of the field along the Y direction; h is the propagation constant of waves in the X direction. It is possible to obtain, using

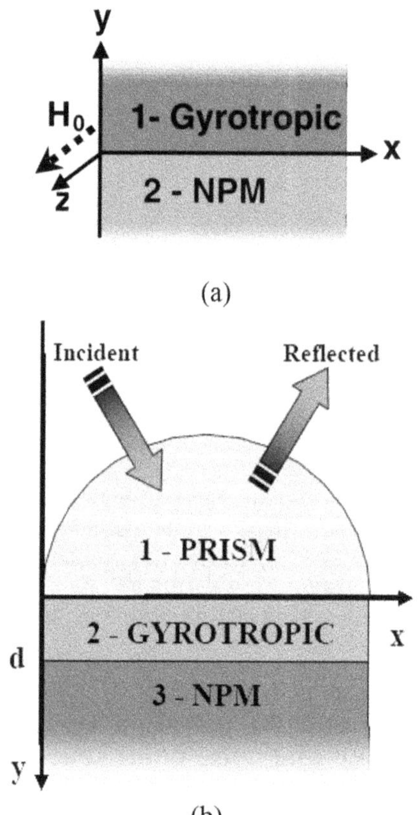

Figure 5.6. (Boardman *et al* 2005b). (a) The system is semi-infinite NPM (MM)—semi-infinite gyrotropic medium. Medium 1 is gyromagnetic or gyroelectric, with the outer field applied in the Z direction. Medium 2 is an isotropic MM with the electrical permittivity and the magnetic permeability ε_2, μ_2. (b) 'Dielectric prism-gyrotropic layer-NPM' system where surface waves can be generated at the interface between the NPM and the gyrotropic layer. The thickness of the gyrotropic layer is equal to d. The reflection coefficient R_{TM} of the electromagnetic wave from such a system is given in appendix C.4, relations (C.24) and (C.25). Reprinted from Boardman *et al* (2005c), copyright IOP Publishing, all rights reserved.

the Maxwell equation, for the cases (1) (gyroelectric waves) and (2) (gyromagnetic waves) wave equations relative to electric or magnetic fields, respectively:

$$(\text{case1})\vec{D} = \widehat{\varepsilon}\vec{E},\ \vec{B} = \mu\vec{H};\quad \nabla\times\nabla\times\vec{E} = -\frac{1}{c^2}\frac{\partial^2}{\partial t^2}(\mu\widehat{\varepsilon}\vec{E}); \tag{5.19}$$

$$(\text{case 2})\vec{D} = \varepsilon\vec{E},\ \vec{B} = \widehat{\mu}\vec{H};\quad \nabla\times\nabla\times\vec{H} = -\frac{1}{c^2}\frac{\partial^2}{\partial t^2}(\widehat{\mu}\varepsilon\vec{H}); \tag{5.20}$$

Electric and magnetic fields for surface polaritons moving along the interface of the structure shown in figure 5.6, are searched in the form

$$\left.\begin{array}{l} E = E_1\exp[ih_x - \omega t + k_1 y] \\ H = H_1\exp[ih_x - \omega t + k_1 y] \end{array}\right\} y < 0;\quad \left.\begin{array}{l} E = E_2\exp[ih_x - \omega t - k_2 y] \\ H = H_2\exp[ih_x - \omega t - k_2 y] \end{array}\right\} y > 0 \tag{5.21}$$

Note that in each of the materials, the field propagating along the interface of the structure decreases exponentially along the normals directed to each of the materials.

Consider approaches to the study of surface polaritons in cases (1) and (2) and then, for example, give more detailed results for surface polaritons in case (2) for the gyromagnetic medium 1 (figure 5.6).

5.5.1.1 Gyroelectric media

For a gyroelectric medium 1, the dielectric constant is as follows

$$\varepsilon = \begin{pmatrix} \varepsilon_{xx} & \varepsilon_{xy} & 0 \\ -\varepsilon_{xy} & \varepsilon_{xx} & 0 \\ 0 & 0 & \varepsilon_{zz} \end{pmatrix} \tag{5.22}$$

Using equations (5.22), (5.19) and (5.20), we get that TE mode E_z, H_x, H_y is not coupled to TM mode H_z, E_x, E_y, which is influenced by the gyrotropy. Using equations (5.18a), (5.18b), (5.21), we obtain the dispersion equation for surface polaritons at the boundary 'gyroelectric medium-metamaterial' in the form (Boardman *et al* 2005c)

$$k_2/\varepsilon_2 + k_1/\varepsilon_V + \left(ik_x\varepsilon_{xy}/\varepsilon_V\varepsilon_{xx}\right) = 0 \tag{5.23}$$

where

$$\begin{aligned}\varepsilon_V = \varepsilon_{xx} + \varepsilon_{xy}^2/\varepsilon_{xx};\ (\omega^2/c^2)\mu_1\varepsilon_V - k_x^2 + k_1^2 = 0;\ (\omega^2/c^2)\mu_2\varepsilon_2 \\ - k_x^2 + k_2^2 = 0;\end{aligned} \tag{5.24}$$

Surface polaritons in the medium 'semi-infinite magnetoplasma-semi-infinite negative phase medium', where the semiconductor n-InSb is accepted as the magnetoplasma, taking into account the corresponding dispersion of the components in the THz range, has been investigated using relation (5.23). Relevant results can be found in Boardman *et al* (2005c) and are not given here.

5.5.1.2 Gyromagnetic medium 1 in figure 5.6

The dispersion equation for TE mode (E_z, H_x, H_y) in this case is obtained by taking into account equation (5.20)–(4.30) and the tensor with components of the form (Boardman *et al* 2005c, Gurevich and Melkov 1996), $\mu_{xx} = \mu_{yy} = \mu$, $\mu_{xy} = -\mu_{yx} = i\mu_a$, where $\mu = (\omega_3^2 - \omega^2)/(\omega_\mathrm{H}^2 - \omega^2)$, $\mu_a = \omega_\mathrm{M}\omega/(\omega_\mathrm{H}^2 - \omega^2)$, $\omega_\mathrm{H} = \gamma H_0$, $\omega_\mathrm{M} = 4\pi\gamma M_0$, $\omega_3^2 = \omega_\mathrm{H}(\omega_\mathrm{H} + \omega_\mathrm{M})$; γ, M_0, H_0 are the gyromagnetic ratio, the saturation magnetization, and the magnetization field, respectively. The obtained dispersion relation for surface polaritons in the system 'semi-infinite ferromagnetic–semi-infinite metamaterial with a negative phase (ferromagnetic/NPM)' has the form

$$k_2/\varepsilon_2 + k_1/\varepsilon_\mathrm{V} + \left(ik_x\varepsilon_{xy}/\varepsilon_\mathrm{V}\varepsilon_{xx}\right) = 0 \tag{5.25}$$

The quantities included in equation (5.25) can be obtained by substituting the corresponding quantities included in equations (5.23) and (5.24), by means of replacement $\varepsilon_{xx;xy;v} \rightarrow \mu_{xx;xy;v}$. According to Shadrivov *et al* (2004), Boardman *et al* (2005d), McCall *et al* (2002), Veselago (1968), and Pendry (2004) the electric permittivity and the magnetic permeability of NPM are, respectively:

$$\varepsilon_2 = 1 - \omega_\mathrm{P}^2/\omega^2,\ \mu_2 = \tilde{F}(\omega^2 - \tilde{\omega}_0^2)/(\omega^2 - \omega_0^2),\ \tilde{F} = 1 - F,\ \tilde{\omega}_0^2 = \omega_0^2/\tilde{F} \tag{5.26}$$

where ω_P, ω_0, F are the characteristic parameters of NPM (Eleftheriades 2005), and in this case dissipative losses are not taken into account. Note that NPM with ε_2, μ_2 in the form (5.26) can be realized, in particular, in the microwave frequency range (Shadrivov *et al* 2004, Boardman *et al* 2005d, McCall *et al* 2002, Veselago 1968, Pendry 2004). The following section and appendix C.3 give a detailed analysis of the dispersion properties of surface magnetic polaritons in the ferromagnetic/NPM system obtained for the first time in Boardman *et al* (2005c).

5.5.2 New types of surface magnetic polaritons in the ferromagnetic system—metamaterial

A detailed analysis of dispersion (figure 5.7(a)–(e)) for electromagnetic waves (EMW)/surface polaritons in the system, shown in figure 5.6 (the details including clarification of the characteristic frequencies indicated on the horizontal axes in figure 5.7(a), are presented in appendix C.3), bring to the onset a summary of the characteristics of the surface waves in the ferromagnetic/NPM system. These results have been compared with these obtained for a 'ferromagnetic/dielectric' system with the same parameters of a ferromagnetic (see, for example, Hartstein *et al* 1973). The results of this comparison are as follows. (1) For the 'ferromagnetic/NPM' system, many branches of the surface waves exist (see figure 5.7(c)–(e)). From the other hand, in the 'ferromagnetic/dielectric' system there is only one dispersion branch. (2) In the 'ferromagnetic/dielectric' system, only surface waves with $V_\mathrm{g} > 0$ propagate, while in the system 'ferromagnetic/NPM' surface waves with both $V_\mathrm{g} > 0$ and $V_\mathrm{g} < 0$ can propagate, with the positive sign of the phase velocity, $V_\mathrm{ph} > 0$ (figure 5.7(b)). (3) For the 'ferromagnetic/NPM' system, the frequencies exist, where $V_\mathrm{g} = 0$, with the finite values of the wavenumber h, (figure 5.7(b) and (e)). (4) The surface waves in

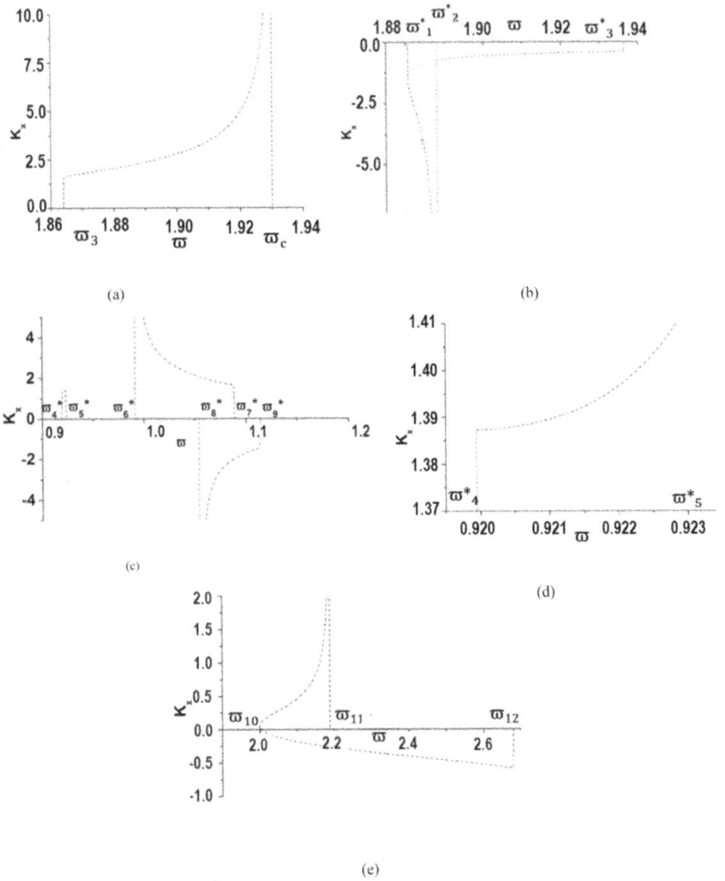

Figure 5.7. (a)–(e) The dispersion in the systems (a) ferromagnetic dielectric; (c)–(e) are ferromagnetic-NPM; $K_x = h/k_0 > 0$, $K_x = h/k_0 > 0$, $\varpi = \omega/\omega_M$; c is speed of light; (b) is the dispersion in the ferromagnetic-NPM system for the parameter set A (see appendix C.3); (c) is the dispersion in the system 'ferromagnetic-NPM' for the set of parameters **B**: the 1st, 2nd and 3rd branches of the surface waves; (d) is the dispersion in the 'ferromagnetic-NPM' system for the parameter set B (see appendix C.3): the 1st branch of the surface waves (the same as in figure 5.7(c), in more detail); (e) is the dispersion in the system 'ferromagnetic-NPM', for the set of parameters B: 4th branch of the surface waves. In the frequency domain (see (b)) in the system 'ferromagnetic-NPM' there are simultaneously direct and backward surface magnetic polaritons, unlike the system 'ferromagnetic-dielectric'.

the 'ferromagnetic/NPM' system strongly depend on the relationship between the parameters of the NPM and the ferromagnetic (compare figure 5.7(b) and figure 5.7(c) and (d)), where the dispersion dependences are shown, which correspond for the set of parameters A and B, determined in appendix C.3). (5) Let us emphasize the peculiarity for the magnetostatic case. For the 'ferromagnetic/dielectric' system, the surface magnetostatic waves (SMSW) with $V_g > 0$, $V_g \to 0$ exist for the definite frequency, $\varpi \to \varpi_2^*$. From the other hand, from the set of parameters B, SMSW are available for three different frequencies, see ϖ_6^*, ϖ_8^*, ϖ_{11}^* in figure 5.7(c) and (e). Again, in distinction to the system 'ferromagnetic/dielectric', in the 'ferromagnetic/NPM' system, SMSW exist with $V_g > 0$, as well as $V_g < 0$, for $\varpi \to \varpi_2^*$, figure 5.7(b) and $\varpi \to \varpi_6^*$, figure 5.7(c) (Boardman *et al* 2005c). The expressions were obtained for the reflection coefficient of incident waves from an ATR type system for TE waves. Some details of this calculation are discussed in appendix C.3. The abbreviation ATR ('attenuated total reflection') refers in this case (Boardman *et al* 2005c) to the system 'dielectric prism-gyrotropic layer-NPM'.

The minima of the reflection coefficient R_{TM} of the electromagnetic wave from the system shown in figure 5.6(b) (see formula (C.25) for R_{TM}, appendix C.4) correspond to the resonant excitation of the surface waves/polaritons (see detailed calculations in Boardman *et al* 2005c). Finding the reflection coefficients (C.24), (C.25) is necessary to determine the Goos–Hänchen shift (Goos and Hänchen 1947) in the reflection of electromagnetic pulses from the system 'dielectric-gyrotropic material-NPM' (Boardman *et al* 2005c).

Suppose that MM (NPM) and dielectric are at $y > 0$ and $y < 0$, respectively. The dispersion equation for TE surface polaritons takes the form $D(\omega, k) = k_b + q/\mu = 0$ where $k_b = [k^2 - \varepsilon_b(\omega/c)^2]^{1/2}$, $q = [k^2 - \varepsilon\mu(\omega/c)^2]^{1/2}$; k, k_b, q are the wavenumber of surface polaritons, and the decrements of the exponential decrease of the field in the directions perpendicular to the interface, in the dielectric and MM, respectively; ε_b and ε,μ are the electric permittivity and the electric permittivity and the magnetic permeability of the dielectric and MM, respectively. The group velocity of TE surface polaritons is equal to $v_g = V_0/V_1$, $V_0 = [1/k_b + 1/(\mu q)]$, $V_1 = (\omega/c^2)[\varepsilon_b/k_b + 1/(\mu q)]$ $-\left(\omega_{pm}^2/\omega^3\mu q c^2\right)\left[\omega_{pe}^2 - 2(qc^2/\mu)\right]$; here it was supposed that $\varepsilon_1 = \varepsilon(\omega) = 1 - \omega_p^2/\omega^2$, $\mu = \mu(\omega) = 1 - \omega_m^2/\omega^2$, where $\omega_{p,\,m}$ are the corresponding 'plasma frequencies' (see Boardman *et al* 2005b and references in this work). Thus, the group velocity sign changes at some frequency where $|\varepsilon_b| = |\varepsilon_c| = |\varepsilon||\mu|$. At this frequency $k_b \to q$, $\mu \to -1$, which means that $|\varepsilon_c| \to |\varepsilon|$. Let us put it $k > 0$. Then, at $\varepsilon_b < \varepsilon_c$, $v_g < 0$, (the surface polaritons are backward waves), while at $\varepsilon_b > \varepsilon_c$, $v_g > 0$, (the surface polaritons are forward waves).

This consideration is necessary for qualitative analysis (after generalization to a nonlinear case) of the influence of nonlinearity on the direction of resonantly excited polaritons in the system 'MM (NPM)—dielectric' and the influence of the nonlinearity on the Goos–Hänchen shift in reflection of electromagnetic pulses (Boardman *et al* 2005b).

5.5.3 Shift of electromagnetic pulses upon reflection from the system 'semi-infinite dielectric–dielectric layer-semi-infinite metamaterial with negative ε, μ'

Details of the derivation of the equations, used for the determination of the features of the effect of electromagnetic momentum shift upon reflection from the system 'semi-infinite dielectric–dielectric layer-semi-infinite MM with negative ε, μ' are considered in appendix C.4. In this section, we will outline and analyze the main results and features of this effect for MMs.

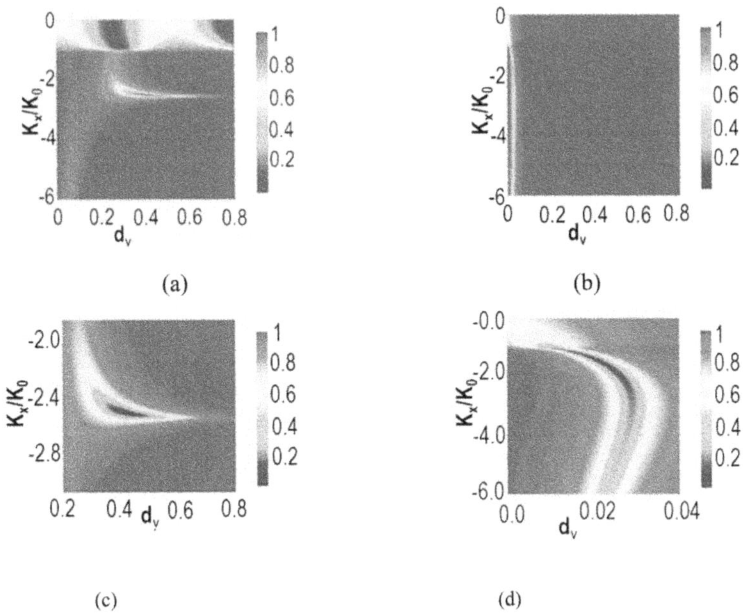

Figure 5.8. (a)–(d) (Boardman *et al* 2005b). The reflection intensity for the TE-polarized plane wave for the prism-YIG-NPM configuration (of ATR type) is calculated. Wavenumbers are normalized to $k_0 = \omega_M/c$. The YIG thickness is given in terms of $d_\lambda = d/\lambda_0$, where λ_0 is the wavelength incident onto the system in a free space. (a) Magnetic field equals to zero. (b) Magnetic field is finite: $\omega_H/\omega_M = 1.43$, $\omega_p/\omega_M = 2.084$, $\omega_0/\omega_M = 1.642$, $\Gamma/\omega_M = 0.001$ for all illustrated cases. (c) and (d) are enlarged versions of (a) and (b). Reprinted from Boardman *et al* (2005c), copyright IOP Publishing, all rights reserved.

Figure 5.8 shows the reflection coefficient of an ATR type system (for details, see appendix C.4) for TE waves as a function of the normalized wavenumber and the thickness of the gyrotropic layer (iron–yttrium garnet (YIG)).

The abbreviation ATR ('attenuated total reflection') refers (Boardman *et al* 2005b) to the arrangement shown in figure 5.8 which consists of a dielectric prism-gyrotropic layer-NPM combination. The minima of the reflection coefficient correspond to the resonant excitation of surface waves/polaritons. One-dimensional dependences of the reflection coefficients on the normalized wavenumber for a certain value of the gyrotropic layer thickness, demonstrate the corresponding minima shown in figure 5.9(a) and (b).

Figure 5.9(c) and (d) illustrate the effect of the Goos–Hänchen shift (see also the caption to figure 5.9(c) and (d)) (Goos Fand Hänchen 1947, Tamir and Boardman 1982, Horowitz and Tamir 1971). The analysis of the dispersion in the corresponding wavenumber intervals (not shown here) shows that the one of the two resonances in figure 5.9, which corresponds to relatively small wavenumbers (figure 5.9), is associated with backward SW (with $V_g V_{ph} < 0$, where $V_{g,\,ph}$ are group and phase velocities, respectively). The resonance in the region of relatively

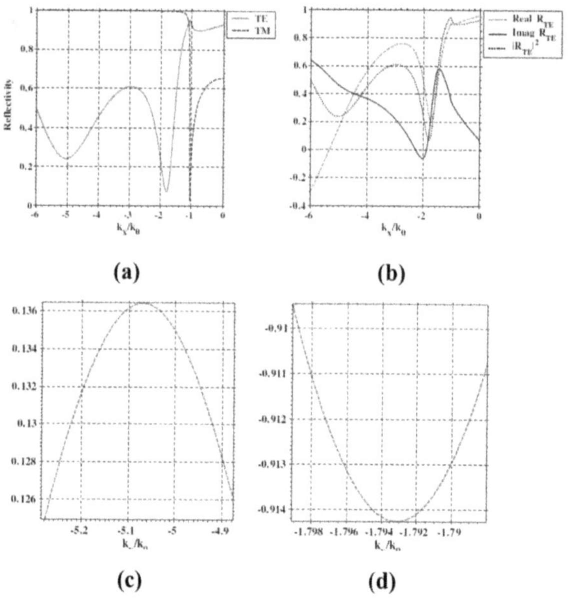

Figure 5.9. (Boardman *et al* 2005b). (a) Calculated TE and TM ATR reflection coefficients for a prism-ferromagnetic (YIG)-NPM configuration. (b) Real and imaginary parts of the TE reflected amplitudes and the TE reflected intensity. Wavenumbers are normalized to $k_0 = \omega_M/c$; the thickness YIG is represented in units $d_\lambda = d/\lambda_0 = 0.025$, where λ_0 is the wavelength of the incident beam in free space. $\omega_H/\omega_M = 1.43$, $\omega/\omega_M = 1.882$, $\omega_p/\omega_M = 2.084$, $\omega_0/\omega_M = 1.642$, $\Gamma/\omega_M = 0.001$. (c, d) Calculations of the S_{GH} factor proportional to the Goos–Hänchen shift for the TE beam falling onto the 'prism-ferromagnetic (ZIG)-NPM' configuration. The wavenumber is normalized to $k_0 = \omega_M/c$. The YIG thickness is represented in units of $d_\lambda = d/\lambda_0 = 0.025$ where λ_0 is the wavelength of the incident beam in a free space. $\omega_H/\omega_M = 1.43$. $\omega/\omega_M = 1.882$, $\omega_p/\omega_M = 2.084$, $\omega_0/\omega_M = 1.642,/\omega_M = 0.001$. (c) For the resonance at large values of wavenumbers. (d) For the resonance at small values of wavenumbers. Reprinted from Boardman et al (2005c), copyright IOP Publishing, all rights reserved.

large values of the wavenumbers (figure 5.9) corresponds to the forward SW. The resonances shown in figure 5.9(a) and (b), correspond to the large positive and negative values of the Goos–Hänchen shift, respectively (see also relations (C.29) and (C.30)). By itself, the effect of the opposite of the Goos–Hänchen shift signs for SW for opposite directions of propagation of the respective SW is not new (Lakhtakia 2004). A new effect presented in our work (Boardman *et al* 2005b) is the possibility of obtaining both positive and negative Goos–Hänchen shifts by the resonant excitation of oppositely directed SWs at the NPM boundary and gyrotropic medium, which have the same wavenumber values belonging to the same dispersion branches and may exist simultaneously.

The role of nonlinearity in the phenomenon of the Goos–Hänchen shift in the layered structure 'NPM-dielectric (air)-nonlinear dielectric (glass)' was investigated in Boardman *et al* (2005b). For the Goos–Hänchen shift, as seen in figure 5.9(c) and (d), it is important to control the sign of the group velocity of surface polaritons on the corresponding interface. In the case under consideration, such a control can be provided, in particular, by the nonlinearity. This possibility is illustrated qualitatively by considering the group velocity of surface polaritons in the system 'semi-infinite dielectric–semi-infinite NPM', and one of these media may be nonlinear (Boardman *et al* 2005b). Suppose the NPM and the dielectric are at $y > 0$ and $y < 0$, respectively. The dispersion equation for TE surface polaritons takes the form $D(\omega, k) = k_b + q/\mu = 0$, where $k_b = [k^2 - \varepsilon_b(\omega^2/c^2)]^{1/2}$, $q = [k^2 - \varepsilon\mu(\omega^2/c^2)]$, k, k_b, q are the wavenumber of surface polaritons, and exponential damping in directions perpendicular to the interface, in the dielectric and MM, respectively; ε_b and ε, μ are the electric permittivity and the electric permittivity and magnetic permeability of the dielectric and MM, respectively. The group velocity TE of surface polaritons is equal to $v_g = V_0/V_1$, where $V_0 = k[1/k_b+1/(\mu q)]$, $V_1 = (\omega/c^2)[\varepsilon_b/k_b + 1/(\mu q)] - (\omega_{pm}^2/\omega^3\mu qc^2)[\omega_{pe}^2 - 2(qc^2/\mu)]$; it is supposed here that $\varepsilon = \varepsilon(\omega) = 1 - \omega_p^2/\omega^2$, $\mu = \mu(\omega) = 1 - \omega_m^2/\omega^2$, where $\omega_{p,\,m}$ are the corresponding 'plasma frequencies' (see Boardman *et al* 2005b and references in this work). Thus, the group velocity sign changes at some frequency, where $|\varepsilon_b| = |\varepsilon_c| = |\varepsilon||\mu|$. At this frequency $k_b \to q$, $\mu \to -1$, which means that $|\varepsilon_c| \to |\varepsilon|$. Let's put it $k > 0$. Then, at $\varepsilon_b < \varepsilon_c$, $v_g < 0$ (the surface polaritons are backward waves), and at $\varepsilon_b > \varepsilon_c$, $v_g > 0$ (the surface polaritons are forward waves). If the quantities $|\varepsilon_b|$ and $|\varepsilon_c|$ are relatively close, and (at least) one of them is nonlinear, then it can be expected that under the action of nonlinearity, the type of waves may change between forward and backward ones.

Accordingly, one can expect the possibility of changing the sign of the Goos–Hänchen shift depending on the 'degree of nonlinearity' when the pulses are reflected from the boundary 'dielectric-NPM/MM'. Numerical calculations in Boardman *et al* (2005b) confirm the qualitative conclusions about the effect of nonlinearity on the Goos–Hänchen shift in the layered NPM-air-dielectric system. Namely, it turned out (see figure 5.3 in Boardman *et al* 2005b) that: (1) in the case of the presence of nonlinearity (in particular, in the dielectric) in the Goos–Hänchen dielectric, it exceeds that in its absence; (2) in the presence of nonlinearity the beam is clearly spatially divided into two beams, of which the 'shadow' is well spatially separated from the main beam and corresponds to the Goos–Hänchen shift; (3) for reflected beams, 'nonlinearity overcomes diffraction', and both reflected beams do not diverge in nonlinear focusing media.

5.6 Conclusions

(1) On the basis of the derived evolution equations, taking into account the effects of magnetic and electrical nonlinearities and diffraction management, a qualitative explanation of the mechanism of effective control of spatial solitons, including breathers, is given by means of an MO effect in a transversely inhomogeneous magnetic field in MO waveguides.

(2) Temporal (pulse) solitons in MMs in the presence of external stationary and transversely inhomogeneous magnetic fields are investigated. A frequency window with negative magnetic permeability and electric permittivity and different signs of group and phase velocities was found. The frequency dependences of the coefficients of the modified normalized NSE are found. The model includes the nonlinear dispersion, diffraction (for two-dimensional pulses), Raman interaction, and higher-order (including third-order) linear dispersion. The frequency dispersion of these normalized coefficients represents the essential properties of nonlinear solitons in MMs. In particular, the standard Drude model for the electric permittivity and magnetic permeability is considered. MO interaction is used as a practical mechanism for controlling the velocity of nonlinear (including one-dimensional) pulses. Practically this effect can be used for data processing systems/signals.

(3) The possibility of effective MO control of bullets in an MM waveguide with a transverse magnetic field (due to the Voigt effect) is shown. The MO interaction leads to a quasi-periodic dependence of the intensity of the bullets on the coordinate along the propagation direction in the waveguide. The most effective stabilization of the bullets in the MM MO waveguide is achieved at the parameters of the waveguide providing a zero (or very small) value of the self-action coefficient (nonlinear dispersion).

(4) Group velocities corresponding to the formation of polaritons in 'dielectric-MM (NPM)' and 'dielectric-ferromagnetic' systems are found. It is shown that in the 'ferromagnetic/NPM' system there are frequency regions where both forward and backward waves can propagate, unlike the 'dielectric-ferromagnetic' system. The coefficients of reflection of electromagnetic waves from the system 'dielectric-ferromagnetic-MM (NPM)' were obtained. These coefficients are necessary for the study of the nontrivial Goos–Hänchen shift, when reflecting electromagnetic pulses from the system 'dielectric-ferromagnetic-MM (NPM)' (Boardman *et al* 2005c). A possibility of nonlinear control of Goos–Hänchen shift in nonlinear dielectric-MM layered media has been proposed.

Appendix C

Appendix C.1 Derivation of evolutionary equation for time domain and layered metamaterial medium with linear and nonlinear effects of higher orders and magnetooptic control

To derive the nonlinear Schrödinger equation (NSE) with higher nonlinear terms for the amplitude of the envelope in a MM waveguide, we use the slowly varying amplitude $A(z, \omega)$ and the 'modal field' distribution $F(y)$ of a planar waveguide. In the presence of nonlinearity in the boundary conditions, the best way is to use the NEELS method (see for example sections 3.1.3, 3.2 and appendix B.2.4 in chapter 3). We will look at the case now when there are no nonlinearities in the boundary conditions, and we will use the standard method for the integrated optics (Agrawal 1989), and for simplification consider the 'slightly waveguiding' structure. The rapid changes are associated with a wavenumber β_0 of the linear waveguiding mode:

$$\tilde{E}_y = A(z, \omega)F(y)e^{i\beta_0 z} \tag{C.1}$$

Using the wave equation (Boardman *et al* 2010b, Boardman and King 2009c, Boardman and Xie 2003, Zvezdin and Kotov 1997), where the separation of variables occurs with the constant $\tilde{\beta}^2$, or with the eigenvalue, we get a set of equations for A, F:

$$\frac{\partial^2 F}{\partial y^2} + \frac{\omega^2}{c^2}\left(\varepsilon(\omega)\mu_{\rm L}(\omega) - \tilde{\beta}^2\right)F = 0,\ \frac{\partial^2 A}{\partial z^2} + 2i\beta_0\frac{\partial A}{\partial z} + \left(\tilde{\beta}^2 - \beta_0^2\right)A = 0 \tag{C.2}$$

$$F\frac{\partial^2 A}{\partial z^2} + 2i\beta_0 F\frac{\partial A}{\partial z} - \beta_0^2 FA = -A\frac{\partial^2 F}{\partial y^2} - \frac{\omega^2}{c^2}\left(\varepsilon(\omega)\mu_{\rm L}(\omega)\right)FA \tag{C.3}$$

The relative electric permittivity and magnetic permeability and nonlinear (Kerr) coefficient for slightly waveguiding (which is considered layered purpose material structure under consideration) are selected as

$$\varepsilon_{\rm L} = \begin{cases} \varepsilon_1 > 0, & y \leqslant -\dfrac{L}{2} \\ \varepsilon_2(\omega), & |y| \leqslant \dfrac{L}{2} \\ \varepsilon_3 > 0, & y \geqslant \dfrac{L}{2} \end{cases} \quad \mu_{\rm L} = \begin{cases} \mu_1 = 1, & y \leqslant -\dfrac{L}{2} \\ \mu_2(\omega), & |y| \leqslant \dfrac{L}{2} \\ \mu_3 = 1, & y \geqslant \dfrac{L}{2} \end{cases}$$
$$\chi^{(3)} = \begin{cases} 0, & y \leqslant -\dfrac{L}{2} \\ \chi^{(3)}, & |y| \leqslant \dfrac{L}{2} \\ 0, & y \geqslant \dfrac{L}{2} \end{cases} \tag{C.4}$$

Neglecting for the term $A\partial^2 F/\partial y^2$ in weakly guiding wave structure, multiplying (C.4) F^* by a complex conjugate value, integrating by y, we obtain an equation for the envelope

$$\frac{\partial^2 A}{\partial z^2} + 2i\beta_0 \frac{\partial A}{\partial z} - \beta_0^2 A + I_1 A + I_2 \mid A \mid^2 A = 0 \tag{C.5}$$

Again due to the slightly waveguiding properties of the system, the integration of the modal fields is expressed approximately in terms of the linear wavenumber, β_L and the effective nonlinear coefficient, $\overline{\chi}^{(3)}$.

$$I_1 = \frac{\dfrac{\omega^2}{c^2}\displaystyle\int_{-\infty}^{\infty} \mid F \mid^2 \varepsilon_\mathrm{L}(\omega)\mu_L(\omega)\mathrm{d}y}{\displaystyle\int_{-\infty}^{\infty} \mid F \mid^2 \mathrm{d}y} \approx \frac{\omega^2}{c^2}\varepsilon_\mathrm{L}(\omega)\mu_\mathrm{L}(\omega)\frac{\displaystyle\int_{-\infty}^{\infty} \mid F \mid^2 \mathrm{d}y}{\displaystyle\int_{-\infty}^{\infty} \mid F \mid^2 \mathrm{d}y} \equiv \beta_\mathrm{L}^2(\omega) \tag{C.6}$$

$$I_2 = \frac{\dfrac{\omega^2}{c^2}\displaystyle\int_{-\infty}^{\infty} \mid F \mid^4 \mu_\mathrm{L}(\omega)\chi^{(3)}\mathrm{d}y}{\displaystyle\int_{-\infty}^{\infty} \mid F \mid^2 \mathrm{d}y} \approx \frac{\omega^2}{c^2}\mu_2(\omega)\chi^{(3)}\frac{\displaystyle\int_{-L/2}^{L/2} \mid F \mid^4 \mathrm{d}y}{\displaystyle\int_{-\infty}^{\infty} \mid F \mid^2 \mathrm{d}y} = \frac{\omega^2}{c^2}\mu_2(\omega)\overline{\chi}^{(3)} \tag{C.7}$$

$$2i\beta_0\frac{\partial A}{\partial z} + \frac{\partial^2 A}{\partial z^2} + \left(\beta_\mathrm{L}^2(\omega) - \beta_0^2\right)A + \frac{\omega^2}{c^2}\mu_2(\omega)\overline{\chi}^{(3)} \mid A \mid^2 A = 0 \tag{C.8}$$

For a pulse with a narrow spectrum and a 'carrier ω_0 frequency', we use the expansions near this central frequency

$$\omega = (\omega - \omega_0) + \omega_0 \Rightarrow \omega_0\left(1 + \frac{i}{\omega_0}\frac{\partial}{\partial t}\right),\ \ \mu_2(\omega) \Rightarrow \left(\mu(\omega_0) + i\frac{\partial \mu}{\partial \omega}\bigg|_{\omega_0}\frac{\partial}{\partial t}\right),$$

$$\beta_\mathrm{L}^2 = \left(\beta_0 + i\beta_1\frac{\partial}{\partial t} - \frac{\beta_2}{2}\frac{\partial^2}{\partial t^2} - \frac{i\beta_3}{6}\frac{\partial^3}{\partial t^3} + \ldots\right)^2 \tag{C.9}$$

After substituting (C.9) in (C.8), go to the moving coordinate frame. The concept of transition from the laboratory frame to the moving coordinate one, taking into account the possibility of the 'soliton motion' with speeds of higher or lower than the group velocity, seems very important for the MO to control solitons in the MM waveguide (layered) media, when this control is associated with the non-stationary bias magnetic fields and magnetizations. (Boardman *et al* 2010b, Boardman and King 2009c, Boardman and Xie 2003, Zvezdin and Kotov 1997). This concept is illustrated accordingly for 'fast' and 'slow' motions using figure C.1.

To transfer to the moving coordinate frame, you need to take into account the transformation

$$z \to z',\ t \to t' - \beta_1 z,\ \beta_1 = \frac{\partial k}{\partial \omega}\bigg|_{\omega_0} = \frac{1}{v_g},\ \beta_n = \frac{\partial^n k}{\partial \omega^n}\bigg|_{\omega_0};\ k(\omega_0) = \beta_0\quad n = 1,2,3\ldots$$

$$\frac{\partial}{\partial t} \Rightarrow \frac{\partial}{\partial t'}\ \frac{\partial}{\partial z} \Rightarrow \frac{\partial}{\partial z'} - \beta_1\frac{\partial}{\partial t'}\ \frac{\partial^2}{\partial z^2} \Rightarrow \frac{\partial^2}{\partial z'^2} - 2\beta_1\frac{\partial^2}{\partial z'\partial t'} + \beta_1^2\frac{\partial^2}{\partial t'^2}$$

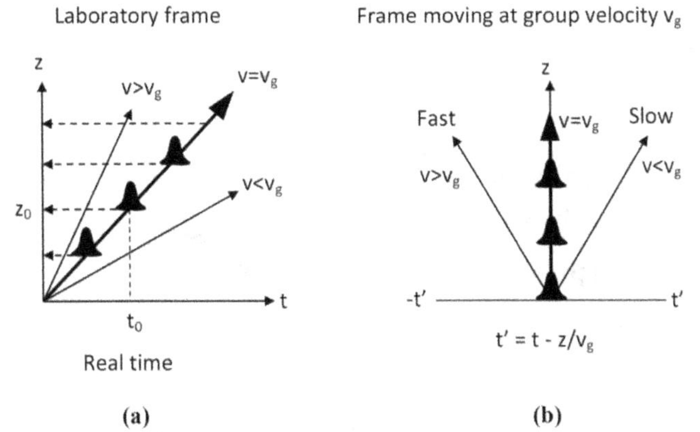

Figure C.1. Coordinate frames moving at a speed of v_g; (a) is the pulse in the laboratory reference frame, and (b) is the pulse in the moving frame. Reprinted from Boardman *et al* 2010b, copyright (2010) with permission from Elsevier.

Here the wavenumber of the wake packet is denoted as k. We take the following orders of magnitudes that are the parts of the evolutionary equation: $\partial/\partial\tilde{z}'$, $\partial^2/\partial t'^{2\sim}$, $\chi^{(3)} \mid A \mid^2$, and, accordingly $\partial^2/\partial t'\partial\tilde{z}'$, $\partial^3/\partial t'^3$. With the same accuracy, neglect the term with $\partial^2/\partial z'^2$; we make transformations when transferring to a moving coordinate frame and an approximate factor and from the equation with an accuracy to the above values (see explanation of the details of the output procedure and reference in Boardman *et al* (2010b), Boardman and King (2009c), Boardman and Xie (2003), and Zvezdin and Kotov (1997)). Note that $z' = L_D Z$, $t' = T_0 T$, $A = \psi/\sqrt{[\omega_0^2/(2 \mid k_0 \mid c^2)]| \mu(\omega_0)| \overline{\chi}^{(3)}}$, where the dispersion length is equal to $L_D = T_0^2/| \beta_2 \mid$ and the value can T_0 be interpreted as the width (duration) of the pulse. We include a frequency-independent term describing the Raman scattering and the nonlinear dispersion (nonlinear coefficient) $\chi^{(3)}$ (Boyd 2003, Boardman *et al* 2010b). As a result, in the absence of the MO effect, we get a normalized evolutionary equation for the envelope (this equation is also written in the section 5.4 as equation (5.8) and repeated here for clarity)

$$
\begin{aligned}
& i\frac{\partial\psi}{\partial Z} - \frac{1}{2}\,\mathrm{sgn}(\beta_2)\frac{\partial^2\psi}{\partial T^2} - i\delta_3\frac{\partial^3\psi}{\partial T^3} \\
& + \mathrm{sgn}(\chi^{(3)})\,\mathrm{sgn}(k')\,\mathrm{sgn}(\mu')\left(|\psi|^2\psi + iS\frac{\partial}{\partial T}(|\psi|^2\psi) - \tau_R\psi\frac{\partial}{\partial T}(|\psi|^2)\right) = 0
\end{aligned}
\tag{C.10}
$$

where $\delta_3 = \dfrac{\beta_3}{6 \mid \beta_2 \mid T_0}$, $S = \dfrac{1}{\omega_0 T_0}\left[2 - \dfrac{\beta_1\omega_0}{\beta_0} + \left(\dfrac{\omega}{\mu}\dfrac{\partial\mu}{\partial\omega}\right)_{\omega_0} + \left(\dfrac{\omega}{\chi^{(3)}}\dfrac{\partial\chi^{(3)}}{\partial\omega}\right)_{\omega_0}\right]$ and $\tau_R = \dfrac{\int_{-\infty}^{\infty} tR(t)\mathrm{d}t}{T_0}$, which uses the response function $R(t)$ that characterizes the Raman interaction (Boardman *et al* 2010b, Agrawal 1989). The nonlinear dispersion, taken into account in (C.10), qualitatively reflects the essential nonlinear dispersion properties of nanostructured nonlinear systems. Equation (C.10) corresponds to the version of equation (3.53) from section 3.3.4, with $\nu = 0$, $k = 0$, $\gamma_0 = 0$, $\nu = 0$.

Appendix C.2 Details of the Lagrange formalism method for qualitative description of magnetooptic control of the purpose of material solitons

The Lagrange function (Whitham 1974) and the average Lagrangian defined as

$$L = \frac{1}{2}\left\{i\left(\psi\frac{\partial\psi^*}{\partial z} - \psi^*\frac{\partial\psi}{\partial z}\right) - \operatorname{sgn}(\beta_2)\left|\frac{\partial\psi}{\partial T}\right|^2 - \operatorname{sgn}(\chi^{(3)})\,|\,\psi\,|^4 - 2v\,|\,\psi\,|^2\right\} \quad \text{(C.11)}$$

$$L = \frac{1}{2}\int_{-\infty}^{\infty}\left\{i\left(\psi\frac{\partial\psi^*}{\partial z} - \psi^*\frac{\partial\psi}{\partial z}\right) - \operatorname{sgn}(\beta_2)\left|\frac{\partial\psi}{\partial T}\right|^2 - \operatorname{sgn}(\chi^{(3)})\,|\,\psi\,|^4 - 2v\,|\,\psi\,|^2\right\}\mathrm{d}T \quad \text{(C.12)}$$

Using the test function of the form (5.16) (section 5.4.3), we get the average Lagrangian in the form of

$$L = -\eta\xi\frac{\partial T_0}{\partial Z} + 2\eta\frac{\partial\phi}{\partial Z} - \operatorname{sgn}(\beta_2)\eta\frac{\xi^2}{4} - \frac{2}{3}\eta^3\left[\frac{1}{2}\operatorname{sgn}(\beta_2) + \operatorname{sgn}(\chi^{(3)})\right] - \eta^2\int_{-\infty}^{\infty} v(T)\operatorname{sech}^2[\eta(T - T_0)]\mathrm{d}T \quad \text{(C.13)}$$

The ‘dynamic’ equations are obtained as ($g = (\eta, \phi, \xi, T_0)$)

$$\frac{\partial}{\partial Z}\frac{\partial L}{\partial\left(\frac{\partial g}{\partial Z}\right)} - \frac{\partial L}{\partial g} = 2\operatorname{Re}\int_{-\infty}^{\infty}\left(R\frac{\partial\psi^*}{\partial g}\right)\mathrm{d}T \quad \text{(C.14)}$$

$$g = \phi\text{:}\quad \frac{\partial\eta}{\partial Z} = 0,\quad \eta = \text{constant} \quad \text{(C.15)}$$

$$g = \xi\text{:}\quad \frac{\partial T_0}{\partial Z} = -\frac{1}{2}\operatorname{sgn}(\beta_2)\xi + \eta^2 S\operatorname{sgn}(\chi^{(3)}) + \delta_3\left(\eta^2 + \frac{3}{4}\xi^2\right) \quad \text{(C.16)}$$

$$\frac{\partial\xi}{\partial Z} = \eta\int_{-\infty}^{\infty} v(T)\frac{\partial}{\partial T_0}\operatorname{sech}^2[\eta(T - T_0)]\mathrm{d}T + \frac{16}{15}\tau_R\eta^4 \quad \text{(C.17)}$$

Thus, we obtain equations (5.16a) and (5.16b).

Appendix C.3 Detailed consideration of dispersion for new types of surface magnetic polaritons in the system 'ferromagnetic-metamaterial' and the 'magnetostatic boundary case'

In this case, the medium occupying the region $y > 0$ in figure 5.6(a) is a ferromagnetic; in the region $y < 0$, NPM will be affected. Taking into account equations (5.25) and (5.24) with redefinition of the corresponding values described after equation (4.25), we obtain for TE mode (with E_z, H_x, H_y components):

$$AK_x^4 + BK_x^2 + C = 0, \tag{C.18}$$

$$\begin{aligned} &K_x = h/k_0, \quad k_0 = \omega/c, \quad K_{x1,2}^2 = (-B \pm \sqrt{D})/2A, \; D = B^2 - 4AC, \\ &A = \left(\mu_2^2 - \mu\mu_{\mathrm{el}}\right)^2 - 4\mu_2^2\mu_{\mathrm{a}}^2, \\ &B = 2\mu_2\left[\mu\left(\mu_2^2 - \mu\mu_{\mathrm{el}}\right)(\varepsilon_2\mu_{\mathrm{el}} - \varepsilon_1\mu_2) + 2\mu_2^2\varepsilon_2\mu_{\mathrm{a}}^2\right], \\ &k_1 = \sqrt{h^2 - k_0^2\varepsilon_1\mu_{\mathrm{el}}}, \quad k_2 = \sqrt{h^2 - k_0^2\varepsilon_2\mu_2} \end{aligned} \tag{C.19}$$

The goal is to demonstrate only one fundamentally new property of the system under consideration, namely 'ferromagnetic/NPM', compared to those known from previous studies of the system 'ferromagnetic-dielectric', described in detail, for example, in Hartstein *et al* (1973). To do this, the parameters of the NPM are selected in such a way that for some frequency to obtain

$$\mu_2 = -1, \quad k_2 = 0, \quad h^2 = -\varepsilon_2 k_0^2 \tag{C.20}$$

$$\mu = \varepsilon_2, \tag{C.21}$$

$$\omega^2 = \omega_{A,B}^2 = \left(\omega_S^2 \pm \sqrt{\omega_S^4 - 8\omega_P^2\omega_H^2}\right)/4 \;\; (\text{"A"} \to \text{" + "}, \text{"B"} \to \text{" - "}) \tag{C.22}$$

$$\begin{aligned} &\omega_S^2 = \omega_P^2 + \omega_H^2 + \omega_3^2, \quad \omega_0^2 = \omega_{0A,B}^2 = (1 - F/2)\omega_{A,B}^2, \\ &\varepsilon_2 < 0, \mu_2 < 0, \quad \omega_P^2 = 1.25\omega_3^2 \end{aligned} \tag{C.23}$$

Different curves for two sets of parameters are shown in figure 5.7. These sets are denoted by 'set A and B', according to the indices 'A, B' in equations (C.21) and (C.22). The difference between dispersion dependences for two option sets (see figure 5.7) demonstrates a strong dependence of the dispersion in the system 'ferromagetic/NPM' on the parameters of NPM.

(I) Set A: $\varpi_H = \omega_H/\omega_M = 1.43$; $\bar{\omega}_3 = \omega_3/\omega_M = 1.8641$; $\bar{\omega}_0 = \omega_0/\omega_M = 1.6423$; $\bar{\omega}_P = \omega_P/\omega_M = 2.0841$; $F = 0.56$; $\bar{\tilde{\omega}}_0 = \tilde{\omega}_0/\omega_M = 2.4759$.

(II) For the set B, different from the set A: $\bar{\omega}_0 = \omega_0/\omega_M = 0.9238$; $\bar{\tilde{\omega}}_0 = \tilde{\omega}_0/\omega_M = 1.3928$. We will find out the characteristic frequencies for different branches of surface waves (SW), and after that it will be possible to make a conclusion about the difference between SW in the 'ferromagnetic/NPM' and 'ferromagnetic/dielectric' systems. For the set of parameters A, there is one branch of SW in the system 'ferromagnetic/NPM'. The dispersion of SW is shown in figure 5.7. Here for $\varpi < \varpi_1^*$ we have $D < 0$, whereas for $\varpi > \varpi_1^*$, it is $D > 0$.

(1) Thus, for $\varpi < \varpi_1^*$ EMW cannot propagate along the X axis in the system shown in figure 5.6. With $\varpi \to \varpi_2^*$, $A \to \infty$. Accordingly, for this frequency we have two solutions equal to (C.19): $|K_x| \to \infty$ and $|K_x| = \sqrt{-C/B}$; both of them correspond to certain points on the dispersion curve shown in figure 5.7.
(2) For $\varpi \to \varpi_3^*$, $\mu_2 \to -1$, $k_2 \to 0$, and k_2^2 is positive or negative for $\varpi < \varpi_3^*$ and $\varpi > \varpi_3^*$, respectively. As a result, we obtain that for a system characterized by a set of parameters A, SW can propagate in the frequency domain $\varpi_1^* < \varpi < \varpi_3^*$, as shown in figure 5.7(b). In figure 5.7(a), the frequencies $\varpi = \varpi_3$ and $\varpi = \varpi_c = \omega_c/\omega_M$, $\omega_c = \omega_H + \omega_M/2$ correspond to the excitation of volumetric and surface unperturbed polaritons (Hartstein *et al* 1973), respectively. The latter one, in fact, represents a limiting case for SMSW (Gurevich and Melkov 1996) in a semi-infinite ferromagnetic. These waves in the magnetostatic case $\varpi \to \varpi_c$ correspond to oscillations rather than waves, because simultaneously $K_x \to \infty$, and the group velocity $V_g \to 0$. The dispersion for the set of the parameters B (see also relation (C.22)) is shown in figure 5.7(c)–(e), and in figure 5.7(d) the fragment of figure 5.7(c) is shown on a larger scale. Four *branches of* SW are obtained.
(3) The first one lies in the frequency interval $\varpi_4^* < \varpi < \varpi_5^*$ (see figure 5.7(a) and (b), where these branches are shown in detail). With $\varpi \to \varpi_4^*$, $k_1^2 \to 0$, $\varpi < \varpi_4^*$, and EMW can only exist as volumetric (non-surface) waves. With $\varpi \to \varpi_5^* \mu_2 \to \infty$, and D changes the sign, so that EMW cannot propagate in the direction of X in the system shown in figure 5.7 at $\varpi > \varpi_5^*$.
(4) The second branch lies in the interval $\varpi_6^* < \varpi < \varpi_7^*$. When $\varpi \to \varpi_6^*$, A and h^2 change the sign, and SW can only propagate when $\varpi > \varpi_6^*$. With $\varpi \to \varpi_7^*$, $\mu_2 \to -1$, $k_2^2 \to 0$, and when $\varpi \to \varpi_7^*$, the waves in environment 2 cannot propagate like SW.
(5) The third branch exists in the interval $\omega_8^* < \varpi < \varpi_9^*$. When $\varpi \to \varpi_8^*$, A and h^2 change the sign, and SW at $h^2 > 0$ can propagate when $\varpi > \varpi_8^*$. With $\varpi \to \varpi_9^*$, D changes the sign, and EMW can propagate when $\varpi < \varpi_8^*$, where $D > 0$ and h^2 is real and positive.
(6) The *Fourth branch* (figure 5.7(e)) *exists* in the interval $\omega_{10}^* < \varpi < \omega_{12}^*$. In the $\omega_{10}^* < \varpi < \varpi_{11}^*$ interval, *there are 'simultaneously'* two solutions for each frequency, in other words, two 'sub-branches' of the fourth branch at $h > 0$ and $h < 0$. At $\varpi \to \varpi_{10}^*$, $D \to 0$, and EMW can propagate along the X direction at $\varpi > \varpi_{10}^*$, where h^2 is real. When $\varpi \to \varpi_{12}^*$, $k_1^2 \to 0$, and with $\varpi > \varpi_{12}^*$ EMW are not surface in the region 1 (ferromagnetic, figure 5.6).

With $\varpi \to \varpi_{11}^{*}$, $A \to 0$ and, for one of the, $|h| \to \infty$. Let us emphasize that only the fourth branch of SW, shown in figure 5.7(e), corresponds to $\mu_2 < 0$; and for $\varepsilon_2 > 0$ and $\varepsilon_2 < 0$ for $\varpi > \varpi_P = 2.0841$ and $\varpi < \varpi_P$, respectively. For all other SW shown in figure 5.7(a)–(d), we have $\mu_2 < 0$, $\varepsilon_2 < 0$.

In the frequency range $\varpi_1^* < \varpi < \varpi_2^*$ (figure 5.7(b)) in the system 'ferromagnetic/NPM', in distinction to the system 'ferromagnetic-dielectric', forward and backward surface magnetic polaritons can be exist *simultaneously*. Thus, there are polaritons of a new type in the system 'ferromagnetic/NPM'.

Appendix C.4 Derivation of equations for the shift of the electromagnetic pulse when reflected from the system 'semi-infinite dielectric layer-semi-infinite metamaterial with negative ε, μ'

Use the approach outlined in Boardman *et al* (2005c), Goos and Hänchen (1947), Tamir and Boardman (1982), Horowitz and Tamir (1971).

The system shown in figure 5.8(b) is uniform along z, and the wave components along the z axis are equal to

$$H_{z1} = \left(A_{1+}\exp\left[ik_{y1}y\right] + A_{1-}\exp\left[-ik_{y1}y\right]\right)\exp\left[i(k_x x - \omega t)\right]$$

$$H_{z2} = \left(A_{2+}\exp\left[ik_{y2}y\right] + A_{2-}\exp\left[-ik_{y2}y\right]\right)\exp\left[i(k_x x - \omega t)\right],$$

$$H_{z3} = A_3\exp\left[ik_{y3}y\right]\exp\left[i(k_x x - \omega t)\right]$$

The tangential components of the electric field are equal to

$$E_{x2} = \left\{-(q_+/\omega\varepsilon_0)A_{2+}\exp\left[ik_{y2}y\right] + (q_-/\omega\varepsilon_0)A_{2-}\exp\left[-ik_{y2}y\right]\right\}\exp\left[i(k_x x - \omega t)\right]$$

$$E_{x1} = -\left(k_{y1}/\omega\varepsilon_0\varepsilon_1\right)\left(A_{1+}\exp\left[ik_{y1}y\right] - A_{1-}\exp\left[-ik_{y1}y\right]\right)\exp\left[i(k_x x - \omega t)\right]$$

$$E_{x3} = -\left(k_{y3}/\omega\varepsilon_0\varepsilon_3\right)A_3\exp\left[ik_{y3}y\right]\exp\left[i(k_x x - \omega t)\right],$$

where $q_+ = (\varepsilon_{zz}k_{y2} + k_x\varepsilon_{xy})/(\varepsilon_{zz}^2 + \varepsilon_{xy}^2)$, $q_- = (\varepsilon_{zz}k_{y2} - k_z\varepsilon_{xy})/(\varepsilon_{zz}^2 + \varepsilon_{xy}^2)$

The complex reflection coefficient from the system 'dielectric prism-gyrotropic material-NPM', shown in figure 5.6(b) on the surface of 'ferrite-NPM' equals

$$R_{\mathrm{TM}} \equiv \frac{A_{1-}}{A_{1+}} = \frac{\left(\frac{k_{y1}}{\varepsilon_1} - q_+\right)\left(\frac{k_{y3}}{\varepsilon_3} + q_-\right)\exp\left[-ik_{y2}d\right] - \left(\frac{k_{y1}}{\varepsilon_1} + q_-\right)\left(\frac{k_{y3}}{\varepsilon_3} - q_+\right)\exp\left[ik_{y2}d\right]}{\left(\frac{k_{y1}}{\varepsilon_1} + q_+\right)\left(\frac{k_{y3}}{\varepsilon_3} + q_-\right)\exp\left[-ik_{y2}d\right] + \left(q_- - \frac{k_{y1}}{\varepsilon_1}\right)\left(\frac{k_{y3}}{\varepsilon_3} - q_+\right)\exp\left[ik_{y2}d\right]} \quad (C.24)$$

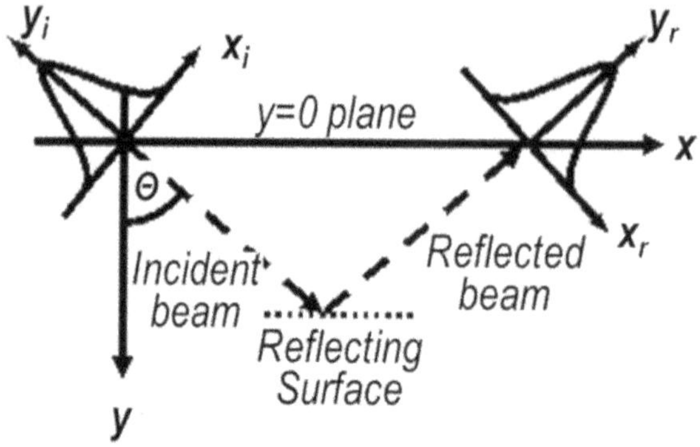

Figure C.2. Gaussian beam under conditions of complete reflection from the surface. This process can be considered as the reflection of a beam of electromagnetic radiation falling upon a plane surface placed at some point below the plane $y = 0$. Reprinted from Boardman *et al* (2005c), copyright IOP Publishing, all rights reserved.

For the system 'dielectric prism-NPM-gyrotropic material', the reflection coefficient is equal to

$$R_{\mathrm{TM}} = \frac{A_{1-}}{A_{1+}} = \frac{\left(\frac{k_{y1}}{\varepsilon_1} - \frac{k_{y2}}{\varepsilon_2}\right)\left(\frac{k_{y2}}{\varepsilon_2} + p\right)\exp\left[-ik_{y2}d\right] + \left(\frac{k_{y1}}{\varepsilon_1} + \frac{k_{y2}}{\varepsilon_2}\right)\left(\frac{k_{y2}}{\varepsilon_2} - p\right)\exp\left[ik_{y2}d\right]}{\left(\frac{k_{y1}}{\varepsilon_1} + \frac{k_{y2}}{\varepsilon_2}\right)\left(\frac{k_{y2}}{\varepsilon_2} + p\right)\exp\left[-ik_{y2}d\right] + \left(\frac{k_{y1}}{\varepsilon_1} - \frac{k_{y2}}{\varepsilon_2}\right)\left(\frac{k_{y2}}{\varepsilon_2} - p\right)\exp\left[ik_{y2}d\right]} \quad \text{(C.25)}$$

where $p = \left(\varepsilon_{xy}k_x + \varepsilon_{yy}k_{y2}\right)/(\varepsilon_{xx}\varepsilon_{yy} - \varepsilon_{xy}\varepsilon_{yz})$.

To determine the features of the Goos–Hänchen shift effect on the surface of 'ferrite-NPM' (Boardman *et al* 2005c), consider the reflection of the Gaussian beam from such a surface, as shown in figure C.2.

The local coordinate frames are (x_i, y_i) for the incident pulse and (x_r, y_r) for the reflected pulse with beginnings in the centers of falling and reflected beams (figure C.2). Between the laboratory (x, y) and local (x_i, y_i) coordinates for the falling beam there are relations $x_1 = x\cos\theta$, $y_1 = x\sin\theta$. If the beam has TM polarization, then consider its H_z component, the shape of which is Gaussian (C.26), and it is assumed that in the local coordinate frame, the beam propagates along the axis y_i.

$$\begin{aligned} H(x_1, y_1) &= \frac{1}{\sqrt{\pi}w}\exp\left\{-\left(\frac{x_1}{w}\right)^2\right\}, \\ H(x, y = 0) &= \frac{1}{\sqrt{\pi}w}\exp\left\{-\left(\frac{x\cos\theta}{w}\right)^2\right\}\exp\left\{ikx\sin\theta\right\} \end{aligned} \quad \text{(C.26)}$$

The second one is the relation in (C.26) describes a projection $y = 0$ in a plane that is parallel to the physical interface. The Fourier transform for the incident pulse is

$H_{inc}(x, y) = (1/2\pi)\int_{-\infty}^{+\infty} \Phi(k_x)\exp\{i[k_x x + k_y y]\}dk_x,\ k_x^2 + k_y^2 = k^2$. After the direct integration $\Phi(k_x)$ can be presented in the form

$$\begin{aligned}\Phi(k_z) &= \int_{-\infty}^{+\infty} H(x, y = 0)\exp\{-ik_z x\}dx \\ &= \exp\{-[(k_x - k\sin\theta)/2\cos\theta]^2\}\cos^{-1}\theta\end{aligned} \tag{C.27}$$

The reflected beam is presented in the form of

$$\begin{aligned}H_{refl}(x, z) = (2\pi\cos\theta)^{-1}\int_{-\infty}^{+\infty} \Gamma(k_x) \\ \exp\{-[(k_1 - k\sin\theta)\cos^{-1}\theta]\}\exp\{i[k_x x - k_z z]\}dk_x\end{aligned} \tag{C.28}$$

In (C.28), $\Gamma(k_z) = R_{TM}$ is a reflection coefficient for a plane wave, θ is a geometric angle of incidence, that is, one corresponding to θ_0 the plane wave propagating along the beam axis. Next, an angle θ_0 is chosen for which the tangential component of the wavenumber is equal to $k_{\perp 0} = k\sin\theta_0$, and then a near vicinity of θ_0 is considered using a Fourier series expansion. This means that in the equation (C.28) is accepted $\Delta k_z = k_z - k\sin\theta_0$, and $\Gamma(k_z) = \Gamma(k_{z0}) + (\partial\Gamma(k_{\perp\partial})/\partial k_z)_{k_z=k_{z0}}\Delta k_\perp + \ldots$, while the y-component of the wavenumber is approximated with accuracy to the second order, namely

$$k_z \approx k_{z0} + \left(\frac{\partial k_z}{\partial k_x}\right)_{k_{z0}}\Delta k_z + \frac{1}{2}\left(\frac{\partial^2 k_z}{\partial k_x^2}\right)_{k_{z0}}(\Delta k_z) + \ldots = k$$

$$\cos\theta_0 - \tan(\theta_0)(\Delta k_x) - \frac{1}{(2k\cos^3(\theta_0))}(\Delta k_x)^2$$

using this approximation in (C.27), along with the corresponding integration, and considering the reflected beam with local coordinates (x_r, y_r), the result is $H_{refl}(x_r) = H_{r0}\exp(-[x_r - i\Delta/2]^2/w^2)\exp(-\Delta^2/4w^2)$. This shows that the 'Goos–Hänchen shift' parallel to the axis x_r is equal to the real part $i\Delta/2$. The last result coincides in the corresponding separate case of the system 'dielectric (quartz)-dielectric (air)-metal' with the corresponding result (Mazur and Djafari 1984). After evaluation of integrals, (C.28) indicates that the shift G is proportional to the combination $\Gamma(k_x)$ and derivative of this value, namely

$$\begin{aligned}G\infty S_{GH} &= \mathrm{Re}\{i[\cos\theta_0\Gamma(\theta_0)]^{-1}(\partial\Gamma/\partial\theta)_{\theta_0}\} \\ &= (\cos\theta_0)^{-1}\,\mathrm{Re}\,[iR_{TM}{}^{-1}(\theta_0)(\partial R_{TM}/\partial\theta)_{\theta_0}]\end{aligned} \tag{C.29}$$

or

$$S_{GH} = \{\cos\theta_0[(R_{TM}^r)^2 + (R_{TM}^t)^2]\}^{-1}[R_{TM}^t(\partial R_{TM}^r/\partial\theta) - R_{TM}^r(\partial R_{TM}^t/\partial\theta)] \tag{C.30}$$

The property to achieve for the shift value G (C.30) 'giant' values (Boardman *et al* 2005c) is associated with the resonance of the excitation of surface polaritons, where the reflection coefficient R_{TM} 'tends to zero'.

Problems to chapter 5

Problem 5.1

Derive the nonlinear evolution equation of the stationary nonlinear Schrödinger equation with both electric and magnetic nonlinearities. Consider a planar MM waveguide; suppose the coordinate x is normal to the waveguide boundaries: waves propagate along z layer and, respectively, the diffraction occurs in y direction; only the linear diffraction is taken in to account. Suppose that the mode structure is (E_x, E_z, H_y), but in fact it is very close to TEM (E_x, H_y) mode.

Obtain the evolution equation for the slowly varying amplitude $A_{\rm H}$ of the H_y component in the dimensional form:

$$i\frac{\partial A_{\rm H}}{\partial z} + \frac{1}{2k_{\rm c}}\frac{\partial^2 A_{\rm H}}{\partial y^2} + \frac{2\pi\omega_{\rm c}^2}{c^2 k_{\rm c}\varepsilon_{\rm c}}\left(\varepsilon_{\rm c}^2\chi_{\rm H} + \mu_{\rm c}^2 k_{\rm E}\right) | A_{\rm H} |^2 A_{\rm H} = 0 \qquad \text{(P5.1)}$$

Here $k_{\rm c}$, $\omega_{\rm c}$ are the carrier wavenumber and the frequency of the electromagnetic wave; $\varepsilon_{\rm c}$, $\mu_{\rm c}$ are corresponding values of the dielectric permittivity and magnetic permeability, $\chi_{\rm H,\,E}$ are the coefficients of magnetic and electric cubic nonlinearities, respectively.

To do this, start Maxwell equations in the form:

$$\mathrm{curl}\vec{H} = \frac{1}{c}\frac{\partial \vec{D}}{\partial t} + \frac{4\pi}{c}\frac{\partial \vec{P}_{\rm NL}}{\partial t} \qquad \text{(P5.2)}$$

$$\mathrm{curl}\vec{E} = -\frac{1}{c}\frac{\partial \vec{B}}{\partial t} - \frac{4\pi}{c}\frac{\partial \vec{\mu}_{\rm NL}}{\partial t} \qquad \text{(P5.3)}$$

In equations (P5.2) and (P5.3), $\vec{P}_{\rm NL}$, $\vec{M}_{\rm NL}$ are electric and magnetic nonlinear polarization, respectively. Use also the relations

$$\mathrm{div}\vec{B} = \mu\,\mathrm{div}\vec{H} + 4\pi\,\mathrm{div}\vec{M}_{\rm NL} = 0 \qquad \text{(P5.4)}$$

Suppose that $P_{{\rm NL}z}$ may be ignored and account for that

$$\mathrm{curl}\left(\frac{\partial \vec{P}_{\rm NL}}{\partial t}\right)_y = \frac{\partial}{\partial z}\frac{\partial P_{{\rm NL}x}}{\partial t} - \frac{\partial}{\partial x}\frac{\partial P_{{\rm NL}z}}{\partial t} \simeq \frac{\partial}{\partial z}\frac{\partial P_{{\rm NL}x}}{\partial t} \qquad \text{(P5.5)}$$

Also use that in the case of quasi-TEM electromagnetic wave,

$$| E_z | \ll | E_x |, \left| H_y \right| \qquad \text{(P5.6)}$$

the dispersion relation for the electromagnetic waves has the form

$$k^2(\omega) \simeq \frac{\omega^2}{c^2}\varepsilon(\omega)\mu(\omega) \qquad \text{(P5.7)}$$

Suppose that the dependence of electromagnetic field on the x coordinate is weak and, respectively

$$\Delta \simeq \frac{\partial^2}{\partial z^2} + \frac{\partial^2}{\partial y^2} \quad \text{(P5.8)}$$

Use the Fourier transform

$$\begin{matrix} \overline{E}(\vec{r}, \omega) = \\ \overline{H}(\vec{r}, \omega) = \end{matrix} \int_{-\infty}^{\infty} \begin{matrix} \overline{E}(\vec{r}, t) \\ \overline{H}(\vec{r}, t) \end{matrix} e^{i\omega t} dt, \quad \begin{matrix} \tilde{\overline{P}}_{NL} = \\ \tilde{\overline{M}}_{NL} = \end{matrix} \int_{-\infty}^{\infty} \begin{matrix} \tilde{\overline{P}}_{NL}(\vec{r}, t) \\ \tilde{\overline{M}}_{NL}(\vec{r}, t) \end{matrix} e^{i\omega t} dt \quad \text{(P5.9)}$$

Applying the curl operation to equation (P5.2) and using (P5.3)–(P5.9), we get, in the approximation of slowly varying ampltudes, equation (P5.1). Then ignore the nonlinear diffraction and neglecting the terms of the corresponding order $| A_H |^2 \dfrac{\partial^2}{\partial y^2} \simeq 0$

$$\frac{\partial^2}{\partial z^2} \ll 2ik_c \frac{\partial}{\partial z} \quad \text{(P5.10)}$$

Take into account that

$$A_H = \frac{k_c}{\mu_c \frac{\omega}{c}} A_E \quad \text{(P5.11)}$$

You get finally the equation (P5.1).

Problem 5.2

Get the modified nonlinear Schrödinger (evolution) equation, accounting for the nonlinear diffraction.

Take into account that $| A_H |^2 \dfrac{\partial^2}{\partial y^2} \neq 0$. Using the same procedure as one user for the solution problem 5.1, get the evolution equation in the form:

$$\begin{aligned} &i\frac{\partial A_H}{\partial z} + \frac{1}{2k_c}\frac{\partial^2 A_H}{\partial y^2} + \frac{2\pi\omega_c}{c^2 k_c}\left(\varepsilon_c \chi_H + \frac{\mu_c^2}{\varepsilon_c}\chi_E\right) | A_H |^2 A_H \\ &+ \frac{2\pi\chi_H}{\mu_c k_c}\frac{\partial^2}{\partial y^2}(| A_H |^2 A) = 0 \end{aligned} \quad \text{(P5.12)}$$

The last term in the left-hand part (P5.12) describes the effect of the nonlinear diffraction with the coefficient

$$K = \frac{2\pi\chi_H}{\mu_c k_c} \quad \text{(P5.13)}$$

Problem 5.3

Use the Drube model for the dependences $\varepsilon(\omega)$ and $\mu(\omega)$ in MM, namely

$$\varepsilon(\omega) = 1 - \frac{\omega_{\text{pe}}^2}{\omega^2},\ \mu(\omega) = 1 - \frac{\omega_{\text{pm}}^2}{\omega^2} \qquad \text{(P5.14)}$$

And put $\Omega \equiv \omega/\omega_{\text{pe}}$, $\omega_{\text{pm}}/\omega_{\text{pe}} = 0{,}6$. Build a dependence of the coefficient of nonlinear diffraction (see the coefficient in the last term in left-hand part of equation (P5.12) formula (P5.13), problem 5.2).

Use relations (P5.13) and (P5.7) and put them in the expression coefficient of the nonlinear diffraction, see equation (P5.12) in problem 5.2.

List of abbreviations

ATR	Attenuated total reflection
CGLE	Complex Ginzburg–Landau equation
EMW	Electromagnetic waves
MO	Magnetooptic
NEELS	Nonlinear evolution equations for layered structures
NPM	Negative phase medium
NSE	Nonlinear Schrödinger equation
PPM	Positive phase medium
SMSW	Surface magnetostatic waves
SW	Surface waves
TE	Transverse electric
TEM	Transverse electric magnetic
TM	Transverse magnetic
YIG	Iron–yttrium garnet

References

Agrawal G 1989 *Nonlinear Fiber Optics* (Boston: Academic) p 323

Argyropoulos C, Estakhri N M, Monticone F and Alu A 2013 Negative refraction, gain and nonlinear effects in hyperbolic metamaterials *J. Opt. Express* **21** 15037–47

Banerjee P P and Nehmetallah G 2007 Spatial and spatiotemporal solitary waves and their stabilization in nonlinear negative index materials *J. Opt. Soc. Am.* B **24** 10

Boardman A D, Marinov K, Pushkarov D I and Shivarova A 2000a Influence of nonlinearly induced diffraction on spatial solitary waves *Opt. Quant. Electron.* **32** 49–62

Boardman A D, Marinov K, Pushkarov D I and Shivarova A 2000b Wave-beam coupling in quadratic nonlinear optical waveguides: Effects of nonlinearly induced diffraction *Phys. Rev. E.* **62** 2871–7

Boardman A D and Xie M 2003 Magnetooptics: A critical review *Introduction to Complex Mediums for Optics and Electromagnetics* ed W S Weiglhofer and A Lakhtakia (Bellingham, WA: SPIE Press) pp 197–219

Boardman A D, King N and Velasco L 2005a Negative refraction perspective *Special Issue: Exotic Electromagnet.* **25** 365–89

Boardman A D, Velasco L, King N and Rapoport Y 2005b Ultra-narrow bright spatial solitons interacting with left-handed surfaces *J. Opt. Soc. Am.* B **22** 1443–52

Boardman A D, King N, Rapoport Y and Velasco L 2005c Gyrotropic impact upon negatively refracting surfaces *New J. Phys.* **7** 1–24

Boardman A D, Egan P, Velasco L and King N 2005d Control of planar nonlinear guided waves and spatial solitons with a left-handed medium *J. Opt. A: Pure Appl. Opt* **7** S57–67

Boardman A D and Marinov K 2006a Electromagnetic energy in a dispersive metamaterial *Phys. Rev.* B **73** 16

Boardman A D and Velasco L 2006b Gyroelectric cubic-quintic dissipative solitons *IEEE J. Select. Top. Quant. Electron.* **12** 388–97

Boardman A D, King N, Mitchell-Thomas R C, Malnev V N and Rapoport Y G 2008 Gain control and diffraction-managed solitons in metamaterials *Metamaterials* **2–3** 145–54

Boardman A D, Egan P, Hess O, Mitchell-Thomas R C and Rapoport Y G 2009a Nonlinear gyroelectric waves in magnetooptics metamaterials *Proc. Second Int. Workshop on Theoretical and Computational Nano-Photonics (TaCoNa-Photonics 2009) (Bad Honnef, (Germany))* vol 1176 pp 10–12

Boardman A D and King N 2009b Magneto-optics and the Kerr effect with ferromagnetic materials *Tutorials in Complex Photonic Media* ed M A Noginov, G Dewar, M W Mc Call and N I Zheludev (Bellingham, WA: SPIE Press)

Boardman A D, Mitchell-Thomas R C, King N J and Rapoport Y G 2010a Bright spatial solitons in controlled negative phase metamaterials *Opt. Commun.* **283** 1585–97

Boardman A D, Hess O, Mitchell-Thomas R C, Rapoport Y G and Velasco L 2010b Temporal solitons in magnetooptic and metamaterial waveguides *Photonics Nanostruct.—Fundam. Appl.* **8** 228–43

Boardman A D, Egan P, Mitchell-Thomas R C, Rapoport Y G and Velasco L 2010c Bright spatial solitons, nonlinear guided waves and complex metamaterial structures *Proc. SPIE-Int. Soc. Opt. Eng.* **7711** 771104

Boardman A D, Egan P, Mitchell-Thomas R C and Rapoport Y G 2010d Solitons, vortices and guided waves in plasmonic metamaterials *Proc. SPIE-Int. Soc. Opt. Eng.* **7757** 775711

Bonod N, Reinisch R and Popov E *et al* 2004 Optimization of surface-plasmon magneto-optical effects *J. Opt. Soc. Am.* B **21** 791–7

Boyd R W 2003 *Nonlinear Optics* 2nd edn (New York: Academic) p 576

Brabec T and Krausz F 1997 Nonlinear optical pulse propagation in the single-cycle regime *Phys. Rev. Lett.* **78** 3282–5

Buttner O, Bauer M, Demokritov S O, Hillebrands B, Kivshar Y S, Grimalsky V V, Rapoport Y and Slavin A N 2000 Linear and nonlinear diffraction of dipolar spin waves in yttrium iron garnet films observed by space-and time-resolved Brillouin light scattering *Phys. Rev.* B **61** 11576–87

D'Aguanno G, Attiucci N and Bloemer M J 2008 Ultra slow light pulses in a nonlinear metamaterial *J. Opt. Soc. Am.* B **25** 1236–43

Davoyan A R and Engheta N 2019 Nonreciprocal emission in magnetized epsilon-near-zero metamaterials *ACS Photonics* **6** 581–6

Eleftheriades G I 2005 ed G I Eleftheriades and K G Balmain *Negative-Refraction Metamaterials: Fundamental Principles and Applications* (New York: IEEE Press, Wiley) p 418

Goos Fand Hänchen 1947 Ein neuer und fundamentaler versuch zur total reflexion *Ann. Phys Lpz* **1** 33–46

Gurevich A G and Melkov G A 1996 *Magnetization Oscillations and Waves* (New York: CRS Press) p 464

Guixin L, Shuang Z and Thomas Z 2017 Nonlinear photonic metasurfaces *Nat. Rev. Mater.* 17010
Hartstein A, Burstein E and Maradudin A A *et al* 1973 Surface polaritons on semi-infinite gyromagnetic media *J. Phys. C: Condens. Matter* **6** 1266–76
Hermann C, Kosobukin V A and Lampel G *et al* 2001 Surface-enhanced magnetooptics in metallic multilayer films *Phys. Rev.* B **64** 235422–33
Horowitz B R and Tamir T 1971 Lateral displacement of a light beam at a dielectric interface *J. Opt. Soc. Am* **61** 586–94
Kadomtsev B B 1988 Collective phenomena in plasma *Science* (Moscow: Nauka) p 304 (in Russian)
Kivshar Y V and Krolikowski W 1995 Lagrangian approach for dark solitons *Opt.Commun.* **114** 353–62
Kochetov B A and Tuz V R 2018 Replication of dissipative vortices modeled by the complex Ginzburg–Landau equation *Phys. Rev.* E **98** 062214
Kochetov B A, Chelpanova O G, Tuz V R and Yakimenko A I 2019 Spontaneous and engineered transformations of topological structures in nonlinear media with gain and loss *Phys. Rev.* E **100** 062202
Kockaert P, Tassin P and Van der Sande G *et al* 2006 Negative diffraction pattern dynamics in nonlinear cavities with left-handed materials *Phys. Rev.* A **74** 033822
Kodera T and Caloz C 2018 Unidirectional loop metamaterials (ULM) as magnetless artificial ferrimagnetic materials: Principles and applications *IEEE Antennas Wireless Propagat. Lett.* **17** 1–5
Lakhtakia A 2004 Positive and negative Goos–Hänchen shifts and negative phase velocity mediums (alias left-handed materials) *Int. J. Electron. Commun.* **58** 229–31
Li P, Yang R and Xu Z 2010 Gray solitary-wave solutions in nonlinear negative-index materials *Phys. Rev.* E **82** 046603–7
Mazur P and Djafari R 1984 Effect of surface polaritons on the lateral displacement of a light beam at a dielectric interface *Phys. Rev.* B **30** 6759–62
McCall M W, Lakhtakia A and Weiglhofer W S 2002 The negative index of refraction demystified *Eur. J. Phys.* **23** 353–9
McGahan W A, He P, Chen L Y and Woollam J A 1991 Optical and magnetooptical characterization of TbFeCo thin-films in Trilayer structures *J. Appl. Phys.* **69** 4568–70
Mihalache D 2012 Linear and nonlinear light bullets: recent theoretical and experimental studies *Rom. J. Phys.* **57** 352–71
Mizumoto T and Naito Y 1982 Nonreciprocal propagation characteristics of YIG thin film *IEEE Trans. Microw. Theory Techn.* **30** 922–5
Pendry J B 2004 Negative refraction *Contemp. Phys.* **45** 191–202
Rapoport Y, Grimalsky V and Zaspel C 2012 Method for the derivation of nonlinear evolution equations in layered structures (NEELS): An example of nonlinear waves in gyrotropic layers *Bull. Kyiv Nat. Taras Shevchenko Univ. Series: Phys.-Math. Sci.* **14/15** 72–6
Rapoport Y G 2013 General method for modeling nonlinear waves in active metamaterial and gyrotropic layered structures *Proc. of Int. Kharkov Symp. on Physics and Engineering of Microwaves, Millimeter and Submillimeter Waves, MSMW'13 (Kharkov, Ukraine)* pp 253–5
Rapoport Y G 2014 General method for deriving the equations of evolution and modeling of nonlinear waves in layered active media with bulk and surface *Bull. Kyiv Nat. Taras Shevchenko Univ. Series: Phys.-Math. Sci.* **1** 281–8 (in Ukrainian)
Remer L, Mohler E, Grill W and Luthi B Nonreciprocity in the optical reflection of magneto-plasmas// *Phys. Rev.* B 1984**30** 3277–82

Remoissenet M, Remer L, Mohler E, Grill W and Luthi B 1984 *Waves Called Solitons* (Berlin: Springer)

Scalora M, Syrchin M S, Akozbek N, Poliakov E Y, D'Aguanno G, Mattiucci N, Bloemer M J and Zheltikov A M 2005 Generalized nonlinear Schrödinger equation for dispersive susceptibility and permeability: Application to negative index materials *Phys. Rev. Lett.* **95** 013902

Shadrivov I V, Sukhorukov A A, Kivshar Y S, Zharov A A, Boardman A D and Egan P 2004 Nonlinear surface waves in left-handed materials *Phys. Rev.* E **69** 016617–25

Simovski C R and Tretyakov S A 2009 Model of isotropic resonant magnetism in the visible range based on core–shell clusters *Phys. Rev.* B **79** 045111

Sugano S and Kojima N 2000 *Magneto-optics* (Berlin: Springer) p 317

Tamir T (ed) 1982 *The Lateral Wave Electromagnetic Surface Modes* ed A D Boardman (Chichester: Wiley) p 768

Torner L and Kartashov Y V 2010 Light bullets in optical tandems *Opt. Lett.* **34** 1129–31

Veselago V G 1968 The electrodynamics of substances with simultaneously negative values of ε and μ *Sov. Phys. Uspekhi* **10** 509–14

Wang H C and She W L 2005 Modulation instability and interaction of non-paraxial beams in self-focusing Kerr media *Opt. Commun.* **254** 145–51

Wang X, Kwon D-H, Werner D H, Khoo I-C, Kildishev A V and Shalaev V M 2007 Tunable optical negative-index metamaterials employing anisotropic liquid crystals *Appl. Phys. Lett.* **91** 143122

Whitham G B 1974 *Linear and Nonlinear Waves* (New York: Wiley) p 630

Wen S, Wang Y, Su W, Xiang Y, Fu X and Fan D 2006a Modulation instability in nonlinear negative-index material *Phys. Rev.* E **73** 36617–22

Wen S, Xiang Y, Su W, Hu Y, Fu X and Fan D 2006b Role of the anomalous self-steepening effect in modulation instability in negative-index material *Opt. Express* **14** 1568–75

Wen S, Xiang Y, Dai Q, Tang Z, Su W and Fan D 2007 Theoretical models for ultrasort electromagnetic pulse propagation in nonlinear metamaterials *Phys. Rev.* A **75** 033815–22

Xiang Y, Wen S, Dai X, Tang Z, Su W and Fan D 2007 Modulation instability induced by nonlinear dispersion in nonlinear metamaterials *J. Opt. Soc. Am.* B **24** 3058–63

Zhang J, Xiang Y, Wen S and Li Y 2014 Controlling self-focusing of ultrashort pulses with anomalous self-steepening in nonlinear negative-index materials *J. Opt. Soc. Am.* B **31** 45–52

Zhang J, Wen S and Xiang Y *et al* 2010 Spatiotemporal electromagnetic soliton and spatial ring formation in nonlinear metamaterials *Phys. Rev.* A **81** 023829–37

Zharov A A and Kurin V V 2007 Giant resonant magneto-optic Kerr effect in nanostructured ferromagnetic metamaterials *J. Appl. Phys.* **102** 123514

Zheng Y, Meng Y and Liu Y 2014 Solitons in Gaussian potential with spatially modulated nonlinearity *Opt. Commun.* **315** 63–8 A69–77

Zvezdin A K and Kotov V A 1997 *Modern Magnetooptics and Magnetooptical Materials* (Boca Raton, FL: CRC Press) p 404

Chapter 6

Parametric interactions of the nonlinear waves in active layered metamaterials and gyrotropic structures

Recent trends in microwave magnetism and superconductivity, namely spintronics, magnonics, magnon caloritronics, physics of magnonic crystals, spin-wave logic, and novel micro- and nano-scale magnetic devices, include also the approaches based on the parametric coupling of 2D pulses in gyrotropic layers (ferrite films (FFs)) (Prokopenko *et al* 2019). The parametric amplification of the 1D magnetostatic waves (MSWs) and wave front conjugation for 1D pulses have been proposed in Melkov and Sholom (1990) and Melkov *et al* (2000), respectively. The authors of the present book introduce, for the first time, magnetic bullets (strongly localized in both longitudinal and transverse direction 2D pulses possessing partially the properties of the solitons) (Grimalsky *et al* 1997, Buttner *et al* 2000), the possibility of the collapse of such pulses (Grimalsky *et al* 1997) and the possibility of parametric amplification and wave front reversal of bullets (Grimalsky *et al* 2000, 2001a,b) and formation of more complicated, in particular ordered multi-bullet structures in wide FFs with parametric pumping (Rapoport 2006). The devices using parametric coupling in ferrite and ferrite-superconductor films are suitable, in particular, for the special signal processing applications including analog ones (Prokopenko *et al* 2019). A lot of fascinating possibilities appear due to combination of the parametric interactions with the unique properties of the metamaterials (MMs). In particular, the combination of the principle of spatio-temporal modulation, which ensures the non-reciprocity of the medium, with parametric interaction leads to the emergence of bandgaps in the forward and backward directions. Namely, such bandgaps appear due to the active parametric interaction of incident waves with space-time harmonics in the medium (Elnaggar and Milford 2018). The use of non-reciprocity in a medium in the presence of spatio-temporal harmonics leads to the creation of

doi:10.1088/978-0-7503-2336-9ch6

many interesting devices, for example, circulators, isolators, nonreciprocal leaky wave antenna, and is potentially interesting for the creation of nonreciprocal metasurfaces (Elnaggar and Milford 2018). Interesting possibilities should be expected when nonlinearity and, accordingly, parametric interaction are combined with artificial unidirectional loop MMs (ULMs) which are in fact magnetless artificial ferrimagnetic materials, (Kodera and Caloz 2018); with locally resonant MM crystals which are artificial materials built from small spatially local resonant inclusions periodically arranged and possessing non-reciprocity transporting down to the subwavelength scale (Zangeneh-Nejad *et al* 2020); as well as using magneto-active media in the regime of epsilon-near-zero MMs; in this case interplay of magnetization and near-zero permittivity leads to a nonreciprocal spiral-like omnidirectional radiation due to a magnetically induced rotating dipole (Davoyan and Engheta 2019). Note that ULMs (Kodera and Caloz 2018) have important advantages compared to ferrites, including a possibility of the multiband operation, ultra broadband and electronic Faraday rotation direction switching. Spontaneous parametric downconversion can be applied for the broadband Purcell enhancement of nonlinear generation of quantum light in MMs with hyperbolic dispersion (Davoyan and Atwater 2018). The nonstationary parametric interaction of waves in negative-index metamaterials is studied and the broadening of the central maximum of the spectrum of the excited signal wave due to the presence of phase modulation of the exciting idler pulse is shown in Kasumova and Amirov (2019). As shown in Kruk *et al* (2019), the combination of bi-anisotropy, nonlinearity and topology provides tunable and non-reciprocal parametric photon generation in the devices of topological MM nano-photonics. In Li Guixin *et al* (2017), parametric nonlinear transformations of frequency, modulation and optical switching, using periodic and aperiodic metasurfaces, are considered. Metasurfaces can provide efficient harmonic and sum frequency generation or parametric downconversion. They can also be used for quantum optics. Entangled photons can be generated on nonlinear metasurfaces using parametric downconversion processes in nonlinear optical crystals. Note that in quantum optical communication, entangled photons play a crucial role (Li Guixin *et al* 2017).

In chapter 6, we consider new nonlinear wave structures in gyrotropic layers/FFs controlled by parametric amplification, parametric interaction in bi-anisotropic metamaterial waveguides and parametric interactions and phase conjugation on active two-dimensional chiral metasurfaces with linear and nonlinear Huygens sources.

6.1 Wave structures in layered active gyrotropic media with parametric interaction

6.1.1 Formulation of the problem

Solitons (Lukomsky 1978, Zvezdin and Popkov 1983, Melkov and Serga 1998) and vortices (Boardman *et al* 2005, Rapoport *et al* 2008) formed on MSWs in FFs have interesting prospects in terms of the creation of information processing devices. This trend is now becoming even more relevant due to the emergence and development of ferrite MMs and the corresponding new capabilities for the development of controlled ferrite devices (Zhao *et al* 2009). An active amplifying medium can be used to overcome wave losses, and parametric amplification is a very effective means of amplification and control for electromagnetic waves in gyrotropic layers. The amplification and phase coupling/inversion of the wave front for 1D (one-dimensional) MSWs were proposed in Melkov and Sholom (1990), Melkov and Serga (1998), and Gordon *et al* (1998). A special method for derivation of nonlinear evolution equations in layered media (NEELS) in the presence of parametric 2D MSW was proposed in Grimalsky *et al* (2000, 2001a, 2001b). In the same works, the wave front reversion and parametric amplification for two-dimensional pulses, including highly focused (magnetic bullets), were proposed (Grimalsky *et al* 2001b). Note that the bullets have some properties that are similar to those that resemble regular 1D solitons. Namely, after a central collision of two equal bullets, they are destroyed by collision (figures 6.1(a)–(d)). In contrast, bullets with centers

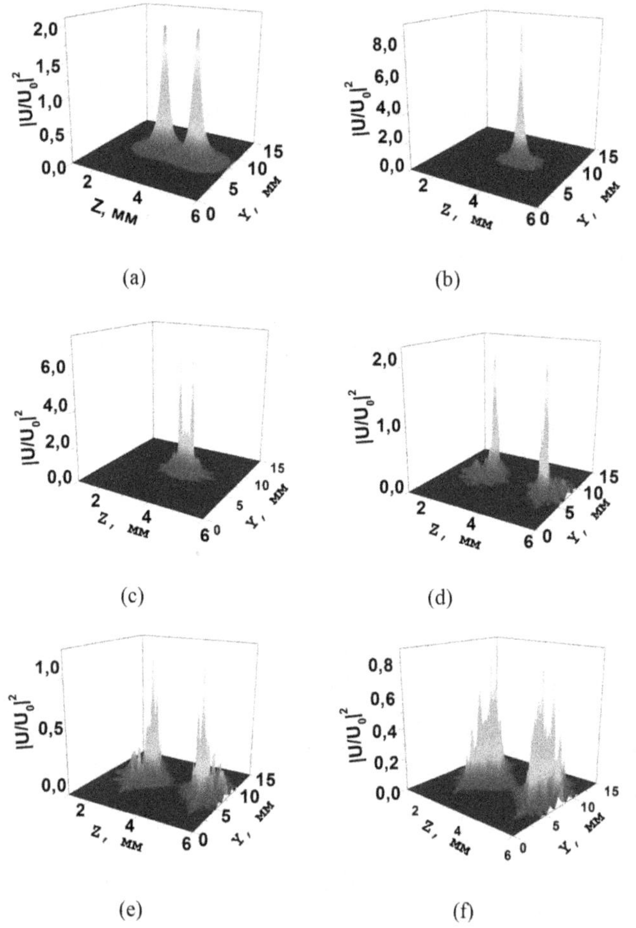

Figure 6.1. Results of numerical simulations for the collision of bullets (a) in a 'wide' FF. (a)–(d) The distance between centers of counter propagating bullets $s = 0.3$ mm; (e) $s = 0.1$ mm. (f) $s = 0$; the bullet propagation time τ from the input to the system ($Z = 0$) is equal to: (a) 82 ns; (b) 100 ns (the effective moment of collision); (c) 108 ns; (d)–(f) 138 ns.

shifted by a distance of about half their width are not destroyed, but retain, in general, their shape (figures 6.1(c)–(f)) (Slavin *et al* 2003). In Serga *et al* (2005) the possibility of forming (strongly focused) bullets for both the incident wave and the counter propagating idle wave was demonstrated. In Serga *et al* (2005), it was noted that the amplitude of the amplified idle pulse was smaller than for the incident wave.

In Serga *et al* (2008), the formation of bullets in 'narrow' FFs was considered (in the absence of parametric amplification). 'Narrow' in this context means that the width of the FF, for the considered power levels, is the magnitude of the inverse increment of the spatial instability in the transverse direction (such an instability leads to self-focusing), and together with self-compression in the longitudinal direction can result in the formation of a bullet on the backward volume MSWs (BVMSWs) (Kivshar and Agraval 2003) in a 'wide' or practically 'infinite' FF (at a scale $k_{\perp}^{'-1}$ where $k_{\perp}'$ is the transverse, with respect to the directions of propagation and the magnetic field, wavenumber). The bullets in narrow FFs, as proposed in Serga *et al* (2008), have an important characteristic, namely their effective width is of about of the width of the FF, so they are formed effectively only by a few transverse modes excited due to the nonlinearity in FF. It is shown (Grimalsky *et al* 2000, 2001a, 2001b, Rapoport 2006a, 2012b, 2014, Rapoport *et al* 2012a) how the effect of parametric interaction in gyrotropic layered nonlinear structures should be included in the NEELS method for the MSWs in FFs. The effects of a controlled formation of 2D nonlinear structures in active nonlinear media will be systematically considered. Bullets of a new type will be presented, namely 'knife-shaped bullets', elongated in the transverse direction. Qualitative mechanisms of the formation of definite active structures will be given and those effects of parametric amplification of solitons, which are due to the peculiarities of the formation of nonlinear structures in narrow and wide FFs, will be analyzed.

6.1.2 Investigations of the three-wave parametric interaction of MSW in gyrotropic layered structures by NEELS method

Consider the propagation of BVMSW pulses (Slavin *et al* 2003) in FFs with a width L in the external magnetic field $\vec{H}_0$ applied in the direction Z in the FF plane. In this case, the pulse propagates in the direction Z of the magnetic field, and the phase and group velocities are directed in opposite directions. In a linear approximation, all spectral components of the input pulse field are proportional to $\exp[i(\omega_1 t - k_z Z)]$, where ω and k_z are the frequency and Z-component of the wave vector, respectively. Consider the case where the pump field $h_0 \exp(2i\omega_1 t)$ is parallel to a constant magnetic field $\vec{H}_0$ (figure 6.2), has a frequency of 2 ω_1 and amplitude h_0 and is considered homogeneous in the scale of the carrier wave in the area where pumping exists $l/2 - L_{\text{pmp}} \leqslant Z \leqslant l/2 + L_{\text{pmp}}$, $\pi/L_{\text{pmp}} \leqslant \Delta k_z \ll k_z$.

Outside the above area, the pump field is considered to be absent. Here Δk_z is the characteristic width of the wave spectrum of the pulse, L_{pmp} is half the length of the region of pumping and parametric amplification of incident and idle waves. Consider the interaction of narrow incident 1 and idle 2 wave packets with the carrier frequencies and wave vectors $\omega_{1,2}$ and $\vec{k}_{1,2}$, respectively, within the approximation of homogeneous pumping within the pumping area. The conditions for

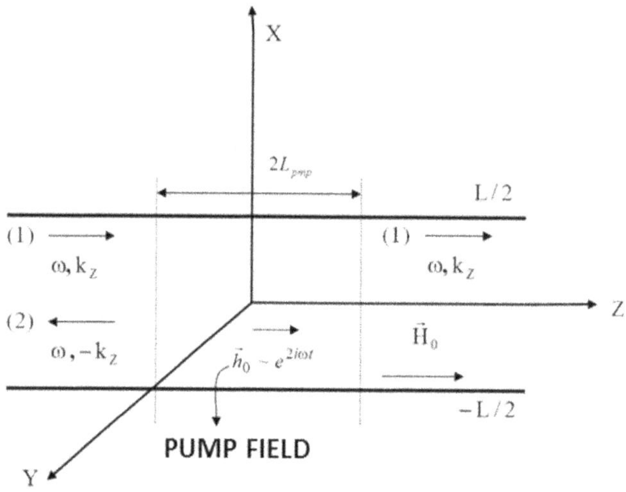

Figure 6.2. The parametric three-wave interaction of the signal pulse and the idle pulse, having carrier frequencies and wavenumbers in the Z direction, equal (ω, k_z) and $(\omega, -k_z)$ respectively. The pump field $\vec{h}_0$ has a frequency 2ω.

parametric coupling in the absence of (nonlinear) phase shift and in approximation (Kalinin and Bayonets 1990) are

$$\Omega = \omega_1 + \omega_2,\ 0 = \vec{k}_1 + \vec{k}_2 \tag{6.1}$$

In chapter 3, sections (3.3.1), (3.3.2), the NEELS method is described in details for BVMSWs in a gyrotropic layer. We will only briefly consider here the version of method NEELS for BVMSW in a gyrotropic layer with parametric coupling. We use the linear properties of waves (Damon and Eshbach 1961, Gurevich and Melkov 1996, Zvezdin and Popkov 1983). Next, we use an approach similar to that described in chapter 3, section 3.1, which is based on the consideration of linear and nonlinear wave systems, for both incident and idle pulses, and in the nonlinear case the effects of self- and mutual action of these waves due to cubic nonlinearity are included, as well as the parametric interaction. For the first time, the corresponding evolution equations were obtained for 2D interacting pulses in Grimalsky *et al* (2000, 2001a, 2001b), Rapoport (2006, 2012b, 2014), Rapoport *et al* (2012a) for a thin FF ($kL \ll 1$). The amplitudes of the envelope incident idling pulses, $U_{1,2}$ have been considered. The corresponding set of equations for the envelopes $U_{1,2}$ takes the form (the idea of the derivation of this system of equation is briefly described in appendix D.1):

$$\begin{aligned}&\frac{\partial U_{1,2}}{\partial t} \pm V_g\frac{\partial U_{1,2}}{\partial z} + i\left(\frac{D}{2}\right)\frac{\partial^2 U_{1,2}}{\partial z^2} + iS\frac{\partial^2 U_{1,2}}{\partial y^2}\\&+ iN(|U_{1,2}|^2 + 2|U_{2,1}|^2)U_{1,2} + \omega_r U_{1,2}\\&+ Vh_{\text{pmp}}(z,\ t)U^*_{2,1} = 0\end{aligned} \tag{6.2}$$

In the system (6.2) the signs «±» correspond to the indices «1,2», $\psi_{1,2}(y, z, t) = \sqrt{\left(m^2_{x1,2} + m^2_{y1,2}\right)/2M_0} = U_{1,2}(y, z, t)\exp\left[i(\omega_{1,2}t - k_{1,2}z)\right], \psi_{1,2}$ are transverse amplitudes (relative to the direction of the bias magnetic field) for wave packets, $U_{1,2}$ are the corresponding envelope amplitudes, which change slowly in time and space. Here, $U_{1,2}^2 \sim \left(m_{x1,2}^2 + m_{y1,2}^2\right)/2M_0^2$, $D = \partial^2\omega/\partial k_z^2$, $S = (1/2)\partial^2\omega/\partial k_y{}^2$, k_y is Y component of the wave vector of BVMSW (which must be taken into account to calculate the diffraction coefficient and to include diffraction effects). As follows from formulas (3.30) and (3.31) from chapter 3, $N \approx (\omega\omega_M/\omega_H)/[\sigma_0^2 + (\mu - 1)^2]$, $V \approx i\gamma\sigma_0/(\omega_H/\omega_M)[\sigma_0^2 + (\mu - 1)^2]$, where D, S, N and $\omega_r = \gamma\Delta H$ are the coefficients of dispersion, diffraction, nonlinearity, and the dissipation parameter proportional to the width of the ferromagnetic resonance line, ΔH, respectively, and V represents the coefficient of parametric coupling. We assume that the absolute values of the group velocity, as well as the coefficients of dispersion, diffraction, nonlinearity, attenuation and pumping are the same for the signal and the idle wave, 1 and 2, respectively.

This follows from the fact that the parametric interaction is degenerate, and at the frequency of homogeneous (in the wavelength scale) pumping 2ω, the frequency and absolute values of the wavenumbers (carriers) of the signal wave and the idle wave are the same (ω and $|k_z|$, respectively). Note that the calculations using the model developed in this paper correspond very well to the results of experimental studies and parametric amplification of magnetic bullets (Serga *et al* 2005). When $D = 0$, $S = 0$, $N = 0$, from the system (6.2) it is possible to obtain the relations of the form (1.8), which alternatively can also be obtained by the energy method (Rapoport 1987, Grimalsky and Rapoport 1995, Rowland 1999).

In figure 6.3, the absolute value of the relative pulse amplitude, ($|U/U_0|$), normalized to the amplitude of the input pulse, U_0, is shown. The amplitude of the input pulses, in turn, can be measured in terms of 'the number of nonlinear thresholds', n_{thr} where $n_{thr} = |U_0/U_{thr}|^2$, $U_{thr} = \sqrt{\omega_r/N} = 2.2 \times 10^{-2}$. Below we will omit the word 'relative' and use the term the 'pulse amplitude' instead.

We express the pump amplitudes in dimensionless values $P = P(Z = l/2)$, where l is the distance between the input and output of the antenna in the direction Z, $P(Z) = -it_{NL}Vh_{pmp}(Z)$, t_{NL} is the time scale we use, in particular $t_{NL} = T = 20$ ns, where T is the duration of a typical pulse, of those to be studied. Finally, the pump field has the form $h_{pmp}(Z) = h_{0pmp}\exp\{-[(Z - l/2)/L_{pmp}]^4\}$, where h_{0pmp} is the field amplitude. As described earlier in this section, in our case there are the inequalities $2L_{pmp} \gg l_z \gg \lambda$, where $l_z = V_gT$ is the wavelength of the wave packet in the direction of propagation, $\lambda = 2\pi/k_{z1}$ is the carrier wavelength, and the inhomogeneity of the pump field within the 'pump area' is not accepted to attention. The distance between the input and output of the system (along the direction Z) is set equal $l = 8$ mm.

6.1.3 Results on the formation of nonlinear active structures in parametric interaction in a gyrotropic medium

The parametric amplification and phase conjugation of pulses were studied for one-dimensional pulses in gyrotropic layer (FF) (Melkov and Sholom 1990, Melkov and Serga 1998, Melkov *et al* 1999) and two-dimensional pulses (Grimalsky *et al* 2000,

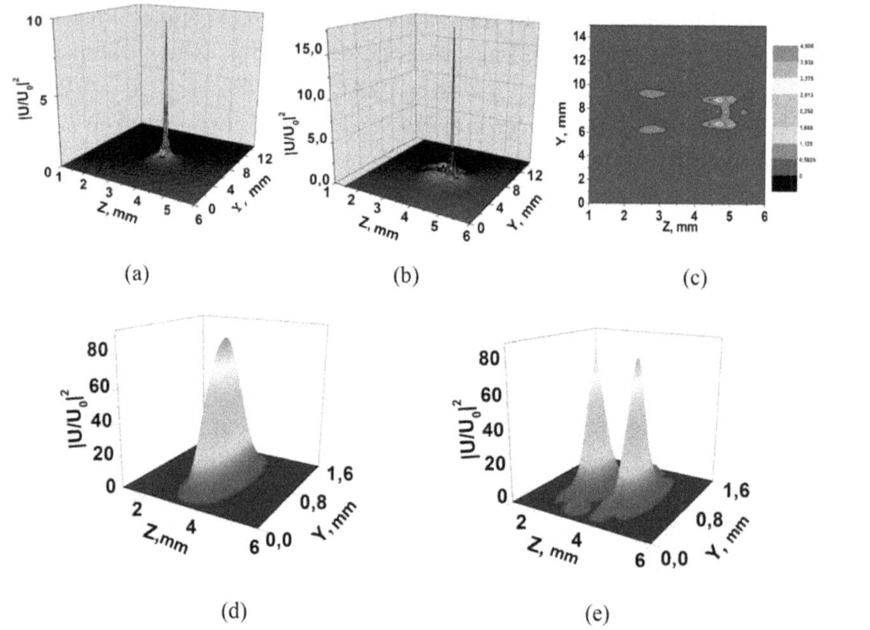

Figure 6.3. (a)–(e) The parametric amplification of magnetic bullets (Grimalsky *et al* 2000, 2001a, 2001b) (a) and (b) and formation of structures (Rapoport 2006a, 2013b, 2014, Rapoport *et al* 2014a) (c) and (d) at parametric interaction of signal pulse and idle pulse in 'wide' (a)–(c) and 'narrow' (d) and (e) FFs; (a)–(c) $P = 2$, $n_{\text{thr}} = 10.42$; (c) and (d) $P = 4$, $n_{\text{thr}} = 0.5$; the moment corresponding to the time of propagation of the signal pulse from the input to the system (a) and (d) is 90 ns; (b) and (e) is 106 ns; (c) is 130 ns; the parametric interaction takes place in the time period from $t_{\text{p1}} = 90$ ns to $t_{\text{p2}} = 110$ ns. In all parts (a)–(e) there are represented both the signal wave and the idle one jointly (see also equation (6.2)).

2001a, 2000b, Rapoport 2006a, 2012b, Rapoport *et al* 2012a, Serga *et al* 2005). This section presents the results of studies of the formation of nonlinear wave structures of BVMSWs in a certain range of input amplitudes (mainly up to 10–12 threshold values) and the pump field (up to values corresponding to $P \sim 6$). Two-dimensional nonlinear effects are studied in a gyrotropic layer with two different values of width (size in the Y direction, figure 6.3), namely for 'wide' and 'narrow' (Demokritov *et al* 2001, Slavin *et al* 2002, Serga *et al* 2008) FF. The corresponding widths are $L_y = 15$ mm and $L_y = 1.6$ mm. The effect of splitting the phase-conjugated pulse in transverse direction in a wide FF with a big enough width input pulse due to the process of formation of a strong enough bullet (figure 6.3(b)) is considered in appendix D.2.

With an amplitude of the order of one threshold of the nonlinearity, half of the wavelength of the modulation instability in the transverse direction in the order of magnitude is (Demokritov *et al* 2001, Slavin *et al* 2002, Serga *et al* 2008) $\lambda_\perp \sim \pi\sqrt{S/(N|\varphi|^2)} \sim 3$ mm. At such amplitudes, the widths of the 'narrow' and 'wide' films are smaller and larger than $\lambda_\perp$, respectively. In the first case, the width of the antenna coincides with the width of the film. In the second case, the width and lateral boundary conditions have a little effect on the formation of nonlinear waves. Accordingly, for a narrow film in the linear case, the entire film is directly excited by the input antenna (Slavin *et al* 2002, Serga *et al* 2008). For rather large amplitudes, transverse modes of higher order are excited nonlinearly due to intermode interaction (Demokritov *et al* 2001, Slavin *et al* 2002, 2003, Serga *et al* 2008). The fundamental difference between the formation of structures in a narrow and wide FFs is that in the first case the development of the modulation instability in the transverse direction is suppressed by limiting the transverse size of the waveguide.

In Grimalsky *et al* (2000, 2001a, 2000b), Rapoport (2006a, 2012b, 2013b, 2014b) the formation of nonlinear structures in an active layered gyrotropic medium with parametric amplification, depending on the input amplitude and amplitude of parametric pumping, has been studied. In particular, the possibility of parametric amplification of magnetic bullets in a wide FF was theoretically shown for the first time (Grimalsky *et al* 2000, 2001a, 2001b). In Rapoport (2006a) the possibility of formation and amplification at parametric interaction of phase-conjugated 'knife-shaped' magnetic bullets (extended in the transverse direction to the direction of propagation) in a narrow FF was shown theoretically (figure 6.4). The peculiarity of

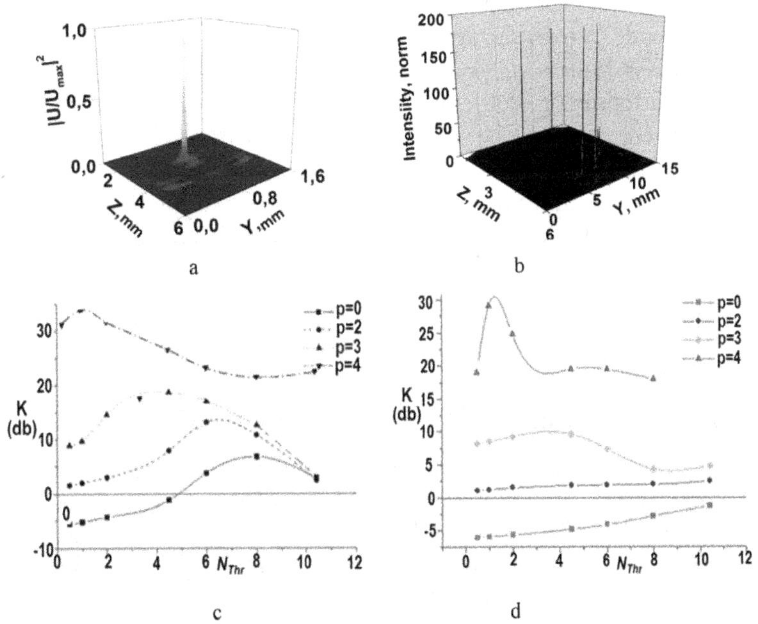

Figure 6.4. (a)–(d) Examples of the formation of new types of structures (a) and (b) (in figure (b), normalized intensity (intensity norm) is used) and the corresponding dependences of the 'transmission factor', determined by the maximum intensity of the signal pulse from 'input' to 'output' of the structure, from the maximum input intensity (c) and (d) in the parametric signal interaction pulse and the idle pulse in 'wide' (c) and 'narrow' (d) FFs; the parameter is the dimensionless amplitude of the pump field *P*; parametric interaction takes place in the period from $t_{p1} = 70$ ns to $t_{p2} = 90$ ns; The 'output' of the system corresponds to the time $t = 130$ ns.

this method of forming knife-shaped bullets is as follows. As can be seen from figure 6.4(a), this bullet is formed on a phase-conjugate wave, whereas at the time under consideration, the bullet was not formed on the direct wave, in contrast to the situations considered for a wide FF (Grimalsky *et al* 2000, 2001a, 2001b, Serga *et al* 2005, Serga 2006). The formation of a bullet from a phase-conjugate wave packet occurs, compared to bullets formed in the absence of parametric interaction, under the action of additional (quasi) linear focusing associated with parametric interaction (Serga *et al* 2005).

In general, the phase-conjugated packet is more focused than the wave packet of the incident wave (Serga *et al* 2005). The wave packet on forward waves strongly feels an influence of lateral walls of a narrow waveguide, and the formation of a bullet of forward waves is suppressed/slowed down. The side walls have a relatively weak effect on the development of the nonlinear structure on the more focused phase-conjugated packet. The result is a pronounced knife-shaped phase-conjugated bullet, while at the same time the bullet on direct waves is absent. This mode of the formation of the phase-conjugate bullet takes place for a relatively weak input direct pulse and a fairly significant intensity of the pumping field, ($|V_0/V_{\text{thr}}|^2 = 1$, $P = 5$, figure 6.4(a)) (Rapoport 2006a). Interestingly, the knife-shaped bullets are noticeably elongated in the transverse direction ($a_y/a_z \sim 4$, where $a_{y,\,z}$ are the characteristic sizes of the bullets in the y and z directions, respectively), and they are bifurcated in the transverse direction. An interesting situation arises. For well-formed bullets of a high intensity in wide gyrotropic layers there is a tendency to equalize the transverse sizes of the BVMSW bullets, $a_y \rightarrow a_z$ the shape of the cross section of the bullet goes to the circular (Buttner *et al* 2000), and in this process the transverse size of the bullet is relatively reduced.

By means of simulations the novel nonlinear structures were found in an active gyrotropic layered medium that are formed under gradual increase of the amplitude of the input pulse and the pumping field in a wide gyrotropic medium. These structures develop before finishing the pumping pulse, especially the structures with four bullets (two bullets are on the base of the incident pulse and two are on the base of the idler wave pulse, see figure 6.4(b)).

6.2 Nonlinear waves in the layered bi-anisotropic metamaterials

Consider evolutional equations for the nonlinear electromagnetic waves with parametric interaction in the layers of bi-anisotropic MMs on the basis of developing the method NEELS. Nonlinear evolution equations are derived using a developed general method for waves in a layered system with a moderate nonlinearity (Boardman *et al* 2009, Rapoport *et al* 2006b, Rapoport 2013b, 2014b, Rapoport and Boardman 2015b). The developed method (see chapter 3 and equation (3.18)) can be applied to bi-anisotropic structures in the presence of both bulk and surface nonlinearities. The set of evolution equations obtained by modifying this method taking into account the parametric interaction of opposing pulses has the form

$$\frac{\partial A_{1,2}}{\partial t} \pm V_{g1,2}\frac{\partial A_{1,2}}{\partial y} + i\frac{V_{g1,2}}{2k_{1,2}}\frac{\partial^2 A_{1,2}}{\partial z^2} + \frac{i}{2}\frac{\partial^2 \omega}{\partial k_y^2}\frac{\partial^2 A_{1,2}}{\partial y^2}$$
$$-\frac{i\omega}{W_{01,2}}\int_{-\infty}^{\infty}\left(\vec{f}^{\,*}_{E1,2}\vec{P}_{A1,2} + \vec{f}^{\,*}_{H1,2}\vec{M}_{A1,2}\right)\mathrm{d}x + \gamma_{1,2}A_{1,2} = i\omega Q_0 A_3 A^*_{2,1};\ W_{01} \tag{6.3}$$
$$= \frac{c}{4\pi V_{g1}}\int_{-\infty}^{\infty}\mathrm{Re}\left(\vec{f}_{E1}\times\vec{f}^{\,*}_{H1}\right)_y \mathrm{d}x$$

Here, $A_{1,2}V_{g1,2}$, and $\gamma_{1,2}$ are slowly varying enveloping amplitudes, group velocities, and dissipation coefficients for two counter propagating waves, $\vec{f}_{E1,2}$, $f_{H1,2}$ are functions describing the spatial structure/spatial distribution in the direction transverse to the propagation direction for the electric and magnetic fields of linear modes, respectively,

$$\begin{pmatrix}\vec{E}_{1,2}\\ \vec{H}_{1,2}\end{pmatrix} = A_{1,2}\begin{pmatrix}\vec{f}_{E1,2}\\ \vec{f}_{H1,2}\end{pmatrix}e^{i(\omega t \mp k_{1,2}y)}, \tag{6.4}$$

In equation(6.4) ω is the carrier frequency (the same for both interacting waves), Q_0 is the value that determines the coupling coefficient for interacting waves, $\vec{P}_{A1,2}$, $\vec{M}_{A1,2}$ are the corresponding nonlinear polarizations. Note that these nonlinear polarizations can belong to a wide class of nonlinear functions, the corresponding nonlinearities are not limited to cubic or quadratic, and the specific form of the corresponding equations is not limited to the Schrödinger type. The obtained set of equations (6.3) provides the possibility of modeling counter propagating waves taking into account the quadratic nonlinearity (and parametric amplification) and self-action effects determined by nonlinearity, which belongs to a fairly wide class of nonlinear functions.

In section 2.3 in chapter 2 (equations (2.82)–(2.84)) it was described the modeling procedure from the study of the characteristics of a single bi-anisotropic metaparticle to the homogenization of the layered active bi-anisotropic MM. This approach, added by the set of equations (6.3), leads to the consideration of the parametric amplification of two-dimensional quasi-soliton pulses in the bi-anisotropic frequency waveguide. Using simulations (Boardman *et al* 2009, Rapoport *et al* 2006b, Rapoport 2013b, 2014b, 2015a, Rapoport and Boardman 2015b), we have shown the fundamental possibility of the parametric amplification of (2 + 1) (spatio-temporal) quasi-soliton pulses in bi-anisotropic layered media with the defocusing nonlinearity. This is qualitatively demonstrated in figures 6.5(a) and (b).

Evaluate the effectiveness of parametric amplification (Boardman *et al* 2009, Rapoport 2014, Rapoport and Boardman 2015b). Using (6.3), we obtain, in order of magnitude, when, and neglecting the effects of self-action

$$\partial A_{1,2}/\partial t \sim \mathrm{PAR}^{(0)}_{\mathrm{TH}}\left(\omega A^*_{2,1}\right);\ \mathrm{PAR}^{(0)}_{TH} = iQ_0A_3;$$
$$\mathrm{PAR}^{(0)}_{\mathrm{TH}} \sim [(1-4\varpi^2)|1-\varpi^2|^{1/2}(1+L_1/L)]^{-1}8\pi\nu^2_{\mathrm{res}}\chi\varpi^4 I_0(S\omega_{\mathrm{res}}/c\varphi_k)A_3 \tag{6.5}$$

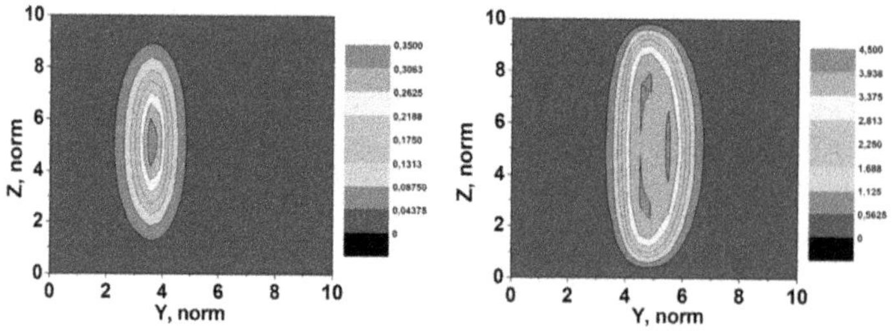

(a) (b)

Figure 6.5. (a) and (b) The parametric processes in the waveguide structure shown in figure 2.8(c); (a) spatial distribution of the intensity of the incident/signal pulse before the impact of parametric pumping at double frequency; (b) spatial distribution of the total intensity of the 'signal' pulse and the counter idle pulse, which is excited as a result of the process of parametric interaction, with a moderate value of the pump amplitude. Coordinates (Y, Z) and intensities are given in relative units. The effects of self-action in the presence of effective cubic defocusing nonlinearity are taken into account.

In (6.5), $\varpi = \omega/\omega_{\text{res}}$, $v = cl/(S\omega_{\text{res}})$, $\chi = NS^2/(c^2L)$, $L_1 = 4\pi NS^2/\left(3\mu_{\text{host}}c^2\right)$, $I_0 = 2/3$, $\varphi_k = 0.5B$ is the value accepted as the 'scale' of the electric potential, μ_{host} is the magnetic permeability of the host medium into which the metaparticles are introduced. Choose for estimates the following parameters of EMW and Ω-particles (see figure 2.8(b) in chapter 2). $\omega_{\text{res}} \sim 1.25 \times 10^{11}$ s^{-1}, $\varpi \sim 1$, $|1 - \varpi^2| \sim 10^{-2}$, $l_0 \equiv S\omega_{\text{res}}/c \sim 3 \times 10^{-2}$ cm, $r_0 \sim 0.05$ cm, $S = \pi r_0^2 \sim 7.5 \times 10^{-3}$ A m^{-1}, $N \sim 200$ cm^{-3}, $A_3 \sim 1.5 \times 10^{-2}$ A m^{-1}, which corresponds to the strength of the magnetic field of pumping ~ 2×10^{-4} Oe. This pumping is quite moderate and generally it is necessary to take into account the saturation of amplification. Note, as shown in experimental and theoretical works with the parametric amplification of MSWs (Gordon *et al* 1998), a pumping field can be used that reaches at least 10 Oe. At the above parameters, including such a very moderate pump amplitude, and $|A_1| \sim |A_2|$ (when two pulses with amplitudes of the same order are launched), the value of the effective 'gain time' t_{char} in the order of magnitude is the same as the period of the envelope of the signal pulse, $\omega t_{\text{char}} \sim 1$. This means that the parametric amplification of pulses in MM bi-anisotropic media with metamolecules loaded on nonlinear elements, such as diodes, can be very effective.

For the transition to higher frequencies (in the millimeter, THz, or optical ranges), it seems promising to use a more complex form of bi-anisotropic particles (Lheurette *et al* 2007, Tretyakov *et al* 1994), as well as as active elements, resonant tunnel diodes (Liu *et al* 2008) (at THz or submm waves) or quantum dots or other active structures (for the optical range) (Boardman *et al* 2011, Zhukovsky *et al* 2005).

6.3 Parametric interactions and phase conjugation on active two-dimensional chiral metamaterial surfaces with linear and nonlinear Huygens sources

This section shows the possibility of an almost perfect phase conjugation on an electrically thin nonlinear layer/metasurface with chiral metaparticles.

6.3.1 Justification and formulation of the problem

The process of wave phase conjugation has been studied for EMWs from microwave (ultrahigh frequencies, GHz) to the optical range. Phase conjugation has been demonstrated in both isotropic (Zel'Dovich *et al* 1985) and anisotropic (ferrite) media (Grimalsky *et al* 2000, 2001a). In Maslovski and Tretyakov (2003) it was proposed to use the phase conjugation of the electromagnetic field to create 'ideal lenses'. The Huygens linear sources have been used in linear mode as efficient metamaterial antennas, without (or with very small) **image distortion** (Jin and Ziolkowski 2010). A new idea (Maslovski and Tretyakov 2003, Maslovski *et al* 2011) is the use of an active surface with nonlinear metaparticles, 'working' as a Huygens source in both linear and nonlinear modes. Huygens sources provide a field structure similar to that in a plane wave, which makes the 'reflection' impossible.

Such sources are provided by an appropriate combination of effective magnetic and electric dipoles. In the long run, this approach may lead to the implementation of superlens without reflection based on three-wave interactions, as well as other applications based on electrically thin nonlinear metalayers/metasurfaces without any reflection of signal and phase-conjugated waves. In this section, we show the fundamental possibility of developing a nonlinear active surface with metaparticles that make up both linear and nonlinear Huygens sources. This will be shown by a specific example of a normally incident signal wave (some discussion of ways to solve this problem for arbitrary angles of incidence of an electromagnetic wave on a metasurface can be found in Maslovski and Tretyakov (2003)).

The processes of the three-wave interaction are considered as the basis of phase conjugation on active surfaces without a reflection of signal and phase-conjugate waves. In the case of the interaction of three waves, the following condition is satisfied: $\omega_p = \omega + \omega_{pc}$. Here ω_p, ω, ω_{pc} are the frequencies of the pump wave, the incident wave/signal, and the phase-coupled wave, respectively. Alternatively, the voltage applied to nonlinear elements possessing quadratic nonlinearity is used for the parametric pumping. These nonlinear elements are embedded in active particles placed on the metasurface. An important question is which of the two possible regimes of the phase conjugation is more appropriate. One is the degenerate interaction when $\omega = \omega_{pe} = \omega_p/2$ (Zel'Dovich *et al* 1985, Maslovski and Tretyakov 2003, Grimalsky *et al* 2001a), and the other is the non-degenerate interaction with different values of ω and ω_{pc} (but $\omega \approx \omega_{pc}$). In Rapoport *et al* (2013a) it was proved, using the Krylov–Bogolyubov method (Landa 1996), that the only possible regime for solving the problem of non-reflective phase conjugation is a non-degenerate coupling (Katko *et al* 2010). It is shown that in the case of the degenerate interaction, the parametric instability is inevitable, which practically makes such a phase conjugation regime impossible.

6.3.2 Conditions of phase conjugation without reflection

Consider the conditions that must be satisfied by electrical and magnetic surface polarizations to provide a phase conjugation for a plane wave with components (Ex, Hy) or (Hx, Ey) that falls normally from $z = -\infty$ to the plane $z = 0$. When the chiral metaparticles are used on the active surface, both modes (Ex, Hy) and (Hx, Ey) will be excited and present, generally speaking, in the fields of both the reflected wave and the transmitted one. Nevertheless, since the corresponding surface polarizations $\vec{P}^{(e)} = (0, P_y^{(e)}, 0)$, $\vec{P}^{(m)} = (P_x^{(m)}, 0,0)$, and $\vec{P}^{(e)} = (P_x^{(e)}, 0,0)$, $\vec{P}^{(m)} = (0, P_y^{(m)}, 0)$, are mutually orthogonal, we can consider the conditions that surface polarizations must provide a reflectionless conjugation of the wave front separately for each of the two modes. Denoting the amplitudes of the electric fields of incident, reflected and transmitted waves as, A_{E_i}, $A_{E_{ir}}$, $A_{E_{ti}}$, respectively, it is possible to obtain and record the conditions of ideal (and without reflection) phase conjugation (see appendix D.4) in the form

$$A_{Er}^{(\omega_{pc})} = 0, \; A_{Et}^{(\omega_{pc})} = B_p A_{Ei}^{(\omega_s)*} \tag{6.6}$$

In relations (6.6), B_p is complex constant, '*' means complex conjugation; the upper indices '$\omega_{s;pc}$' correspond to the frequencies at which the fields are considered. For a non-degenerate interaction, amplitudes A_{Et}, A_{Ei}^* refer to a transmitted wave (with a conjugate phase) and an incident (signal) wave, respectively, having different frequencies. The consideration of the incidence, reflection and transmission with the appropriate boundary conditions (see also appendix D.4), gives the conditions of the absence of reflection, at a given frequency, in the form

$$P_y^{(m)} = P_x^{(e)} \text{ or } P_x^{(m)} = -P_y^{(e)} \tag{6.7}$$

for incident mode (E_x, H_y) and surface polarization, $\vec{P}^{(e)} = (P_x^{(e)}, 0,0)$; $\vec{P}^{(m)} = (0, P_y^{(m)}, 0)$ and incident mode (H_x, E_y) and surface polarization, $\vec{P}^{(e)} = (0, P_y^{(e)}, 0)$; $\vec{P}^{(m)} = (P_x^{(m)}, 0,0)$, respectively. In the case of using a metasurface with an elementary cell in the form of a pair of orthogonal chiral particles (figure 6.6(a)), both surface polarizations are excited, even if only one of the above modes falls on the active surface. Note that: (1) in the case of a non-degenerate interaction, the incident phase-conjugate wave is absent, while the possible transmitted wave and the reflected wave at the frequency ω_{pc} are excited and are proportional to the nonlinear surface polarization at the corresponding frequency; (2) conditions of the form (6.7) to ensure a reflection-free phase-conjugation should be satisfied for both signal and nonlinearly excited phase-conjugate waves at appropriate frequencies. The latter, in turn, is excited by mixing, on nonlinear elements (diodes/capacitors) embedded in metaparticles, incident wave fields and pumping applied to a nonlinear element (with quadratic nonlinearity). Thus, in the case of non-degenerate communications, the second condition from (6.6) is satisfied 'automatically'.

To ensure a reflectionless phase conjugation, the first of the conditions of (6.6) must be satisfied for both signal and phase-conjugate waves. Accordingly, the first and/or second of the formulas. (6.7) must also be performed for both frequencies ω, ω_{pe}. Which one of the first, second, or both of the conditions (6.7) must be met depends on the polarization of the incident wave and whether or not the metaparticles used are chiral. Conditions, including (6.7) and others, providing no- or relatively small reflection on both the fundamental frequency (the linear effect) and the frequency of the phase-conjugate wave (the nonlinear effect), as well as the efficiency of converting

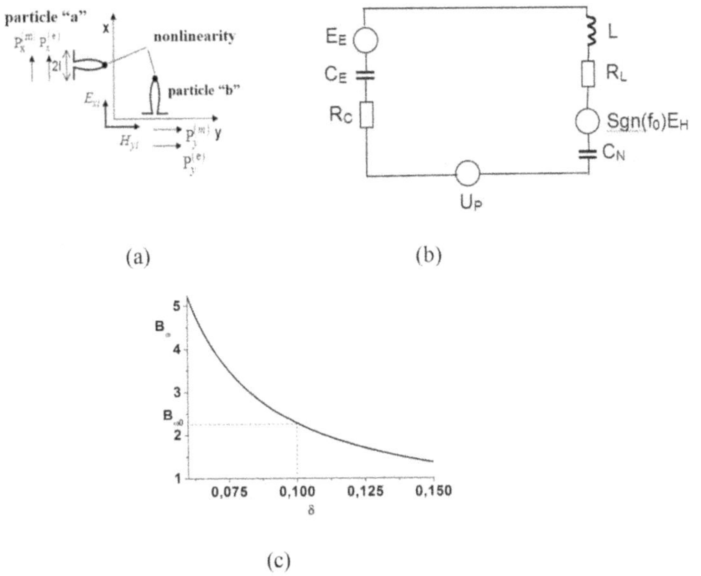

Figure 6.6. (a)–(c) An elementary cell of a metasurface with chiral particles (with a magnetic loop lying in the plane xy) (a) and an equivalent circuit describing the metaparticle (b); $P_{x,\,y}^{(e,\,m)}$ are surface polarizations excited in chiral particles ***a*** and ***b*** by incident waves with electric fields (E_{xi}, E_{yi}); $E_{E,\,H}$; C_E; L, and $R_{C,\,L}$ denote the electromotive forces associated with the electric dipole and the magnetic loop, the capacitance associated with the electric dipole (stem) of the chiral particle, the inductance of the magnetic loop and the effective resistances that describe the dissipative losses in the magnetic and electric dipoles, respectively; C_N, U_P describe a nonlinear capacitance (using quadratic nonlinearity) embedded in the chiral particle and the amplitude of the pump voltage at the frequency ω_p, respectively; (c) is the frequency dependence of the conversion factor B_p (see equation (D.15)) in the vicinity of the parametric resonance (but under the condition of nondegeneracy, $\omega \neq \omega_{pc}$). Reprinted from Rapoport et al (2013a) by permission of the publisher (Taylor & Francis Ltd, http://www.tandfonline.com).

the incident wave into a phase-conjugate, are considered briefly in appendix D.4 (see the relations (D.10)–(D.15)) and in more detail in Rapoport *et al* (2013a).

6.3.3 Simulation results

The 'elementary cell' of chiral metaparticles, namely, Ω-particles, and the corresponding equivalent scheme are shown in figure 6.6(a) and (b), respectively. The frequency dependence of the conversion factor of the signal wave into the phase-conjugated wave B_p (see appendix D.4, including the relation (D.15)) in the vicinity of the parametric resonance (but under the condition of nondegeneracy, $\omega \neq \omega_{pc}$) is shown in figure 6.6(c). Analysis of the condition (6.7) and the equivalent circuit shown in figure 6.6(c) shows that a phase conjugation with the zero reflection of the signal wave and the small 'reflection' (the generation in the direction opposite to the incident signal wave) of order $|(\omega - \omega_{pc})/\omega| \sim 0.1$, at the frequency of the phase-coupled wave ω_{pc}, close to but different from the frequency of the signal wave ω_{pc}, is possible. Note that the 'reflection' at the frequency ω_{pc} is determined relative to the amplitude of the passing wave, instead of the amplitude of the incident wave, which is zero for the phase-coupled wave.

The conditions for such a 'non-reflective' phase conjugation are found in terms of the parameters of the metaparticles shown in figure 6.6(a). The estimates for the signal frequency $\omega \sim 2\pi \cdot 14 \times 10^{10}$ s^{-1} show that the efficiency of converting the energy of the incident phase-conjugate wave $|B_p|^2$ may be of the order of 10^{-5} (see appendix D.4). This value is not yet optimized, but it is quite sufficient for experimental observation. The optimization is possible with further modification of the metaparticles, which is not the subject of this work. More details are included in appendix D.4.

6.4 Conclusions to chapter 6

The possibility of the parametric amplification of magnetic bullets in wide FFs using parallel pumping is shown. For the first time, the possibility of forming and parametric amplification of knife-shaped phase-conjugated bullets in the absence of bullets on direct waves in narrow FFs was shown. This process can occur for a relatively weak input direct pulse amplitude and a significant intensity of the pumping field.

A new 'idler wave splitting' effect was found. It appeared on the basis of nonlinear phase mismatch in the presence of a strong incident wave. The effect is achieved at moderate values of the amplitudes of the input waves and moderate values of the parametric pumping intensities. The moderate amplitude of the input wave means that it is sufficient to form a magnetic bullet before starting the parametric interaction. It was found that for relatively weak values of the input pulse amplitudes and increased values of the pumping intensities, the integral energies of the incident pulse and the idle one tend to be equal after the end of the pumping action. The system 'forgets' the initial conditions, and the amplitudes of the counter propagating pulses (incident and idle), after the end of the parametric interaction, usually have magnitudes of the same order. It is shown that new nonlinear structures are formed by the amplitude of the input pulses and large pump fields in the 'wide' ferrite area. Such a structure can develop after finishing the pumping, consisting of a structure of four strong bullets (two based on the incident wave and two based on the idler one). The features of the transformation of the amplitude of the incident pulse during the passage of the gyrotropic layer during the action of the pump field were found; they appear in connection with the development of nonlinear structures in a wide and narrow gyrotropic layer . The amplitude transformation coefficient increases with increasing amplitude in the range of values of the normalized pump field $P = 0$–3 and the amplitude of the incident pulse, which corresponds to 8–10 thresholds for a narrow gyrotropic layer in contrast to a wide gyrotropic layer . This is explained by the fact that for a narrow gyrotropic layer it forms nonlinear structures in the transverse direction not as intense as for a wide gyrotropic layer.

6.4.1 A new method of nonlinear homogenization for a bi-anisotropic medium has been developed. It is used for a special case of Ω-particles with nonlinear loading, located parallel to each other. For the case of GHz frequencies, a single metaparticle is analyzed based on Kirchhoff's laws for currents and voltages, taking into account nonlinear elements excited by the total action of local electromagnetic fields (including action from other metaparticles in the Lorentz–Lorenz approximation) and internal sources that can be effectively enclosed in metaparticle loads.

6.4.2 A method for deriving evolutionary macroscopic equations for the amplitudes of envelope waves in layered systems in the presence of bulk and surface nonlinearities has been developed, for the first time, for media with bi-anisotropic layers.

6.4.3 A special case of a layered bi-anisotropic medium with metaparticles containing nonlinear loads that allow both self-influence and parametric interaction of nonlinear counter propagating waves is considered. The possibilities of self-influence, cross-modulation, second harmonic generation, parametric interaction for counter propagating waves in layered bi-anisotropic structure are accounted for. The surface nonlinearity and linear loss or amplification are taken into account as well. The physical reason for the effects mentioned above could be, in particular the presence of the effective negative conductivity in the loading of metaparticles.

6.4.4 The possibility of close to the ideal phase conjugation is shown. Close to the ideal phase conjugation is achieved with the use of an active metasurface with a normal incidence of a signal wave and a non-degenerate three-wave interaction, with a small inequality of the frequencies of the incident and phase-conjugate waves. The use of linear and nonlinear Huygens sources for both linear (signal) and nonlinear (phase-coupled) waves causes zero or very small values of the reflection coefficient for both waves simultaneously. It is possible to achieve the value of energy transformation of an incident to a phase-conjugate EMW of order 10^{-5} that is sufficient for experimental demonstration.

Appendix D.1 Formation of nonlinear structures in 'wide' film and 'narrow' active gyrotropic wave

We assume that a wave with index '1' is a nonlinear wave at a frequency propagating in the structure considered in section 6.1.2 (figure 6.2). We use the index '2' for a nonlinear idle wave, excited in the process of parametric interaction. Then we obtain for the parametric interaction of counter propagating waves in the presence of parametric pumping a set of evolution equations for counter propagating waves in the form of (3.30) (section 3.3.1); for a specific system and type of waves considered in section 6.1.2 (also figure 6.2) this set of equations takes the form (6.2).

Appendix D.2 Splitting of the peak of the 'idle' pulse in the transverse direction during parametric interaction

Consider the relationship for the amplitude (envelope) $U_2(t,\ Y,\ Z)$ of the idle wave

$$\mathrm{d}U_2/\mathrm{d}t + \Gamma U_2 = -Vh_{\mathrm{pmp}}U_{10}^{*} \tag{D.1}$$

where a rough approximation is used for qualitative estimations; we focus only on the terms that describe the parametric interaction and are quadratic; the effects of the cubic nonlinearity and linear diffraction, and wave dispersion are not considered. We consider the input amplitude of the signal wave to be moderate, but sufficient for the formation of a pronounced magnetic bullet before the beginning of the parametric interaction, and the amplitude of the parametric pumping is relatively small. In a qualitative consideration, the change in the amplitude of the signal during the parametric interaction is neglected.

We present the amplitude of the signal wave U_{10} in the form $U_{10} \sim U_{100}\mathrm{e}^{-i\Delta\omega_{\mathrm{NL}}t}$, $\Delta\omega_{\mathrm{NL}} = \alpha|U_{100}|^2$, where α is a nonlinear coefficient in the equation for the amplitude of the wave (envelope) of the signal U_{10} (not written here), and consider the small values of time from the beginning of the parametric interaction. We obtain from (D.1) after averaging over time, given that $U_2(t=0)=0$, $\langle(\mathrm{d}U_2/\mathrm{d}t + \Gamma U_2)\rangle \sim \left\langle Vh_{\mathrm{pmp}}U_{10}^{*}\mathrm{e}^{-i\Delta\omega_{\mathrm{NL}}t}\right\rangle$, or, in order of magnitude,

$$\left|U_2\right| \sim \left|Vh_{\mathrm{pmp}}U_{10}/\Delta\omega_{\mathrm{NL}}\right| \sim \left|Vh_{\mathrm{pmp}}/(\alpha U_{10})\right| \sim \left|U_{10}\right|^{-1} \tag{D.2}$$

where the ‘averaging time’ T disappears from this approximate result. The relation (D.2) explains the decrease in the amplitude in the center of the ‘idle’ pulse, at $y = W/2$ where the signal wave has a maximum as shown in figures 6.3(a) and (b). Note that figures 6.3(b) and (c) correspond to the moment of time around the ‘signal bullet’ shown in figure 6.3(a), already destroyed and transformed into a set of transverse and longitudinal (relative to the direction of propagation) peaks. The ‘single’ wave has two transverse peaks with a decrease inside (figure 6.3(c)), which is described by the qualitatively presented above theory.

Appendix D.3 The tendency to equalize the integral amplitudes of the signal and the ‘idle’ pulse at relatively small values of the amplitude of the input pulse and large values of the pump amplitude in the ‘narrow’ FF

Combine both equation (6.2) for each of the amplitudes $U_{1,2}$, with the corresponding complex equations for $U_{1,2}^{*}$ and multiply the equation for U_1 by U_1^{*}, the equation for U_2 by U_2^{*}, etc. After summing up the obtained equations and integration over the spatial region (over y and z) occupied by the incident (signal) and the ‘idle’ pulse, both of which propagate in the ferrite layer (film), we obtain the following set of equations for $W_{1,2} = \iint |U_{1,2}|^2\mathrm{d}y\mathrm{d}z$:

$$\partial W_1/\partial t + 2\Gamma W_1 + Vh_{\mathrm{pmp}}\iint (U_1^{*}U_2^{*} + U_1U_2)\mathrm{d}y\mathrm{d}z = 0 \tag{D.3}$$

$$\partial W_2/\partial t + 2\Gamma W_2 + Vh_{\mathrm{pmp}}\iint (U_1^{*}U_2^{*} + U_1U_2)\mathrm{d}y\mathrm{d}z = 0 \tag{D.4}$$

The solution (D.3) is represented as:

$$W_1 = W_{10}\exp(-2\Gamma t) + \int_0^t \exp[-2\Gamma(t-t')]F(t')\mathrm{d}t' \quad \text{(D.5)}$$

$$W_2 = W_{20}\exp(-2\Gamma t) - \int_0^t \exp[-2\Gamma(t-t')]F(t')\mathrm{d}t' \quad \text{(D.6)}$$

where $W_{1,20} = W_{1,2}(t=0)$, $F(t') = Vh_{\mathrm{pmp}}\iint (U_1^* U_2^* + U_1 U_2)\mathrm{d}y\mathrm{d}z$

For long enough time and amplitude of pumping h_{pmp}, the second terms in the right parts (D.5) become much larger than the corresponding first terms, and we obtain from (D.5) and (D.6) that $W_1 \sim W_2$ (figure 6.3(d); see also figure 6.3(e) and the signature to these figures). The system 'forgets' the initial conditions, and the integral intensities of the corresponding pulses after their interaction under the above conditions tend to equalize. This tendency can be observed in a wide FF, but in a wide film where the transverse instability is not suppressed, the pulses can split more easily than in a narrow film, and the tendency to equalize the integral intensities of interacting waves at appropriate parameters may become indistinct.

Appendix D.4 Almost reflectionless phase conjugation and Huygence sources

For nondegenerate three-wave interaction (see sections 6.3.1 and 6.3.2) with different values of signal and phase-coupled waves, respectively ω_s and ω_{pc} (Katko *et al* 2010), $\omega_s \neq \omega_{pc}$ (but $\omega_s \approx \omega_{pc}$). The electromagnetic field of signal (incident) and complex conjugate waves is proportional

$$\begin{pmatrix} E_x \\ H_y \end{pmatrix}_{i,r,t}^{(\omega_{s;pc})} = \begin{pmatrix} A_{Ex} \\ A_{Hy} \end{pmatrix}_{i,r,t}^{(\omega_{s;pc})} \mathrm{e}^{i\left(\omega_{s;pc}t - k_{zi,r,t}^{(\omega_{s;pc})}z\right)} \text{ and } \begin{pmatrix} H_x \\ E_y \end{pmatrix}_{i,r,t}^{(\omega_{s;pc})} = \begin{pmatrix} A_{Hx} \\ A_{Ey} \end{pmatrix}_{i,r,t}^{(\omega_{s;pc})} \mathrm{e}^{i\left(\omega_{s;pc}t - k_{zi,r,t}^{(\omega_{s;pc})}z\right)} \quad \text{(D.7)}$$

for (E_x, H_y) and (H_x, E_y) modes of incident waves (the metasurface with chiral AM lies in the XY plane, figure 6.6(a)). In equation (D.7), the indices 'i, r, t' correspond to the incident, reflected and passing waves, respectively; the superscripts '$\omega_{s;pc}$' correspond to the frequencies at which the fields are considered; for the corresponding components of the wavenumbers we have $k_{zi}^{(\omega_{s;pc})} = k_{zt}^{(\omega_{s;pc})} = k_0^{(\omega_{s;pc})} = \omega_{s;pc}/c$, $k_{zr}^{(\omega_{s;pc})} = -k_0^{(\omega_{s;pc})}$. Using Maxwell's equation and boundary conditions on the surface $z = 0$, we obtain conditions (6.7) no reflection at a given frequency, $\vec{n} = \vec{z}_0$ is the normal to the active surface $z = 0$, $\vec{j}^{(e,m)}$, $\vec{P}^{(e,m)}$ is the surface current density and polarization, respectively; indices 1.2 correspond to the values of the fields on both sides of the active surface. The absence of 'reflection' at the frequency of the phase-coupled wave and the relationship between the 'passing'/'emitted' phase-coupled wave and the signal wave are clearly reflected in the relations (6.6).

The relationship between voltage u_N and the charge q_N for a nonlinear element in figure 6.6(b) has the form $u_N = \alpha q_N + \beta q_n^2$, α and β are linear coefficient and quadratic nonlinearity coefficient, respectively. For a nonlinear differential capacitance C_N of a parametric diode (Ionkin 1972, Migulin *et al* 1988, Kalinin and Bayonets 1990) we have $C_N(u_N) = \dfrac{dq_N}{du_N} = (1 - u_N/\varphi_0)^{-k}C_0$, $q_N = \int_0^{u_N} C_N(u)du$. Here C_0 is the differential capacitance in the linear approximation, φ_0 is the contact potential difference of the p-n junction in the diode used as a nonlinear capacitor, $k = 1/2$ (Kalinin and Bayonets 1990, Migulin *et al* 1988), $\alpha = 1/C_0$, $\beta = -1/4C_0^2\varphi_0$ (Ionkin 1972). The potential φ_0 is further used as a scale in voltage normalization; the corresponding scale of the charge and the electric field are defined as $q_0 = C_0\varphi_0$ and $E_0 = \varphi_0/l_0$, respectively, where l_0 is the length of the 'tendrils'/half-length of the electric dipole included in the chiral metaparticle, figure 6.6(a). The following field components can excite metaparticles, depending on the polarization of the signal wave incident on the system:

$$\overline{E}_{\text{ed}} = \begin{cases} \overline{E}_x, & \text{for particle "}a\text{"} \\ \overline{E}_y, & \text{for particle "}b\text{"} \end{cases}, \quad \overline{H}_{\text{md}} = \begin{cases} \overline{H}_x, & \text{for particle "}a\text{"} \\ \overline{H}_y, & \text{for particle "}b\text{"} \end{cases} \tag{D.8}$$

In the quasilinear (parametric) approximation we obtain the equation for the charge on the nonlinear capacity (figure 6.6(b)) (Rapoport *et al* 2013a):

$$d^2\overline{q}_N/d\tau^2 + \overline{q}_N + r_0 d\overline{q}_N/d\tau = E_E + E_H + \theta_0\overline{u}_p - v_{NL}\overline{q}_N^2; \tag{D.9}$$

For the values included in (D.9), there are relations:

$$\begin{aligned}
&\overline{q}_N = q_N/q_0,\ \tau = \omega_{\text{eff}}t,\ t - \text{time},\ \omega_{\text{eff}} = 1/\sqrt{C_{\text{eff}}L},\ C_{\text{eff}}^{-1} \equiv C_E^{-1} + \alpha = C_E^{-1} + C_0^{-1},\\
&r_0 = \omega_{\text{eff}}(R_C + R_L)C_{\text{eff}},\ E_E = \theta_0\overline{E}_{\text{ed}},\ E_H = -f_0\, d\overline{H}_{\text{md}}/d\tau,\\
&\overline{u}_p = u_p/\varphi_0, f_0 = \operatorname{sgn}(f_0)\varsigma_0\theta_0,\ \varsigma_0 = S\omega_{\text{eff}}/(cl_0),\ \theta_0 = C_{\text{eff}}/C_0,\\
&v_{\text{NL}} = C_{\text{eff}}q_0\beta = -C_{\text{eff}}/(4C_0)
\end{aligned}$$

In equation (D.9), u_p is the amplitude of the parametric voltage at the frequency ω_p, applied to the nonlinear element in the circuit in figure 6.6(b), $|f_0| = 1$, $\operatorname{sgn}(f_0)$ defines the 'sign of rotation' ('left' or 'right') of the spiral/chiral particle. As follows from ((D.8) and (D.9)) and figure 6.6(a), for polarization (E_x, H_y), the chiral particles 'a' and 'b' are linearly excited at the 'signal' frequency ω_s by an external electric field E_x (corresponding electromotive force is equal) and magnetic field H_y, (corresponding electromotive force is equal $-f_0 d\overline{H}_y/d\tau$). For polarization (H_x, E_y), the corresponding electromotive forces are equal to $-f_0 d\overline{H}_x/d\tau$ and $\theta_0\overline{E}_y$, respectively. Suppose, without loss of generality, that $\operatorname{sgn}(f_{0a}) = 1$. Hereinafter, the indices 'a', 'b' refer to the corresponding metaparticles shown in figure 6.6(a). As shown in Rapoport *et al* (2013a), the first and second relations (6.7) for the signal frequency ω_s are reduced to the conditions

$$[\varsigma_{0a}\theta_{0a}l_{0a}/\varsigma_{0b}\theta_{0b}l_{0b}] = \mathrm{sgn}(f_{0b}),\ [(\varsigma_{0b}\varpi_b)^{-2}\theta_{0a}l_{0a}/(\theta_{0b}l_{0b})] = \mathrm{sgn}(f_{0b}) \tag{D.10}$$

in accordance. If the signs f_{0a} and f_{0b} are the same, then the sizes and characteristic frequencies of the particles 'a' and 'b' are equivalent, namely $\omega_{\mathrm{eff}a} = \omega_{\mathrm{eff}b}$, $\varsigma_{0a} = \varsigma_{0b} \equiv \varsigma_0$, $\theta_{0a} = \theta_{0b} \equiv \theta_0$, $l_{0a} = l_{0b} \equiv l_0$, $\varpi_{\mathrm{eff}a} = \varpi_{\mathrm{eff}b} = \varpi_{\mathrm{eff}}$, $\varpi_{pa} = \varpi_{pb} = \varpi_p$. Using the same, in absolute value, the amplitudes of the pumping voltages for the particles 'a' and 'b', $|\bar{u}_{pb}| = |\bar{u}_{pa}|$, the relationship (D.10) is reduced to $(\varsigma_0\varpi_s) = 1$, which means

$$S\omega_s/(cl_0) = 1,\ \mathrm{sgn}(f_{0b}) = 1 \tag{D.12}$$

Conditions (6.7) for a nonlinear phase-conjugate wave at frequency ω_{pc} are reduced to very simple relations (Rapoport 2012b)

$$\varpi_{pc}/\varpi_s = |\bar{u}_{pb}| = |\bar{u}_{pa}|,\ \ \varpi_{pc}/\varpi_s = |\bar{u}_{pa}| = |\bar{u}_{pb}|,\ \mathrm{sgn}(\bar{u}_{pb}) = -\mathrm{sgn}(\bar{u}_{pa}) \tag{D.13}$$

Given the above condition $|\bar{u}_{pa}| = |\bar{u}_{pb}|$, (D.13) is reduced to the form

$$\varpi_{pc}/\varpi_s = 1,\quad \mathrm{sgn}(\bar{u}_{pb}) = -\mathrm{sgn}(\bar{u}_{pa}) \tag{D.14}$$

Since the frequency ϖ_s is slightly different from ϖ_{pc}, the first equation from (D.14) can be satisfied only approximately. If you provide the exact satisfaction of equation (D.12), the reflection of the signal wave will be absent. To characterize the 'equivalent reflection' of a phase-coupled wave (in the absence of an incident phase-coupled wave), we introduce an effective reflection coefficient in the form $R_{\mathrm{ph.c.}} = (A_{\mathrm{ph.c.}})_{\mathrm{reflected}}/(A_{\mathrm{ph.c.}})_{\mathrm{transmitted}}$, where $(A_{\mathrm{ph.c.}})_{\mathrm{reflected}}$ and $(A_{\mathrm{ph.c.}})_{\mathrm{transmitted}}$ are the amplitudes of the 'reflected' and phase-coupled waves propagating in the '$+z$' direction, respectively. In order of magnitude, $|R_{\mathrm{ph.c.}}| \sim |(\varpi_s - \varpi_{pc})/\varpi_s| \sim 0.1$. For the amplitude of the phase-conjugate wave taking into account the rotation of the plane of polarization, we obtain $\vec{E}_t \approx B_p E_{xi}^*(\vec{x}_0 - i\vec{y}_0)$, where $\vec{x}_0$ and $\vec{y}_0$ are the unit vectors in the x and y directions, respectively,

$$B_p = i\left[\left(1 - \varpi_p^2\right)\left(1 - \varpi_{pc}^2\right)\left(1 - \varpi_s^2\right)\right]^{-1}\varsigma_0\theta_0^2 v_{\mathrm{NL}}F\bar{u}_{pa} = i\varsigma_0\theta_0^2 v_{\mathrm{NL}}F\bar{u}_{pa}B_\omega \tag{D.15}$$

where $B_\omega \equiv [(1 - \varpi_p^2)(1 - \varpi_{pc}^2)(1 - \varpi_s^2)]^{-1}$, $F = 4\pi N_s l_0^3 C_0/S$. N_s, l_0 and S measured in cm^{-2}, cm and cm^2 respectively, C_0 measured in pF. It was considered for the nonlinearity coefficients of particles 'a' and 'b' $v_{\mathrm{NL}a} = v_{\mathrm{NL}b} = v_{\mathrm{NL}}$. We neglect energy losses. If $\varpi_p/2 = 1 + \delta_p$, $\varpi_s = 1 + \delta$, $\varpi_{pc} = 1 - \delta$, then the frequency factor B_ω is a function of frequency tuning δ_p and δ. In figure 6.6(c), shows the dependence of the 'frequency factor' B_ω on the frequency detuning δ at a given value δ_p. Put for numerical estimates (Rapoport *et al* 2013a) $|\bar{u}_{pa}| \sim 0.1$. Assuming $\omega_s \sim 2\pi \cdot 14 \times 10^{10}\ \mathrm{s}^{-1}$, $S = \pi a_0^2$, $a_0 \sim l_0 \sim 0.2$ cm, $N_s \sim 4\ \mathrm{cm}^{-2}$, $\varphi_0 \sim 0.5$ V, $\delta \sim 0.1$, we obtain the ratio of energy flows P_t and P phase-conjugate and incident waves $P_t/P_i = |E_t|^2/|E_i|^2 = |B_p|^2 \sim 0.25 \cdot 10^{-6}$. Note that due to the strong dependence of the ratio P_t/P_i on the value C_{eff}/C_0, for values of $C_{\mathrm{eff}}/C_0 \sim 1$, the ratio P_t/P_i can reach the order of magnitude 10^{-5}. Although the system is not yet optimized, these values $|B_p|^2$ are already sufficient for the experimental implementation of the phase conjugation effect without reflection.

Problems to chapter 6

Problem 6.1 Derive parametric term in equations (6.2).

The way of solution:

To solve this problem, use equations (3.24)–(3.27) (chapter 3) and obtain the system of equations for parametric coupling of counter propagating BMSW (figure 6.2). Use the system of Bloch equations in the form

$$\begin{aligned}\frac{\partial m_{x1}^{(1)}}{\partial t} &= \omega_H m_{y1} - \frac{\omega_M}{4\pi}\tilde{h}_{y1} - \gamma\left(h_{y1}m_{z1} - h_{z1}m_{y1}\right)^{(1)} \\ &- \gamma\left(h_{y1}m_{z1} + h_{y1}m_{y1} - \tilde{h}_{z1}m_{y1} - \tilde{h}_{z1}m_{y1}\right)^{(1)} + \gamma h_{z\mathrm{pmp}}m_y\end{aligned} \tag{P6.1}$$

$$\begin{aligned}\frac{\partial m_{y1}^{(1)}}{\partial t} &= -\omega_H m_{x1} + \frac{\omega_M}{4\pi}\tilde{h}_{x1} + \gamma(\tilde{h}_{x1}m_{z1} + \tilde{h}_{z1}m_{x1})^{(1)} \\ &+ \gamma(\tilde{h}_{x1}m_{z1} + h_{x1}m_{z1} + h_{z1}m_{x1} + h_{z1}m_{x1})^{(1)} - \gamma h_{z\mathrm{pmp}}m_x\end{aligned} \tag{P6.2}$$

$$\vec{h}_m = h_{z\mathrm{pmp}}\mathrm{e}^{zi\omega t}\vec{e}_z \tag{P6.3}$$

In equations (P6.1)–(P6.3), the denotations described in chapter 3 and sections (3.3.1) and (3.3.2) are used; the upper index (1) in (P6.1) and (P6.2) corresponds to the values $\sim\exp[i(\omega t - kz)]$; corresponding system of equations for the values $\sim\exp[i(\omega t + kz)]$ is added; h_{pmp} is the amplitude of pumping waves. Applying the procedure of the method NEELS, as described in section 3.3.1, obtain the differential relation

$$\begin{aligned}&\frac{C}{4\pi}div\left\{[\vec{e}_2^{\,*}\vec{h}_1] + [\vec{e}_1\vec{h}_2^{\,*}]\right\} + \frac{\partial}{\partial t}\left\{\frac{1}{4\pi}(\varepsilon\vec{e}_1\vec{e}_2^{\,*} + \vec{h}_1\vec{h}_2^{\,*}) + \right. \\ &\left.\frac{4\pi\omega_H}{\omega_M}(m_{x1}m_{x2}^* + m_{y1}m_{y2}^*)\right\} = \frac{4\pi f}{\omega_M}\left[\frac{\partial m_{y2}^*}{\partial t}(\tilde{h}_{y1}m_{z1} - h_{z1}^*m_{y1})^{(1)} + \right. \\ &\left. + \frac{\partial m_{x2}^*}{\partial t}(\tilde{h}_{x1}m_{z1} - \tilde{h}_{z1}m_{x1})\right] - \left[h_{z2}^*\frac{\partial m_{z2}}{\partial t}\right] + \gamma(h_{y2}^*m_{x1} - h_{x2}^*m_{y1})h_{zpmp}\end{aligned} \tag{P6.4}$$

Suppose, for simplicity that the system is symmetrical with respect to the axis x (direction to the normal to FF). Accounting for the linear dispersion equation for the main mode of BVMSW (Gurevich and Melkov 1996) and omitting, writing the evolution equation, all the terms besides the time derivative of the slowly varying amplitude of BVMSW 1 (propagating in positive Z direction, figure 6.2) and the term describing the parametric coupling, one can obtain

$$\frac{\partial U_1}{\partial t} + \ldots = -i\frac{\sigma_0}{\frac{\omega_H}{\omega_M}[\sigma_0{}^2 + (\mu - 1)^2]}\gamma h_{z\text{pmp}} U_2^* \tag{P6.5}$$

The term in the right-hand side of (P6.5) describes the parametric coupling, namely an influence of the idle wave 2 on the incident wave 1. The similar equation can be written for the time derivative of the slowly varying amplitude A_2 of the idle wave 2. The coefficient in the right-hand part in equation (P6.5) plays a role of the coefficient V in equations (6.2).

Problem 6.2 Derive in detail the conditions, to which surface polarization densities p_x, m_y should correspond, to provide phase conjugation for the plane wave (E_x, H_y), which is incident **normally** from $z = -\infty$ to the plane $z = 0$, where the surface polarization is located (see section 6.3.2 and appendix D.4). Therefore, if the amplitudes of incident, reflected and transmitted electric waves/field are A_{Ei}, A_{Er}, A_{Et}, respectively, we should have

$$A_{Er} = 0, \; A_{Et} = C_0 A_{Ei}{}^* \tag{P6.7}$$

where C_0 is arbitrary complex constant.

The way of solution:

Suppose, semi-infinite regions $z < 0$, $z > 0$ are homogeneous magneto-dielectrics, with corresponding constants ε, μ. Put

$$\begin{pmatrix} E_x \\ H_y \end{pmatrix}_{i,r,t} = \begin{pmatrix} A_E \\ A_H \end{pmatrix}_{i,r,t} e^{i(\omega t - k_z z)} \tag{P6.8}$$

where indexes i,r,t correspond to incident, reflected and transmitted waves, respectively. Using Maxwell equations in the 'source free' regions $z < 0$, $z > 0$

$$\nabla \times \vec{H} = \frac{\varepsilon}{c}\frac{\partial \vec{E}}{\partial t}, \; \nabla \times \vec{E} = -\frac{\mu}{c}\frac{\partial \vec{H}}{\partial t}, \tag{P6.9a}$$

one can get

$$A_{Hi,\,t} = \sqrt{\frac{\varepsilon}{\mu}} A_{Ei,\,t}, \; A_{Hr} = -\sqrt{\frac{\varepsilon}{\mu}} A_{Er}, \; k_0 = \frac{\omega}{c}, \; k_z = k = \sqrt{\varepsilon\mu}\, k_0 \tag{P6.9b}$$

Boundary conditions on the interface $z = 0$ between media 1 ($z < 0$) and 2 ($z > 0$) has the standard form

$$\vec{n} \times (\vec{E}_2 - \vec{E}_1) = -\frac{4\pi}{c}\vec{j}_{ts}^{(m)} = -\frac{4\pi}{c}\frac{\partial \vec{P}_{ts}^{(m)}}{\partial t} \tag{P6.10a}$$

$$\vec{n} \times (\vec{H}_2 - \vec{H}_1) = -\frac{4\pi}{c}\vec{j}_{ts}^{(e)} = -\frac{4\pi}{c}\frac{\partial \vec{P}_{ts}^{(e)}}{\partial t} \tag{P6.10b}$$

Here

$\vec{n} = \vec{e}_z, \vec{j}_{ts}^{(e,m)}$ are equivalent electric and magnetic (tangential) surface currents, $\vec{P}_{ts}^{(e,m)}$ are corresponding (tangential) surface polarizations,

$$\partial/\partial t = i\omega, \tag{P6.11a}$$

$$\vec{P}_{ts}^{(e)} = (P_{sx}^{(e)}, 0,0), \quad \vec{P}_{ts}^{(m)} = \left(0, P_{sy}^{(m)}, 0\right) \tag{P6.11b}$$

Using equations (P6.11a, P6.11b) and (P6.9b), it is possible to reduce (P6.10a, P6.10b) to

$$A_{Et} - A_{Er} = A_{Ei} - \frac{4\pi}{c}\frac{\partial P_{sy}^{(m)}}{\partial t} \tag{P6.12a}$$

$$A_{Et} + A_{Er} = A_{Ei} - \frac{4\pi}{c}\sqrt{\frac{\mu}{\varepsilon}}\frac{\partial P_{sy}^{(m)}}{\partial t} \tag{P6.12b}$$

As follows from (P6.12a and P6.12b),

$$A_{Et} = A_{Ei} - \frac{1}{2}\frac{4\pi}{c}\left(\sqrt{\frac{\mu}{\varepsilon}}\frac{\partial P_{sx}^{(e)}}{\partial t} + \frac{\partial P_{sy}^{(m)}}{\partial t}\right), \tag{P6.13a}$$

$$A_{Er} = \frac{1}{2}\frac{4\pi}{c}\left(-\sqrt{\frac{\mu}{\varepsilon}}\frac{\partial P_{sx}^{(e)}}{\partial t} + \frac{\partial P_{sy}^{(m)}}{\partial t}\right) \tag{P6.13b}$$

Finally, conditions of 'reflectionless complex conjugation' (P6.7), accounting for (P6.11a) and (P6.13a, P6.13b) lead to

$$P_{sy}^{(m)} = \sqrt{\frac{\mu}{\varepsilon}}P_{sx}^{(e)} \tag{P6.14a}$$

$$\frac{4\pi}{c}\frac{\partial P_{sy}^{(m)}}{\partial t} = A_{Ei} - C_0 A_{Ei}^{*} \tag{P6.14b}$$

Then, if $C_0 = 1$, (P6.14b) reduces to

$$\frac{4\pi}{c}\frac{\partial P_{sy}^{(m)}}{\partial t} = \frac{4\pi}{c}i\omega P_{sy}^{(m)} = 2iA_{Ei}^{''},$$

where $A_{Ei} = A_{Ei}^{'} + iA_{Ei}^{'}$, or

$$P_{sy}^{(m)} = \frac{c}{2\pi\omega}A_{Ei}^{''} \tag{P6.15}$$

As follows from (P6.15) and (P6.14a), for the real amplitude of the incident wave, a solution is $P_{sy}^{(m)} = 0$, $P_{sx}^{(e)} = 0$, which physically does correspond to the non-distorted propagation of plane wave in normal direction through the plane $z = 0$ in the absence of any surface polarization/currents.

Problem 6.3 Present corresponding (see problem 6.2) results for **oblique incidence** of a plane wave on interface $z = 0$.

The way of solution:
An incident wave has the components : $(E_{ix,}H_{iy}, E_{iz})$, where

$$E_{ix} = A_{Ei}e^{i(\omega t - k_z z - k_x x)}, \quad E_{ix} = A_{Ei}e^{i(\omega t - k_z z - k_x x)}, \quad A_{Ei} = \frac{k_z}{k_0 \varepsilon} A_{Hi} = \cos\varphi \sqrt{\frac{\mu}{\varepsilon}} A_{Hi},$$

$$\cos\varphi = k_z/k, \quad k = k_0\sqrt{\varepsilon\mu}$$

Then, let us consider the real amplitude of the incident wave and therefore put $A_{Ei} = \mathrm{Re}(A_{Ei}) = A_{Ei}^{'}(A_{Hi} = \mathrm{Re}(A_{Hi}) = A_{Hi}^{'})$,

Again, the consideration of all (incident, reflected and transmitted) waves, along with conditions for 'reflectionless phase conjugation' in the form

$$H_{\text{reflected}} = 0, \ \mathrm{H}_{\text{transmitted}(z=+0)} = \left[H_{\text{incident}(z=-0)}\right]^* \tag{P6.16}$$

gives the following outcomes:

$$P_{sx}^{(e)} = \frac{1}{\cos\varphi}\sqrt{\frac{\varepsilon}{\mu}} P_{sy}^{(m)},$$

$$P_{sx}^{(e)} = -\frac{c}{2\pi\omega}\frac{1}{\cos\varphi}\sqrt{\frac{\varepsilon}{\mu}} \sin(k_x x) A_{Ei} = -\frac{c}{2\pi\omega} \sin(k_x x) A_{Hi}$$

Again, for normal incidence ($\cos\varphi = 1$, $k_x = 0$), conditions (P6.16) of 'reflectionless phase conjugation' require $P_{sx}^{(e)} = 0$, $P_{sy}^{(m)} = 0$.

List of abbreviations

BVMSWs	Backward volume magnetostatic waves
FFs	Ferrite films
MSWs	Magnetostatic waves
NEELS	Nonlinear evolution equations in layered media
ULMs	Unidirectional loop metamaterials

References

Boardman A D, Grimalsky V V, Kivshar Y, Koshevaya S V, Lapine M, Litchinitser M, Malnev V N, Noginov M, Rapoport Y G and Shalaev V M 2011 Active and tunable metamaterials *Laser Photonics Rev.* **5** 287–307

Boardman A D, Grimalsky V V, Ivanov B, Koshevaya S V, Velasko L, Zaspel C E and Rapoport Y G 2005 Excitation of vortices using linear and nonlinear magnetostatic waves *Phys. Rev. E.* **71** 026614–24

Boardman A D, Mitchell-Thomas R and Rapoport Y G 2009 Weakly nonlinear waves in layered bi-anisotropic *Proc. of Third Int. Congress on Adv. Electromagn. Materials in Microwaves and Optics: Metamaterials (London UK)* 495–7

Buttner O, Bauer M, Demokritov S O, Hillebrands B, Kivshar Y S, Grimalsky V V, Rapoport Y and Slavin A N 2000 Linear and nonlinear diffraction of dipolar spin waves in yttrium iron garnet films observed by space- and time-resolved Brillouin light *Phys. Rev. B.* **61** 11576–87

Damon R W and Eshbach J R 1961 Magnetostatic modes of a ferromagnetic slab *J. Phys. Chem. Solids.* **19** 308–20

Davoyanм A R and Engheta N 2019 Nonreciprocal emission in magnetized epsilon-near-zero metamaterials *ACS Photonics* **6** 581–6

Davoyan A and Atwater H 2018 Quantum nonlinear light emission in metamaterials: Broadband Purcell enhancement of parametric downconversion *Optica* **5** 608–11

Demokritov S O, Hillebrands B and Slavin A N 2001 Brillouin light scattering studies of confined spin waves: linear and nonlinear confinement *Phys. Reports* **348** 441–89

Elnaggar S Y and Milford G N 2018 Controlling non-reciprocity using enhanced Brillouin scattering *IEEE Trans. Antennas Propagation* **66** 3500–11

Gordon A L, Melkov G A and Serga A A *et al* 1998 Phase conjugation of linear signals and solitons of magnetostatic waves *ZETP Lett.* **67** 913–6

Grimalsky V V and Rapoport Y G 1995 Convolution of magnetostatic waves in ferrite films *J. Magn. Magn. Mater.* **140–4** 2195–6

Grimalsky V, Rapoport Y and Slavin A N 1997 Nonlinear diffraction of magnetostatic waves in ferrite films *J. Phys. IV France* **7** PC1-393–4

Grimalsky V V, Rapoport Y G, Slavin A N and Zaspel C E 2000 Parametric amplification and wave front conjugation of 2D pulse in ferromagnetic film *Proc. of Meeting of the APS (Minneapolis, MN)* Z25-5 1035

Grimalsky V V, Rapoport Y G, Zaspel C E and Slavin A N 2001a Parametric amplification of two-dimensional dipolar spin wave pulses in ferrite films *Proc. of the 8th Int. Conf. On Ferrites (Japan Society of Powder and Powder Metallurgy, Kyoto and Tokyo 2000)* pp 921–3

Grimalsky V V, Mantha J H, Rapoport Y G, Slavin A N and Zaspel C E 2001b Numerical models of amplification and wave front reversal of two-dimensional spin wave packets in magnetic films *Materials Science Forum, Kyiv (Ukraine), 8th Europ. Magn. Mater. and Appl. Conf. (EMMA 2000)* (Zurich: TransTech Publications) pp 377–80

Gurevich A G and Melkov G A 1996 *Magnetization Oscillations and Waves* (New York: CRS Press) p 464

Ionkin P A 1972 *Principels engeneering electrophysics* (Moscow: Visshaya shkola) p 439 (in Russian)

Jin P and Ziolkowski R W 2010 Metamaterial-inspired, electrically small Huygens sources *IEEE Anten. Wireless Propag. Lett.* **9** 501–5

Kalinin B A and Shtikov V V 1990 On the possibility of reversing the front of radio waves in the artificial nonlinear media *J. Commun. Techol. Electron.* **35** 2275–81 (in Russian)

Kasumova R J and Amirov S S 2019 Frequency transformation of ultrafast laser pulses in metamaterials *Superlatt. Microstruct.* **126** 49–56

Katko A R, Gu S and Barrett J P *et al* 2010 Phase conjugation and negative refraction using nonlinear active metamaterials *Phys. Rev. Lett.* **105** 123905

Kivshar Y S and Agraval G P 2003 *Optical solitons From Fibers to Photonic Crystals* (New York: Academic) p 643

Kodera T and Caloz C 2018 Unidirectional loop metamaterials (ULM) as magnetless artificial ferrimagnetic materials: principles and applications *IEEE Antennas Wireless Propagation Lett.* **17** 1–5

Kruk S, Poddubny A, Smirnova D, Wang L, Slobozhanyuk A, Shorokhov A, Kravchenko I, Luther-Davies B and Kivshar Y 2019 Nonlinear light generation in topological nanostructures *Nat. Nanotechnol.* **14** 126–30

Landa P S 1996 *Nonlinear Oscillations and Waves in Dynamical Systems* (Dordrecht: Kluwer) p 683

Lheurette E, Vanbesien O and Lippens D 2007 Double negative media using interconnected-type metallic particles *J. Microwave Opt. Technol. Lett.* **49** 84–90

Li G, Zhang S and Zentgraf T 2017 Nonlinear photonic metasurfaces *Nat. Rev. Mater.* **2** 17010

Liu N, Kaiserand S and Giessen H 2008 Magnetoinductive and electroinductive coupling in plasmonic metamaterial molecules *Adv. Mater.* **20** 4521–5

Lukomsky V P 1978 Nonlinear magnetostatic waves in ferromagnetic plates *Ukr. J. Phys.* **23** 134–9

Maslovski S, Rapoport Y and Tretyakov S 2011 Perfect lensing with phase-conjugating surfaces: Approaching practical realization *Proc. of Intern. Conf. 'Days of diffraction'2011' (St. Petersburg (Russia))* pp 143–4

Maslovski S and Tretyakov S 2003 Phase conjugation and perfect lensing *J. Appl. Phys.* **94** 4241–3

Melkov G A and Sholom S V 1990 Amplification of surface magnetostatic waves by a parametric pump *Sov. Phys. Tech. Phys.* **35** 943–6

Melkov G A and Serga A A 1998 Nonlinear parametric excitation of spin waves *Frontiers in Magnetism of Reduced Dimension Systems* ed V G Bar'yakhtar (Dordrecht: Kluwer) pp 555–78

Melkov G A, Serga A A, Tiberkevich V S, Oliynyk A N, Bagada A V and Slavin A N 1999 Parametric interaction of spin wave pulse with localized non-stationary pumping: amplification and phase conjugation *IEEE Trans. Magn* **35** 3157–9

Melkov G A, Serga A A, Tiberkevich V S, Oliynyk A N and Slavin A N 2000 Wave front reversal of a dipolar spin-wave pulse in a nonstationary three-wave parametric interaction *Phys. Rev. Lett.* **84** 3438

Migulin V V, Medvedev V I, Mustel E P and Parigin V N 1988 *Fundamentals of Oscilation Theory* (Moscow: Nauka) p 480 (in Russian)

Prokopenko O V *et al* 2019 Recent trends in microwave magnetism and superconductivi *Ukr. J. Phys.* **64** 888–926

Rapoport Y G 1987 Contribution of the ε tensor components responsible for nonlinear magneto-optical effects and gyromagnetic effects to light scattering by a magnetostatic wave in a gyromagnetic waveguide *Regional Conf. Spin-Wave Phenomena Of Microwave Electronics (Krasnodar)* pp 193–4 Abstracts (in Russian)

Rapoport Y G 2006a Formation of structures under parametric coupling of two-dimensional nonlinear pulses of magnetostatic waves in magnetic films *IEEE Proc. of the 16th Int. Crimean Conf. 'Microwave and Communication Technology' CriMiCo'2006 (Sevastopol, Crimea, Ukraine)* pp 629–30 IEEE Catalog No. 06EX1376

Rapoport Y G, Boardman A D, Kanevskiy V I, Malnev V N, King N J and Velasco L 2006b Modelling new active media based on metamaterials with artificial *IEEE Proc. of the 16th Int. Crimean Conf. 'Microwave and Communication Technology' CriMiCo'2006 (Sevastopol, Crimea, Ukraine)* pp 671–2 IEEE Catalog No. 06EX1376

Rapoport Y G 2006 Formation of structures under parametric coupling of two-dimensional nonlinear pulses of magnetostatic waves in magnetic films *Proc. IEEE 16th Int. Crimean*

Conf. 'Microwave & Telecommunication Technology', CriMiCo'2006 vol 2, Article No. 4023416 *(Sevastopol, Crimea, Ukraine, Sept.11–5)* pp 629–30 Catalog No. 06EX1376.

Rapoport Y G, Boardman A D, Grimalsky V V, Koshevaya S V, Zaspel C E and Ivanov B A 2008 Nonlinear vortex generation by forward volume magnetostatic waves *Electromagnetic, Magnetostatic and Exchange—Interaction Vortices in Confined Magnetic Structures* ed E O Kamenetskii (Kerala, India: Transworld Research Network) pp 29–44

Rapoport Y, Grimalsky V and Zaspel C 2012a Method for the derivation of nonlinear evolution equations in layered structures (NEELS): An example of nonlinear waves in gyrotropic layers *Bull. Univ. Kyiv Phys.* **14/15** 72–6

Rapoport Y G 2012b General method for modeling nonlinear waves in layered structures of diferent physical nature including bi-anisotropic and active metamaterials *Progress in Electromagn Research Symp. PIERS (Moscow Russia)* pp 18–9 Abstracts

Rapoport Y G, Tretyakov S and Maslovski S 2013a Nonlinear active Huygens metasurfaces for reflectionless phase conjugation of electromagnetic waves in electrically thin layers *J. Electromagn. Waves Appl.* **27** 1309–28

Rapoport Y G 2013b General method for modeling nonlinear waves in active metamaterial and gyrotropic layered structures *Proc. of Int. Kharkov Symp. on Phys. and Engineering of Microwaves, Millimeter and Submillimeter Waves MSMW'13 (Kharkov Ukraine)* pp 253–5

Rapoport Y G, Grimalsky V V, Boardman A D and Malnev V N 2014a Controlling nonlinear wave structures in layered metamaterial, gyrotropic and active media *Proc. of IEEE 34th Int. Sci. Conf. Electronics and Nanotechnology* (Kyiv Ukraine: ELNANO) pp 46–50

Rapoport Y G 2014b General method for deriving the equations of evolution and modeling of nonlinear waves in layered active media with bulk and surface *Bull.of Kyiv Nat. Taras Shevchenko Univ. Series: Phys.-Math. Sci.* **1** 281–8 (in Ukrainian)

Rapoport Y G 2015a Modeling 'From properties of a metaparticle (MP) to characteristics of nonlinear waves in layered active bi-anisotropic metamaterials (BIAM)' *Ukr.–Germ. Symp.on Phys. and Chem. of Nanostructures and on Nanobiotechnology (Kyiv Ukraine)* p 159

Rapoport Y G and Boardman A D 2015b Modeling 'from the characteristics of the metaparticle to the characteristics of nonlinear waves in layered active bi-anisotropic metamaterials with bulk and surface nonlinearities' *Bull. of Kyiv Nat. Taras Shevchenko Univ. Series: Physics* **3** 207–12 (in Ukrainian)

Rowland D R 1999 Conservation law for multimoded nonlinear optical waveguide interactions and its physical interpretation *Phys. Rev.* E **59** 7141–7

Serga A A, Kostylev M P and Hillebrands B 2008 Formation of guided-wave bullets in ferromagnetic film stripes *Phys. Rev. Lett.* **101** 137204

Serga A A, Hillebrands B, Demokritov S O, Slavin A N, Wierzbikicki P, Vasyuchka V, Dzyapko J and Chumak F 2005 Parametric generation of forward and phase-conjugated spin-wave bullets in magnetic films *Phys. Rev. Lett.* **94** 167202–5

Serga O O 2006 Parametric interaction of spin waves and oscillations with nonstationary local pumping *Abstract of the dissertation for the degree of Doctor of Physical and Mathematical Sciences* Kiev p 36 (in Russian)

Slavin A N, Büttner O, Bauer M, Demokritov S O, Hillebrands B, Kostylev M P, Kalinikos B A, Grimalsky V V and Rapoport Y 2003 Collision properties of quasi-one-dimensional spin wave solitons and two-dimensional spin wave bullets *Chaos* **13** 693–701

Slavin A N, Demokritov S O and Hillebrands B 2002 Nonlinear spinwaves in one and two-dimensional magnetic waveguides *Spin Dynamics in Confined Magnetic Structures* (Topics in Applied Physics vol 83) ed B Hillebrands and K Ounadjela (Berlin: Springer) pp 35–66

Tretyakov S A, Kharina T G, Simovski K R and Pavlov A A 1994 Frequency dispersion in chiral and omega media: An approximate theoretical model *Antennas and Propagation Society Int. Symp., AP S. Digest* vol 2 pp 722–5

Zangeneh-Nejad F, Kaina N, Yves S, Lemoult F, Lerosey G and Fleury R 2020 Non-reciprocal manipulation of subwavelength fields in locally-resonant metamaterial crystals *IEEE Trans. Antennas Propagation* 1726–32

Zel'Dovich V Ya, Pilipetsky N P and Shkunov V V 1985 Principles of phase conjugation *Springer Ser. Opt. Sci.* **42** 228

Zhao H, Zhou J, Kang L and Zhao Q 2009 Tunable two-dimensional left-handed material consisting of ferrite rods and metallic wires *Opt. Express* **17** 13373–80

Zhukovsky S V, Ozel T, Gaponik N, Eychmuller A, Lavrinenko A V, Demir H V and Gaponenko S V 2005 Hyperbolic metamaterials based on quantum-dot plasmon-resonator nanocomposites *Opt. Lett.* **30** 3210–2

Zvezdin A K and Popkov A K 1983 Contribution to the nonlinear theory of magnetostatic spin waves *Zh. Eksp. Teor. Fiz* **2** 606–15

IOP Publishing

Waves in Nonlinear Layered Metamaterials, Gyrotropic and Plasma Media

Yuriy Rapoport and Vladimir Grimalsky

Chapter 7

Formation propagation, and control of bullets in metamaterial waveguides with higher-order nonlinear effects and magnetooptic interaction

7.1 Introduction

Bullets are highly localized two-dimensional (Büttner *et al* 2000) or three-dimensional (Kivshar and Agrawal 2003) soliton-like (Kivshar and Agrawal 2003, Akhmediev and Ankiewicz 2005) spatio-temporal wave packets that can propagate, retaining their shape, at a distance sufficient for observation, and practical use (for example, for signal processing devices (Serga *et al* 2006)) and, accordingly, for a sufficient time.

For nonlinear bullets (Büttner *et al* 2000, Kivshar and Agrawal 2003, Balakin *et al* 2017), the existence and propagation with the retention of shape for a sufficiently long time is ensured by the balance between nonlinearity, dispersion, diffraction, etc and wave scattering in appropriate environments. Along with such bullets, linear light bullets (Abdollahpour *et al* 2010, Li *et al* 2019) can also exist, for example, based on the temporal and spatial distribution of Airy or 2D plasmon bullets on specially designed metasurfaces (Karalis and Joannopoulos 2019). Nonlinear bullets based on Airy pulses are also possible (Fattal *et al* 2011, Abdollahpour *et al* 2010). The bullets can be useful for analog signal processing (Serga *et al* 2006), light control in combined positive and negative waveguide arrays (Ali *et al* 2021), tomography, microscopy, and signal transmission over long distances, as well as in biological applications (Abdollahpour *et al* 2010), etc.

In this chapter, we present the investigations of self-compression due to nonlinearity and linear and nonlinear dispersion and diffraction on the modulation instability (MI) and the formation of bullets (Grimal'sky *et al* 2000, Rapoport *et al* 2004, 2012, Rapoport 2013, Boardman *et al* 2010, Slavin *et al* 2003, Boardman *et al* 2011). As media for the wave propagation, there are considered the metamaterial (MM)

doi:10.1088/978-0-7503-2336-9ch7

waveguides with the 'negative phase' (NP) (Boardman *et al* 2009b, Rapoport 2006, Buttner *et al* 2000, Malomed *et al* 2005), or alternating waveguide wave propagation, which formed from materials with 'positive' and 'negative' phases. As a result the diffraction management is provided (Eisenberg *et al* 2000, Boardman *et al* 2010a, Boardman *et al* 2010b). The diffraction management is based on using waveguide structures that are periodic in the direction of wave propagation, see chapter 5 and figure 5.1. The spatial period of the structures is much less than the wavelength, so averaging is possible of the parameters of the waveguiding media. The proper parameters of the elementary cell provide the control of the coefficient of the wave diffraction within wide intervals of its value.

To point out the principal effects, the diffraction in the layered structures that include the layers of MMs is considered by means of two simplified methods. The first one is close to that in the unlimited MM in the approximation of the weak waveguiding dispersion, whereas the second method uses qualitative simulations like in a plane metal waveguide. More complex approaches do not change the general wave dynamics qualitatively. With the help of these approximations, the coefficients of the normalized evolution equation for the magnetooptical control of bullets are determined. For a stabilization of bullets, the higher nonlinearities in the MM waveguide should be taken into account. Different modes of formation and distribution of bullets occur; the modes are investigated as a general physical problem without specific considerations of a certain MM dispersion.

7.2 Instabilities of bullets in the metamaterial waveguides with the influence of the higher-order nonlinear effects

In Rapoport *et al* (2012) and (2014a), a simultaneous influence of the nonlinear dispersion and diffraction on the propagation of pulses in a nonlinear gyrotropic layered medium was investigated. It was first shown that the nonlinear dispersion results in the formation of complex nonlinear structures in the transverse direction. The nonlinear diffraction, in turn, influences the formation of structures in the longitudinal direction, for example, the formation of additional peaks, (see also section 4.3). The papers mentioned above summarized the results of investigations of the effects of the self-compression due to cubic nonlinearity and wave linear and nonlinear dipersion and diffraction on MI (see also appendix E and figure E.3 in this appendix). The formation of bullets is investigated in an MM waveguide (see figure E.3) with a negative phase MM, or a waveguide with alternating (in the direction of wave propagation) materials possessing positive and negative phases. In such media the diffraction control, or the diffraction management, (Eisenberg *et al* 2000, Akhmediev and Ankiewitcz 2005) is provided. In this case, the diffraction and other coefficients in the nonlinear evolution equation for an envelope in the structure with layers of MMs can be considered to be close to that in unbounded MMs in the weak waveguide dispersion approximation. Alternatively, one can use for the determination of the above mentioned coefficients the model of planar metal MM waveguide. In appendix E figure E.2, the frequency dependences of the normalized values of the coefficients of linear $(\beta_{2\text{norm}})$ and nonlinear (S_{norm}) dispersion coefficients are shown for the simplest model of a planar metallized waveguide with a finite normalized distance between metal boundaries, $L_{QP} = 4$. A nonlinear wave is investigated in a planar waveguide structure with a nonlinear wave propagating in the Z direction, with the Y axis normal to the MM layer and the X axis lying in the plane of wave propagation. It can be shown (see Rapoport *et al* (2014a), Boardman *et al* (2010a) and equation (3.53) in section 3.3.4) that the evolution equation for the slowly varying amplitude of a wave packet can be represented as follows in the dimensionless form:

$$\left[i\frac{\partial}{\partial\bar{\xi}} + q\frac{\partial^2}{\partial\bar{\xi}^2} - \frac{1}{2}\operatorname{sgn}(\beta_2)F\frac{\partial^2}{\partial\bar{\tau}^2} + \frac{1}{2}\operatorname{sgn}(k_c')\frac{L_{\mathrm{Disp}}}{L_{\mathrm{Difr}}}D\frac{\partial^2}{\partial\bar{x}^2} + i\gamma_0\right]\psi$$
$$+ \operatorname{sgn}(\tilde{C}_{\mathrm{nl}})|\psi|^2\psi + Q|\psi|^4\psi\operatorname{sgn}(\tilde{C}_{\mathrm{nl}}) + i\operatorname{sgn}(\tilde{C}_{\mathrm{nl}})b_\tau\frac{\partial}{\partial\bar{\tau}}(|\psi|^2\psi) \tag{7.1}$$
$$+ NL_{\mathrm{Difr}}\frac{\partial^2}{\partial\bar{x}^2}(|\psi|^2\psi) - \nu\psi = 0$$

The equations of the type of (7.1) can be obtained for the X component of the electric or magnetic field with modes possessing the components either (E_x, H_y, H_z) (H-, or TE-mode) or (H_x, E_y, E_z) (E-, or TH-mode) respectively. Here, the dimensionless coordinates and time are related to the corresponding dimensional values as follows: $\xi = L_{\mathrm{Disp}}\bar{\xi}$, $x = \bar{x}D_0$, $\tau = T_0\bar{\tau}$, where L_{Disp}, D_0 and T_0 are the corresponding spatial and temporal scales, $\xi = z$, $\tau = t - z/|V_{\mathrm{gc}}|$. Here t and τ are time in the laboratory frame and time in the moving coordinate frame, or the running time, respectively, V_{gc} is the group velocity at the carrier frequency of the wave packet, $L_{\mathrm{Disp}} = T_0^2/|\beta_2|$, $\beta_2 = (\partial^2k/\partial\omega^2)_{\omega_c}$ is the dispersion coefficient. The values ω_c, k_c are the carrier frequency and the wavenumber, the coefficient F, $0 \leqslant F \leqslant 1$, describes the effect of controlling the dispersion. The coefficients $\mathrm{NL}_{\mathrm{difr}}$ and $(-\nu)$ in the left-hand parts of equation (7.1) describe the nonlinear diffraction effect and transverse magnetooptic (Voigt) one (Rapoport *et al* 2014a), respectively. The dimensional amplitudes A are either for the electric field or for magnetic one, depending on whether equation (7.1) is derived for either the electric E_x field or the magnetic one H_x for the modes (E_x, H_y, H_z) or (H_x, E_y, E_z), respectively. These amplitudes are normalized in such a way that

$$A = \psi A_0 \tag{7.2}$$

Here $L_{\mathrm{Disp}}|\tilde{C}_{\mathrm{nl}}|A_0^2 = 1$; A_0 and ψ are the normalization coefficient and the dimensionless amplitude, respectively. $\tilde{C}_{\mathrm{nl}}$ and Q are the effective coefficients that determine the cubic and fifth order nonlinearity, respectively. The dimensionless coefficients q, D, γ_0, b_τ, and $\mathrm{NL}_{\mathrm{Difr}}$ describe the effects of the non-paraxiality, diffraction (with a possibility of the diffraction control), the dissipative (linear) energy loss or gain, the self-compression coefficient, i.e. the nonlinear dispersion, and the nonlinear diffraction, respectively. These dimensionless coefficients are related to the corresponding dimensional coefficients for the two cases of nonlinear waves in systems with electric and magnetic nonlinearities, respectively. Here the corresponding formulas are presented for one of these cases, namely for the mode with components (E_x, H_y, H_z) (H-, or TE-mode) in an MM medium with the electric nonlinearity. In this case it is assumed that the longitudinal magnetic component is $|H_z| \ll |H_y|$, and the nonlinear electric polarization $\vec{P}_{\mathrm{NL}}$ is defined as

$$P_{\mathrm{NLx}} = (\chi_{E_2}|A_{\mathrm{Ex}}|^2 + \chi_{E_4}|A_{E_x}|^4)A_{E_x}, \tag{7.3}$$

where A_{E_x} is the amplitude of the electric component E_x, thus $A = A_{E_x}$ is the amplitude included in equation (7.2) (Rapoport 2013, 2014b, Rapoport *et al* 2014a). The dimensionless coefficients included in equation (7.1) are related to the corresponding dimensional coefficients as follows: $\tilde{C}_{\mathrm{nl}} = (2\pi\omega_{\mathrm{c}}^2/k_{\mathrm{c}}'c^2)\mu_{\mathrm{c}}'\chi_{E_2}$, so $\mathrm{sgn}(\tilde{C}_{\mathrm{nl}}) = \mathrm{sgn}(k_{\mathrm{c}}')\mathrm{sgn}(\mu_{\mathrm{c}}')\mathrm{sgn}(\chi_{E_2}) = \mathrm{sgn}(\chi_{E_2})$, $\mu_{\mathrm{c}}' = \mathrm{Re}(\mu_{\mathrm{c}})$ is the real part of the linear magnetic permeability at the carrier frequency $\omega = \omega_{\mathrm{c}}$, $q = (2k_{\mathrm{c}}'L_{\mathrm{Disp}})^{-1}$, $\mathrm{NL}_{\mathrm{Difr}} = \mathrm{sgn}(\chi_{E_2})\mathrm{sgn}(k_{\mathrm{c}}')\mathrm{sgn}(\varepsilon_{\mathrm{c}}')(|k_{\mathrm{c}}'|^2 D_0^2)^{-1} = \mathrm{sgn}(\chi_{E_2})(|k_{\mathrm{c}}'|L_{\mathrm{Difr}})^{-1}$,
$\varepsilon_{\mathrm{c}}' = \mathrm{Re}(\varepsilon_{\mathrm{c}}) = \mathrm{Re}(\varepsilon|_{\omega=\omega_{\mathrm{c}}})$ is the real part of the linear electric permittivity at the carrier frequency, $\gamma_0 = (\omega_{\mathrm{c}}/c)^2(L_{\mathrm{Disp}}/2k_{\mathrm{c}}')(\varepsilon_{\mathrm{c}}'\mu_{\mathrm{c}}'' + \varepsilon_{\mathrm{c}}''\mu_{\mathrm{c}}')$ is the decrement of losses or increment of gain. The increment of MI for two-dimensional perturbations was obtained by using equation (7.1) (see appendix E and figure E.1). It is shown that, under other equivalent conditions, the nonlinear dispersion and diffraction both influence MI increments. But the results of MI in figure E.1 correspond to the initial quasilinear stage of this instability (Rapoport *et al* 2014a, Rapoport 2014b). As follows from numerical simulations using equation (7.1), the development of the corresponding nonlinear instability (Rapoport *et al* 2014a, Rapoport 2014b) involves a rather complex interaction of transverse and longitudinal perturbations of the corresponding nonlinear structures (see appendix E including figure E.1).

The simplest model, namely the layered planar metallized waveguide, is considered. For the general case of a layered medium, the nonlinear coefficients can be found using the general NEELS method. Consider the case where the signs of the coefficients in (7.1) are chosen so that the wave bullets can be formed (or the wave collapse could develop), namely they are $\mathrm{sgn}(k_z') < 0$, $\mathrm{sgn}(\tilde{C}_{\mathrm{nl}}) < 0 \cdot [-\mathrm{sgn}(\beta_2)] < 0$, (or all signs are opposite). Note that the calculations of the frequency-dependent coefficients of evolution equation (3.53) from chapter 3 (or (7.1)) (Rapoport *et al* 2014a), the examples of which are shown in appendix E (figure E.2), are very important for revealing the regimes of the most efficient magnetooptical stabilization of bullets in the magnetooptical waveguide proposed in (Rapoport *et al* 2014a). This issue will be briefly discussed in section 7.3.

Based on the calculations, the results of which are given in appendix E (see in particular figure E.3), we can point out the following conclusions. In the absence of the diffraction and the dispersion management (at standard values of linear dispersion and diffraction) and a sufficient value of the input amplitude, the bullet is formed at the normalized distance $X \sim 2$, and then, after propagation by a distance of about 1, it is destroyed. The evolution of bullets in MMs is determined by the competition between two types of, generally speaking, nonlinear instabilities in the presence of energy losses. The first type of instability is the one that causes the formation of the bullet and leads to a tendency of increasing the amplitude and reducing the cross-section of the bullet. As the amplitude increases, this type of instability can shift from linear stage to a nonlinear one. In the absence of the linear diffraction control, or the diffraction management, but in the presence of the self-compression and nonlinear diffraction of the positive sign, the bullet collapses, then it splits in the longitudinal direction, but then, unlike in the situation where higher nonlinearities are absent, the larger one of the formed peaks propagates for a relatively large distance, of order of $X \sim 6$, as a pronounced quasi-stable bullet. In the presence of the linear diffraction control, in particular at a sufficiently small linear diffraction coefficient, the following phenomena occur:

(1) the instability of the first type does not stop and leads to a catastrophic collapse and the tendency of the increasing amplitude of the wave to infinity, even with a relatively moderate value of the input amplitude;
(2) at the normal value of the linear diffraction, the corresponding collapse can be stopped taking into account the dispersion of nonlinearity and dissipation;
(3) in the presence of both the self-compression and nonlinear diffraction, and unsaturated nonlinearity, there is some minimum value of the diffraction coefficient (D) for which the 'second type' instability, which tends to split the bullet, is predominant, leading to splitting of pulses and ensuring a finite amplitude of the bullet.

It is important to note that the nonlinear diffraction significantly influences the formation of nonlinear structures in the longitudinal direction. The presence of the dispersion of nonlinearity causes in the nonlinear waveguide the arrangement of peaks of different intensities when a multi-soliton structure with many peaks is formed (Agrawal 2012). The saturation of nonlinearity can prevent, for a given set of parameters, the catastrophic development of instability of the first type and, therefore, promote quasi-stabilization of the bullets.

7.3 Stabilization of bullets in periodical and magnetooptic metamaterial waveguides

Consider the possibility of stabilization of bullets in a periodic medium in the presence of the nonlinear diffraction and dispersion, i.e. the self-compression. The existence of such a possibility in the presence of higher nonlinearities is shown for the first time.

In figure 7.1 the results are shown of comparing the propagation of space-time nonlinear (2 + 1) pulses in periodic and non-periodic media. As can be seen from figure 7.1, the bullets are stabilized at a fairly large distance in a periodic medium using the diffraction control, namely the creation of (phenomenological MM) a waveguide with the periodic diffraction coefficient. Figure 7.1 shows the normalized distance traveled by space-time pulses (bullets) from the input to the system $Z = 3.5$, but the simulations demonstrate that the bullets in such a system are stored at least up to the distance $Z = 12$, and there are the nonlinear diffraction and dispersion, see

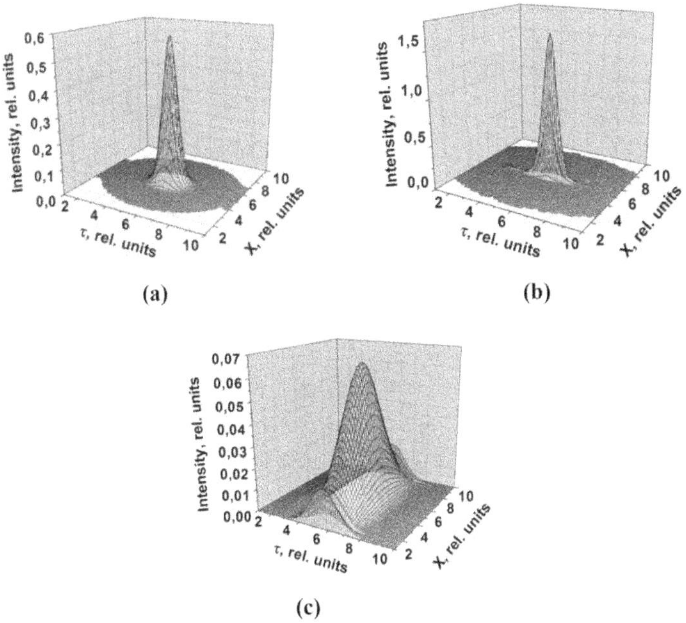

Figure 7.1. (a–c) Possibility of stabilization of bullets in the periodic medium in the presence of higher nonlinearities (nonlinear diffraction and dispersion). The spatio-temporal structures are shown of the pulses for the corresponding normalized distances traveled by the pulses from the origin: (a) is for $Z = 0$; (b) is for $Z = 3.5$; in (b) the medium is periodic with a normalized period $\Delta Z = 1$; in (c) the medium is not periodic. Copyright IEEE (2014), reprinted with permission from Rapoport *et al* (2014a).

the upper pair of figures. As can be seen from figure 7.1(c), in the absence of periodicity, under all other conditions, as for the upper pair, at $Z = 3.5$, the bullets already destroy. Thus, the diffraction control, or construction of the periodic values of diffraction coefficient, allows one to stabilize the bullets also in the presence of the nonlinear diffraction and dispersion.

Consider the magnetooptical control and stabilization of the bullets in an MM waveguide near a frequency with a value close to zero of the nonlinear dispersion coefficient (nonlinear self-steepening). Figure 7.2 illustrates the magnetooptical control (Apostolov *et al* 2016) and the stabilization of bullets in an MM waveguide in the presence of magnetooptical control by means of a transverse magnetooptical effect. The carrier frequency of the pulses corresponds to the zero coefficient of the nonlinear dispersion $S = 0$. A characteristic dispersion was chosen for the MM medium that filled the model planar nonlinear waveguide.

From figure 7.2 it is seen that the dependence of the maximum intensity (2 + 1) of space-time strongly focused solitons or bullets on the longitudinal coordinate is oscillatory. The simulation was performed for the following normalized parameters: input amplitude $A_{\text{inp}} = 1.08$, the normalized width of the MM waveguide is equal to $L_{\text{PQ}} = 4$, $T_0\omega = 18$, $T_0/T = 3$, $D_0k = 90$. Here T_0, T, λ are the duration of the pulse at the input of the system, the period corresponding to the carrier frequency, and the wavelength. After forming the bullets we get $D_{0\text{eff}}/\lambda \sim (3 - 4)$, where $D_{0\text{eff}}$ is the

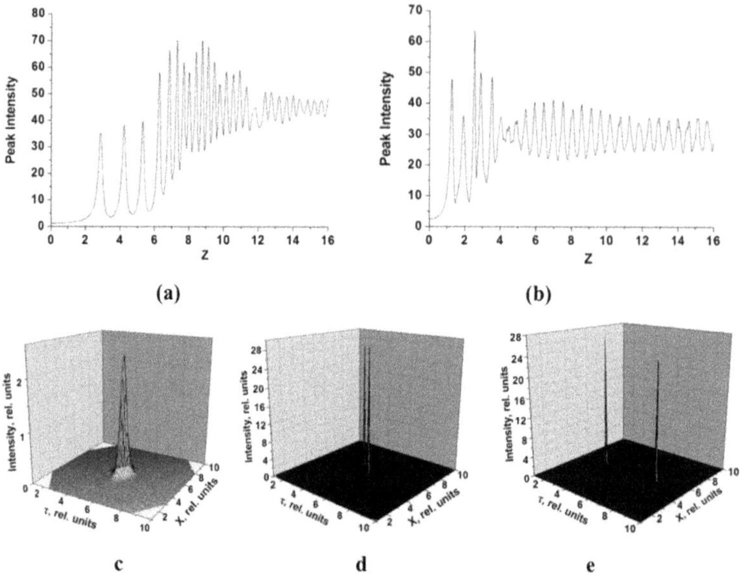

Figure 7.2. (a–e) The oscillatory dependence of the peak intensity of the bullets in the MM waveguide on the longitudinal coordinate and the presence of magnetooptical control. (a), (b) are the dependence of the normalized peak intensities on the normalized distance Z, traveled by the pulse from the input of the system for the normalized magnetooptic parameter ν equal to 0 and 20, respectively; (c–e) are the spatio-temporal two-dimensional structures of the pulses for $\nu = 20$ and the corresponding distance Z equal to 0.1, 3. And the frequency corresponds, taking into account the waveguide dispersion, to the zero coefficient of nonlinear dispersion $S = 0$. Here $D_0 k = 90$; D_0, k are the transverse size of the beam at the input of the system and the value of the carrier wavenumber, respectively; T_0, T, λ are the pulse duration at the input of the system, the period corresponding to the carrier frequency, and the wavelength corresponding to the carrier wavenumber, respectively. (a) and (b) copyright IEEE (2014), reprinted with permission from Rapoport *et al* (2014a) .

characteristic transverse pulse width after the nonlinear self-compression in the magnetooptical waveguide. The normalized carrier frequency of the wave packet $\overline{\omega} \approx 0.4854$ was chosen near the point on the dispersion curve (not shown here), where the coefficient of nonlinear dispersion S tends to zero.

It is shown (Apostolov *et al* 2016) that when the value of the nonlinear dispersion coefficient S is different from zero, the dependence of the value of the bullet intensity at the maximum on the longitudinal coordinate is a smooth function and has a pronounced maximum. In contrast, in the presence of the magnetooptical interaction with a positive constant and near the frequency point to one where $S = 0$, there is a quasi-stabilization of the bullets, and the dependence of the maximum intensity on the distance Z becomes quasi-periodic. When the sign of the magnetooptical interaction changes, the quasi-stabilization of the bullets is replaced by their destabilization.

Finally, the bullets can be stabilized in the periodic and active MM waveguides even in the media with gain, as shown in figure 7.3. The normalized parameters of such a bullet are presented in the caption to figure 7.3. A possibility of the 'stabilization of amplification of the bullet' in the periodic active MM media is shown in the present chapter. The proposed effectively controllable nonlinear

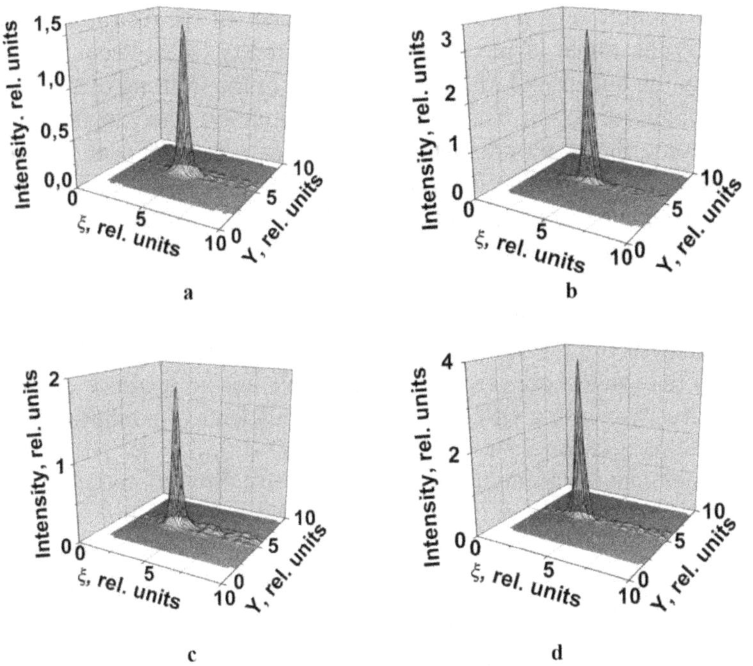

Figure 7.3. (a–d) 'Stabilization of amplification' of the bullets in a model layered active MM waveguiding media with a periodic coefficient of the linear diffraction (diffraction management). The values of normalized distance Z, passing by the pulse from the input of the system: (a) is for $Z = 9$; (b) is for $Z = 9.5$; (c) is for $Z = 13$; (d) is for $Z = 13.5$. The normalized coefficient of self-steepening (the nonlinear dispersion) is $S = -0.133$, the coefficient of nonlinear diffraction is $K = 0.003$. Here ξ is the normalized value of time in the moving coordinate frame. The 'diffraction management' (Eisenberg *et al* 2000, Boardman *et al* 2010a, Boardman *et al* 2010b) is used, and the media is periodical; the coefficient of diffraction is proportional to $D(z) = \pm 1$; the normalized period is $\Delta z = 1$; (a and b) are for the pulse passing the 9th period of the periodic media; (c and d) are for the 13th period. (a)–(d) copyright IEEE (2014), reprinted with permission from Rapoport *et al* (2014a).

structures are a good perspective for signal processing, space communications, and other applications for information technologies.

7.4 Conclusions for chapter 7

The formation and propagation of a bullet in a model MM waveguide with magnetooptical control are investigated using the advanced version of the NEELS method. The possibility of stabilizing the bullets due to the magnetooptical control is pointed out. The effect of the 'gain stabilization' in layered media with the diffraction management is established. The possible applications can be useful for signal processing, and space communication, from microwaves to optics. Finally, the following conclusions are put forward.

(1) The evolution of magnetic bullets/(2 + 1) space-time solitons in MMs is determined by the competition between two types of nonlinear instabilities and dissipative losses. The instability of the first type causes the formation of a bullet and tends to lead to an increase in the amplitude and decrease in

the cross-section of the bullet. In the absence of the diffraction control and in a certain range of parameters, the instability of the second type tends to destroy the bullets and manifests primarily in the splitting of the bullets in the longitudinal direction. Then, unlike in the situation where higher nonlinearities are absent, the larger one of the forming peaks extends over a relatively long distance, as a pronounced quasi-stable bullet. The nonlinear diffraction has a non-trivial and significant influence on the formation of nonlinear structures in the longitudinal direction. The presence of the self-steepening causes in the nonlinear waveguide the arrangement of peaks of different intensities when a multi-soliton structure with many peaks is formed.

(2) The analysis of (2 + 1) (space-time) solitons/quasisolitons in MM waveguides shows that bullets are possible, for example, in the presence of the magnetic nonlinearity, where a set of signs of the coefficients of nonlinear equations are realized as a necessary condition: $k < 0$, $D < 0$, $\beta_2 > 0$, $\chi^{(3)} < 0$. Here k is the longitudinal wavenumber, D is the effective diffraction coefficient, β_2 is the dispersion coefficient, $\chi^{(3)}$ is the nonlinear Kerr coefficient.

Appendix E Results of the modeling formation and propagation of the bullets in metamaterial waveguides accounting for influence of higher-order nonlinearities

The dimensionless coefficients included in the evolutional equation (7.1) are equal to:

$$\tilde{C}_{\rm nl} = \frac{2\pi\omega_{\rm c}^2}{k_{\rm c}'c^2}\mu_{\rm c}'\chi_{E_2},\ \text{therefore}\ \mathrm{sgn}(\tilde{C}_{\rm nl}) = \mathrm{sgn}(k_{\rm c}')\mathrm{sgn}(\mu_{\rm c}')\mathrm{sgn}(\chi_{E_2}) = \mathrm{sgn}(\chi_{E_2}),$$

$$\tilde{C}_{\rm nl} = \frac{2\pi\omega_{\rm c}^2}{k_{\rm c}'c^2}\mu_{\rm c}'\chi_{E_2},\ \text{so}\ \mathrm{sgn}(\tilde{C}_{\rm nl}) = \mathrm{sgn}(k_{\rm c}')\mathrm{sgn}(\mu_{\rm c}')\mathrm{sgn}(\chi_{E_2}) = \mathrm{sgn}(\chi_{E_2}),$$

where $\mu_{\rm c}' = \mathrm{Re}(\mu_{\rm c})$ is the real part of the linear magnetic permeability at the carrier frequency $\omega_{\rm c}$, $\mu_{\rm c} \equiv \mu|_{\omega=\omega_{\rm c}}$, $Q = \frac{\chi_{E_4}}{\chi_{E_2}}A_0^2$, $q = \frac{1}{2k_{\rm c}'L_{\rm Disp}}$,
$\mathrm{NL_{Difr}} = \mathrm{sgn}(\chi_{E_2})\mathrm{sgn}(k_{\rm c}')\mathrm{sgn}(\varepsilon_{\rm c}')\frac{1}{|k_{\rm c}'|^2 D_0^2} = \mathrm{sgn}(\chi_{E_2})\frac{1}{|k_{\rm c}'|L_{\rm Difr}}$,
$\varepsilon_{\rm c}' = \mathrm{Re}(\varepsilon_{\rm c}) = \mathrm{Re}(\varepsilon|_{\omega=\omega_{\rm c}})$ is the real part of the linear electric permittivity at the carrier frequency, $\gamma_0 = \left(\frac{\omega_{\rm c}}{c}\right)^2\frac{L_{\rm Disp}}{2k_{\rm c}'}(\varepsilon_{\rm c}'\mu_{\rm c}'' + \varepsilon_{\rm c}''\mu_{\rm c}')$.

The increment of two-dimensional MI is obtained by a 'quasilinearization' (which takes into account higher-order nonlinearities, in particular the nonlinear diffraction and dispersion), equation (7.1), illustrated in figure E.1. As can be seen from figure E.1, other factors being important equally, namely the nonlinear dispersion and diffraction, affect 'symmetrically' the increments of MI. But the results given in this section represent the beginning of the investigations of equation (7.1) for instability (Rapoport *et al* 2014a, Rapoport 2014b), and are obtained on the base of the 'linearized' analysis of two-dimensional MI. As will be shown below, the development of the corresponding nonlinear (quasi-soliton) instability (Rapoport *et al* 2014a, Rapoport 2014b) involves a rather complex interaction of transverse and longitudinal perturbations of the corresponding nonlinear structures.

An example of frequency dependences (normalized ones) (Boardman *et al* 2010a, Boardman *et al* 2009a) of the coefficients of the evolution equation, in particular the linear dispersion coefficient (β_{2norm}) and the self-compression (nonlinear dispersion) (S_{norm}), is shown in figure E.2.

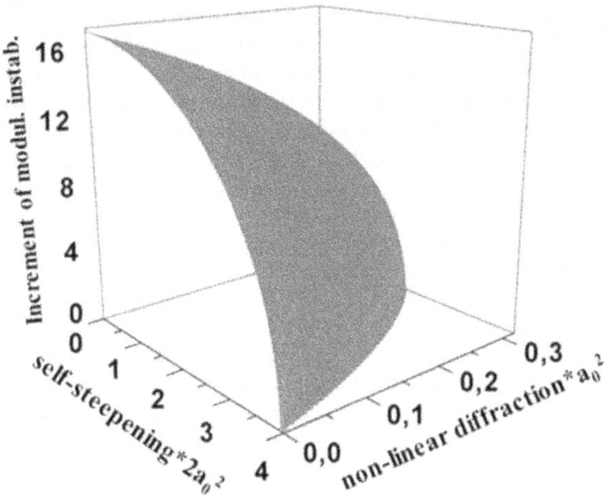

Figure E.1. The dependence of the increment of MI on nonlinear effects of higher orders (nonlinear diffraction and dispersion) that exist simultaneously.

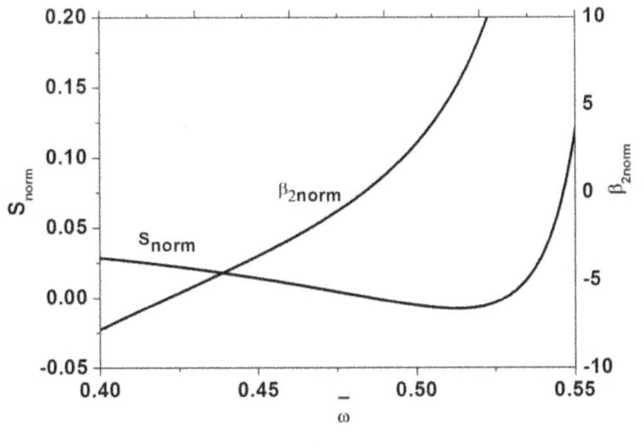

Figure E.2. Frequency dependences of normalized values of linear coefficients (β_{2norm}) and nonlinear (S_{norm}) dispersions for the simplest model of the MM planar metallized waveguide with a finite normalized distance between metal boundaries, $L_{QP} = 4$.

The simplest model layered planar metallized waveguide with a finite distance between the metal boundaries is considered. For the general case of a layered medium, the nonlinear coefficients can be found using the general NEELS method (see section 3) for MM layered systems. Note that the value of the coefficient S_{norm} (figure E.2) and the signs of the coefficients in the normalized evolution equation are very important for solving the problem of stabilization of bullets.

We simulated the formation and propagation of 2D bullet, which are described by the evolution equation of the form (7.1) (Rapoport *et al* 2014a, Rapoport 2014b, Ballav and Chowdhury 2006, Basu 1997). Consider the case when the signs of the coefficients in (7.1) are chosen so that bullets can be formed (or the wave collapse), namely, $\mathrm{sgn}(k_c') < 0$, $\mathrm{sgn}(\tilde{C}_{\mathrm{nl}}) < 0$, $[-\mathrm{sgn}(\beta_2)] < 0$, (or all signs are opposite). One of the results of the numerical simulations is shown in figure E.3. In this case, the decay of the bullets is mainly due to the development of structures like splitting of the pulses in the longitudinal direction. At the same time, a wide structure with many small peaks also develops.

Based on the simulations, including the results illustrated in figure E.3, and a series of other numerical and analytical calculations (not given), it is possible to draw the conclusions set forth in sections 7.2 and 7.4 of this chapter.

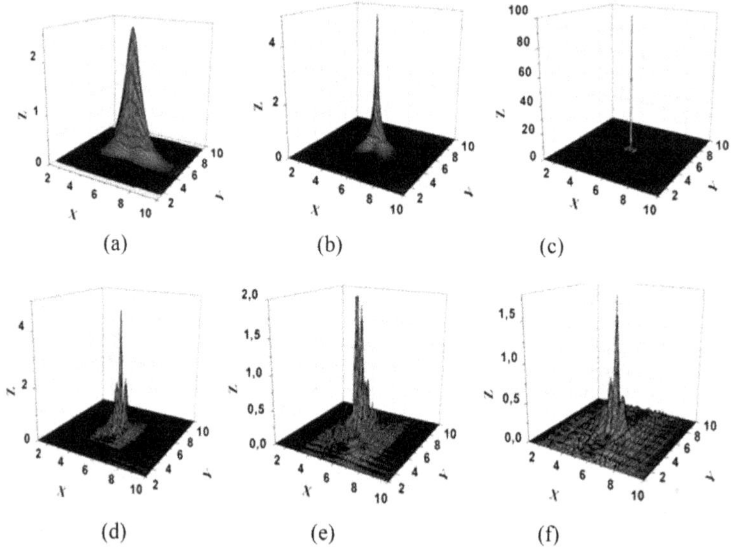

Figure E.3. (a–f) Formation and disintegration (due to nonlinear 'instability of the second type') of bullets in a model MM waveguide with the parameters: the linear diffraction control coefficient $D_{\mathrm{diffr}} = 1$, the linear dispersion control coefficient $D_{\mathrm{disp}} = 1$, the nonlinear dispersion coefficient (self-compression) $S = +0$, the nonlinear diffraction coefficient $\mathrm{NL}_{\mathrm{difr}} = 0.003$, the 'number of thresholds' (measuring the input wave intensity) is $K_{\mathrm{thr}} = 60$; (a) is for $Z = 1$; (b) is for $Z = 3$; (c) is for $Z = 4$; (d) is for $Z = 5$; (e) is for $Z = 6$; (f) is for $Z = 8$. The normalized values of time in the 'running' coordinate frame, the transverse coordinate, and wave intensity, respectively, are plotted on the X, Y, and Z axes. (a), (c) and (f) copyright (2014) IEEE. Reprinted with permission from Rapoport *et al* (2014a).

Problems to chapter 7

Problem 7.1

Derive the dispersion equation for 3-layer waveguide in the form:

$$k_0L = \frac{1}{\sqrt{\mu_1\varepsilon_1 - n_{\text{eff}}^2}}[\arctan(F(n_{\text{eff}})) + m\pi]$$

The central layer possesses the electric permittivity ε_1 and magnetic permeability μ_1, the lower half-space possesses the tensor of permittivity $\hat{\varepsilon}_3 = \begin{pmatrix} \varepsilon_{xx} & ig_3 & 0 \\ -ig_3 & \varepsilon_{yy} & 0 \\ 0 & 0 & \varepsilon_{zz} \end{pmatrix}$ and the permeability μ_3, and the upper half-space possesses $\hat{\varepsilon}_2 = \begin{pmatrix} \varepsilon_{xx} & 0 & 0 \\ 0 & \varepsilon_{yy} & 0 \\ 0 & 0 & \varepsilon_{zz} \end{pmatrix}$ and $\mu_2 = \mu_3$,

Where m is an integer value and corresponds to the number of the mode in the structure under consideration

$$F(n_{\text{eff}}) = \frac{Q_{12} + Q_{13}}{1 - Q_{12}Q_{13}}$$

$$Q_{12} = \frac{q_2}{\tilde{q}_1}\frac{\varepsilon_1}{\varepsilon_2};\ Q_{13} = \frac{(\varepsilon_3 q_3 - g_3 k_y)\varepsilon_1}{(\varepsilon_3^2 - g_3^2)\tilde{q}_1};\ \tilde{q}_1 = \sqrt{\mu_1\varepsilon_1 k_0^2 - k_y^2};\ q_2 = \sqrt{k_y^2 - \mu_2\varepsilon_2 k_0^2};$$

$$k_y/k_0 \equiv n_{\text{eff}};$$

Problem 7.2

Describe qualitatively the linear magnetooptic control of dispersion and group velocity V_g/delay time in the delay line for the 'slow light', see problem 7.1. Suppose that the magnetooptic parameter g_3 can be taken with large enough absolute value. Draw qualitatively the dependence $n_{\text{eff}}(k_0L)$. Draw qualitatively also the dependence $KyL(k_0L)$. Demonstrate that it is possible to get $V_g \sim d(k_0L)/d(k_0L) = 0$ at some relatively small value of k_0L.

Problem to section 7.2

Draw qualitatively also the dependence $k_yL\ (k_0L)$.

Problem 7.3.

Describe qualitatively the control of the (cubic) nonlinearity over the group velocity in the magnetooptic delay line. See also problems 7.1, 7.2 for the corresponding linear case.

Suppose that the delay line/layered structure can be considered in the quasilinear approximation. In the computations below the magnetooptic effect is not included, while in principle, the mixing of MO (magnetooptic)—NL (nonlinear) control can be also considered. Suppose that the materials in layers 2, 3 can be made nonlinear using other (pumping) waveguiding mode with a smaller frequency and strong amplitude E_{pump}. If we work with mode TM_3, this can be, say, the mode TM_2 or TE mode at a smaller frequency. Due to the pumping wave, the permittivity can be changed, such as, for example,

$$\varepsilon_{xxNL} = \varepsilon_{xx} + a_{NL}|E|^2,\ \varepsilon_{yyNL} = \varepsilon_{yy} + a_{NL}|E|^2,\ \varepsilon_{zzNL} = \varepsilon_{zz} + a_{NL}|E|^2,$$

where a_{NL} is a nonlinear coefficient. Respectively to a main (signal) wave (mode TM_3 in our case), the structure is quasilinear, namely this mode as itself propagates as a linear wave, but within a structure, where $\widehat{\varepsilon}$ is changed due to nonlinearity. Illustrate the possibility of a nonlinear control over the group velocity, including the region of light stopping ($V_g = 0$), by means of the qualitative dependence $V_g(k_0L)$ for different values of the parameter ($a_{\text{NL}}|E_{\text{pump}}|^2$).

List of abbreviations

MI	Modulation instability
MMs	Metamaterials
NL	Nonlinear
NP	Negative phase
NEELS	Nonlinear evolution equation for layered structures
TE	Transverse electric

References

Abdollahpour D, Suntsov S, Papazoglou D G and Tzortzakis S 2010 Spatiotemporal Airy light bullets in the linear and nonlinear regimes *Phys. Rev. Lett.* **105** 253901

Agrawal G 2012 *Nonlinear Fiber Optics* (New York: Academic)

Akhmediev N and Ankiewicz A (eds) 2005 *Dissipative Solitons* (Berlin: Springer)

Ali S A K, Maimistov A I, Porsezian K, Govindarajan A and Lakshmanan M 2021 Modulational instability in a non-Kerr photonic Lieb lattice with metamaterials *Phys. Rev.* A **103** 013517

Apostolov S S, Maizelis Z A and Yampol'skii V A *et al* 2016 Transmission of THz waves through layered superconductors controlled by a dc magnetic field *Phys. Rev.* B **94** 024513

Balakin A A, Mironov V A and Skobelev S A 2017 Self-action of Bessel wave packets in a system of coupled light guides and formation of light bullets *J. Exp. Theor. Phys.* **124** 49–56

Ballav M and Chowdhury A R 2006 On a study of diffraction and dispersion managed soliton in a cylindrical media *Progr. Electromagn. Res., PIER* **63** 33–50

Basu B 1997 Generalized Rayleigh–Taylor instability in the presence of timedependent Equilibrium *J. Geophys. Res.* **102** 17305–12

Boardman A D, Egan P, Hess O, Mitchell-Thomas R C and Rapoport Y G 2009a Nonlinear gyroelectric waves in magnetooptics metamaterials *Proc. Second Int. Workshop on Theoretical and Computational Nano–Photonics (TaCoNa–Photonics 2009) (Bad Honnef, Germany)* vol 1176 pp 10–12

Boardman A D and Mitchell-Thomas R 2009b Rapoport Yu G Weakly nonlinear waves in layered bi–anisotropic *Proc. of Third Int. Congress on Adv. Electromagn. Materials in Microwaves and Optics: Metamaterials (London, UK)* pp 495–97

Boardman A D, Mitchell-Thomas R C, King N J and Rapoport Y G 2010b Bright spatial solitons in controlled negative phase metamaterials *Opt. Commun.* **283** 1585–97

Boardman A D, Hess O, Mitchell-Thomas R C, Rapoport Y G and Velasco L 2010a Temporal solitons in magnetooptic and metamaterial waveguides *Photonics Nanostr.-Fundam. Appl.* **8** 228–43

Boardman A D, Grimalsky V V, Kivshar Y, Koshevaya S V, Lapine M, Litchinitser M, Malnev V N, Noginov M, Rapoport Y G and Shalaev V M 2011 Active and tunable metamaterials *Laser Photonics Rev.* **5** 287–307

Büttner O, Bauer M, Demokritov S O, Hillebrands B, Kivshar Yu S, Grimalsky V V, Rapoport Y and Slavin A N 2000 Linear and nonlinear diffraction of dipolar spin waves in yttrium iron garnet films observed by space–and time–resolved Brillouin light *Phys. Rev. B.* **61** 11576–87

Eisenberg H S, Silberberg Y, Morandotti R and Aitchison J S 2000 Diffraction management *Phys. Rev. Lett.* **85** 1863–66

Fattal Y, Rudnick A and Marom D M 2011 Soliton shedding from Airy pulses in Kerr media *Opt. Express* **19** 17298

Grimal'sky V, Rapoport Yu, Zaspel C and Slavin A N 2000 Numerical modeling of wave front reversal for two-dimensional spin wave packets in magnetic films *Abstr. of the 8th Europ. Magn. Materials and Applications Conf. (Kiev, Ukraine)* We-OB01 pp 38–40

Karalis A and Joannopoulos J D 2019 Plasmonic metasurface 'bullets' and other 'moving objects': Spatiotemporal dispersion cancellation for linear passive subwavelength slow light *Phys. Rev. Lett.* **123** 067403

Kivshar Y S and Agrawal G P 2003 *Optical Solitons: From Fibers to Photonic Crystals* (New York: Academic)

Li H, Hao W, Lin X and chen L 2019 Circularly polarized Airy beam generation with hyperbolic metamaterials at telecom wavelengths *IEEE 18th Int. Conf. on Optical Communications and Networks (ICOCN)* 1–3

Malomed B A, Mihalache D, Wise F and Torner L 2005 Spatiotemporal optical solitons *J. Opt. B: Quantum Semiclass. Opt.* **7** R53–72

Rapoport Y G, Zaspel C E, Grimalsky V V and Sanchez-Mondragon J 2004 Nonlinear Lorentz Lemma with the influence of exchange interaction and propagation of the magnetostatic waves with higher diffraction and dispersion *IEEE Proc. of the 14th Int. Crimean Conf. 'Microwave and Communication Technology' CriMiCo'2004 (Sevastopol, Crimea, Ukraine)* Catalog No.:04EX843 pp 361–63

Rapoport Y G 2006 Formation of structures under parametric coupling of two-dimensional nonlinear pulses of magnetostatic waves in magnetic films *IEEE Proc. of the 16th Int. Crimean Conf. 'Microwave and Communication Technology' CriMiCo'2006 (Sevastopol, Crimea, Ukraine)* Catalog No.:06EX1376 pp 629–30

Rapoport Y, Grimalsky V and Zaspel C 2012 Method for the derivation of nonlinear evolution equations in layered structures (NEELS): an example of nonlinear waves in gyrotropic layers *Bull. Univ. Kyiv Phys.* **14/15** 72–6

Rapoport Y G 2013 General method for modeling nonlinear waves in active metamaterial and gyrotropic layered structures *Proc. of Int. Kharkov Symp. on Phys. and Engineering of Microwaves, Millimeter and Submillimeter Waves MSMW'13 (Kharkov, Ukraine)* pp 253–55

Rapoport Y G, Grimalsky V V, Boardman A D and Malnev V N 2014a Controlling nonlinear wave structures in layered metamaterial, gyrotropic and active media *Proc. of IEEE 34th Int. Sci. Conf. Electronics and Nanotechnology ELNANO (Kyiv, Ukraine)* pp 46–50

Rapoport Y G 2014b General method for deriving the equations of evolution and modeling of nonlinear waves in layered active media with bulk and surface *Bull. Kyiv Nat. Taras Shevchenko Univ. Series: Phys.-Math. Sci.* **1** 281–88 (in Ukrainian)

Serga A A, Kostylev M P, Kalinikos B A, Demokritov S O, Hillebrands B and Benner H 2006 Parametric generation of soliton-like spin wave pulses in ferromagnetic thin-film ring resonators *J. Exp. Theor. Phys.* **102** 497–508

Slavin A N, Büttner O, Bauer M, Demokritov S O, Hillebrands B, Kostylev M P, Kalinikos B A, Grimalsky V V and Rapoport Y 2003 Collision properties of quasi-one-dimensional spin wave solitons and two-dimensional spin wave bullets *Chaos* **13** 693–701

IOP Publishing

Waves in Nonlinear Layered Metamaterials, Gyrotropic and Plasma Media

Yuriy Rapoport and Vladimir Grimalsky

Chapter 8

Giant double-resonant second harmonic generation in the multilayered dielectric–graphene metamaterials

Excitation of the second harmonic of THz radiation is investigated theoretically in the planar multilayered structure dielectric–graphene–dielectric–graphene…. The case of the oblique incidence of the s-polarized fundamental wave is studied, where the electric field is parallel to the interfaces, and generation of the *p*-type second harmonic wave occurs. The original concept is proposed to employ the double resonance arrangement for the effective generation of the second harmonic. The double-resonant case can be realized when a high-permittivity dielectric is at the input of the structure and the vacuum is at the output. High efficiency is demonstrated; the second harmonic reflectance coefficient is ≥ 0.01 under realistic values of the collision frequency in graphene $>10^{12}$ s^{-1}. Such a great efficiency, which is four–five orders of magnitude higher than reported for the graphene–dielectric structures previously, is proposed for the first time. To compute the nonlinear surface currents two approaches were used, the kinetic and the hydrodynamic. A qualitative agreement between two approaches, proven in the present modeling, ensures an applicability of the results.

8.1 Introduction

The nonlinear electromagnetic processes in graphene, namely self-action, electromagnetic and lattice-electromagnetic discrete solitons, envelope solitons, nonlinear mixing and harmonic generation, increasing nonlinearity using plasmon resonances etc, are the subject of growing interest. These nonlinear phenomena cover frequency ranges from microwave to optics (Dragoman and Dragoman 2009, Garcia de Abajo 2014, Mikhailov 2007, 2013, 2014, Cox and Garcia de Abajo 2015, Zhang *et al* 2011, Hendry *et al* 2010, Dong *et al* 2013, Nesterov *et al* 2013, Kryuchkov and Kukhar 2014,

doi:10.1088/978-0-7503-2336-9ch8

Bludov *et al* 2015, Smirnova and Kivshar 2014, Sekwao and Leburton 2015, Christensen and Mortensen 2015, Li *et al* 2013, Paul *et al* 2013). Because the graphene is a center-symmetrical material and no geometrical asymmetry is induced in the media (Zheludev and Emel'yanov 2004, Suprun and Shmeleva 2014), the second harmonic can be generated either due to the quadratic nonlinearity of the Lorentz force, i.e. due to an influence of the magnetic field of an electromagnetic wave on carriers, or due to the excitation of the component of 2D electron concentration at the double frequency. In Mikhailov (2011) the second harmonic generation in graphene was investigated by means of the quantum kinetic equation for the density matrix. It was shown that the efficiency of the generation of the second harmonic is about one order of magnitude higher than in GaAs quantum wells with typical experimental parameters. Under conditions of a 2D plasmon resonance the amplitude of the second harmonic is further increased by several orders of magnitude and may become comparable with the amplitude of the external electric field. Namely, it could be principally realized under the plasmon resonance, when the wavenumber of the external electric field is equal to the wavenumber of the 2D plasmon at the fundamental frequency ω and, in turn, the wavenumber of the second harmonic of the external electric field is equal to the wavenumber of the 2D plasmon at 2ω.

The second harmonic generation under the scattering from a spherical graphene nanoparticle was considered in Smirnova and Kivshar (2014).

In Glazov (2011) the second harmonic generation in graphene was theoretically studied based on phenomenological analysis and symmetry arguments. It is demonstrated that the second harmonic generation in ideal graphene samples is possible only when the oblique incidence of the electromagnetic wave takes place thus removing the inverse symmetry. A microscopic theory has been developed for the classical regime of light–electrons interactions, when the photon energy is much lower than the characteristic energy of the charge carriers. In Dean and van Driel (2009) the optical second harmonic generation of fundamental pulses of the 800 nm wavelength and 150 fs duration is observed from exfoliated graphene and multilayer graphitic films mounted on an oxidized silicon(001) substrate. In Bykov *et al* (2012) the optical second harmonic generation is studied from multilayer graphene films in the presence of a dc electric current flowing in the sample plane. The graphene layers are manufactured by the chemical vapor deposition technique and deposited on an oxidized Si(001) substrate. The second harmonic intensity from the graphene layer is found to be negligible in the absence of the direct current, while it increases dramatically with the application of such a current. The current-induced change of the second harmonic generation intensity from graphene/SiO_2/Si(001) rises linearly with the current amplitude. The observed effect is explained in terms of the interference of the second harmonic radiation reflected from the Si surface and that induced by the dc current in the multilayer graphene structure. The resonant phenomena in the case of strong nonlinearity under the self-action were investigated in Rapoport *et al* (2013, 2014a), namely the sharp threshold transition between regimes of 'reflection' and 'transmission', when the input amplitude of a THz pulse passing through the multilayered dielectric–graphene–dielectric... system is changed.

The estimations based on papers (Mikhailov 2011, Smirnova and Kivshar 2014) demonstrated that in non-resonant cases the transformation into the second harmonic is of low efficiency for the electromagnetic intensities, that is $<10^{-5}$. By the transformation into the second harmonic, we understand here the relation between intensities of the second and fundamental harmonics. The plasmon frequency resonance proposed in Mikhailov (2011) could be realized only in the quasi-stationary case, when the wavenumbers of plasmons are at least one order greater than the wavenumbers of incident electromagnetic waves. Therefore, for effective second harmonic generation a new strategy should be identified.

In spite of the presence of the experiments on harmonic generations in structures with graphene (Paul *et al* 2013), effectiveness of this process is relatively small. In this chapter, we propose a way of dramatic enhancement of such effectiveness up to values of the order of 0.01 and even higher, in a perspective with proper optimization. Our theoretical estimations may be a basis of a corresponding experimental implementation of the proposed great enhancement of the second harmonic generation.

Our approach of providing nonlinear signal operations for the THz range follows from the rather severe conditions for the wave propagation, in particular at room temperatures. Namely, for typical THz frequencies and the electron collision frequency, or dissipation loss rate, of the order of 10^{12}–10^{13} s^{-1}, respectively, effective propagation along the graphene layers is practically impossible. In this case, to provide effective nonlinear operation, we need to enhance the nonlinearity by, e.g. resonant conditions. Plasmonic resonances were exploited, for example in Smirnova and Kivshar (2014) and Sekwao and Leburton (2015). In the structure 'semi-bounded plasma–dielectric' the theory of a giant second harmonic generation with the nonlinear surface plasmons, under the condition of temporal plasmonic resonance with the second harmonic was proposed first in Grimalsky *et al* (1996) and then developed farther in Grimalsky and Rapoport (1998) and Rapoport and Grimalsky (2011), respectively. Namely the resonance condition was exploited $2\omega \approx \omega_p$, or $\varepsilon(2\omega) \approx 0$, in the presence of surface and volume nonlinearities. Plasmonic resonances were exploited also for example in Smirnova and Kivshar (2014) and Sekwao and Leburton (2015). Instead of this in the present chapter, another, much more effective strategy, is put forward, in particular for the second harmonic generation in the multilayered dielectric–graphene structure. We propose, to our best knowledge for the first time, to use the double resonance both for the fundamental and second harmonics of the electromagnetic waves.

In the present chapter, an effective resonant second harmonic generation of THz waves in the range 0.5–3 THz in the multilayered 'dielectric of high-permittivity–dielectric–graphene–dielectric–...–vacuum' system is demonstrated. The double resonance approach is determined as follows. (1) The double resonance means that there is realized simultaneously the minimum of the reflection of the first harmonic and the maximum of the effective reflection of the second harmonic. (2) This approach is based on the idea of accumulation of the fields at the fundamental and double frequencies inside a system in a neighborhood of the graphene layers. At the same time, effective sources that provide the reflection on the main harmonic and effective reflection of the second harmonic, placed on the first graphene layers, have minimum and maximum values, respectively. As a result the proper conditions for the reflections of the first harmonic and effective reflection of the second harmonic, outlined above in item (1) are provided. (3) The case of different polarizations of the first and second harmonics is under investigation. It is shown that the second harmonic possesses the polarization different from that of the fundamental harmonic. (4) The most effective regime of the resonant character of the system is realized, adjusting the incidence angle of the fundamental wave near the angle of total internal reflection. As a result, the efficiency of the transformation into the second harmonic can reach extreme values $\geqslant 0.01$, comparatively to $\sim 10^{-6}$ in the case of SHG, based on plasmonic/concentration nonlinearity in Smirnova and Kivshar (2014) and Mikhailov (2011).

In section 8.2 the formulation of the problem and the basic equations are presented. The expressions for the nonlinear surface current in graphene are obtained. The more detailed quasi-classical kinetic approach uses the expansion of the Boltzmann distribution function near the non-shifted Fermi distribution (Mikhailov 2011, Smirnova and Kivshar 2014). The hydrodynamic approach uses the nonlinear hydrodynamic equations for the electron concentration and momentum. These two approaches add each other. In the general nonlinear case, when the perturbations are applied, the results of these two approaches may be qualitatively, or at least quantitatively, different. The quantitative differences when using perturbations may be due to the following fact. In the kinetics the expansion near the non-shifted Fermi function is used (Mikhailov 2011, Smirnova and Kivshar 2014), whereas the hydrodynamic equations are related to the Fermi function shifted by the average momentum (Svintsov *et al* 2012). The corresponding difference will be reflected also in the nonlinear characteristics, in particular for second harmonic generation, which is of interest in the present chapter. There are no guarantees which of the two above indicated approaches would give a more accurate result for the second harmonic in the general case. So, the comparison of results can be useful. A qualitative agreement between two approaches ensures an applicability of results. Section 8.3 presents results of simulation of the stationary nonlinear reflection into the second harmonic in the layered structure a dielectric with a high-permittivity–dielectric–graphene–dielectric–….–vacuum. The details of the double resonance concept are presented in section 8.3. The transformation into the second harmonic facilitated by the double-resonant reflection is demonstrated, with the simplest optimization of the incidence angle and thickness of the first dielectric layer. Namely, in the resonant case the minimum of the reflection coefficient of the first harmonic coincides with the maximum of one for the second harmonic. The optimum values of the incidence angle for the fundamental harmonic are close to the limiting angle for the total internal reflection for the incidence from a dielectric of a high permittivity to the internal dielectric. Discussion and conclusions are presented in section 8.4.

8.2 Basic equations

Consider graphene sheets with 2D equilibrium concentration $n_{20} = (10^{10}\text{–}10^{12})\ \text{cm}^{-2}$. When investigating the dynamics of 2D electron gas in the presence of the THz electromagnetic field in the absence of interband transitions, it is possible to use either the quasi-classical kinetic method (Dragoman and Dragoman 2009, Garcia de Abajo 2014) or the simple hydrodynamic approach (Svintsov *et al* 2012). The quantum approach based on the equation for the density matrix is important in the presence of the interband transitions, but in our case it yields the same results as the kinetic method based on the quasi-classical Boltzmann equation (Cox *et al* 2016).

An oblique incidence of s-polarized THz electromagnetic wave onto the planar layered structure dielectric–graphene–dielectric–…–dielectric is considered. The OZ axis is aligned perpendicularly to the surfaces of the layers, see figure 8.1(a). The incident electromagnetic wave has (H_x, E_y, H_z) components at the frequency ω. The incidence angle with respect to the normal is Θ_i. Under the oblique incidence the H_z component of the magnetic field exerts the Lorentz force on free electrons. As a result, the excited second harmonic possesses the complementary p-polarization (E_x, H_y, E_z) different from the polarization of the fundamental harmonic. This is in a clear distinction from Mikhailov (2011) and Smirnova and Kivshar (2014), where the first and the second harmonics possess the same p-polarizations (E_x, H_y, E_z).

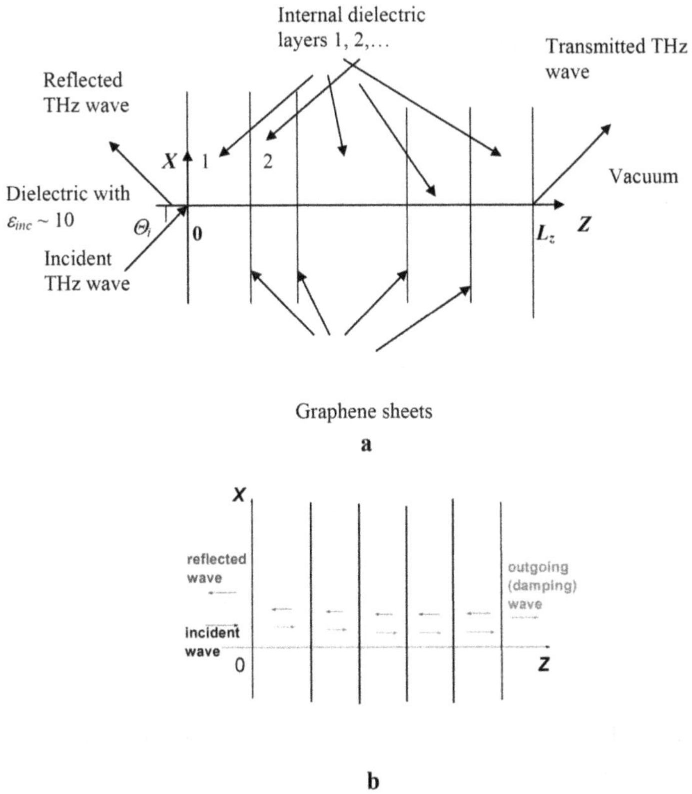

Figure 8.1. Geometry of the problem and the concept of partial waves; (a) is layered structure proposed for the giant second harmonic generation. The graphene sheets are between the internal dielectric layers (1, 2, …) with permittivities $\varepsilon \sim 4$. At the input $z < 0$ there is a high-permittivity dielectric with $\varepsilon_{\rm Inc.} \sim 10$, at the output $z > L_z$ there is vacuum with $\varepsilon = 1$. Reprinted from Rapoport *et al* (2017), copyright IOP Publishing, all rights reserved; (b) is partial waves in the layers. The direction of propagation is presented with respect to OZ axis. For the first harmonic, the amplitude of the incident wave is known. For the second harmonic this amplitude is 0, but there are the current sources at the graphene sheets.

The stationary case of nonlinear scattering is considered, where only two frequencies are present in the system, ω and 2ω.

Electromagnetic waves (EMWs) obey the Maxwell equations:

$$\nabla \times \vec{H} = \frac{\varepsilon}{c}\frac{\partial \vec{E}}{\partial t}; \quad \nabla \times \vec{E} = -\frac{1}{c}\frac{\partial \vec{H}}{\partial t} \tag{8.1}$$

Here ε is a permittivity of dielectric layers. The frequency dispersion is neglected, but our simulations have demonstrated that the resonant effects occur also under a weak frequency dispersion. Within each dielectric layer the EMWs have the same component of the wave vector k_x: $k_x \equiv k_{ix} = k_i \sin \Theta_i$, where $k_i = \omega/c(\varepsilon_{\mathrm{Inc.}})^{1/2}$ is the wavenumber of the incident EMW, $\varepsilon_{\mathrm{Inc}}$ is the permittivity of the medium at $z < 0$. The dielectric layers are assumed linear, whereas the nonlinearity is due to 2D electron gas of the graphene sheets.

Electric field E_y and components of the magnetic field H_x, H_z of the fundamental harmonic are:

$$\begin{aligned} &E_y = \frac{1}{2}E_1(z)\exp(i\phi_1) + \mathrm{c.c.}, \quad \phi_1 = \omega t - k_{ix}x; \\ &H_x = -\frac{ic}{2\omega}\left(\frac{\partial E_1(z)}{\partial z}\exp(i\phi_1) + \mathrm{c.c.}\right) \equiv \frac{1}{2}H_{1x}\exp(i\phi_1) + \mathrm{c.c.}; \\ &H_z = \frac{ck_{ix}}{\omega}E_y. \end{aligned} \tag{8.2a}$$

Within each dielectric layer the electromagnetic field is the superposition of the forward and backward waves.

Expressions for the electric field components E_x, E_z and the magnetic field H_y of the induced second harmonic are:

$$\begin{aligned} &E_x = \frac{1}{2}E_2(z)\exp(i\phi_2) + \mathrm{c.c.}, \quad \phi_2 = 2\omega t - k_{ix}x; \\ &H_y = -\frac{ic}{2\omega}\left(1 - \frac{k_{ix}^2c^2}{\omega^2\varepsilon}\right)^{-1}\left(\frac{\partial E_1(z)}{\partial z}\exp(i\phi_1) + \mathrm{c.c.}\right); \\ &E_z = \frac{k_{ix}c}{\omega\varepsilon}H_y. \end{aligned} \tag{8.2b}$$

8.2.1 Kinetic approach

In the kinetic method the Boltzmann equation for the one particle distribution function $f(t,x,p_x,p_y)$ takes a form:

$$\begin{aligned} &\frac{\partial f}{\partial t} + v_x\frac{\partial f}{\partial x} + e\left(E_x + \frac{1}{c}v_yH_z\right)\frac{\partial f}{\partial p_x} + eE_y\frac{\partial f}{\partial p_y} = -\gamma(f - f_0), \\ &\vec{v} \equiv \frac{\partial E(p)}{\partial \vec{p}} = v_{\mathrm{F}}\frac{\vec{p}}{p}, \quad E(p) = v_{\mathrm{F}}p, \quad p \equiv (p_x^2 + p_y^2)^{1/2} \end{aligned} \tag{8.3}$$

Here $\vec{v}$ is the electron velocity, $\vec{p}$ is the quasi-momentum, $E(p)$ is the electron energy, γ is the electron collision frequency.

For solving equation (8.3) the standard perturbation method is used. Namely, the trial solution is searched as (Smirnova and Kivshar 2014):

$$f = f_0(p_x, p_y) + \frac{1}{2}(F_1(x, p_x, p_y) \cdot \exp(i\phi_1) + F_2(x, p_x, p_y) \cdot \exp(i2\phi_1) + \text{c.c.}) \quad (8.4)$$

Here f_0 is the equilibrium Fermi distribution function $f_0(p_x, p_y) = (1 + \exp(v_F(p - p_F)/k_B T))^{-1} \approx \Theta(p_F - p)$, $F_{1,2}$ are the amplitudes of the distribution function at the first and the second harmonics; E_F is the Fermi energy, $p_F = E_F/v_F$ is the Fermi-momentum.

From equations (8.3) and (8.4) the expressions for $F_{1,2}$ are:

$$\begin{aligned} F_1 &\approx \frac{i}{\omega - i\gamma}\left(1 + \frac{k_x v_x}{\omega}\right) e E_1 \frac{\partial f_0}{\partial p_y}; \\ F_2 &\approx \frac{ie}{2\omega - i\gamma} E_{2x} \frac{\partial f_0}{\partial p_x} - \frac{e^2}{2(2\omega - i\gamma)(\omega - i\gamma)c} E_1 H_z v_y \frac{\partial^2 f_0}{\partial p_x \partial p_y} \end{aligned} \quad (8.5)$$

Here $E_1 \equiv E_1(z)$ is the amplitude of the fundamental harmonic and $E_{2x} \equiv E_{2x}(z)$ is the amplitude of the second harmonic, see equation (8.2).

The amplitude of the surface current density at the second harmonic $I_{\text{sx}}^{(2)}$ can be expressed as:

$$\begin{aligned} I_{\text{sx}}^{(2)} = 4e\sum_{\vec{p}} v_x F_2(x, p_x, p_y) &= \frac{2ie^2}{\omega\left(1 - \frac{i\gamma}{2\omega}\right)} E_{2x} \sum_{\vec{p}} v_x \frac{\partial f_0}{\partial p_x} \\ &- \frac{e^3}{\omega^2\left(1 - \frac{i\gamma}{2\omega}\right)\left(1 - \frac{i\gamma}{\omega}\right)} E_1 H_{1z} \sum_{\vec{p}} v_x v_y \frac{\partial^2 f_0}{\partial p_x \partial p_y} \end{aligned} \quad (8.6)$$

The summing up in equation (8.6) is done over the first Brillouin zone and can be replaced by the integration (Smirnova and Kivshar 2014):

$$\sum_{\vec{p}} \ldots = \frac{1}{4\pi^2\hbar^2} \int_{-\infty}^{+\infty} \int_{-\infty}^{+\infty} \ldots \, \mathrm{d}p_x \mathrm{d}p_y. \quad (8.7)$$

Due to the step-like shape of the Fermi distribution function $f_0(p_x, p_y)$ the integration in (8.7) can be extended into the total plane $-\infty < p_{x,y} < +\infty$ and thus fulfilled analytically. As a result, the expression for the amplitude of the surface current density at the second harmonic $I_{\text{sx}}^{(2)}$ is:

$$I_{\text{sx}}^{(2)} = -\frac{ie^2}{2\omega\left(1 - \frac{i\gamma}{2\omega}\right)} \frac{v_F p_F}{\pi\hbar^2} E_{2x} - \frac{e^3}{8\omega^2 c\left(1 - \frac{i\gamma}{2\omega}\right)\left(1 - \frac{i\gamma}{\omega}\right)} \frac{{v_F}^2}{\pi\hbar^2} E_1 H_{1z} \quad (8.8)$$

And, using equation (8.2a), it is possible to get:

$$I_{\mathrm{sx}}^{(2)} = -\frac{ie^2}{2\omega\left(1 - \dfrac{i\gamma}{2\omega}\right)}\frac{v_{\mathrm{F}}p_{\mathrm{F}}}{\pi\hbar^2}E_{2x} - \frac{e^3k_{ix}}{8\omega^3\left(1 - \dfrac{i\gamma}{2\omega}\right)\left(1 - \dfrac{i\gamma}{\omega}\right)}\frac{{v_{\mathrm{F}}}^2}{\pi\hbar^2}{E_1}^2 \tag{8.9}$$

The density of the surface current at the second harmonic for the incident s-polarized electromagnetic wave is of the same order as in Smirnova and Kivshar (2014), equation (17) in that paper, for the p-polarized incident wave.

It is also possible to get the analytical expression for the density of surface current under the finite temperatures $T \neq 0$, the structure is just the same as equation (8.9) but the coefficients are slightly different and possess a more complicated form. The coefficient near the nonlinear term is 20%–30% smaller at T = 150–300 K when compared with the limiting case $T = 0$.

Equation (8.9) is a relation between the surface current at the second harmonic, the electric field at the second harmonic, and the known field of the fundamental harmonic. In the following subsection the expression for the density of the surface current is also obtained by a simper hydrodynamic approach.

8.2.2 Hydrodynamic approach

As an alternative route we use also a simplified hydrodynamic approach, based on the determination of the effective nonlinear current sources by introducing the quasiparticles with the effective mass m_{g}^*. This approach was used for the nonlinear electromagnetic phenomena in a multilayered dielectric–graphene metamaterial in Katsnelson (2008). In the hydrodynamic approach the electron dynamics of 2D electron gas in the graphene sheets is described by the following equation for the average electron quasi-momentum $\vec{p}$, or for average electron velocity $\vec{v}$:

$$\frac{\mathrm{d}\vec{p}}{\mathrm{d}t} = e\vec{E} + \frac{e}{c}\vec{v} \times \vec{H} - \gamma\vec{p}, \quad \vec{p} = m_{\mathrm{g}}^*\vec{v} \tag{8.10}$$

Here collision frequency $\gamma \sim 10^{12}$–10^{13} s^{-1} is assumed, the same as in the kinetic approach, equation (8.3).

In the hydrodynamic approach the expression for effective electron mass m_{g}^* is obtained from the comparison of the first term of the current density in kinetics, equation (8.9), and in the hydrodynamics:

$$\begin{aligned} &(2i\omega + \gamma)m_{\mathrm{g}}^* V_x^{(2)} = eE_{2x} + \text{non-linear term}; \\ &I_{\mathrm{sx}}^{(2)} \equiv en_{20}V_x^{(2)} \approx -\frac{ie^2n_{20}E_{2x}}{(2\omega - i\gamma)m_{\mathrm{g}}^*} \end{aligned} \tag{8.11}$$

Here $V_x^{(2)}$ is the amplitude of the electron velocity at the second harmonic frequency. As a result, the expression for the effective mass is:

$$m_{\mathrm{g}}^* = \frac{p_{\mathrm{F}}}{v_{\mathrm{F}}} \sim {n_{20}}^{1/2}; \quad p_{\mathrm{F}} = \hbar(\pi n_{20})^{1/2}; \quad E_{\mathrm{F}} = v_{\mathrm{F}}p_{\mathrm{F}.} \tag{8.12}$$

This effective mass is about $m_g^* \approx 0.01 m_e$ and depends on n_{20}.

We consider the case when the electric field of the second harmonic is smaller than the field of the fundamental one. Thus, the inverse influence of the second harmonic on the fundamental wave is neglected. The expression for the electron velocity at the fundamental harmonic in graphene is:

$$v_y^{(1)} = -\frac{ieE_1(z)}{2m_g^*(\omega - i\gamma)}\exp(i\phi_1) + \text{c.c.} \tag{8.13}$$

It is possible to search the electric field of the second harmonic as:

$$E_x^{(2)} = \frac{1}{2}E_2(z)\exp(i\phi_2) + \text{c.c.}; \quad \phi_2 \equiv 2\phi_1 = 2\omega t - 2k_{ix}x; \tag{8.14}$$

The equation for the electron velocity on the double frequency is:

$$\frac{\partial v_x^{(2)}}{\partial t} + \gamma v_x^{(2)} = \frac{e}{m_g^*}E_x^{(2)} + \frac{e}{m_g^* c}(v_y H_z)^{(2)} \tag{8.15}$$

Note that the nonlinearity is due to the Lorentz force, as in the kinetic approach.

The solution for $v_x^{(2)}$ is searched as:

$$v_x^{(2)} = \frac{1}{2}V_x^{(2)}\exp(i\phi_2) + \text{c.c.} \tag{8.16}$$

The solution of equation(8.15) in the nonlinear case is:

$$V_x^{(2)} = -\frac{iE_{2x}}{m_g^*(2\omega - i\gamma)} - \frac{e^2 k_{ix} E_1^2}{2(m_g^*)^2\omega(\omega - i\gamma)(2\omega - i\gamma)} \tag{8.17}$$

Therefore, the expression for the surface electric current at the second harmonic is:

$$I_{\text{sx}}^{(2)} \equiv en_{20}V_x^{(2)} = -\frac{ien_{20}E_{2x}}{m_g^*(2\omega - i\gamma)} - \frac{e^3 v_F^2 k_{ix} E_1^2}{2\pi\hbar^2\omega(\omega - i\gamma)(2\omega - i\gamma)} \tag{8.18}$$

Equation (8.18) or equation (8.9) are used below in the boundary conditions for the magnetic field, see section 8.2.3, to calculate in a self-consistent manner the distribution of the field at the second harmonic $E_{2x}(z)$.

Note that the similar expression for the nonlinear term proportional to E_1^2 has been obtained from the more complicated kinetic approach, equation (8.9). The difference is in the coefficient near the nonlinear term, which is two times smaller in the kinetic approach. Surprisingly, the expression for the effective mass m_g^* was obtained from comparison of the linear terms in the surface current, but the expression for the nonlinear term is also close.

One can see that the nonlinear coefficient does not depend on the value of the 2D electron concentration n_{20} (Mikhailov 2011). Therefore, it is preferable to use smaller values of n_{20}. The more rigorous quantum kinetic approach (Mikhailov 2011) points out the limits of the hydrodynamic method. Namely, condition

$E_F \gg \hbar\omega$ should be satisfied to prevent interband transitions (Falkovsky and Varlamov 2007). This limits the possible values of $n_{20} \geqslant 10^{10}$ cm^{-2} when the frequency is within THz range. Thus, the optimum values of n_{20} are of about 10^{10} cm^{-2}–10^{11} cm^{-2}, where the influence of dissipation within the graphene sheets on propagation of the first harmonic is still not essential.

8.2.3 Boundary conditions at the graphene sheets

The boundary conditions for the second harmonic of EMW at each graphene sheet are:

$$\begin{aligned} E_x|_{\rm II} - E_x|_{\rm I} &= 0; \\ H_y|_{\rm II} - H_y|_{\rm I} &= \frac{4\pi}{c} i_{sx} \end{aligned} \tag{8.19}$$

The tangential components of the magnetic field H_y and H_x are discontinuous due to the surface current. The boundary condition for the fundamental harmonic E_y can be reduced to:

$$\left.\frac{\partial E_1}{\partial z}\right|_{\rm II} - \left.\frac{\partial E_1}{\partial z}\right|_{\rm I} = \frac{4e^2\omega E_F}{c^2(\omega - i\gamma)\hbar} E_1 \tag{8.20}$$

Correspondingly, the boundary condition for the second harmonic can be rewritten as:

$$\begin{aligned} &\left(1 - \frac{k_{ix}{}^2}{\omega^2\varepsilon_{\rm II}}\right)^{-1} \left.\frac{\partial E_{2x}}{\partial z}\right|_{\rm II} - \left(1 - \frac{k_{ix}{}^2}{\omega^2\varepsilon_{\rm I}}\right)^{-1} \left.\frac{\partial E_{2x}}{\partial z}\right|_{\rm I} \\ &= \frac{8e^2\omega E_F}{c^2(2\omega - i\gamma)\hbar}\left[E_{2x} - \frac{iek_{ix}E_1^2}{2m_g^*\omega(\omega - i\gamma)}\right] \end{aligned} \tag{8.21}$$

The coefficient near the nonlinear term is written under the hydrodynamic approach. Under the kinetic approach this coefficient is two times smaller.

The boundary conditions at the input $z = 0$ and output $z = L_z$ are as follows. For the fundamental harmonic there exists the incident wave of a known amplitude at $z = 0$, whereas at $z > L_z$ there is only the outgoing wave in vacuum, which has an imaginary wavenumber and damps in z-direction. For the second harmonic the incident wave is absent at $z < 0$, whereas there exists the damping outgoing wave in the vacuum at $z > L_z$. Within each internal dielectric layer there are two partial waves that propagate in opposite directions with respect to OZ axis, see figure 8.1(b).

The standard matrix method to compute the electric fields in the structure is applied. Initially the distribution of the electric field for the first harmonic is calculated, then this distribution is used for the simulations of the second harmonic in the presence of the surface currents in all graphene sheets, as seen from equation (8.21).

8.3 Double resonant reflection and nonlinear scattering into second harmonic: simulations

Initially, excitation of the second harmonic of E_x polarization was considered in a simple geometry, where the oblique incidence of the fundamental harmonic of E_y polarization occurred from the vacuum onto the graphene sheets. The nonlinear scattering with a low efficiency $R_{2\omega} < 10^{-5}$ at the input intensities ~1 MW cm^{-2} takes place in this case, which is in good agreement with the earlier results (Smirnova and Kivshar 2014). Therefore, the optimization of the nonlinear scattering process by implying resonant geometries is required. The coefficient of reflection of the fundamental wave is determined as usually: $R_\omega = |E_{1R}/E_{10}|^2$, where E_{1R} is the amplitude of the reflected wave at $z = 0$, E_{10} is the amplitude of the incident wave. The efficiency of scattering into the second harmonic is estimated as the effective corresponding reflection coefficient $R_{2\omega} = |E_{2x}(z = 0)/E_{10}|^2$.

The best results have been obtained when at the input of the structure $z < 0$ there is a dielectric with large permittivity $\varepsilon_{\text{Inc.}} \geqslant 10$ and at the output $z > L_z$ there is vacuum of the permittivity 1, see figure 8.1(a). In this case the total internal reflection of electromagnetic waves at the output occurs at some incident angles and the structure behaves like a resonator. Namely, the total internal reflection $\Theta_i = \arcsin((\varepsilon/\varepsilon_{\text{Inc.}})^{1/2}) = 35.8°$ for the incidence of an electromagnetic wave from a dielectric medium with $\varepsilon_{\text{Inc.}} = 11.7$ onto a dielectric medium with $\varepsilon = 4$. An incident angle for the fundamental harmonic is near this total reflection angle. The reflection coefficient for the second harmonic $R_{2\omega}$ can reach high values $\geqslant 0.01$, when the number of dielectric layers is 5–10 and the electron collision frequency is $\gamma = 2 \times 10^{12}$–$5 \times 10^{12}$ s^{-1}, as our simulations have demonstrated. At higher values $\gamma > 5 \times 10^{12}$ s^{-1} it is better to use fewer dielectric layers, 2–4. At smaller values of $\gamma < 10^{12}$ s^{-1} the reflection coefficient may reach even higher values $R_{2\omega} > 0.1$, but in this case the inverse influence of the second harmonic on the fundamental one should be considered and this needs special investigation.

Initially, in the simulations the thicknesses of all internal dielectric layers were the same. The simplest optimization of the efficiency of the scattering into the second harmonic was done by means of variation of the thickness $l_1 \sim 5$–500 μm of the first dielectric layer and incidence angle $\Theta_i \sim 0 - \pi/2$. The thicknesses of other internal layers are preserved as constant.

Typical results of simulations are presented in figures 8.2–8.7. The incident amplitude is $E_{10} = 20$ kV cm^{-1} in the hydrodynamic approach, or $E_{10} = 30$ kV cm^{-1} in the kinetic approach. For figures 8.2–8.7 the parameters of the structure are: the dielectric permittivity at $z < 0$ is $\varepsilon_{\text{Inc.}} = 11.7$, intrinsic silicon, the permittivity of the internal dielectric layers $\varepsilon = 4$, like SiO_2, their thicknesses are $l_2 = l_3 = \ldots = 80$ μm; at $z > L_z$ there is vacuum. The 2D electron concentration is $n_{20} = 2.5 \times 10^{10}$ cm^{-2}. In figure 8.2 the resonant frequency of the fundamental harmonic is $\omega_1 = 6.52 \times 10^{12}$ s^{-1}.

Figures 8.2–8.7 demonstrate application of the outlined above the double resonance concept. Indeed, under the chosen parameters the resonant reflection *minima* for the fundamental harmonic coincide with the corresponding resonant *maxima* of the 'effective reflection' into the second harmonic. This means the

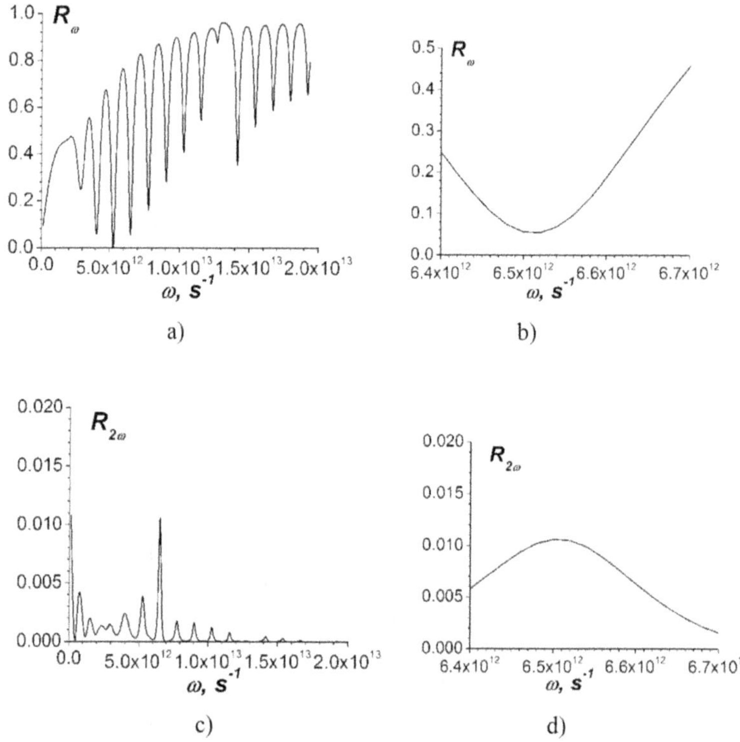

Figure 8.2. Dependences of the linear R_ω (parts (a) and (b)) and nonlinear $R_{2\omega}$ (parts (c) and (d)) reflection coefficients on fundamental frequencyω for the electron collision frequency $\gamma = 2 \times 10^{12}\ \mathrm{s}^{-1}$. The left panels are general pictures, the right panels are detailed views near the resonant frequency. The incidence angle is $\Theta_i = 31.5°$, the thickness of the first dielectric layer is $l_1 = 91\ \mu\mathrm{m}$, other dielectric layers are $l_2 = l_3 = \dots 80\ \mu\mathrm{m}$. There are 10 internal dielectric layers and 9 graphene sheets between them. Reprinted from Rapoport *et al* (2017), copyright IOP Publishing, all rights reserved.

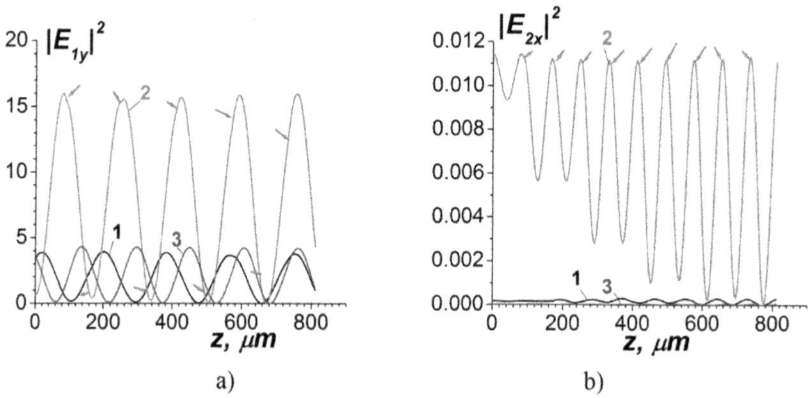

Figure 8.3. Profiles of the first, part (a), and the second harmonics, part (b). Curve 1 is for the frequency $\omega = 6 \times 10^{12}\ \mathrm{s}^{-1}$, curve 2 is for $\omega = 6.52 \times 10^{12}\ \mathrm{s}^{-1}$, the resonant case, curve 3 is for $\omega = 7 \times 10^{12}\ \mathrm{s}^{-1}$. The data are as in figure 8.2, the collision frequency is $\gamma = 2 \times 10^{12}\ \mathrm{s}^{-1}$. The positions of 9 graphene sheets are marked by arrows. Reprinted from Rapoport *et al* (2017), copyright IOP Publishing, all rights reserved.

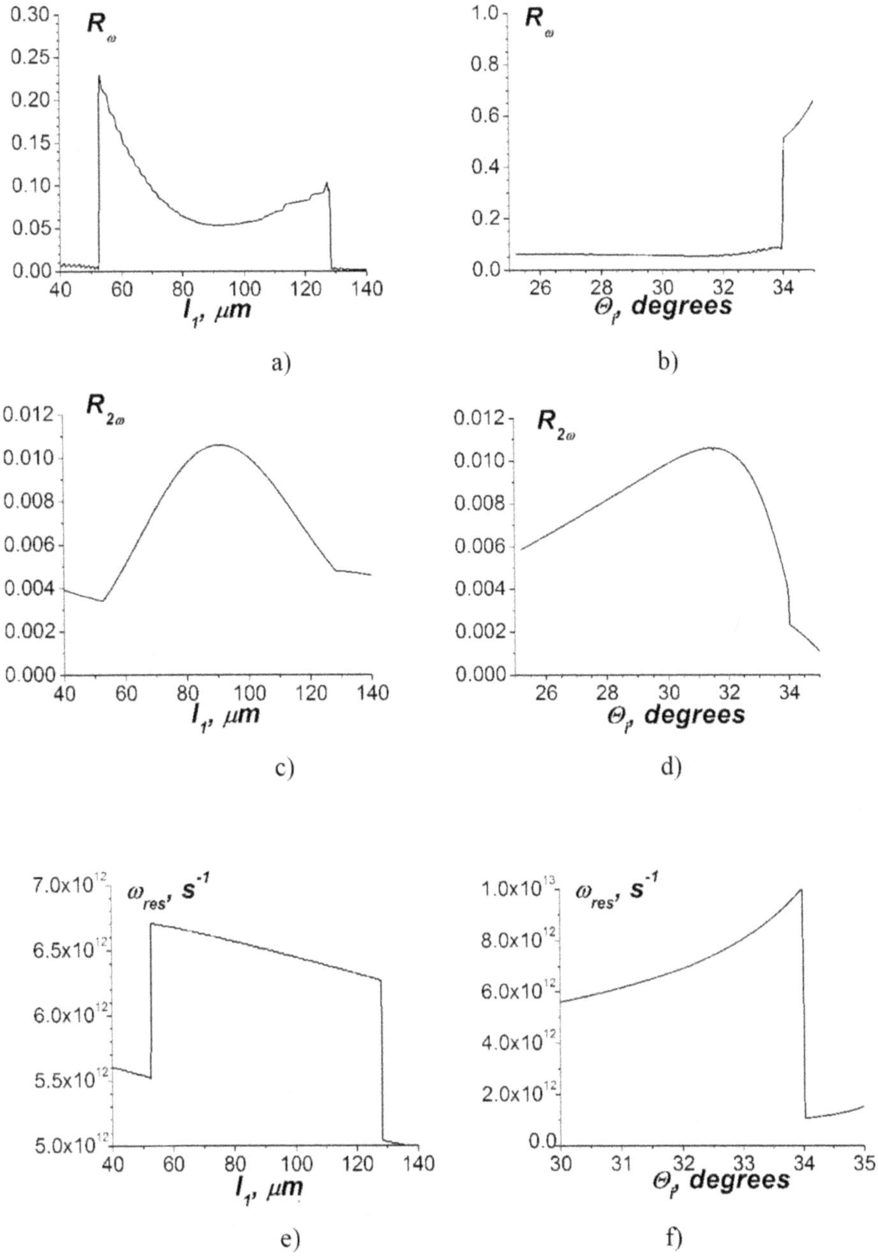

Figure 8.4. The resonant dependences of the maximum value of the linear reflection coefficient R_ω and the nonlinear reflection coefficient $R_{2\omega}$ on the thickness of the first internal dielectric layer l_1, parts (a) and (c), and on the incidence angle Θ_i, parts (b) and (d). The dependences of the resonant frequencies, where the maxima of $R_{2\omega}$ occur, are given in parts (e) and (f). The data are as in figure 8.2, the collision frequency is $\gamma = 2 \times 10^{12}$ s^{-1}. In parts (a) and (c) the incidence angle is $\Theta_i = 31.5°$; in parts (b) and (d) the thickness of the first internal dielectric layer is $l_1 = 91$ μm, other dielectric layers are $l_2 = l_3 = \ldots$ 80 μm. Reprinted from Rapoport *et al* (2017), copyright IOP Publishing, all rights reserved.

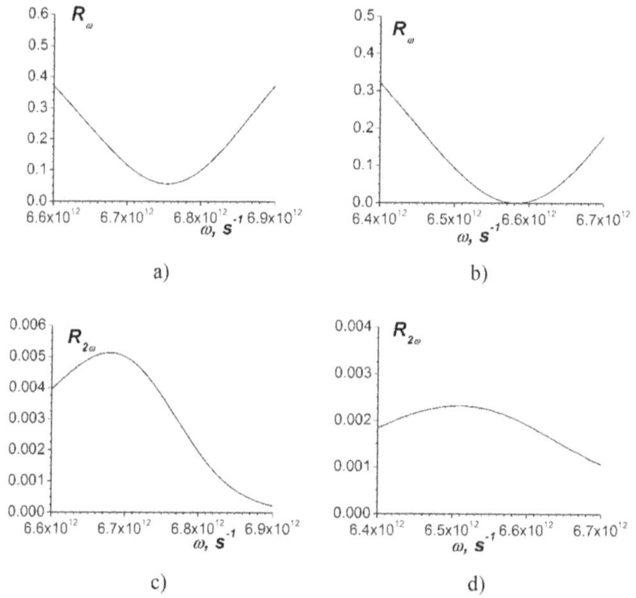

Figure 8.5. Detailed dependences of the linear R_ω, parts (a) and (b) and nonlinear $R_{2\omega}$, parts (c) and (d), reflection coefficients on fundamental frequency ω for the electron collision frequency $\gamma = 2 \times 10^{12}\ \mathrm{s}^{-1}$. A single parameter is different from those under the double resonance as in figure 8.2. Parts (a) and (c) are when the optimum incidence angle is $\Theta_i = 31.5°$, but the thickness of the first internal dielectric layer is $l_1 = 60\ \mu\mathrm{m}$, other dielectric layers are $l_2 = l_3 = \ldots\ 80\ \mu\mathrm{m}$. Parts (b) and (d) are for the optimum thickness of the first internal dielectric layer $l_1 = 91\ \mu\mathrm{m}$, but the incidence angle is $\Theta_i = 33°$. There are 10 internal dielectric layers and 9 graphene sheets between them, as in figure 8.2. Reprinted from Rapoport *et al* (2017),

resonant excitation of the second harmonic in the chosen structure. When linear reflection R_ω is minimal, the electric field within the structure increases ~5 times, compared with the electric field at frequencies far from the resonant value.

The distributions of the squares of amplitudes along the structure are depicted in figure 8.3. In the double-resonant case at the frequency $\omega = 6.52 \times 10^{12}\ \mathrm{s}^{-1}$, curves 2, the values of the electric field $|E_{1y}|$ of the fundamental harmonic at the graphene layers are near the maxima, as well as the values of the electric field $|E_{2x}|$ of the excited second harmonic. Also, one can see that at the resonant frequency the accumulation of the energy of the fundamental harmonic takes place within the system. The positions of the graphene sheets are marked by arrows. One can see that the simple optimization, where the thicknesses of the second, the third etc dielectric layers are the same, is quite rough, because some graphene sheets are near the minima of the first harmonic.

In figure 8.4 the dependences of the maximum values of nonlinear reflection coefficient $R_{2\omega}$ on the thickness of the first dielectric layer l_1 are depicted, when incidence angle Θ_i is the optimal, and on incidence angle Θ_i, when thickness l_1 is optimal. The parameters correspond to those in figure 8.2, i.e. the collision frequency is $\gamma = 2 \times 10^{12}\ \mathrm{s}^{-1}$. It is seen that these dependences possess the resonant character. To reach the resonance value of the reflection the accuracy in l_1 should be

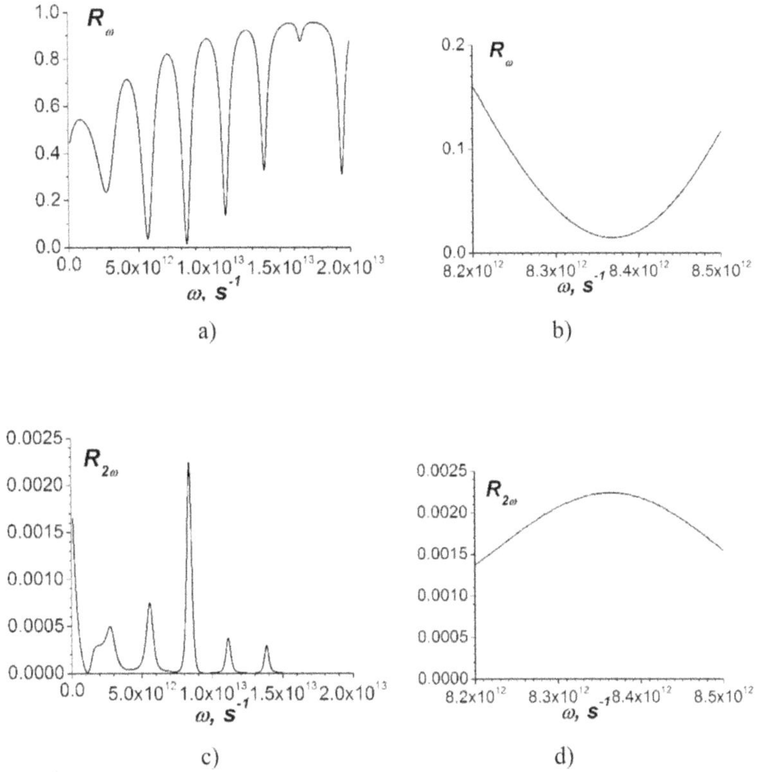

Figure 8.6. Dependences of the linear R_ω (parts (a) and (b)) and nonlinear $R_{2\omega}$ (parts (c) and (d)) reflection coefficients on fundamental frequency ω for electron collision frequency $\gamma = 5 \times 10^{12}$ s^{-1}. The left panels are general pictures, the right panels are detailed views near resonant frequency $\omega_1 = 8.38 \times 10^{12}$ s^{-1}. The incidence angle is $\Theta_i = 33.3°$, the thickness of the first internal dielectric layer is $l_1 = 86$ μm, other dielectric layers are $l_2 = l_3 = \ldots$ 80 μm. There are 6 internal dielectric layers, 5 graphene sheets between them. Reprinted from Rapoport *et al* (2017), copyright IOP Publishing, all rights reserved.

within ±20 μm and in incidence angle Θ_i should be ±1°. Thus, to provide the resonant conditions the incident beam should be transversely wide. The maximum values of $R_{2\omega}$ are realized at different frequencies of the fundamental harmonic when the values of l_1 or Θ_I vary, as seen from figure 8.4, parts (c) and (d), but the difference is within 20%.

When the thickness of the first dielectric layer l_1 or incidence angle Θ_i, differ from the optimal values, the maxima of the reflection coefficients for the second harmonic $R_{2\omega}$ and the minima ones for fundamental harmonic R_ω do not coincide, as seen from figure 8.5. In figure 8.5, parts (a) and (c), the thickness of the first internal dielectric layer differs from the optimum value but the incidence angle is the same as in figure 8.2. The minimum of reflection at the fundamental harmonic is at frequency $\omega_1 = 6.76 \times 10^{12}$ s^{-1}, the maximum of reflection at the second harmonic is at $\omega_1 = 6.67 \times 10^{12}$ s^{-1}. In figure 8.5, parts (b) and (d) the value of l_1 is the same as in figure 8.2, whereas the incidence angle is not optimal. The minimum of reflection at

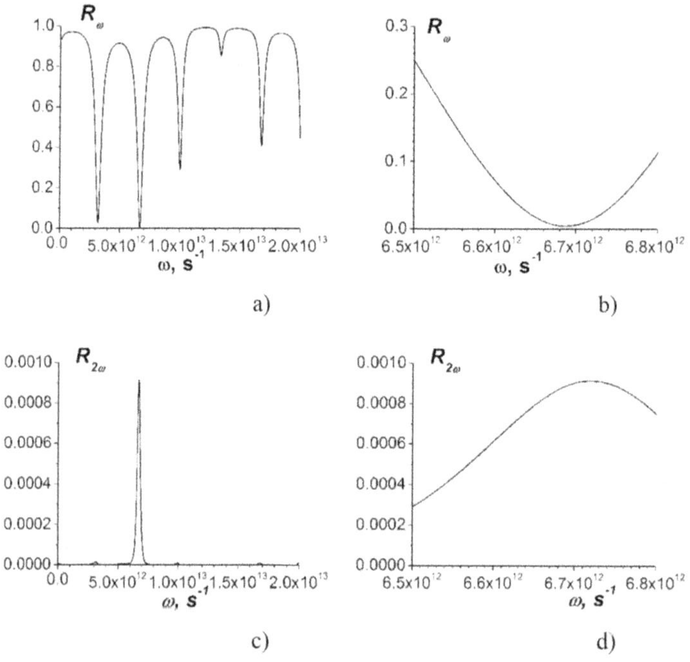

Figure 8.7. Dependences of the linear R_ω (parts (a) and (b)) and nonlinear $R_{2\omega}$ (parts (c) and (d)) reflection coefficients on fundamental frequency ω for the electron collision frequency $\gamma = 1 \times 10^{13}\ \mathrm{s}^{-1}$. The left panels are general pictures, the right panels are detailed views near the resonant frequency $\omega_1 = 6.68 \times 10^{12}\ \mathrm{s}^{-1}$. The incidence angle is $\Theta_i = 34.8°$, the thickness of the first dielectric layer is $l_1 = 480\ \mu\mathrm{m}$, the second dielectric layer is $l_2 = 150\ \mu\mathrm{m}$. There are 2 internal dielectric layers, 1 graphene sheet between them. Reprinted from Rapoport *et al* (2017),

the fundamental harmonic is at frequency $\omega_1 = 6.58 \times 10^{12}\ \mathrm{s}^{-1}$, the maximum of reflection at the second harmonic is at $\omega_1 = 6.50 \times 10^{12}\ \mathrm{s}^{-1}$.

The resonant values of the incidence angle and the thickness of the first dielectric layer are $\Theta_i = 31.5°$, $l_1 = 90\ \mu\mathrm{m}$ for collision frequency $\gamma = 2 \times 10^{12}\ \mathrm{s}^{-1}$, see figure 8.2. Correspondingly, these resonant values for $\gamma = 5 \times 10^{12}\ \mathrm{s}^{-1}$ are $\Theta_i = 33.3°$, $l_1 = 86\ \mu\mathrm{m}$. see figure 8.6. One can see from figures 8.2, 8.6 and 8.7, parts (b) and (d), that both linear reflection coefficient R_ω, parts (b), and the reflection coefficient into the second harmonic $R_{2\omega}$, parts (d), demonstrate the resonant character. Moreover, the resonant values of the incidence angle are close to the limiting angle of the total internal reflection $\Theta_i = \arcsin((\varepsilon/\varepsilon_{\mathrm{Inc.}})^{1/2}) = 35.8°$ for the incidence of an electromagnetic wave from a medium with $\varepsilon_{\mathrm{Inc.}} = 11.7$ onto a medium with $\varepsilon = 4$. The efficiency in $R_{2\omega}$ depends essentially on the graphene collision frequency, or the dissipation coefficient, γ. The resonance conditions are encountered at all realistic values of dissipation in graphene sheets $\gamma > 10^{12}\ \mathrm{s}^{-1}$.

At higher values of collision frequency $\gamma > 5 \times 10^{12}\ \mathrm{s}^{-1}$ it is better to use 1–3 graphene sheets within the structure, because of the high dissipation of the incident wave in the structures. The results of simulations for two internal dielectric layers, $l_1 = 480\ \mu\mathrm{m}$, $l_2 = 150\ \mu\mathrm{m}$, and a single graphene sheet, $n_{20} = 2.5 \times 10^{10}\ \mathrm{cm}^{-2}$, are given in figure 8.7. One can see from figure 8.6, part (b), that the reflection into the second harmonic can reach relatively high values $R_{2\omega} \sim 0.001$ even at such high value of the collision frequency $\gamma = 1 \times 10^{13}$ s^{-1}. The resonant frequency is $\omega_1 = 6.68 \times 10^{12}$ s^{-1}. In this case, figure 8.7, both thicknesses l_1 and l_2 were optimized to provide the maximum value of the second harmonic under the double-resonant conditions. Moreover, the resonant properties of the structure are preserved here, when the frequency of the fundamental harmonic $\omega \sim 10^{13}$ s^{-1} is comparable with the graphene collision frequency γ. The use of several graphene sheets does not result in an essential increase of the efficiency of the nonlinear scattering here.

Note that the resonant minima of the reflection coefficient for fundamental harmonic R_ω decrease under increase of collision frequency γ, compare figures 8.2(b), 8.6(b) and 8.7(b). At the same time the non-resonant values of R_ω are practically the same. Therefore, this layered structure can be an effective resonant absorber in the THz range.

When the incidence angle Θ_i is chosen small and the total internal reflection is absent, the resonant phenomena also occur, see figure 8.8. Again in the double-resonant

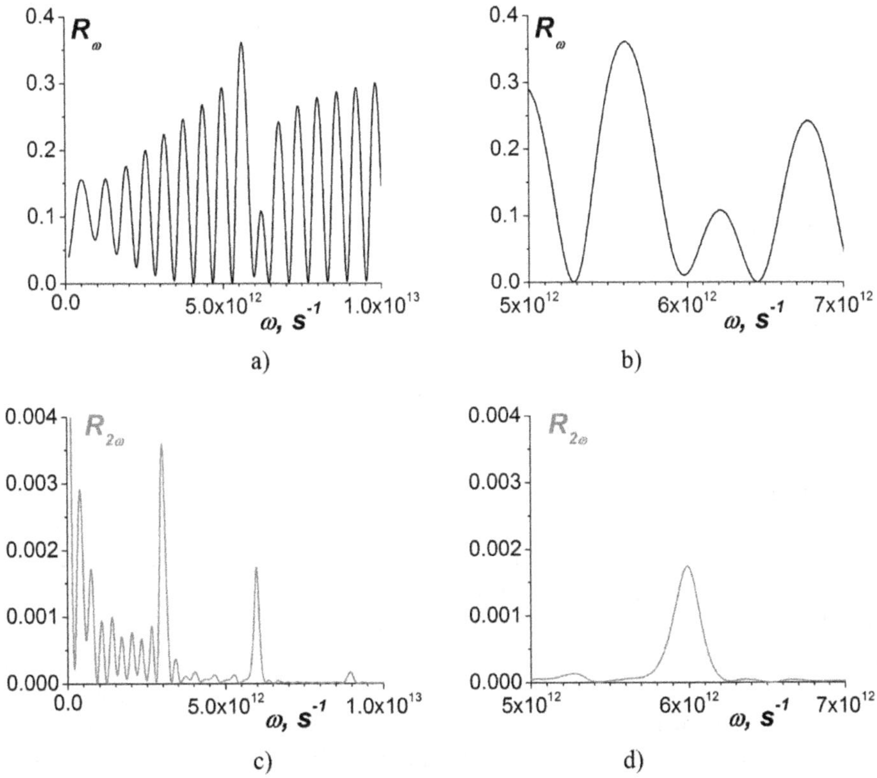

Figure 8.8. The case of almost normal incidence onto the structure, $\Theta_i = 4.8°$, $l_1 = 57$ μm is resonant for this angle, it is different from figure 8.2. The dissipation is $\gamma = 2 \times 10^{12}$ s^{-1}, $l_2 = l_3 = \ldots = l_{10} = 80$ μm, these parameters are as in figure 8.2. The resonant frequency for the optimum transformation is $\omega = 6 \times 10^{12}$ s^{-1} here. Parts (a) and (b) are the dependences of the reflection coefficients of the first harmonic on frequency ω, parts (c) and (d) are the reflection coefficients for the second harmonic. Reprinted from Rapoport *et al* (2017), copyright IOP Publishing, all rights reserved.

case the minimum of the reflection of the first harmonic coincides with the maximum of the reflection for the second harmonic. But the efficiency of the scattering into the second harmonic is one order smaller than in the case of higher incidence angles, compare with figure 8.2.

Thus, the resonant mechanism of the scattering into the second harmonic results in the accumulation of the EM energy within the layered structure on the fundamental frequency, where the minimum of reflection of the fundamental harmonic takes place, and the subsequent excitation of the second harmonic, where the maximum of reflection occurs. Because the steady 2D electron concentration in the graphene sheets can be controlled by the applied bias voltage, there is a possibility to control the nonlinear reflection coefficient too (Arezoomandan *et al* 2014).

8.4 Discussion and conclusions

The resonant nonlinear scattering into the second harmonic of THz radiation has been theoretically demonstrated in planar structures that include graphene sheets and intermediate dielectric layers. The oblique incidence of the input wave onto the structure is necessary to provide the nonlinear scattering into the second harmonic. The different polarizations of the fundamental and the second harmonics are considered. To compute the nonlinear surface currents two approaches have been used, quasi-classical kinetic and hydrodynamic. The obtained expressions for the nonlinear currents are different in coefficients only of the second harmonic. Some quantitative differences when using perturbations may be explained as follows. In the kinetics the expansion near the non-shifted Fermi function is used (Mikhailov 2011, Smirnova and Kivshar 2014). In the hydrodynamic approach the equations are related to the Fermi function shifted by the average momentum (Svintsov *et al* 2012). The comparison of the results obtained by means of both methods is useful. At least, a qualitative agreement between two approaches, proven in the present modeling ensures an applicability of the results.

A high efficiency of nonlinear scattering can be obtained in the resonant structure, where there is a dielectric of a high permittivity at the input and vacuum at the output. The incidence angle of the impinging wave should be close to the value of the total internal reflection angle at the input interface. The original double resonance concept for the effective generation of the second harmonic has been proposed and explored. Namely, under the chosen parameters the resonant reflection *minima* for the fundamental harmonic coincide with the corresponding resonant *maxima* of the reflection coefficient for the second harmonic. This approach is based on the idea of accumulation of the fields of the fundamental and double frequencies inside a system in a neighborhood of the graphene layers. At the same time, effective sources that provide the reflection on the main harmonic and effective reflection of the second harmonic, placed on the first graphene layers, have minimum and maximum values, respectively. As a result, the proper conditions for the reflections of the first and effective reflection of the second harmonics, outlined above are provided.

The resonant properties of the layered structures manifest at relatively high collision frequencies in graphene, when the fundamental frequency is comparable to or even smaller than the collision frequency. The nonlinear reflection coefficients for the intensity of the second harmonic in this double-resonant case are $>10^3$ times higher than for non-resonant cases. The resonant nonlinear scattering can be practically exploited for the frequency multipliers in the THz range.

The efficiency of the second harmonic generation in double-resonant graphene–dielectric structures is 4–5 orders of value larger than reported previously.

Because the resonant nonlinear reflection into the second harmonic takes place also at high collision frequencies in graphene, it is possible to expect the same effect when the interband transitions are important and the linear and nonlinear conduction possesses the dissipative character.

Further work will be connected with the joint electromagnetic-resonant effects with controllable electron waves and effects of graphene solid-state metamaterial electron optics (GMEO) (Rapoport *et al* 2012, 2014b, 2014e, 2014f), using the numerical methods based on the combined nonlinear multimode spectral-finite difference technique developed in Rapoport *et al* (2012, 2014b, 2014c, 2014d, 2014e, 2014f).

List of abbreviations

Graphene solid-state metamaterial electron optics (GMEO)

Problems to chapter 8

8.1. Estimate the decrease of the coupling coefficient for the generation of the second harmonic under finite temperatures, i.e. the generalization of equation (8.9). The stationary distribution function f_0 should be taken in the form of the Fermi distribution.

8.2. Investigate a possibility to generate higher harmonics for another kind of polarization, i.e. E_x, H_y, E_z, p-polarization of the first harmonic. In this case the equation for variable 2D concentration should be taken into account $\partial n_2/\partial t + \partial(n_2 v_x)/\partial x = 0$ where $n_2 = n_{20} + \tilde{n}_2$, so both the first harmonic and the second one possess p-polarization. Also the hydrodynamic equation for the electron velocity is $\partial v_x/\partial t + v_x\, (\partial v_x/\partial x) + \gamma v_x = eE_x$.

8.3. Investigate the wave mixing, when in the input there are two EM waves of different frequencies ω_1, ω_2. As the result, the EM is generated at the sum frequency $\omega = \omega_1 + \omega_2$.

References

Arezoomandan S, Yang K and Sensale-Rodriguez B 2014 Graphene-based electrically reconfigurable deep-subwavelength metamaterials for active control of THz light propagation *Appl. Phys.* A **117** 423–6

Bludov Yu V, Smirnova D A, Kivshar Yu S, Peres N M R and Vasilevskiy M I 2015 Discrete solitons in graphene metamaterials *Phys. Rev.* B **91** 045424

Bykov A Y, Murzina T V, Rybin M G and Obraztsova E D 2012 Second harmonic generation in multilayer graphene induced by direct electric current *Phys. Rev.* B **85** 121413(R)

Christensen T, Yan W, Jauho A-P, Wubs M and Mortensen N A 2015 Kerr nonlinearity and plasmonic bistability in graphene nanoribbons *Phys. Rev.* B **92** 121407

Cox J D and Javier Garcia de Abajo F 2015 Plasmon-enhanced nonlinear wave mixing in nanostructured graphene *ACS Photon.* **2** 306–12

Cox J D, Silveiro J D and Garcia de Abajo F J 2016 Quantum effects in the nonlinear response of graphene plasmons *ACS Nano* **10** 1995–2003

Dean J J and van Driel H M 2009 Second harmonic generation from graphene and graphitic films *Appl. Phys. Lett.* **95** 261910

Dong H, Conti C, Marini A and Biancalana F 2013 Terahertz relativistic spatial solitons in doped graphene metamaterials *J. Phys.* B **46** 155401

Dragoman M and Dragoman D 2009 Graphene-based quantum electronics *Progr. Quant. Electron.* **33** 165–214

Falkovsky L A and Varlamov A A 2007 Space-time dispersion of graphene conductivity *Eur. Phys. J.* B **56** 281–4

Garcia de Abajo F J 2014 Graphene plasmonics: challenges and opportunities *ACS Photon.* **1** 135–52

Glazov M M 2011 Second harmonic generation in graphene *JETP Lett.* **93** 366–71

Grimalsky V V, Kotsarenko N Y and Rapoport Y G 1996 Nonlinear surface waves in electronic plasma *Proc. of the 23rd European Physical Society Conf. on Controlled Fusion and Plasma Physics (Kyiv, Ukraine)* 396

Grimalsky V V and Rapoport Y G 1998 Modulational instability of surface plasma waves in the second-harmonic resonance region *Plasma Phys. Reports* **24** 980–2

Hendry E, Hale P J, Moger J, Savchenko A K and Mikhailov S A 2010 Coherent nonlinear optical response of graphene *Phys. Rev. Lett.* **105** 097401

Katsnelson M I 2008 Optical properties of graphene: the Fermi-liquid approach *Europhys. Lett.* **84** 37001

Kryuchkov S V and Kukhar E I 2014 Solitary electromagnetic waves in a graphene superlattice under influence of high-frequency electric field *Superlatt. Microstruct.* **70** 70–81

Li Y, Rao Y, Mak K F, You Y, Wang S, Dean C R and Heinz T F 2013 Probing symmetry properties of few-layer MoS_2 and h-BN by optical second-harmonic generation *Nano Lett.* **13** 3329–33

Mikhailov S A 2007 Non-linear electromagnetic response of graphene *Europhys. Lett.* **79** 27002

Mikhailov S A 2011 Theory of the giant plasmon-enhanced second-harmonic generation in graphene and semiconductor two-dimensional electron system *Phys. Rev.* B **84** 045432

Mikhailov S A 2013 Graphene-based voltage-tunable coherent terahertz emitter *Phys. Rev.* B **87** 115405

Mikhailov S A 2014 Nonlinear electromagnetic response of a uniform electron gas *Phys. Rev. Lett.* **113** 027405

Nesterov M L, Bravo-Abad J, Nikitin A Y, Garcia-Vidal F J and Martin-Moreno L 2013 Graphene supports the propagation of subwavelength optical solitons *Laser Photon. Rev.* **7** L7–11

Paul M J *et al* 2013 High-field terahertz response of graphene *New J. Phys.* **15** 085019

Rapoport Y U G and Grimalsky V V 2011 Nonlinear surface 2D plasmons and giant second harmonic generation *Proc. Days on Diffraction Int. Conf. 2011* (Piscataway, NJ: IEEE) 168–73

Rapoport Y U G, Grimalsky V V and Nefedov I S 2012 Graphene as electron wave density metamarerial and modeling 2D electron dynamics *Proc. of the XXXII Intern. Sci. Conf. Electronics and Nanotechnology, (ELNANO) (Kyiv, Ukraine)* (Piscataway, NJ: IEEE) 86–7

Rapoport Y, Grimalsky V, Iorsh I, Kalinich N, Koshevaya S, Castrejon-Martinez C and Kivshar Y S 2013 Nonlinear reshaping of THz pulses with graphene metamaterials *JETP Lett.* **98** 561–4

Rapoport Y G, Grimalsky V V, Castrejon-M C, Koshevaya S V and Kivshar Y S 2014a Nonlinear spatiotemporal focusing of terahertz pulses in the structures with graphene layers *Proc. of the XXXIV Int. Scientific Conf. Electronics and Nanotechnology (ELNANO), Kyiv, Ukraine* (Piscataway, NJ: IEEE) 31–4

Rapoport Y G, Grimalsky V V, Castrejon-M C and Koshevaya S V 2014b 2D Electron and electromagnetic waves in graphene metamaterials. Graphene solid-state electron optics *Proc. of the 15th Int. Conf. on Mathematical Methods in Electromagnetic Theory (Dnipropetrovsk, Ukraine)* 46 IEEE AP/MTT/ED/GRS/NPS/AES/EMB East Ukraine Joint chapter)

Rapoport Y, Selivanov Y, Ivchenko V, Grimalsky V, Tkachenko E and Fedun V 2014c Excitation of planetary electromagnetic waves in the inhomogeneous ionosphere *Ann. Geophys.* **32** 449–63

Rapoport Y G, Boardman A D, Grimalsky V V, Ivchenko V M and Kalinich N 2014d Strong nonlinear focusing of light in nonlinearly controlled electromagnetic active metamaterial field concentrators *J. Opt.* **16** 0552029

Rapoport Y, Grimalsky V, Koshevaya S and Castrejon-Martinez C 2014e 2D electron dynamics in single layer graphene with spin-orbital interaction and resonator-like external fields *Proc. of the 29th Int. Conf. on Microelectronics, (MIEL) 2014 (Kyiv, Ukraine)* (Piscataway, NJ: IEEE) 201–04

Rapoport Y G, Grimalsky V V, Koshevaya S V, Boardman A D and Malnev V N 2014f New method for modeling nonlinear hyperbolic concentrators *Proc. of the XXXIV Int. Scientific Conf. Electronics and Nanotechnology (ELNANO) 2014 (Kyiv, Ukraine)* (Piscataway, NJ: IEEE) 35–8

Rapoport Y, Grimalsky V, Lavrinenko A V and Boardman A 2017 Double resonant excitation of the second harmonic of terahertz raditation in dielectric-graphene layered metamaterials *J. Optics* **19** 095104

Sekwao S and Leburton J-P 2015 Terahertz harmonic generation in graphene *Appl. Phys. Lett.* **106** 063109

Smirnova D and Kivshar Y S 2014 Second-harmonic generation in subwavelength graphene waveguides *Phys. Rev.* B **90** 165433

Smirnova D A, Shadrivov I V, Miroshnichenko A E, Smirnov A I and Kivshar Y S 2014 Second-harmonic generation by a graphene nanoparticle *Phys. Rev.* B **90** 035412

Suprun A D and Shmeleva L V 2014 The centrally-symmetric solutions of electronic excitations of semiconductors in the conditions of relativistic like degeneracy of dynamical properties *Funct. Mater.* **21** 69–79

Svintsov D, Vyurkov V, Yurchenko S, Otsuji T and Ryzhii V 2012 Hydrodynamic model for electron-hole plasma in graphene *J. Appl. Phys.* **111** 083715

Zhang Q, Lin K and Luo Y 2011 Laser-launched evanescent surface plasmon polariton field utilized as a direct coherent pumping source to generate emitted nonlinear four-wave mixing radiation *Opt. Express* **19** 4991–5001

Zheludev N I and Emel'yanov V I 2004 Phase matched second harmonic generation from nanostructured metallic surfaces *J. Opt. A: Pure Appl. Opt.* **6** 26–8

IOP Publishing

Waves in Nonlinear Layered Metamaterials, Gyrotropic and Plasma Media

Yuriy Rapoport and Vladimir Grimalsky

Chapter 9

Nonlinear transformation optics and field concentration

9.1 Introduction to metamaterial transformations and geometrical optics mapping onto full-wave nonlinear solutions. Impact of nonlinear wave transformations on the design of realistic devices

An important problem is creating inhomogeneous media with desired properties for linear and nonlinear propagation of electromagnetic waves. These properties can be extreme concentrating electromagnetic energy or an opposite case like electromagnetic cloaking (Rahm *et al* 2008). To reach this goal, a general method based on the form invariant transformation of the Maxwell equations, or transformation optics (Zhang *et al* 2019, Yang *et al* 2019, McCall *et al* 2018, Gilarlue and Hadi 2019), can be applied including nonlinear transformational optics (Bergman *et al* 2011, Sklan and Baowen 2018). From the physical view, there is an analogy between the electromagnetic wave propagation in a gravitation field and in non-uniform media.

A possibility of analog modeling of geophysics and cosmology processes using metamaterials and linear and nonlinear electromagnetic transformational optics are illustrated in Narimanov and Kildishev (2009; see also references in this work).

The desire for energy concentrators is partly driven by their role in solar cells, but also by the need to generate high electromagnetic field intensities (Rahm *et al* 2008, Kildishev and Shalaev 2011, Narimanov and Kildishev 2009). In the beginning, it seemed natural to reach for metamaterials and to engage in a form of the linear transformation optics (Rahm *et al* 2008). This direction does need some care (Boardman *et al* 2011a) because the use of resonators in metamaterial manufacture can lead, directly, to high losses and to very narrow frequency band operation. In addition, the classical use of linear transformation optics, incorporating the form-

doi:10.1088/978-0-7503-2336-9ch9

invariance of the Maxwell equations, leads to the need for both dielectric and magnetic properties (Rahm *et al* 2008). Given this direction, it is natural to ask about the possibility of developing a new form of nonlinear transformation optics that might give access to spectacular high intensities and novel behaviour. In this connection, a preliminary report (Boardman *et al* 2011b) has emerged involving complex geometrical optics (CGO) and this will be developed rigorously here. Secondly, a full nonlinear transformation optics approach has appeared (Bergman *et al* 2011, Sklan and Baowen 2018) but it is limited to *weak* nonlinearities. It is possible, however, to proceed towards highly nonlinear concentrator designs, without going down the classical transformation route, with the aid of a graded-index, *completely non-magnetic* form of material (Narimanov and Kildishev 2009) that can be designed as an analogue to the Kepler problem concerning the capture of bodies by a gravitational field. Such a graded-index material approach, with cylindrical symmetry, is also adopted here, but with the very important, and realistic, added novel feature of a nonlinear core as the energy capture region. In this way, high intensities are put under very realistic control and versatility is generated, especially in a region where it is inevitable that the medium may become extremely nonlinear. It will be shown here that a superconcentrator can be created that will lead not only to very high levels of field intensity but to the discovery of a new kind of nonlinear switching into a superfocused state: designated here as a 'hot spot'. New methods of analysis are deployed to model this fascinating phenomenon because it exceeds, by far, the final state of any weakly nonlinear systems. It is not the kind of outcome arising from the usual formal transformation optics theory. However, it does point to the need to try and develop a new approach (Boardman *et al* 2011b), of the kind governed by the invariant form of the Maxwell equations.

The emphasis is upon the goal of reaching a super-high field concentration by means of a simple practical realization. To be quite specific, the new methods are designed to seek out strongly nonlinear effects in a system that has small dimensions. It remains to be seen if future research can discover the strongly nonlinear transformation generalizations (which is also allowed by the method, designed in the present work) that are required for a classic metamaterial route. In this work, an inhomogeneous, outer, *linear* medium is used for the initial stage of electromagnetic field capture and then a *nonlinear* homogeneous inner core is used for the final capture. This arrangement is sketched in figure 9.1 that contains the cross-section of the cylindrical concentrator used in this investigation.

The coordinate frame is defined as (r,θ,z) and the cylinder axis coincides with the z-axis. The linear external region ($r > R_0$) is occupied by a homogenous medium, with a relative permittivity of ε_0, the inner core ($r \leqslant R_c$) is both nonlinear and homogeneous. The latter also has the additional saturable nonlinear permittivity $\delta\varepsilon_{NL}$ that is shown. The outer cylindrical shell, ($R_c \leqslant r \leqslant R_0$) is special because it is both inhomogeneous and linear. The inhomogeneous permittivity increases with decreasing the radius, thus ensuring light capture through graded-index means (Narimanov and Kildishev 2009). For treating wave propagation in such a media, geometrical optics is used (Gershman *et al* 1984). Namely, a new version of CGO (Bravo-Ortega and Glassor 1991, Peeters 1996, Timofeev 2005) was developed here.

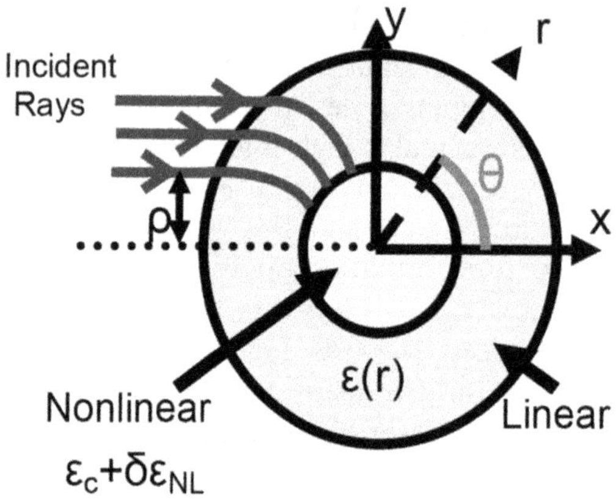

Figure 9.1. (r, θ) Cross-section of cylindrical nonlinear electromagnetic energy concentrator. Linear external cylinder ($R_c \leqslant r \leqslant R_0$). Nonlinear inner cylinder ($r \leqslant R_c$). Three incident complex rays. ρ is an impact distance. Reprinted from Rapoport *et al* (2014).

The complex eikonal describes both amplitude and phase of the electric field for a given ray and moreover, as it is explained below, even propagation of beams of finite transverse widths can be included in consideration. Each beam includes several rays here.

This type of model permits a very accurate description of the energy profiles that are to be admitted to the nonlinear core. For the inner nonlinear region the appropriate technique to deploy is called full-wave nonlinear electrodynamics.

9.2 Inhomogeneous dielectric permittivity and the wave equation

As an example, three complex rays are partially sketched in figure 9.1 and the polarization of the rays is set equal to (H_r, H_θ, E_Z). In order to capture electromagnetic energy a form of graded-index is needed. Ideally an isotropic form of permittivity is required that will capture incoming electromagnetic energy and deposit it into the core of the concentrator. The creation of such a concentrator, which will both capture energy and not engage in reflections that will diminish its effectiveness, can be done using a straightforward form of radially dependent permittivity distribution, to which can be added the possibility of the core becoming nonlinear and the further possibility of absorption. The practical form of the relative permittivity distribution can therefore be written very simply as

$$\varepsilon(r) = \begin{cases} \varepsilon_0, \; r > R_0 \\ \varepsilon_0 \dfrac{R_0^2}{r^2} + \dfrac{i\gamma}{\dfrac{R_0^2}{R_c^2} - 1}\left(\dfrac{R_0^2}{r^2} - 1\right), \; R_c < r < R_0 \\ \varepsilon_c + \delta\varepsilon_{\mathrm{NL}} + i\gamma, \; r < R_c \end{cases} \tag{9.1a}$$

where the elementary form $\varepsilon_c = \varepsilon_0 \frac{R_0^2}{R_c^2}$ simply ensures that, as $r \to R_c$, there is complete continuity across the interface. The nonlinear correction to the relative permittivity of the core is set to the saturable Kerr form $\delta\varepsilon_{NL} = \alpha|E|^2/(1 + \beta|E|^2)$, where α, β are constants that vary from one material to another. Finally, γ is a measure of any possible absorptive loss. The contact with a previously used simple form (Narimanov and Kildishev 2009) of permittivity is achieved by setting $\gamma = 0$, and setting $\delta\varepsilon_{NL} = 0$, in order to stay in the linear domain. As emphasized earlier, it is easy to see that, in a linear environment, each electric permittivity has been set to merge into the other as the appropriate radius is reached. This is still the case here, with the assumption that the nonlinear permittivity contribution in the core does not modify the reflections significantly at the boundary.

In order to simulate this type of structure, an original method has been developed that is a new version of CGO, within the inhomogeneous region $R_c \leqslant r \leqslant R_0$.

This is then matched to a full-wave solution for the nonlinear electrodynamic region $r \leqslant R_c$. Inclusion of the saturation, characterized by β, is very important, because any superfocusing will lead to such strong field concentrations that the core will move rapidly out of the weakly nonlinear regime and beyond the scope of the recently reported (Bergman *et al* 2011) nonlinear transformation optics. Figure 9.1 shows beams being incident upon the concentrator and the outcomes due to selecting one, two or three beams simultaneously are investigated. In each case, all the beams have equal initial amplitudes. Each beam is modeled as a central ray with some angular distribution of the field around it. In order to introduce the new method of calculation, consider the transverse electric mode (H_r, H_φ, E_z) and the corresponding 'full-wave' equation, namely

$$\frac{1}{r}\frac{\partial}{\partial r}\left(r\frac{\partial E_z}{\partial r}\right) + \frac{1}{r^2}\frac{\partial^2 E_z}{\partial \theta^2} + k_0^2\varepsilon(r)E_z = 0 \tag{9.1b}$$

where

$$k_0 = \frac{\omega}{c}, \tag{9.1c}$$

ω is the angular frequency and c is the velocity of light in free space,

$$\theta = \arctan\left(\frac{x}{y}\right) \equiv \tan^{-1}\left(\frac{x}{y}\right), \quad r = \sqrt{x^2 + y^2}, \tag{9.1d}$$

(see figure 9.1). For these structures, described, in the case of full-wave consideration, by equation (9.1a), the new version of CGO is developed.

9.3 'Ordinary' geometrical optics

Recall first what is the 'ordinary' geometrical optics, and then come to the new version CGO, presented in the present chapter. Both 'ordinary' geometrical optics (Gershman *et al* 1984) and the new version of CGO, presented in this section, use eikonal (phase) $S = S(r, \theta, t)$(depending on whether losses are present or absent). The introduction of any version of geometrical optics, requires that

$$E_z = Ae^{iS(r,\theta,t)} \tag{9.1e}$$

in the region $R_c \leqslant r \leqslant R_0$, where $S(r, \theta)$ is the eikonal connected to the image formation. Respectively, effective 'wavenumbers' and frequency are determined as

$$p_r = \frac{\partial S}{\partial r}, p_\theta = \frac{\partial S}{\partial \theta}, \omega = -\frac{\partial S}{\partial t} \tag{9.1f}$$

and, respectively,

$$dS = p_r dr + p_\theta d\theta - \omega dt \tag{9.1g}$$

Generally speaking, application of (9.1e) leads to the dispersion equation in the form

$$D(p_r, p_\theta, r, \theta, \omega) = 0 \tag{9.2}$$

and for the particular problem and structure (figure 9.1) considered in the present chapter, the function D is of the form

$$D = p_r^2 - \frac{ip_r}{r} + \frac{1}{r^2}p_\theta^2 - k_0^2\varepsilon(r) = 0, \tag{9.3}$$

which is the 'geometrical optics' correspondence to 'full-wave' equation (9.1b).

We will distinguish between 'ordinary' geometrical optics and the new version of CGO, presented in this chapter. 'Ordinary' geometrical optics can include either real (Gershman *et al* 1984) or imaginary (Bravo-Ortega and Glassor 1991) eikonal, depending on whether losses are included or not, respectively. Note also, that CGO was introduced before with *real* spatial coordinates, where complex eikonal was used to describe change of amplitude due to diffraction divergence of rays (Peeters 1996). In developing the present version of CGO, we follow the ideas of (Ginzburg 1961). Generally speaking, determination of equivalent 'pulses' and frequency in lossy inhomogeneous media is not unambiguous, and it is hard even to determine in such a medium unambiguously an energy flow. This results finally in existence of few different versions of 'geometrical optics', in particular 'CGO' (Bravo-Ortega and Glassor 1991, Peeters 1996, Timofeev 2005, Ginzburg 1961, Berczynski 2011). Nevertheless, for the case of slowly varying and weakly dissipative media, the following conclusion was made in the classical book (Ginzburg 1961): the signal propagation is, even in this case, still characterized by the 'group velocity' $\vec{v}_g$ which has, approximately, the form

$$\vec{v}_g \approx \frac{\partial \omega}{\partial \vec{p}\,'}; \vec{p} = \vec{p}\,' + i\vec{p}\,'', \vec{p}\,' \equiv \mathrm{Re}(\vec{p}), \vec{p}\,'' = \mathrm{Im}(\vec{p}) \tag{9.4a}$$

We will show relationship (9.4a) can be presented, and, approximately, in the form

$$\vec{v}_g \approx -\mathrm{Re}\left(\frac{D_{\vec{p}}}{D_\omega}\right) \tag{9.4b}$$

In general case, 'group velocity' could not be determined unambiguously for an inhomogeneous and lossy media, therefore sign '≈' emphasizes that the relations (9.4a) and (9.4b) keep the sense of 'signal' or 'pulse' velocity only in the case, when both inhomogeneity and dissipation are weak, and, in addition, the dispersion of dissipation is small enough in the 'characteristic frequency range' (Ginzburg 1961). For the stationary problem (for which the numerical calculations will be presented below) the last restriction should not be important, but general analytical formulas presented in this and the next sections are valid even for pulse propagation, therefore we will suppose that all the conditions listed above, are satisfied.

Hereafter denotations like $D_{\vec{k}}$ etc mean

$$D_{\vec{k}} \equiv \frac{\partial D}{\partial \vec{k}} \tag{9.4c}$$

Details of the 'new version of CGO' will be considered in the next section, and now we only note that this 'new version' is determined by two conditions, never satisfied simultaneously (to our best knowledge) in previous versions of geometrical optics/CGO before. These conditions are: (1) despite that eikonal S is complex, spatial coordinates should be real. Otherwise (Bravo-Ortega and Glassor 1991) we are faced with extra additional procedure of analytical continuation of complex function, to come back from complex to real space after all computations. Instead, we require from the very beginning that the coordinates are real, and potentially the ambiguous procedure of analytical continuation and 'rebuilding' real coordinates is avoided. (2) Group velocity should reduce to the formula (9.4a), where we follow classical book (Ginzburg 1961). Unification of these two features determines the novelty of the new version of CGO.

Let us consider first 'ordinary' geometrical optics, 'real' for losses and 'complex' for lossy media. We will refer to the 'ordinary geometrical optics' as to the method, where we preserve approximately the same expression for the 'group velocity', as for lossless and weakly inhomogeneous media, namely $\vec{v}_g = \dfrac{\partial \omega}{\partial \vec{p}}$. In this case, as we will show (not only for the dispersion equation in the form (9.3), but for rather general form of D), the equations for coordinates are

$$\frac{\mathrm{d}r}{\mathrm{d}t} = -\frac{D_{p_r}}{D_\omega}, \quad \frac{\mathrm{d}\theta}{\mathrm{d}t} = -\frac{D_{p_\theta}}{D_\omega} \tag{9.5}$$

We present below the derivation of equation (9.5).

Let us obtain first the expressions for group velocity, and to do this, we consider variation of D, which can be presented as $D = D(\vec{p}, \vec{r}, \omega)$ due to variation of ω. Respectively, consider an expansion:

$$\omega = \omega_0 + \delta\omega, \quad \vec{p} = \vec{k}_0 + \delta\vec{p}; \quad D(\omega_0, \vec{p}_0) = 0 \tag{9.6}$$

Using the Taylor formula, one can get:

$$D_\omega \delta\omega + D_{\vec{p}} \delta\vec{p} = 0 \tag{9.7a}$$

Or (for x-component, for instance):

$$\delta\omega + \frac{D_{kx}}{D_{\omega}}\delta p_x = 0 \tag{9.7b}$$

In accordance to (9.7b), the component of the group velocity is:

$$\frac{dx}{dt} = v_{gx} = \frac{\delta\omega}{\delta k_x} \tag{9.7c}$$

Accounting for (9.7b), we obtain:

$$v_{gx} = -\frac{D_{p_x}}{D_{\omega}} \tag{9.8a}$$

Similarly to (9.7b) and (9.7c), (9.8a), we obtain:

$$\frac{dy}{dt} = v_{gy} = \frac{\delta\omega}{\delta p_y} = -\frac{D_{p_y}}{D_{\omega}} \tag{9.8b}$$

In accordance with (9.7c), the trajectory is:

$$\frac{d\vec{r}}{dt} = \vec{v}_g = -\frac{D_{\vec{p}}}{D_{\omega}} \tag{9.8c}$$

As is seen from (9.7c), (9.8a)–(9.8c), accounting for (9.2), for any particular form of dispersion equation, such as ((9.3), for example), both $\vec{v}_g$ and coordinates $\vec{r}$ become complex for a lossy media (with complex ε). Nevertheless, we derive first sequentially the whole set of equations for 'ordinary' geometrical optics, including equation (9.5) for components of a 'pulse' $\vec{p}$ and after that present the new version of CGO, free of the above mentioned disadvantage. To finish the derivation for 'ordinary' geometrical optics, let us consider the equation for $r = \sqrt{x^2 + y^2}$:

$$\frac{dr}{dt} = \frac{x\frac{dx}{dt} + y\frac{dy}{dt}}{\sqrt{x^2 + y^2}} = \frac{x}{r}\frac{dx}{dt} + \frac{y}{r}\frac{dy}{dt} \tag{9.9a}$$

Putting (9.7c), (9.8a) and (9.8b) into (9.9a) we get

$$\frac{dr}{dt} = -\frac{\cos\theta D_{p_x} + \sin\theta D_{p_y}}{D_{\omega}} \tag{9.9b}$$

In accordance with the definition (9.1f), and accounting for that $x = r\cos\theta$, $y = r\sin\theta$, we obtain from (9.9b)

$$p_r = \left(\frac{\partial S}{\partial r}\right)_{\theta} = \left(\frac{\partial S}{\partial x}\right)_{y}\left(\frac{\partial x}{\partial r}\right)_{\theta} + \left(\frac{\partial S}{\partial y}\right)_{x}\left(\frac{\partial y}{\partial r}\right)_{\theta} = p_x\cos\theta + p_y\sin\theta \tag{9.10a}$$

Similarly to (9.10a), one can get

$$p_\theta = \left(\frac{\partial S}{\partial \theta}\right)_r = \left(\frac{\partial S}{\partial x}\right)_y\left(\frac{\partial x}{\partial \theta}\right)_r + \left(\frac{\partial S}{\partial y}\right)_x\left(\frac{\partial y}{\partial \theta}\right)_r = -p_x r\cos\theta + p_y r\cos\theta \qquad (9.10b)$$

As is seen from (9.10a) and (9.10b),

$$p_x = p_r\cos\theta - \frac{p_\theta}{r}\sin\theta \qquad (9.11a)$$

$$p_y = p_r\sin\theta + \frac{p_\theta}{r}\cos\theta \qquad (9.11b)$$

Accounting for that

$$D_{p_r} = D_{p_x}\left(\frac{\partial p_x}{\partial p_r}\right)_{p_\theta} + D_{p_y}\left(\frac{\partial p_y}{\partial p_r}\right)_{p_\theta} \qquad (9.12a)$$

and using (9.11a) and (9.11b), one can get from (9.12a)

$$D_{p_r} = D_{p_x}\cos\theta + D_{p_y}\sin\theta \qquad (9.12b)$$

Comparing (9.11b) with (9.9b), we obtain the first of equation (9.5).
To obtain the second of equation (9.5), let us note that

$$D_{p_\theta} = D_{p_x}\left(\frac{\partial p_x}{\partial p_\theta}\right)_{p_r} + D_{p_y}\left(\frac{\partial p_y}{\partial p_\theta}\right)_{p_r} \qquad (9.12c)$$

Using (9.11a) and (9.11b), (9.12a) can be rewritten as

$$D_{p_\theta} = -D_{p_x}\frac{\sin\theta}{r} + D_{p_y}\frac{\cos\theta}{r} \qquad (9.12d)$$

On the other hand, it follows from (9.1d), that

$$\frac{\mathrm{d}\theta}{\cos^2\theta} = \frac{x\mathrm{d}y - y\mathrm{d}x}{x^2} \qquad (9.13)$$

Therefore,

$$\frac{\mathrm{d}\theta}{\mathrm{d}t} = \frac{\cos\theta}{r}v_y - \frac{\sin\theta}{r}v_x \qquad (9.14a)$$

Accounting for (9.8a) and (9.11b), present (9.14a) as

$$\frac{\mathrm{d}\theta}{\mathrm{d}t} = -\left(\frac{\cos\theta}{r}\frac{D_{p_y}}{D_\omega} - \frac{\sin\theta}{r}\frac{D_{p_x}}{D_\omega}\right) \qquad (9.14b)$$

Then, comparing (9.14b) with (9.12d), we obtain, finally, the second of equation (9.5).
To derive equations for the 'moments' $p_{r,\,\theta}$, let us write

$$\frac{\mathrm{d}p_r}{\mathrm{d}t} = \frac{\partial p_r}{\partial r}\frac{\mathrm{d}r}{\mathrm{d}t} + \frac{\partial p_r}{\partial \theta}\frac{\mathrm{d}\theta}{\mathrm{d}t} \tag{9.15a}$$

$$\frac{\mathrm{d}p_\theta}{\mathrm{d}t} = \frac{\partial p_\theta}{\partial r}\frac{\mathrm{d}r}{\mathrm{d}t} + \frac{\partial p_\theta}{\partial \theta}\frac{\mathrm{d}\theta}{\mathrm{d}t} \tag{9.15b}$$

Let us also use the relations, which follow directly from equation (9.2), namely:

$$\frac{\mathrm{d}D}{\mathrm{d}r} = D_{p_r}\frac{\partial p_r}{\partial r} + D_{p_\theta}\frac{\partial p_\theta}{\partial r} + D_r = 0 \tag{9.16a}$$

$$\frac{\mathrm{d}D}{\mathrm{d}\theta} = D_{p_r}\frac{\partial p_r}{\partial \theta} + D_{p_\theta}\frac{\partial p_\theta}{\partial \theta} + D_\theta = 0 \tag{9.16b}$$

As follows from (9.5), equation (9.15a) reduces to

$$\frac{\mathrm{d}p_r}{\mathrm{d}t} = -\frac{1}{D_\omega}\left[D_{p_r}\frac{\mathrm{d}p_r}{\mathrm{d}r} + D_{p_\theta}\frac{\mathrm{d}p_r}{\mathrm{d}\theta}\right] \tag{9.17a}$$

Using definition (9.1), we can see that

$$\frac{\partial p_r}{\partial \theta} = \frac{\partial^2 S}{\partial \theta \partial r} = \frac{\partial p_\theta}{\partial r} \tag{9.17b}$$

and putting (9.17b) into (9.17a), we present the last relationship in the form

$$\frac{\mathrm{d}p_r}{\mathrm{d}t} = -\frac{1}{D_\omega}\left[D_{p_r}\frac{\mathrm{d}p_r}{\mathrm{d}r} + D_{p_\theta}\frac{\mathrm{d}p_\theta}{\mathrm{d}r}\right] \tag{9.17c}$$

Hence, accounting for (9.16a), we present (9.17c) in the form

$$\frac{\mathrm{d}p_r}{\mathrm{d}t} = \frac{D_r}{D_\omega} \tag{9.18}$$

which is the final 'equation of motion' for the pulse component p_r.
Similarly to (9.17c), it is possible to obtain from (9.15b), (9.5)

$$\frac{\mathrm{d}p_\theta}{\mathrm{d}t} = -\frac{1}{D_\omega}\left[D_{p_r}\frac{\mathrm{d}p_\theta}{\mathrm{d}r} + D_{p_\theta}\frac{\mathrm{d}p_\theta}{\mathrm{d}\theta}\right], \tag{9.19a}$$

which reduces, with using (9.17b), to the form

$$\frac{\mathrm{d}p_\theta}{\mathrm{d}t} = -\frac{1}{D_\omega}\left[D_{p_r}\frac{\mathrm{d}p_r}{\mathrm{d}\theta} + D_{p_\theta}\frac{\mathrm{d}p_\theta}{\mathrm{d}\theta}\right] \tag{9.19b}$$

Finally, accounting for (9.16b), we rewrite (9.19b) in the form

$$\frac{\mathrm{d}p_\theta}{\mathrm{d}t} = \frac{D_\theta}{D_\omega} \tag{9.19c}$$

The set of equations (9.5), (9.18) and (9.19c) is the final set of 'ordinary geometrical optics' equations. Note again, that this set of equations implies both complex coordinates (Bravo-Ortega and Glassor 1991) and 'pulses' $\vec{p}$ for a (weakly) lossy medium. Now we come to the new version of CGO.

9.4 New CGO techniques

Now we will present 'the new version of CGO', which provides, as was already mentioned, both 'real coordinates' $\vec{r}$ and proper expression(s) for 'group velocity' (9.4a) and (9.4b). Let us first explain, how (9.4b) follows from (9.4a).

Consider an expansion (with real ω):

$$\omega = \omega_0 + \delta\omega, \quad \vec{p} = \vec{p}_0 + \delta\vec{p} \tag{9.20}$$

Using the Taylor formula, one can get:

$$D_\omega \delta\omega + D_{\vec{k}}(\delta\vec{p}\,' + i\delta\vec{p}\,'') = 0 \tag{9.21a}$$

Or (for x-component only, for instance):

$$\delta\omega + \frac{D_{p_x}}{D_\omega}\delta p_x' + i\frac{D_{p_x}}{D_\omega}\delta p_x'' = 0 \tag{9.21b}$$

But the component of the group velocity is, in accordance with (9.4a):

$$v_{gx} = \frac{\delta\omega}{\delta p_x'} \tag{9.22a}$$

equation (9.21b) yields

$$\frac{\delta\omega}{\delta p_x'} + \mathrm{Re}\left(\frac{D_{p_x}}{D_\omega}\right) + i\mathrm{Im}\left(\frac{D_{p_x}}{D_\omega}\right) + i\mathrm{Re}\left(\frac{D_{p_x}}{D_\omega}\right)\frac{\delta p_x''}{\delta p_x'} \approx 0 \tag{9.22b}$$

Here a higher-order term $\sim \mathrm{Im}\left(\frac{D_{kx}}{D_\omega}\right)\frac{\delta p_x''}{\delta p_x'}$ is omitted. Separating the real and imaginary parts in (9.22b), we obtain:

$$v_{gx} \approx -\mathrm{Re}\left(\frac{D_{p_x}}{D_\omega}\right) \tag{9.22c}$$

Note that all this is valid when $\left|\mathrm{Im}\left(\frac{D_{p_x}}{D_\omega}\right)\right| \ll \left|\mathrm{Re}\left(\frac{D_{p_x}}{D_\omega}\right)\right|$, $|\delta p_x''| \ll |\delta p_x'|$. In the case of essential complexity (as was stated in Ginzburg 1961), the meaning of the group velocity fails.

Therefore, in this method, the trajectory is, instead of (9.8c):

$$\frac{d\vec{r}}{dt} = -\mathrm{Re}\left(\frac{D_{\vec{p}}}{D_\omega}\right) \tag{9.23}$$

It is possible to get from (9.23), instead of (9.5)

$$\frac{\mathrm{d}r}{\mathrm{d}t} = -\mathrm{Re}\left(\frac{D_{p_r}}{D_\omega}\right), \quad \frac{\mathrm{d}\theta}{\mathrm{d}t} = -\mathrm{Re}\left(\frac{D_{p_\theta}}{D_\omega}\right) \tag{9.24}$$

Note that $p_{r,\theta}$ are determined by formulas (9.1f). For example, the first formula from (9.24) can be obtained using (9.1c), (9.23), (9.11b) and

$$\begin{aligned}\frac{\mathrm{d}r}{\mathrm{d}t} &= \frac{x}{r}\mathrm{Re}\left(\frac{D_{p_x}}{D_\omega}\right) + \frac{y}{r}\mathrm{Re}\left(\frac{D_{p_y}}{D_\omega}\right) = -\mathrm{Re}\left(\frac{D_{p_x}\cos\theta + D_{p_y}\sin\theta}{D_\omega}\right)\\ &= -\mathrm{Re}\left(\frac{D_{p_r}}{D_\omega}\right)\end{aligned} \tag{9.25}$$

The second equation from (9.24) can be obtained, using (9.23), in the way, similar to the derivation of the second equation from (9.5). Consider now the derivation of the analogs of equations (9.18) and (9.19c) for the components of the momentum $p_{r,\theta}$. Let us use, again, relationships (9.15a),(9.15b) and (9.16a) and (9.16b), along with equation (9.17b). Similarly to (9.17c) and (9.18), we obtain

$$\frac{\mathrm{d}p_r}{\mathrm{d}t} = -\frac{\partial p_r}{\partial r}\mathrm{Re}\left(\frac{D_{p_r}}{D_\omega}\right) - \frac{\partial p_r}{\partial \theta}\mathrm{Re}\left(\frac{D_{p_\theta}}{D_\omega}\right) \tag{9.26a}$$

Similarly to (9.19a)–(9.19c), using (9.16b), it is possible to get

$$\frac{\mathrm{d}p_\theta}{\mathrm{d}t} = \frac{D_\theta}{D_\omega} + i\frac{\partial p_\theta}{\partial r}\mathrm{Im}\left(\frac{D_{p_r}}{D_\omega}\right) + i\frac{\partial p_\theta}{\partial \theta}\mathrm{Im}\left(\frac{D_{p_\theta}}{D_\omega}\right) \tag{9.26b}$$

As is seen from equations (9.26a) and (9.26b), there is *no* closed set of equations for p_r, p_θ! This result is rather unexpected 'beforehand'. But, when the media is independent of θ, i.e.

$$D_\theta = 0, \ \frac{\partial p_r}{\partial \theta} = \frac{\partial p_\theta}{\partial \theta} \tag{9.27a}$$

and (as follows from (9.17b))

$$\frac{\partial p_\theta}{\partial r} = \frac{\partial p_r}{\partial \theta} = 0 \tag{9.27b}$$

we get from (9.26b)

$$\frac{\mathrm{d}p_\theta}{\mathrm{d}t} = 0, \quad p_\theta = \mathrm{const} \tag{9.28a}$$

Then, we get from (9.26a), (9.27b) and (9.16a)

$$\frac{\mathrm{d}p_r}{\mathrm{d}t} = -\frac{\partial p_r}{\partial r}\mathrm{Re}\left(\frac{D_{p_r}}{D_\omega}\right) = \frac{D_r}{D_{p_r}}\mathrm{Re}\left(\frac{D_{p_r}}{D_\omega}\right) \tag{9.28b}$$

Equations (9.24), (9.28a) and (9.28b) constitute the whole set of equations for the 'new version of geometrical optics'.

9.5 Formulas of CGO for the particular system shown in figure 9.1

As was mentioned above, we consider a system shown in figure 9.1, as energy/field concentrator. The necessity to collect 'the beams', generally speaking, incident on the system from different directions, the most adequate method for modelling waves/light in the external linear inhomogeneous region shown in figure 9.1 ($R_c \leqslant r \leqslant R_0$) is CGO. Because we use the principle of 'superconcentration', after passing the linear region, a wave meets nonlinear inner region $r \leqslant R_c$. Moreover, as will be shown below, we search the new strongly nonlinear effect in this region. Therefore full-wave electromagnetic solution is used in the region $r \leqslant R_c$, where any sort of 'weak nonlinearity approximation' is not used. Therefore, we need matching of the full-wave nonlinear solution with CGO at the boundary $r = R_c$. We will concentrate on full-wave nonlinear solution and its matching with CGO in the next sections. In the present section, let us consider how to use the formulas for CGO, derived above, for the weakly inhomogeneous and weakly lossy linear region $R_c \leqslant r \leqslant R_0$,

The parameters of the medium shown in figure 9.1 do not depend on the angle θ, the dispersion equation is of the form (9.3), therefore

$$D_\theta = 0, \quad p_\theta = \text{const}, \tag{9.29a}$$

in accordance with (9.27a) and (9.28a). Respectively, we should use in the system of equations (9.24), (9.28a) and (9.28b). Derivatives included in the equations of CGO have the form, in accordance with (9.2a):

$$D_{p_r} = 2p_r - \frac{i}{r},\ D_{p_\theta} = \frac{2p_\theta}{r^2},\ D_r = \frac{ip_r}{r^2} - \frac{2}{r^3}p_\theta - k_0^2\frac{\mathrm{d}\varepsilon}{\mathrm{d}r} \tag{9.29b}$$

Therefore all 'right-hand parts' in the set of equations (9.24), (9.28a) and (9.28b) are known. To determine finally this system of equations, we need to add proper boundary (initial) conditions at the 'beginning of a ray', which correspond to the boundary of the external cylinder $r = R_0$. We are not interested in the propagation of the wave along rays in the homogeneous medium which surrounds the external cylinder (at $r > R_0$) in the approximation of geometrical optics, because such a propagation happens without any distortion, and only phase changes along a ray. Then, due to continuity of (linear) ε at the interface $r = R_0$, reflection at this interface is absent. Therefore we start our modeling at the point, where the ray incidents on the interface, and this point ('beginning' of the ray) corresponds to $t = 0$. Let us specify p_θ first. As we will see, p_θ is determined by the impact distance ρ (see figure 9.2).

Indeed, p_θ is determined by the dependence of the derivative of phase (eikonal) by the coordinate, in accordance with the second equation of (9.1a) and the first equation of (9.1f). If, at $r > R_0$ electric field is proportional to $E_z \sim \exp[i(kx - \omega t)]$, $k = \sqrt{\varepsilon}k_0$, then

$$S = kx - \omega t = kr\cos\theta - \omega t \tag{9.30a}$$

$$p_\theta = \frac{\partial S}{\partial \theta} = -kr\sin\theta = -k\rho \tag{9.30b}$$

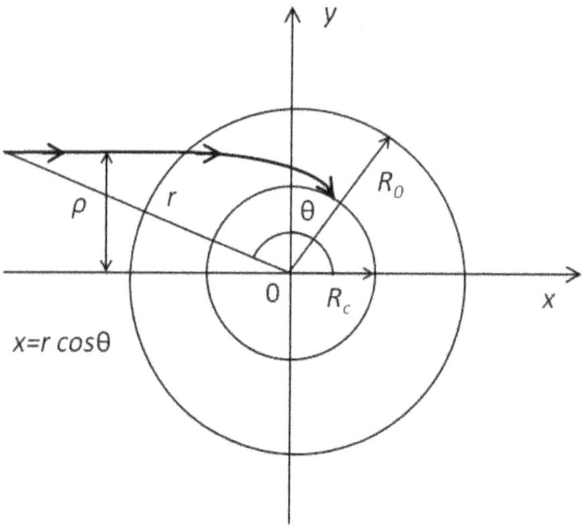

Figure 9.2. Incidence of pulse on the field concentrator; ρ is impact distance and θ azimuth component. $r \leqslant R_c$ is internal nonlinear region of the field concentrator; $R_c \leqslant r \leqslant R_0$ is external linear region of the field concentrator.

In spite of p_θ being determined on a ray in the homogeneous region ($r > R_0$) surrounding the external cylinder, in accordance with the relation (9.29a), we get the same value of p_θ also for the rays inside the external cylinder ($R_c \leqslant r \leqslant R_0$)

$$p_\theta = -k\rho = \text{const} \tag{9.30c}$$

where $\rho \equiv y(t = 0)$ (see figure 9.1) is the impact distance for the incident ray in the surrounding homogeneous linear region $r > R_0$ (figures 9.1 and 9.2). Let us consider boundary/initial conditions at $t = 0$ ($r = R_0$). We can say about 'boundary/initial' conditions for the following reason. It is quite reasonable, in the course of making numerical geometrical optics modeling for the propagation of the field along the ray inside linear inhomogeneous region $R_c \leqslant r \leqslant R_0$, to start with $t = 0$ at the 'beginning of the ray' inside this region, at $r = R_0$. Then, from the point of view of formal solution of the system of equations (9.24), (9.28a) and (9.28b) with independent variable t(time) we should have, of course initial boundary condition at $t = 0$. At the same time, we consider the propagation along the corresponding ray. From this (and only this) sense we have a right to say also about corresponding 'boundary' conditions at the point corresponding to 'the beginning' of a ray, namely $r = R_0$.

An initial condition for p_r at $t = 0$ is (see also problem 9.1)

$$p_{r_{1,2}} = \frac{i}{2R_0} - \sqrt{k_0^2\varepsilon(R_0) - \frac{p_\theta^2}{{R_0}^2} - \frac{1}{4{R_0}^2}} \tag{9.31}$$

Finally, the system of equations (9.24), (9.28a) and (9.28b) with the boundary conditions (9.31) is a closed mathematical formulation for the CGO describing the field propagation inside the internal linear cylinder shown in figure 9.1.

Let us note two important points. Firstly, the first term in relation (9.31) (see also relation (P9.1) in problem 9.1) has a clear physical sense. Namely, in accordance with equation (9.1e), the imaginary term in (P9.2) (see problem 9.1) describes the increase of amplitude (because this means the presence of imaginary part in the phase, i.e. in the eikonal) due to linear concentration of the field, when rays move towards the center of the cylindrical system. Secondly, as is seen from relation (P9.1) (see problem 9.1), the used approximation loses a sense, when $r \to 0$. But this is a natural requirement for an applicability of (complex) geometrical optics, because for the 'trajectories' with small enough radius, no approximation of 'slowly varying media' (on the scale of wavelength) can be applicable.

Let us also emphasize a possibility of using the (complex) geometrical optics approach for describing not only 'rays', but also 'beams', and we will consider below the propagation of the 'beams'. Indeed, we can easily extend the geometrical optics from only separate rays to beams. There are two ways to describe beams in geometrical optics approximation. The first is just to launch a lot of rays, to consider them in the present approximation independently, and to 'collect' the field on the 'output' of the system. In this chapter, we use the other way to consider a narrow beam. Namely, we do not 'trace' a lot of rays, from which a beam under consideration does consist. Instead, we prescribe to a beam some bell (Gauss)-like distribution of amplitude around its center, similarly to the approach of Peeters (1996). We can include in the present theory the diffraction (the 'next' approximation after geometrical optics), but would not do this, because this does not change the results qualitatively. Therefore, we use the 'aberrationless geometrical optics' (Gershman *et al* 1984, Weinberg 1962), while the field propagates along the rays without any distortion, caused by diffraction effect. Then, instead of tracing a lot of rays, we make a numerical solution and tracing *only* for the *central* ray of a beam using the method described above. Then the eikonal approach is used. Actually, we characterize an eikonal of the central ray and, as is demonstrated below, we will be able to find '*the new*' distribution of the eikonal of the beam around the central ray, using the beam narrowness. Having all characteristics of the 'central ray' and knowing the (complex) eikonal distribution in this (narrow) beam around the central ray, we finally characterize the beam on the 'output' of the system 'as a whole', and therefore the problem of describing the beam propagation in the approximation of CGO is solved in a rather elegant way. Then, having field distribution at $r = R_c$, we will be able to match this electromagnetic field with the corresponding field inside the inner nonlinear region $r \leqslant R_c$.

Let us describe some details concerning modeling a beam. First, let us consider only one, 'central' ray of a beam under consideration. We should make a note concerning an 'initial phase (eikonal)' of the central ray. Let us first introduce 'the reference plane', from which the phase will be counted, namely the plane A′A ($x = -R_0$). If we consider different beams incident on the concentrator (in other words, on the 'external cylinder surface, $r = R_0$) with different 'impact distance' (characterizing, in fact, the 'central ray' of the particular beam) their (central) rays

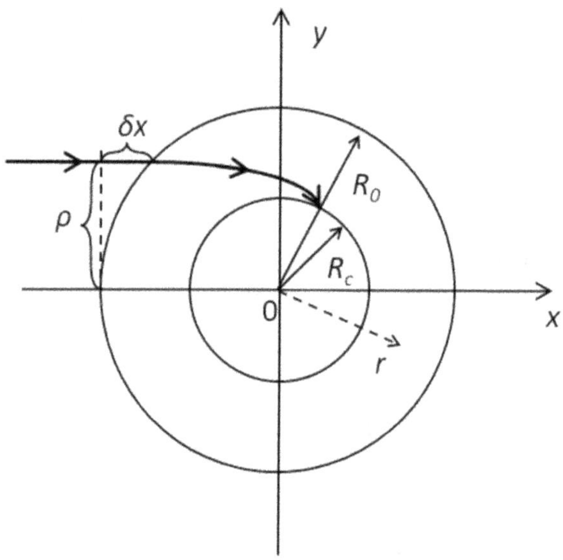

Figure 9.3. Eikonal for the central ray of the beam. $r \leqslant R_c$ is internal nonlinear region of the field concentrator; $R_c \leqslant r \leqslant R_0$ is external linear region of the field concentrator.

would have different phase, respectively, to the reference plane. As is seen from figure 9.3, such a phase (eikonal) $S(r = R_0)$ for a given central ray is equal to

$$S(r = R_0) = k\delta x = k\left(R_0 - \sqrt{R_0^2 - \rho^2}\right) \tag{9.32a}$$

Then, dropping the part kR_0 of a phase, common for all the (central) rays, we get for the 'initial eikonal'

$$S(r = R_0) \to S(r = R_0) - kR_0 = -k\sqrt{R_0^2 - \rho^2} \tag{9.32b}$$

To find the corresponding eikonal at the 'end' of the considered ray, at $r = R_c$, let us integrate the 'phase differential' given by equation (9.1g) along the ray trajectory (again, we are talking about the 'central ray' now). We consider the stationary media and stationary wave propagation in this media, so we can safely drop in the expression (9.1g) the term proportional to dt. Accounting for also (9.30c), we get finally

$$S(r = R_c) = S(r = R_0) + \int_{R_0}^{R_c} p_r \mathrm{d}r + p_\theta[\theta(r = R_c) - \theta(r = R_0)] \tag{9.32c}$$

Note that integral $\int_{R_0}^{R_c} p_r \mathrm{d}r$ is calculated along the ray trajectory.

The field distribution E_{CGO}, obtained in the CGO approximation, inside a beam ray is described at $r = R_0$ and $r = R_c$, by the following functions:

at $r = R_0$,

$$E_{\mathrm{CGO}}(r = R_0) = E_0[F_0(\theta)]_{r=R_0} \exp\left[iS(r = R_0) + ip_\theta(\theta - \theta_0)\right]; \tag{9.33a}$$

at $r = R_c$,

$$E_{CGO}(r = R_c) = E_0[F_0(\theta)]_{r=R_c} \exp\left[iS(r = R_c) + ip_\theta(\theta - \theta_c) \right] \tag{9.33b}$$

Here θ_0, θ_c are the angles θ, corresponding to the 'beginning' and 'end' of the 'central ray', at $r = R_0$ and $r = R_c$, respectively. The functions

$$[F_0(\theta)]_{\theta_{0,c}} = \exp\left[-\left(\frac{\theta - \theta_{0,c}}{\delta\theta_{0,c}} \right)^2 \right] \tag{9.33c}$$

describe the Gauss distribution of the field amplitude inside a beam around the 'central ray' at $r = R_{0;c}$, respectively with 'effective angular width' $\delta\theta_{0,c}$, respectively ($\delta\theta_{0,c} \ll \pi$, what corresponds to a narrow beam). The phases inside the exp include two terms. The first is the phase of the 'central ray', different at the 'beginning of the ray' ($r = R_0$, see formula (9.33b)) and at the 'end of the ray' ($r = R_0$, formula (9.33c)). The second term in each of the exponents describes the 'extra' phase difference across a beam relative to the central ray. This 'extra' difference of the phases arises due to difference of beam propagation direction from the local normal to the surface $r =$ const, see figure 9.4.

Because any beam under consideration is narrow, we present the corresponding extra phase as a Taylor series expansion, where we leave only the first term. At the

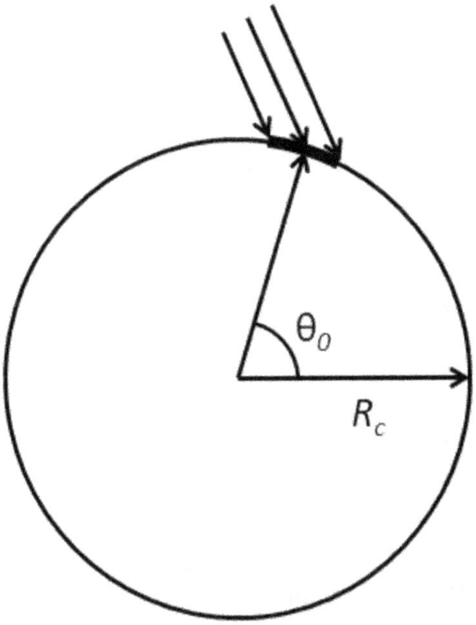

Figure 9.4. Oblique incidence (in the case $p_\theta \neq 0$) of a ray from the external linear region ($R_c \leqslant r \leqslant R_0$) on the boundary $r = R_c$ between external and internal ($r \leqslant R_c$) regions of field concentrator. The 'extra' phase difference arises due to the difference of beam propagation direction from the local normal to the surface $r =$ const. Both the central ray of the beam and two additional rays are presented.

any of the surfaces $r = R_0$ or $r = R_c$, for any 'part of a beam', radius is the same at the 'point of incidence'. Therefore, differences between corresponding phases 'inside a beam' could not be connected with the terms like $\sim p_r dr$, which could be included, in principle, in the phase differential (see equation (9.19)). Indeed, in this case $dr = 0$, because the points are lying on the same radius. Therefore, 'the extra phase' could contain only terms connected with p_θ, and only 'the lower order term' among them is left in the second phase term included in exponents in equations (9.33a) and (9.33b). Note also, that in the case of beam incidence, normal to the surface $r = \text{const}$ ($r = R_c$), in other words, for beam, propagating along a radius, we would get $d\theta/dt = 0$. Respectively, in accordance with (9.24), (9.3) and (9.30c) $p_\theta = 0$ and $\rho = 0$ (see figure 9.1). In this case 'extra phase difference', equal to $p_\theta(\theta - \theta_c)$ (see (9.33b)) would also disappear. Note that this is possible only for a beam with central ray going along the radius 'from the very beginning' (already when it incidences from the region $r > R_0$), otherwise we would not get $\rho = 0$ (the case $\rho \neq 0$ is shown in figure 9.1).

We will use the following approximation: that the angle distributions are the same for the field distributions inside a beam at $r = R_0$ and $r = R_c$ (see equations (9.33a) and (9.33c)), so

$$\delta\theta_c = \delta\theta_0 \tag{9.33d}$$

Indeed, effective angle width is the same for the beams at $r = R_0$ and $r = R_c$ and the value p_θ is constant along any given ray, including 'central ray of a beam', for which a value p_θ, included in equations (9.33a) and (9.33c) is determined. This is connected with the absence of any aberration/distortion and using diffractionless approximation (Peeters 1996). Note that the absence of any distortion, reflected by equations (9.33a) and (9.33b), was checked directly in the special numerical experiment, for our medium with particular choice the dependence of $\varepsilon(r)$. In this experiment, couples of rays were launched at $r = R_0$ under some small angle to each other. It was checked, that such an 'angle inclination' was the same (with a large accuracy) for the corresponding couple of rays at $r = R_c$, as it was at $r = R_0$. This approach and the results of the numerical experiment correspond to the aberrationless approximation, in particular for Gaussian beams (Prokhorov 1988, Balakina *et al* 2008).

Finally, relation (9.33b) allows one to match the field at both sides of the interface $r = R_c$. In the next sections, we will consider the way of presentation of full-wave nonlinear solution in the region $r \leqslant R_c$ and mention a method of treatment of the obtained set of coupled equations for a lot of nonlinearly interacting electromagnetic modes.

9.6 Electromagnetic field inside an internal nonlinear region $r \leqslant R_c$

The structure under investigation (figure 9.1) provides 'superfocusing', as a result of sequential linear and nonlinear focusing in the external and internal cylindrical regions, respectively, even with a possibility of 'hot spot' formation, as will be shown below. Due to such a strong focusing, instead of ordinary Ker nonlinearity, the saturating Kerr nonlinearity is used, to provide nonlinear field convergence, in the region $r \leqslant R_c$. The corresponding nonlinear the electric permittivity has the form

$$\delta\varepsilon_{\text{NL}} = \alpha|E|^2/(1 + \beta|E|^2) \tag{9.34}$$

(see also figure 9.1), where α, β are the constants. Inclusion of the saturation (characterized by β) is important, because we expect that 'superfocusing' will lead to a strong energy concentration. Inside the nonlinear cylinder $r \leqslant R_c$, where the full-wave field is described by equations (9.1b) and (9.1c), the solution is searched in the form

$$E_z = \sum_{m=-N/2+1}^{N/2} A_m(r)\mathrm{e}^{-im\theta} \tag{9.35}$$

(N is the number of the Fourier modes accounted for). The nonlinear part of dielectric constant (see relationship (9.34)) is a function of all nonlinear harmonic amplitudes A_m. At the same time, $\delta\varepsilon_{\text{NL}} = \delta\varepsilon_{\text{NL}}(r, \theta)$ can be presented, for any given r, in the form of expansion into Fourier series by angle harmonics, $\exp(-il\theta)$

$$\delta\varepsilon_{\text{NL}}(r, \theta) = \sum_l \delta\varepsilon_l(r)\mathrm{e}^{-il\theta}, \tag{9.36}$$

while coefficients $\delta\varepsilon_l(r)$ are nonlinear and depend on all Fourier field harmonic amplitudes A_m. The nonlinear wave equation for the amplitude of mth harmonic takes the form

$$\frac{1}{r}\frac{\mathrm{d}}{\mathrm{d}r}\left(r\frac{\mathrm{d}A_m}{\mathrm{d}r}\right) - \frac{m^2}{r^2}A_m + k_0^2\varepsilon(r)A_m + k_0^2\sum_l \delta\varepsilon_l(r)A_{m-l} = 0 \tag{9.37a}$$

or

$$\frac{1}{r}\frac{\mathrm{d}}{\mathrm{d}r}\left(r\frac{\mathrm{d}A_m}{\mathrm{d}r}\right) - \frac{m^2}{r^2}A_m + k_0^2[\varepsilon(r) + \sum_l \delta\varepsilon_l(r)]A_m + k_0^2 \sum_{\substack{l=-N/2+1 \\ (l\neq 0)}}^{N/2} \delta\varepsilon_l(r)(A_{m-l} - A_m) = 0 \tag{9.37b}$$

Briefly, equations (9.37a) and (9.37b) can be presented in the form

$$\frac{1}{r}\frac{\mathrm{d}}{\mathrm{d}r}\left(r\frac{\mathrm{d}A_m}{\mathrm{d}r}\right) - \frac{m^2}{r^2}A_m + k_0^2\varepsilon(r)A_m + k_0^2F_m(r) = 0;\; m = -(N/2 - 1), \ldots, N/2 \tag{9.37c}$$

where the components $F_m(r)$ include all nonlinear terms,

$$F_m(r) = \sum_l \delta\varepsilon_l(r)A_{m-l} = \sum_l \delta\varepsilon_l(r)A_m + \sum_{\substack{l=-N/2+1 \\ (l\neq 0)}}^{N/2} \delta\varepsilon_l(r)(A_{m-l} - A_m) \tag{9.37d}$$

and the system of equations (9.37c) describes the effects of both self- and cross-interactions for all the modes $A_m(r)$.

Boundary conditions at $r = 0$ have the form that is typical for the cylindrical functions. In other words, peculiarities of the modes at $r \to 0$ even in nonlinear case still reflect the peculiarities of the cylindrical functions of the corresponding orders:

$$\frac{dA_0}{dr}\,|_{r=0} = 0, \quad A_m\,|_{r=0} = 0\ (m \neq 0) \tag{9.38}$$

To show this, it is enough to rewrite equation (9.37b) in the form

$$r^2\frac{\mathrm{d}^2A_m}{\mathrm{d}r^2} + r\frac{\mathrm{d}A_m}{\mathrm{d}r} - m^2A_m + r^2{k_0}^2[\varepsilon(r)A_m + F_m(r)] = 0 \tag{9.39}$$

and to consider the asymptotics for

$$r \to 0 \tag{9.40a}$$

We are looking for the solution with all $A_m(r)$(and, respectively, $F_m(r)$) finite for $r \to 0$. The analysis of corresponding asymptotics, finite at $r \to 0$, shows that (see more details in section (10.2) of chapter 10 including problem 10.1).

$$\frac{dA_0}{dr}\,|_{r\to 0} \to \frac{d\zeta_0}{dr}\,|_{r=0} = 0; \; A_m\,|_{r\to 0} \to \zeta_m\,\big|_{r=0} = 0\ (m \neq 0) \tag{9.40b}$$

Asymptotic relations (9.40b) prove the validity of the boundary conditions (9.38). To determine entirely the nonlinear full-wave solution of the set of equations (9.39), we need to add to the conditions (9.38), proper boundary conditions at the boundary $r = R_c$ of the internal (nonlinear) and external (linear) cylindrical regions (figure 9.1). The procedure of matching full-wave nonlinear solution with CGO solution implies the derivation of the proper boundary conditions at $r = R_c$. Such a procedure is described in the next section.

9.7 Matching 'full-wave' and 'CGO' solutions and possible applications

Before describing the matching of CGO and full-wave solutions, let us make the following note. Of course, CGO and corresponding boundary conditions obtained using this method, which are the subject of the present section, include some approximation. On the other hand, even if a purely numerical method is used such as finite element method (Jian-Ming 2014), approximate boundary conditions are needed such as, for example, perfectly matched layer (PML) conditions (Gedney 1996). We are sure that the new method described in the present chapter, which includes CGO, full-wave strongly nonlinear solution and their matching, is the most adequate for the modeling of field/energy concentrator (shown, in particular in figure 9.1). Because we want to collect energy/rays incident on the concentrator from different sides, we use first in linear and homogeneous external cylindrical region ($R_c < r < R_0$) CGO. Then, because we want to provide very strong 'superconcentration', we add nonlinearity in the internal region ($r < R_c$), and, moreover, consider non-weak nonlinearity and get, finally new and strongly nonlinear effect of 'non-linear focusing switching' and 'hot spot formation' (see chapter 10). Then, the necessity of full-wave nonlinear solution in the internal cylinder ($r < R_c$) and its matching with CGO solution is obvious. The advantage of the present method is, again that it is ideally adequate for the considered problem on 'superconcentration' and, at the same time, very physically intuitive. As a result of its application, we get the new physical effect, described in chapter 10.

Let us derive the approximate boundary conditions for matching CGO and full-wave solution at the interface $r = R_c$. Note that we will account for both incidence and reflection of the waves incident on the boundary $r = R_c$ from the side of the linear region $R_c < r < R_0$. To do this, we use the presentation of the field in the incident beam in the close vicinity of the boundary $r = R_c$ inside the region $R_c < r < R_0$, which is denoted as $r = R_c + 0$. We present such a field as a Fourier set by angular harmonics $\sim\exp^{(-im\theta)}$ and, at the same time, apply quasi-planar approximation for the incident and reflected waves.

We could formulate the problem as follows. We know the incident wave from the geometrical optics solution (considered in sections 9.3–9.5). Then, we will introduce the waves, reflected from the boundary $r = R_c$ back to the region $R_c < r < R_0$ and obtain boundary conditions, matching of incident and reflected waves in the region $R_c < r < R_0$ with amplitudes of angular harmonics inside the region $r < R_c$ (see section 9.6). We should also note that the geometrical optics deals with propagation of beam along the ray trajectories, while the full-wave solution in section 9.6 can be interpreted as 'presentation of the wave propagation through Fourier angular harmonics with amplitudes dependent on radius' inside the inner cylinder. To match both types of the solution on the interface $r = R_c$, let us transform the field presentation in the region $r = R_c + 0$ (vicinity of the external cylinder near the interface $r = R_c$) from the 'presentation of field along the ray' also to the presentation 'through Fourier angular harmonics with amplitudes dependent on radius'. For the implementation of this technique, first of all, we identify an incident wave with the field of the incident beam, given by formulas (9.33b) and (9.32c). Because we need to match the fields at $r = R_c$, let us present the corresponding field obtained from geometrical optics approximation (see equations (9.33b) and (9.32c)) as a set of Fourier angular harmonics. For convenience, we include 'constant' part of the phase (i.e. not dependent on θ), namely $[S(r = R_c) - p_\theta \theta_c]$ into complex amplitude E_0. As a result, we get for incident field E_{inc}

$$E_{\text{inc}}(r = R_c) = E_{\text{CGO}}(r = R_c) = E_{00} \exp\left[-\left(\frac{\theta - \theta_c}{\delta\theta_c}\right)^2\right] \exp(ip_\theta \theta)$$

$$= \sum_{m=-\frac{N}{2}+1}^{N/2} E_{im}(r = R_c) e^{-im\theta} \tag{9.41a}$$

$$\text{where } E_{00} = E_0 \exp\{i[S(r = R_c) - p_\theta \theta_c]\} \tag{9.41b}$$

Here the index 'inc' denotes the incident field, E_{im} is an amplitude of mth harmonic of the incident field, N is the number of Fourier harmonics accounted for (as was already mentioned before). Therefore, incident harmonic amplitudes can be determined from equation (9.41a) with given input amplitude $E_{00}(E_0)$.

An idea of matching CGO and full-wave solution (and derivation of corresponding boundary conditions) at $r = R_c$, is based on the 'local quasi-planar' approximation for the cylindrical Fourier modes at $r = R_c + 0$ (in the external cylinder near the interface between the regions). Using such an approximation, we can write *the same* field (9.41a), obtained before 'from CGO', again, but now from the approximate 'quasi-planar presentation' of full-wave equation (9.1b). Indeed, the quasi-planar approximation full-wave solution for the field can be presented in the form

$$E(r, \theta) = \sum_{m=-\frac{N}{2}+1}^{N/2} \left\{ E_{im}^{(\mathrm{EM})}(r = R_c)\mathrm{e}^{-im\theta+ik_{im}(r-R_c)} + E_{rm}^{(\mathrm{EM})}(r = R_c)\mathrm{e}^{-im\theta+ik_{rm}(r-R_c)} \right\} \tag{9.42}$$

Here $E_{im,\,rm}^{(\mathrm{EM})}$ are amplitudes of mth field harmonics of incident and reflected (on and from) the interface $r = R_c$, respectively, based on 'full-wave' (electromagnetic) quasi-planar approximation. Excluding an amplitude of the reflected wave, we obtain, finally, effective boundary conditions at $r = R_c$ in the form (see also problems 9.2 and 9.3):

$$A_m + \frac{i}{k_{rm}}\frac{\mathrm{d}A_m}{\mathrm{d}r} = \left(1 - \frac{k_{im}}{k_{rm}}\right)E_{im}(r = R_c) \tag{9.43}$$

In equation (9.43),

$$k_{im,\,rm} = \frac{i}{2R_c} \pm \sqrt{k_0^2\varepsilon(r = R_c) - \frac{m^2}{R_c^2} - \frac{1}{4R_c^2}} \tag{9.44}$$

The system of equation (9.37c), altogether with boundary conditions (9.38) and (9.43) concludes the formulation of the problem for full-wave nonlinear solution in the inner cylindrical region $r \leqslant R_c$, while for an incident amplitude we can find an amplitude $E_{im}(r = R_c)$ is determined through the input amplitude $E_{00}(E_0)$, using equation (9.41a).

Then, the set of equations (9.37a) for the nonlinear Fourier modes A_m, with obtained boundary conditions is solved by means of especially developed multimode nonlinear spectrum iteration-sweep method (MNSISM).

9.8 Superfocusing combining linear and nonlinear media to create new forms of energy capture and field concentration

The results of nonlinear wave diffraction and interference in the inner cylinder ($r \leqslant R_c$) are shown, using two equal amplitude incident beams as an example, in chapter 10, section 10.2.

9.9 Conclusions

A new method is proposed for modeling nonlinear energy concentrator structure, where the inner homogeneous region is nonlinear. This method includes matching of complex geometrical optic and full-wave solutions. For the structures under consideration, the 'nonlinear superfocusing' is used, which includes both linear and nonlinear focusing. The calculations show a possibility of strong field focusing and even an occurrence of 'hot spots'. It is also shown that the new effects of nonlinear switching of focusing and hot spot formation happen, when input amplitude exceeds some threshold value. The numerical estimations adopt the doped *n*-Si. The value of the corresponding cubic permittivity $\chi^{(3)}$ is of order of 10^{-7}–10^{-8} e.s.u (Sutherland 1996). The electromagnetic wave (with intensity of order ~1–5 GW cm^{-2}) is necessary for the observing the nonlinear phenomena in the field concentrator with the internal nonlinear region of the radius $R_c = 10$ μm. Respectively, the intensity of the electromagnetic wave incident on the external cylinder of the field concentrator should be of order (100–500) MW cm^{-2}. Applications based on electromagnetic or acoustic metamaterial structures may be prospective for nonlinear versions of energy harvesting systems, such as solar cells, subwavelength imaging, antenna-based sensing including field-enhanced microscopy, systems for feeding optical fibers, frequency mixing and high-harmonic generation, new types of optical nonlinear switches, reduction of (strong) acoustic noise. The proposed method which includes matching of full-wave nonlinear solution and CGO is rather general and, besides electromagnetic and acoustic metamaterials, can be used for modeling 'natural' inhomogeneous, active and strongly nonlinear media, such as, for example, laboratory and space plasma (Tao and Bortnik 2010, Demekhov *et al* 2003). The developed technique may be useful for active space experiments (Sauvaud *et al* 2005), where the processes of wave coupling in the nonlinear system 'magnetosphere–ionosphere' will be studied, while spatial scales in different regions of a system differ by many orders of magnitude and matching of CGO and full-wave electromagnetic solutions is very useful. Details concerning nonlinear superfocusing in acoustic metamaterials and anisotropic metamaterials are the subject of strong interest for future investigation.

List of abbreviations

CGO	Complex geometrical optics
MNSISM	Multimode nonlinear spectrum iteration-sweep method
PML	Perfectly matched layer

Problems to chapter 9

Problem 9.1

Obtain the boundary condition for the system of equations (9.24), (9.28a) and (9.28b).

Take into account the explanation after equation (9.30c) concerning the 'boundary/initial condition'. Then let us write initial conditions.

At $t = 0$,

$$r = R_0,\ \theta = \pi - \arcsin\left(\frac{\rho}{R_0}\right) \equiv \pi - \sin^{-1}\left(\frac{\rho}{R_0}\right) \tag{P9.1}$$

(see also figures 9.1 and 9.2). To find the initial condition for p_r, let us first write the general expression for p_r, using equation (9.3):

$$p_{r_{1,2}} = \frac{i}{2r} \pm \sqrt{k_0^2\varepsilon(r) - \frac{p_\theta^2}{r^2} - \frac{1}{4r^2}} \tag{P9.2}$$

Accounting for (9.1b), and that we are interested in the wave propagating toward the center of the system (figure 9.1), we conclude that the pulse is p_{r_2} with sign '−' in equation (P9.2). Respectively, obtain the relation (9.31).

Problem 9.2

Obtain effective wavenumber $k_{im,rm}$ of the incident (onto the field concentration) and reflected (from the field concentrator) quasi-planar harmonics at $r = R_c$.

Put the field (9.42) into the 'full-wave' electromagnetic equation (9.2a). This yields equation

$$k_{im,rm}^2 - \frac{ik_{im,rm}}{r} + \frac{1}{r^2}m^2 - k_0^2\varepsilon(r = R_c) = 0 \tag{P9.3}$$

Obtain the wavenumbers $k_{im,rm}$ as the roots of the quadratic equation (P9.3).

Problem 9.3

Obtain the system of two equations for continuity of tangential components of electric and magnetic field, in quasi-planar approximation, at $r = R_c$. Then, excluding from the obtained system of equations, an amplitude of reflected wave, obtain the boundary condition at $r = R_c$, which connects an amplitude of incident wave and its derivative by radius.

As follows from equation (P9.3),

$$k_{im,rm} = \frac{i}{2R_c} \pm \sqrt{k_0^2\varepsilon(r = R_c) - \frac{m^2}{R_c^2} - \frac{1}{4R_c^2}} \tag{P9.4}$$

The choice of sign '−' in equation (P9.4) for incident wave corresponds to the propagation of this wave to the 'side of decreased r'; the choice of sign '+' for reflected wave corresponds to the propagation of this wave to the 'side of increased r'. The imaginary part of $k_{im,rm}$ corresponds to imaginary part of complex eikonals, describing increasing of amplitude, while the waves propagate towards the center of the system, in the direction, where dielectric permittivity increases. Then, equating incident angular harmonic amplitudes, we get

$$E_{im}^{(EM)}(r = R_c) = E_{im}(r = R_c) \tag{P9.5}$$

Matching the fields (9.35) at $r = R_c - 0$ with the field (9.42a) in the region $r = R_c + 0$ (i.e. at the interface $r = R_c$), basing on the continuity of the E and $\frac{dE}{dr}$ (which corresponds to the continuity of the electric and magnetic field components), obtain the system of equations

$$E_{im}^{(EM)}(r = R_c) + E_{rm}^{(EM)}(r = R_c) = A_m \tag{P9.6}$$

$$ik_{im}E_{im}^{(EM)}(r = R_c) + ik_{rm}E_{rm}^{(EM)}(r = R_c) = \frac{dA_m}{dr} \tag{P9.7}$$

Finally, excluding an amplitude of reflected wave from the set of equations (P9.6) and (P9.7), get the boundary condition in the form (9.43).

References

Balakina A A, Balakin M A, Permitin G V and Smirnov A I 2008 Influence of dissipation on propagation of wave beams in inhomogeneous, anisotropic and gyrotropic media (in Russian) *Phys. Plasmas* **34** 533–47

Berczynski P 2011 Complex geometrical optics of nonlinear inhomogeneous fibres *J. Opt.* **13** 035707

Bergman L, Alitalo P and Tretyakov S A 2011 Nonlinear transformation optics and engineering of the Kerr effect *Phys. Rev.* B **84** 205103

Boardman A D, Grimalsky V V, Kivshar Y S, Koshevaya S V, Lapine M, Litchinitser N M, Malnev V N, Noginov M, Rapoport Y G and Shalaev V M 2011a Active and tunable metamaterials *Laser Photonics Rev.* **5** 287–307

Boardman A, Egan P, Mitchell-Thomas R and McCall M 2011b Integrated linear and nonlinear metaphotonics *Proc. Metamaterials' 2011 Fifth Int. Cong. on Advanced Electromagnetic Materials in Microwaves and Optics (Barcelona, 10–15 October 2011)*

Bravo-Ortega A and Glassor A H 1991 Theory and application of complex geometric optics in inhomogeneous magnetized plasmas *Phys. Fluids* B **3** 528–38

Demekhov A G, Trakhtengerts V Y, Mogilevsky M M and Zelenyi L M 2003 Current problems in studies of magnetospheric cyclotron masers and new space project 'Resonance' *Adv. Space Res.* **32** 355–74

Gedney S D 1996 An anisotropic perfectly matched layer-absorbing medium for the truncation of FDTD lattices *IEEE Trans. Antennas Propag.* **44** 1630–9

Gershman B N, Erykhimov L M and Yashin Y Y 1984 *Wave Phenomena in the Ionosphere and Space Plasma (in Russian)* (Moscow: Nauka)

Gilarlue M M and Hadi B S 2019 Photonic crystal waveguide crossing based on transformation optics *Opt. Commun.* **450** 308–15

Ginzburg V L 1961 *Propagation of Electromagnetic Waves in Plasma* (Gordon: Breach Science)

Jian-Ming J 2014 *The Finite Element Method in Electromagnetics* (New York: Wiley)

Kildishev A V and Shalaev V M 2011 Transformation optics and metamaterials *Usp. Fiz. Nauk* **181** 59–70 (English Transl.: 2011 *Uspekhi Fizicheskih Nauk* **54**)

McCall M, Pendry J B, Galdi V, Lai Y, Horsley S A R, Li J, Zhu J, Mitchell-Thomas R C and Quevedo-Teruel O 2018 Roadmap on transformation optics *J. Opt.* **20** 063001

Miller M A, Sorokin Y M and Stepanov N S 1977 Covariance of Maxwell equations and comparison of electrodynamic systems *Sov. Phys. Usp.* **20** 264–72

Narimanov E E and Kildishev A V 2009 Optical black hole: Broadband omnidirectional light absorber *Appl. Phys. Lett.* **95** 041106

Peeters A G 1996 Extension of the ray equations of geometric optics to include diffraction effects *Phys. Plasmas* **3** 4386–95

Prokhorov A M 1988 *Physical Encyclopedia in 5 Volumes (in Russian)* (Moscow: Soviet Enciclopedia)

Rahm M, Schurig D, Roberts D A, Cummer S A, Smith D R and Pendry J B 2008 Design of electromagnetic cloaks and concentrators using form-invariant coordinate transformations of Maxwell's equations *Photonics Nanostruct. Fundam. Appl.* **6** 87–95

Rapoport Y G, Boardman A D, Grimalsky V V, Ivchenko V M and Kalinich N 2014 Strong nonlinear focusing of light in nonlinearly controlled electromagnetic active metamaterial field concentrators *J. Optics* **16** 055202

Sauvaud J-A, Fedorov A O, Pichkhadze K M and Goroshkov I N 2005 Perspectives of the studies of wave-particle interactions in the inner magnetosphere: RESONANCE project *Geophys. Res. Abstracts* **7** 05599

Sklan S R and Li B 2018 A unified approach to nonlinear transformation materials *Sci. Rep.* **8** 4436

Sutherland R L 1996 *Handbook of Nonlinear Optics* (New York: Marcel Dekker)

Tao X and Bortnik J 2010 Nonlinear interactions between relativistic radiation belt electrons and oblique whistler mode waves *Nonlin. Processes Geophys.* **17** 599–604

Tikhonov A N and SaMarskii A A 1977 *Equations of Mathematical Physics (in Russian)* (Moscow: Nauka (Science))

Timofeev A V 2005 Geometrical optics and diffraction phenomena *Uspekhi Fizicheskih Nauk* **175** 637–41 (in Russian) (English Translation: 2005 *Uspekhi Fizicheskih Nauk* **48**)

Weinberg S 1962 Eikonal method in magnetohydrodynamics *Phys. Rev.* **126** 1899–909

Yang W, Jichun L, Yungqing H and He B 2019 Developing finite element methods for simulating transformation optics devices with metamaterials *Commun. Comput. Phys.* **25** 135–54

Zhang J, Pendry J B and Luo Y 2019 Transformation optics from macroscopic to nanoscale regimes: A review *Adv. Photonics* **1** 014001

Yuriy Rapoport and Vladimir Grimalsky

Chapter 10

Wave processes in controlled and active metamaterials and plasma-like media in the presence of resonance and strong nonlinearity

Strong field localization is used in nonlinear and plasma resonant and active media, including nanoplasmonic (Krasavin *et al* 2019), superconducting (Schalch *et al* 2019), hyperbolic (Bhardwaj *et al* 2020) metamaterials (MMs), and ultralight MMs. based on carbon aerogels (Xie *et al* 2018), metasurfaces (Pertsch and Kivshar 2020), and such applications include microwave (Ruisheng Yang *et al* 2021, Siddiqui and Mohra 2017), THz (Yang *et al* 2021, Schalch *et al* 2019, Qu *et al* 2020) and optical (Pertsch and Kivshar 2020, Krasavin *et al* 2019) frequency ranges. Such approaches to the physics of highly nonlinear MMs and plasma-like media pave the ways for applications in the form of ultrasensitive sensors, light control using light, built-in optical data processing, optical communication networks and laser applications (Krasavin *et al* 2019), nonlinear meta-devices, on metasurfaces with spatially controlled nonlinear fields such as nonlinear deflectors, nonlinear vortex beam generators, nonlinear metalens and nonlinear holographic devices (Pertsch and Kivshar 2020), etc.

This chapter discusses wave interactions and self-action with strong and controlled nonlinearity, including strongly nonlinear controlled active and resonant plasma-like and MM media.

10.1 Conditions for transition to the mode of strong nonlinearity during the generation of a giant localized surface plasmonic second harmonic

In chapter 4 we considered the interaction of the first and the second harmonics of the surface plasmons in the system semi-infinite plasma–vacuum (or dielectric),

doi:10.1088/978-0-7503-2336-9ch10

figure 4.1. It is shown that the most interesting effects are in the case when the second harmonic is near the plasma resonance. The consideration is within the framework of equation (4.1). The new effect called the giant generation of the second harmonic is pointed out. The nonlinearity, both quadratic and cubic, is assumed as moderate there. In this section the situation is considered when the condition of the moderate nonlinearity of surface plasmons ceases to be valid.

Figure 10.1 presents the dynamics of the main and second harmonics in the presence of almost constant pumping at the input to the system, when $x = 0$ (namely, the input signal/fundamental harmonic in the shape of a very long pulse). Relevant input data are shown in the captions to the figure. Figure 10.1 is given for the case when both cubic and quadratic nonlinearities are present, and the sign of the total nonlinearity coefficient corresponds to the sign of the surface nonlinearity. A very large resonant second harmonic is obtained (figure 10.1). As can be seen from the comparison of figure 10.1 with figures 4.2(a)–(d), increasing (with other equivalent input parameters) the input amplitude by only 25% leads to an increase in the ratio of the maximum intensities of the second and first harmonics by about two times. Note that the calculations presented in figure 10.1, already, strictly speaking, go beyond the conditions of real smallness of at least one of the small parameters, described in appendix B.2.2 in chapter 3, namely (χ_2), which determines the relative value of the second harmonic of the plasma concentration. Therefore, figure 10.1 is in fact an illustration of the trend of transition of the generation of the giant second harmonic (considered in section 3.2 of chapter 3 in the mode of weak nonlinearity) to the mode of strong resonant nonlinearity. Relevant calculations, strictly speaking, should be performed not within the NEELS method, but with another method that does not require the condition of moderation of nonlinearity and which is beyond the scope of this work.

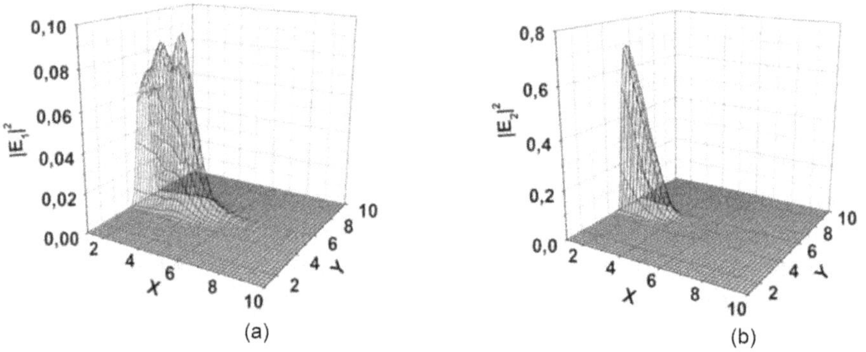

Figure 10.1. Dependence of the normalized amplitude of the first (a) and second (b) harmonics on the normalized coordinates x, y for the value of the normalized time $T = 4.5$, which has passed since the beginning of the action of the pump pulse (fundamental, or main, harmonic) at the input ($x = 0$) There are both quadratic and cubic nonlinearities. The pump field (main harmonic at the input to the system) at $x = 0$ is a step-type function that decreases so slowly over time that it can be considered approximately stationary; $E_1(x = 0) \approx \text{const} = 0.25$, (a) and (b) are the normalized frequency detuning (from frequency $\omega_p/2$ for the main harmonic) and the decrement decay for the first and second harmonics $(-\Delta\omega) = \gamma_{1,\,2} = 0.1$.

10.2 Nonlinear electromagnetic waves in metamaterial field concentrators

The strong field concentration in a nonlinear MM electromagnetic 'black hole' (the inhomogeneous mesoscale MM field concentrator with combined linear–nonlinear focusing, figure 10.2) was studied. The problem of the concentration of electromagnetic energy in the infrared range is very important. A linear MM infrared concentrator has been proposed in Narimanov and Kildishev (2009). It was shown that by applying the principles of (MM) transformation optics, using an analogy with the Kepler's problem, it is possible to create an optical concentrator of a linear field for incident rays with arbitrary directions of propagation. This section presents the main results of a series of works (Boardman *et al* 2011, Rapoport and Grimalsky 2011, Rapoport *et al* 2012a, 2012b, 2013d, 2014b), where a new type of MM concentrators is proposed and studied namely 'nonlinear electromagnetic black holes'. An attempt (Bergman *et al* 2011) to develop an approach of completely nonlinear transformation optics is limited by the case of a weak nonlinearity. The theory (Bergman *et al* 2011) is unsuitable for new and very interesting applications of the strongly nonlinear phenomenon of the focusing jump and hot spot formation (Boardman *et al* 2011, Rapoport and Grimalsky 2011, Rapoport *et al* 2012a, 2012b, 2014a, 2014b). The reason is that in our case the inverse problem and the determination of equivalent electromagnetic (dielectric permittivity and/or magnetic permeability), which are integral parts of the method of transformation optics based on a weak nonlinearity, has no solution, in the general case of considerable (moderate or significant) nonlinearity. Therefore, the work uses a solution obtained

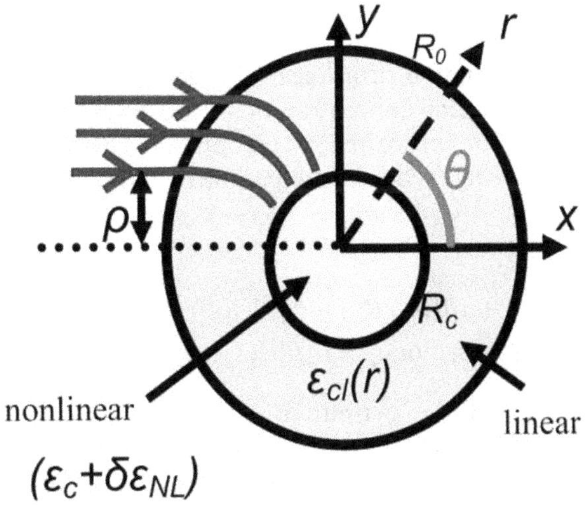

Figure 10.2. (r, θ)—Cross-section of a cylindrical nonlinear electromagnetic field concentrator. For convenience, figure 9.1 from chapter 9 is repeated here. The linear outer cylinder occupies the area $R_c \leqslant r \leqslant R_0$. The nonlinear inner cylinder occupies the region $r \leqslant R_c$. The three rays incident on the system are shown. The impact parameter (impact distance) for the beam falling on the system is denoted as ρ. Reprinted from Rapoport et al (2014b), copyright IOP Publishing, all rights reserved.

on the basis of the principle of the linear transformation optics in fact as a zero (linear) approximation.

The structure (figure 10.2), built on this principle, is considered. A heterogeneous, external, linear environment ($R_c \leqslant r \leqslant R_0$, figure 10.2) used for the pre-field concentration, and a homogeneous inner region with focusing nonlinearity ($r \leqslant R_c$, figure 10.2) used for the final capture and strong concentration of the electromagnetic field. Figure 10.2, shows a cross-section of a cylindrical concentrator used in this study. The coordinate system is defined as (r, θ, z), and the axis of the cylinder coincides with the axis z. The linear outer area ($r > R_0$) includes a homogeneous medium with a relative dielectric permittivity ε_0, the inner core ($r \leqslant R_c$) is both nonlinear and homogeneous (in the absence of nonlinearity). The additional nonlinear dielectric permittivity is present in the inner core $\delta\varepsilon_{NL}$, it is saturated, which is shown in figure 10.2. The outer cylindrical shell, ($R_c \leqslant r \leqslant R_0$) is both inhomogeneous and linear. The heterogeneous permittivity in the area $R_c \leqslant r \leqslant R_0$ increases with decreasing radius, thereby providing light capturing through a gradient of refractive index (Narimanov and Kildishev 2009). To simulate the propagation of waves in such media, the geometric optics is used (Gershman *et al* 1984). In our work, we developed a new version of the complex geometric optics (CGO), different from previously known versions of the corresponding method (see, for example, Bravo-Ortega and Glassor 1991, Berczynski 2011). In our version, the complex eikonal describes both the amplitudes and phases of the electric field for a given beam and, moreover, as explained below, can be considered to include even the change of amplitude due to change of the area of a cross-section of a beam at its distribution. In figure 10.2 is shown schematically the case when three rays fall onto the concentrator. Considered is the (H_r, H_θ, E_z) polarization of the incident wave.

To realize the maximum concentration of the electromagnetic energy (or the high field intensity in the focusing area) in the absence of reflections (at least in the linear approximation), a combination of a linear concentration in the external gradient medium with the subsequent concentration in the internal inhomogeneous and homogeneous (in the linear approximation) is optimal. This achieves a very effective focusing, i.e. 'superfocusing'. The electric permittivity is chosen for this purpose in the form

$$\varepsilon(r) = \begin{cases} \varepsilon_0, \ r > R_0 \\ \varepsilon_0(R_0^2/r^2) + i\gamma[(R_0^2/R_c^2) - 1]^{-1}[(R_0^2/r^2) - 1], \ R_c < r < R_0 \\ \varepsilon_c + \delta\varepsilon_{NL} + i\gamma; \ \delta\varepsilon_{NL} = \alpha|E|^2/(1 + \beta|E|^2), \ r < R_c \end{cases} \tag{10.1}$$

where the dependence of the permittivity is in the form $\varepsilon_c = \varepsilon_0(R_0^2/R_c^2)$ that guarantees the realization of the condition that when $r \to R_c$, the dielectric permittivity is continuous (and it is continuous in the system as a whole); the nonlinear part is taken in the form of the Kerr nonlinearity, but with saturation, and α, β are constant values (depending on the material), γ is a value of dissipative losses. The inclusion of saturation (characterized by the value β) is important because we expect superfocusing to result in a strong concentration of energy. The transition to that previously considered for a linear concentrator electric permittivity

(Narimanov and Kildishev 2009) occurs when we put $\gamma = 0$, $\delta\varepsilon_{\rm NL} = 0$. In addition to the continuity of the electric permittivity in the linear approximation, the small reflection of the incident waves is guaranteed by the fact that the nonlinearity is relatively smoothly 'switched on' from the boundary $r = R_c$ between the linear and nonlinear domains. In the linear domain $R_c \leqslant r \leqslant R_0$ the developed method of CGO is used. In the inner area $r \leqslant R_c$ a 'full-wave' solution of the Maxwell nonlinear equations is provided. Special boundary conditions are derived for 'sewing together' both solutions (i.e. satisfying the corresponding boundary conditions) at the boundary $r = R_c$. Each of the beams incident onto the structure of a nonlinear concentrator is presented by a central beam with some angular distribution of the field around the 'center' of such a beam. In order to present a new method of calculation, consider the transverse electric mode (H_r, H_θ, E_Z) and accordingly the 'full-wave' equation, namely

$$\frac{1}{r}\frac{\partial}{\partial r}\left(r\frac{\partial E_z}{\partial r}\right) + \frac{1}{r^2}\frac{\partial^2 E_z}{\partial \theta^2} + k_0^2\varepsilon(r)E_z = 0 \qquad (10.2)$$

where $k_0 = \omega/c$, ω is the angular frequency, c is the speed of light in a free space. According to the CGO method, the electric field is represented as $E_z = Ae^{iS(r,\,\theta)}$, in the area $R_c \leqslant r \leqslant R_0$, where $S(r,\theta)$ is the complex eikonal. The components of the effective momentum are defined as

$$p_r = \partial S/\partial r,\ p_\theta = \partial S/\partial\theta,\ {\rm d}S = p_r{\rm d}r + p_\theta{\rm d}\theta \qquad (10.3)$$

Based on (10.2) and (10.3), you can get the variance equation

$$D = p_r^2 - ip_r/r + p_\theta^2/r^2 - k_0^2\varepsilon(r) = 0 \qquad (10.4)$$

The CGO equations, describing the propagation of rays, have the form

$$\begin{aligned} &{\rm d}r/{\rm d}t = -{\rm Re}(D_{p_r}/D_\omega),\ {\rm d}\theta/{\rm d}t = -{\rm Re}(D_{p\theta}/D_\omega),\\ &{\rm d}p_r/{\rm d}t = (D_r/D_{p_r}){\rm Re}(D_{p_r}/D_\omega),\ p_\theta = {\rm const} = -k_0\rho \end{aligned} \qquad (10.5)$$

where D_r, D_{p_r}, D_{p_θ}, D_ω denote the corresponding partial derivatives with respect to the functions included in the left part of the dispersion equation, ρ is the impact parameter for the beam, as shown in figure 10.2. Different versions of CGO can be found in the literature, for example in Bravo-Ortega and Glassor (1991) and Berczynski (2011), but our proposed new version of CGO has two obvious advantages. First, whereas the eikonal and the components of the pulse p_r are complex, the spatial coordinates are real. This approach is more physical than the use of complex coordinates, as in Bravo-Ortega and Glassor (1991). Second, the method proposed in this work gives an expression for the group velocity (Ginzburg 1961) in the form $\vec{v}_g = \delta\omega/\delta({\rm Re}\,\vec{k})$. This is the form of the group velocity for an environment with weak dissipative losses and nonlinearity, as recommended in Ginzburg (1961). You can show using definitions $\vec{v}_g$ and equation (10.5) that the group velocity can be expressed in the form $\vec{v}_g = -{\rm Re}(D_{\vec{k}}/D_\omega)$. Inside the nonlinear cylinder $r \leqslant R_c$, a full-wave solution is sought and a nonlinear addition to the dielectric permittivity is presented (see (10.1)) in forms, respectively

$$E_z = \sum_{m=-N/2+1}^{N/2} A_m(r)e^{-im\theta}; \quad \delta\varepsilon_{NL}(r, \theta) = \sum_l \delta\varepsilon_l(r)e^{-il\theta}; \tag{10.6}$$

(N is the number of Fourier harmonics that are taken into account). The nonlinear part of the electric permittivity $\delta\varepsilon_{NL} = \delta\varepsilon_{NL}(r, \theta)$ is a function of the amplitudes of all nonlinear harmonics A_m, and, as can be seen from (10.6), for any given value r, is represented in the form of a Fourier series expansion by the angular harmonics, $\exp(-il\theta)$. The coefficients $\delta\varepsilon_l(r)$ are nonlinear and depend on all amplitudes of Fourier harmonics A_m. The nonlinear wave equation for the amplitude of the mth harmonic can be represented in the following equivalent forms (we repeat hereafter for the sake of the convenience, equation (9.37a) and some other details from chapter 9):

$$\frac{1}{r}\frac{d}{dr}\left(r\frac{dA_m}{dr}\right) - \frac{m^2}{r^2}A_m + k_0^2\varepsilon(r)A_m + k_0^2\sum_l \delta\varepsilon_l(r)A_{m-l} = 0 \tag{10.7}$$

In particular, the set of equation (10.7) describes the effects of both self-action and interaction for all modes $A_m(r)$. Boundary conditions at $r = 0$ have a structure that is typical for cylindrical geometries. In other words, the features of the behavior at $r \to 0$, even in the presence of nonlinearity, reflect the behavior of cylindrical functions of corresponding orders

$$\frac{dA_0}{dr}\Big|_{r=0} = 0, \;\; A_m\Big|_{r=0} = 0 (m \neq 0) \tag{10.8}$$

To show this, it is enough to rewrite equation (10.7) in the form $r^2(d^2A_m/dr^2) + r(dA_m/dr) - m^2A_m + r^2k_0^2[\varepsilon(r)A_m + F_m(r)] = 0$,

$$F_m(r) = \sum_l \delta\varepsilon(r)A_{m-l} \tag{10.9}$$

and consider the asymptotics for

$$r \to 0. \tag{10.10}$$

The analysis of the corresponding asymptotics for fields, finite at $r \to 0$ demonstrates that

$$\frac{dA_0}{dr}\Big|_{r\to 0} \to \frac{d\zeta_0}{dr}\Big|_{r=0} = 0; \;\; A_m\Big|_{r\to 0} \to \zeta_m\Big|_{r=0} = 0 (m \neq 0) \tag{10.11}$$

The asymptotic relations (10.11) (see also (10.9) and (10.10)) prove the validity of the boundary conditions (10.8). To determine a completely nonlinear full-wave solution of the set of equation (10.9), we must add to the conditions (10.8) the appropriate boundary conditions at the boundary $r = R_c$ between the inner (nonlinear) and outer (linear) cylindrical regions (figure 10.2). The special procedure of matching of the full-wave nonlinear solution (FWNS) with the solution by the method CGO (method CGO-FWNS) results in the appropriate conditions at the boundary $r = R_c$. The concept of 'sewing together' solutions based on the method of CGO and full-wave solution at $r = R_c$, based on the local quasi-planar approximation for cylindrical Fourier modes. At $r = R_c$ the field distribution corresponding to each beam has the form

$$E = E_0 \exp\left[-\left(\frac{\theta - \theta_c}{\delta\theta_c}\right)^2\right] \times \exp[iS(r = R_c) + ip_\theta(\theta - \theta_c)) \tag{10.12}$$

where $\delta\theta_0$ is the angular width of the beam, and direct calculations show that the angular width is the same for the beam when falling as onto the outer ($r = R_0$) and internal ($r = R_c$) cylinders, in the diffractionless approximation of the geometric optics. Diffraction effects in a linear inhomogeneous medium can be included in this model, but they will not neither qualitatively nor essentially quantitatively change the results of this work obtained in the non-diffraction approximation for the outer region of the concentrator (figure 10.2). The analysis shows that for a system with a cylindrical (angular) symmetry, we have $p_\theta = \text{const}$, and this value is determined by the impact parameter shown in figure 10.2 (these conditions are already actually used when writing down the set of equation (10.5)). $S(r = R_c)$ is the eikonal (phase) of the central beam of the beam when it falls onto the inner cylinder, $\theta(r = R_0) = \theta_0$ is the angle corresponding to the center of the beam when it falls onto the outer cylinder ($r = R_0$). In the exponent in the equation (10.12), the term $p_\theta(\theta - \theta_c)$ is an additional phase of the peripheral part of the beam when it falls onto the inner cylinder ($r = R_c$), relative to the central ray of this beam. At $r = R_c$, continuity is used E, $\partial E/\partial r$. The corresponding boundary conditions are reduced to the form $A_m + (i/k_{rm})(dA_m/dr) = [1 - (k_{im}/k_{rm})]E_{im}(r = R_c)$ Here $E_{im}(r = R_c)$ is the amplitude of the mth Fourier harmonic for the incident wave at the boundary $r = R_c$, $k_{im,\,rm} = (i/2R_c) \pm [k_0^2\varepsilon(r = R_c) - (m^2/R_c^2) - (1/4R_c^2)]^{1/2}$, $\varepsilon(r = R_c)$ is the dielectric permittivity at the boundary $r = R_c$.

Consider the possibility of controlling the process, in particular with a nonlinear beam, and the result of the nonlinear concentration, as well as the formation of a 'hot spot' that occurs when the amplitudes of incident beams exceed a certain threshold, as shown in Rapoport *et al* (2014a, 2014b). Consider nonlinear wave structures, figures 10.3, 10.4, in the central (inner) part of the concentrator $r \leqslant R_c$ (see figure 10.3). To illustrate the possibility of the nonlinear control, we will gradually increase the amplitudes of each of the two incident beams (their amplitudes are the same). The input normalized amplitudes for the signal and control beams are given in the captions to figures 10.3, 10.4. The amplitude of the input beams is chosen so that their values are below the threshold value. When the input amplitude of the incident beams gradually increases, still not reaching the threshold value, the shape and amplitude of the nonlinear structure formed in the core area of the structure also gradually changes (figures 10.3(a)–(c)). At the same time, the maximum of the nonlinear structure is located at some point inside the nonlinear inner area. Then, when the input amplitude of the incident beams becomes

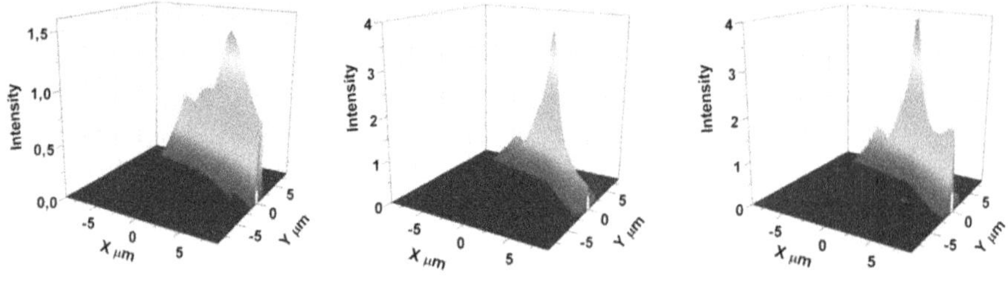

(a) (b) (c)

Figure 10.3. The preparation for the jump of the focus point in an inhomogeneous isotropic mesoscale MM field concentrator (figure 10.2) (Rapoport *et al* 2014b); the internal nonlinear parts of the concentrator are shown (in the area $r \leqslant R_c$, figure 10.2); the number of incident beams is 2; (a) is without nonlinearity; in (b) and (c) the nonlinearity is present; (a) and (b) is the normalized amplitude of each of the two input beams $A_{\rm inp} = 0.444$ $A_{\rm inp} = 0.444$; (b) $A_{\rm inp} = 0.4445$, which is close to but less than the threshold. Reprinted from Rapoport *et al* (2014b), copyright IOP Publishing, all rights reserved.

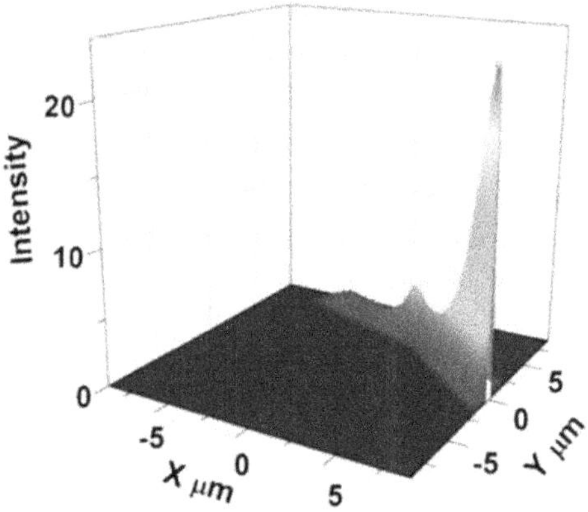

Figure 10.4. The focus point jump and hot spot formation near the point when the normalized amplitude of each of the two incident beams exceeds the 'threshold' value (which is in the range between 0.4445 and 0.4520); the normalized amplitude of the input of each of the two incident beams is 0.452, and all other parameters are the same as for the cases illustrated in figure 10.3. Reprinted from Rapoport *et al* (2014b), copyright IOP Publishing, all rights reserved.

close to a certain threshold value, the amplitude of the nonlinear structure begins to increase in the immediate vicinity of a certain point ($x = -R_c$, $y = 0$) located in the X axis and on the boundary between the nonlinear inner and outer linear regions (figure 10.4). Physically, this point is interpreted (Boardman *et al* 2011, Rapoport *et al* 2011, 2012a, 2014b) as a self-localized strongly nonlinear resonator formed at the boundary between the linear and nonlinear regions of the field concentrator. With a further increase in the input amplitude of the incident beams (namely $A_{\rm inp} = 0.4445$) hot spot formation manifests itself in the form of the formation of a highly localized maximum of relatively high field intensity (a self-localized resonator/hot spot) near the point ($x = -R_c$, $y = 0$).

Figure 10.4 corresponds to the situation when the input amplitude of the signal beam exceeds the threshold level (see the captions to figures 10.3 and 10.4). The new strongly nonlinear phenomenon found is that: (1) the focus point 'jumps' when the value of the input amplitude of a certain threshold level is exceeded, which depends on the structure, position, and the number of incident beams. A similar phenomenon has been shown to occur in the presence of two, three, four, or five parallel incident signal beams. (2) There is a tendency to form a 'hot spot' or a 'spot' with a size of the order of the single wavelength and relatively high intensity (figure 10.4).

Interestingly, for a beam propagating in the direction of the X axis and falling onto the system (figure 10.2), there is a tendency for 'jumping' the focus point from some position within the inner nonlinear region $r \leqslant R_c$ (figures 10.3(c), 10.4) to a point (or a position near this point) ($x = -R_c$, $y = 0$), lying at the border between the inner (central) nonlinear and outer linear regions (figure 10.4). Such an effect can be called 'nonlinear focusing switching'. Also, the direct numerical calculations show that if you continue the calculation in the approximation (linear) CGO in the central region $r \leqslant R_c$, then the corresponding beam passes through the same point ($x = -R_c$, $y = 0$, figures 10.4 and 10.2), in which the focus point jumps when the amplitude exceeds the threshold value in the mode of strong nonlinearity.

Thus, we have shown the possibility of the effective control of nonlinear structures and the transition to a mode of the strong nonlinearity with a jump of the focus point and the tendency to form a hot spot, by means of an additional control beam. Numerical studies show that the same trend is observed in other ways to realize that the amplitude of the signal beam exceeds the appropriate threshold level, in addition to increasing the amplitude of the input beam.

This threshold level can be shifted at a constant amplitude of the signal beam, changing the level of the nonlinearity by a method beyond changing the input amplitude of the signal beam (Boardman *et al* 2011, Rapoport and Grimalsky 2011, Rapoport *et al* 2012a, 2012b, 2014b): (1) adding an auxiliary control beam; (2) the placement to the central (nonlinear) region of the concentrator of the active medium and a gradual increase in the linear gain, which compensates for the linear dissipative losses γ and even leads to the resulting gain; (3) a gradual reduction of nonlinear dissipative losses; (4) a reduction of the coefficient β (the saturation coefficient) (see relation (10.1)); (5) increasing the nonlinearity coefficient α.

All these methods lead to the same qualitative result, which is shown in figure 10.4—that is, the jump of the focus point and the formation of a hot spot/hot point. All these results are qualitative, but they correspond to a certain physics, for example, the linear gain can be increased by adding quantum dots or dye molecules (in the optical range) in the central region of the concentrator. As we have shown, the system 'forgets' the prehistory of the process of transition of the system to a state when the input amplitude of the signal exceeds the threshold level.

The result (the tendency to form a hot spot) does not depend on the specific process of transition through the threshold value, so we state that it is the first phenomenon of focus jumping and hot spot formation, which is a reality for the concentrator considered in this section. Nevertheless, it should be noted that the theory used on the base of the local nonlinearity approximation ceases to adequately describe the system during the transition to the above-threshold mode (figure 10.4).

This is reflected by the quasi-chaotic dependence of the output amplitude (amplitude at the point of focus, figure 10.4), i.e., in fact, the lack of numerical convergence itself, in the above-threshold mode. For this mode, it will be necessary to develop a new, nonlocal quantum-electrodynamic nonlinear theory for an active MM field concentrator in the above-threshold mode in subsequent works that go beyond this section. The above-threshold field concentration was also obtained in a nonlinear active field concentrator on hyperbolic MM (Rapoport *et al* 2012b, 2014a, Boardman *et al* 2012).

Let us estimate the intensities required for the observation of the strongly nonlinear phenomena. For the central nonlinear region, you can use doped *n*-Si, where the cubic nonlinear permittivity coefficient $\chi^{(3)}$ is of about $\chi^{(3)\sim}$ $(10^{-7}\text{–}10^{-8})$ e.s.u., abs. units (Sutherland 2003). To observe nonlinear phenomena in an inner nonlinear cylinder of radius $R_c \approx 10$ μm, the intensity of the electromagnetic wave should be ~ 1 GW cm^{-2}. As a result, the intensity of electromagnetic waves incident on the outer cylinder (figure 10.2) should be (10–100) MW cm^{-2}.

10.3 Nonlinear switching effect when electromagnetic waves pass through a multilayer resonant system 'dielectric–graphene'

The passage of THz EMW through layered structures, controlled by nonlinearity and external magnetic field, is of considerable interest and has been studied, in particular for structures with superconductors in Rokhmanova *et al* (2013) and Apostolov *et al* (2016). Consider new nonlinear electromagnetic effects and applications in multilayer dielectric–graphene (DG) structures. Possibilities are known from the literature formation of electromagnetic solitons (Dong *et al* 2013) and generation of harmonics (Smirnova and Kivshar 2014) in the plane of graphene layers. We investigated (Rapoport *et al* 2013a, 2013b, 2013c) the effect of nonlinear switching between transmission and reflection modes for electromagnetic waves propagating in multilayer DG structure (figure 10.5(a)). The linear transmission and reflection characteristics are presented in figure 10.5(b). The electrodynamic nonlinearity of graphene is based on the electronic nonlinearity. Namely, the nonlinear current in the graphene plane is expressed as (Dong *et al* 2013) $i_s \approx i_o F(\psi)$, where $i_o = eV_F n_2$, $F(\psi) = \psi(1+\psi^2)^{-1/2}$, $\psi = eA/(\hbar k_F)$, V_F, n_2, and A are the Fermi velocity, the effective surface concentration of electrons, and the amplitude of the vector potential, respectively, k_F is the Fermi wavenumber. Nonlinear equations with respect to amplitude $C = i\omega A$ of the electric field (ω is wave frequency) with the first derivative with respect to time and the second derivative in the coordinate includes a new structure of the nonlinear term proportional to the multiplier $\mathrm{NL} = 1 - Q^{-1/2}[1 - Q^{-1}(1 - \mu Q^{-1}|C|^2)]$, where $Q = 1 + \alpha|C|^2$ (the coefficients μ, α (Rapoport *et al* 2013a, 2013b, 2013c) are given below). The following quasi-parabolic equations for the amplitude C of the electromagnetic wave in a multilayer

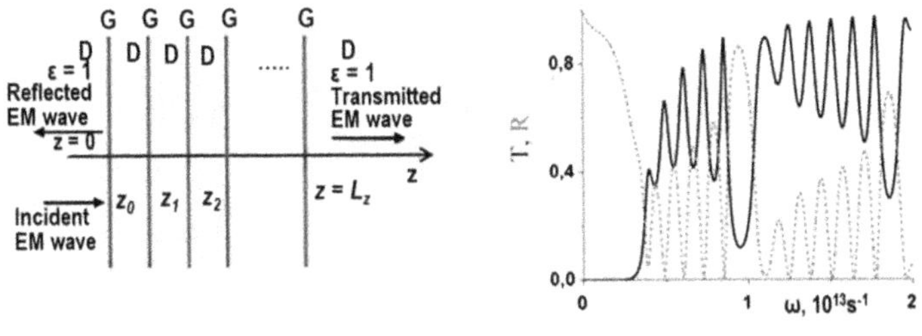

(a) (b)

Figure 10.5. (a) is the layered dielectric–graphene CDG structure; the graphene layers correspond to the coordinates z_0, z_1,... (b) is the dependence of transmission coefficients T, solid line, and reflection R, dotted line, linear THz electromagnetic waves at the frequency for relatively small dissipation, $\nu = 3 \times 10^{11}$ s^{-1}. The frequency dependences of the linear transmission coefficients (T) and reflection (R) of electromagnetic waves are shown in figure 10.5(b). Figure 10.6 presents the results of numerical simulations of nonlinear propagation of THz pulses through the structure. (b) Reprinted from Rapoport *et al* (2013a), copyright (2013) with permission of Springer.

plane-parallel DG system were obtained. For such a system, it is inapplicable to approximate amplitudes that change slowly, at least with respect to their change along the normal to the system layers, because the thickness of nonlinear graphene layers is much smaller than the electromagnetic wavelength in the (THz) range considered:

$$\frac{\partial C}{\partial t} + \frac{ic^2}{2\omega\varepsilon^{(1)}}\left(\frac{\partial^2 C}{\partial z^2} + \Delta_\perp C\right) + \frac{i}{2\omega\varepsilon^{(1)}}\left(\omega^2\varepsilon - \sum_j \omega_{\rm pg}^2 l_n \delta(z - z_j)\left(1 + \frac{i\nu}{\omega}\right)\right)C$$
$$+ \frac{i\omega_{\rm pg}^2 l_n \omega}{2\omega^2\varepsilon^{(1)}}\sum_j\left(1 - Q_{\rm g}^{-1/2}\left(1 - \frac{1}{8}\qquad\frac{e^2|C|^2}{(m^* v_{\rm F}\omega)^2}Q_{\rm g}^{-1}\right)\right)\delta(z - z_j)C \quad (10.13)$$
$$= 0;\ Q_{\rm g} = 1 + \frac{e^2|C|^2}{2(m^* v_{\rm F}\omega)^2};\ \alpha = \frac{e^2|C|^2}{2(m^* v_{\rm F}\omega)^2};\ \mu = \frac{1}{8}\frac{e^2|C|^2}{(m^* v_{\rm F}\omega)^2}$$

The summings up in equation (10.13) are performed at the positions z_j of all graphene layers (figure 10.5(a)); D is the linear electrical induction in dielectric layers; ${\omega_{\rm pg}}^2 l_n = (4\pi e^2 n_{20}/m^*)$. Typical spatial scales l_n used in the simulation for the THz band are equal to $l_n = 10^{-2}$ cm; $v_{\rm F} \approx 10^8$ cm s^{-1} is the characteristic Fermi velocity of electrons in graphene, $p_{\rm F} \equiv \hbar k_{\rm F}$ is Fermi momentum of electrons; A is the vector potential of the electromagnetic wave. An analogue of the effective mass is $m^* = \hbar p_{\rm F}/v_{\rm F} \sim (0.01–0.03)m_{\rm e}$. The electronic nonlinearity is saturable; $\varepsilon^{(1)} = \varepsilon + (\omega/2)(d\varepsilon/d\omega)$ is a correction related to frequency dispersion. This theory takes into account the collisions of electrons whose frequency is close to $\nu = 3 \times 10^{11}–3 \times 10^{12}$ c^{-1}. From equation (10.13) it follows that the nonlinearity in graphene is focusing both in the longitudinal Z and in the transverse directions. Equation (10.13) is supplemented by boundary conditions, which are derived from the condition of continuity of tangential components E_x and H_y. A layered structure surrounded by a free space is considered. At $z < 0$ there are both incident and reflected waves, and at $z > L_z$ there is only passing one. In the parabolic approximation, the boundary conditions are as follows:

$$\frac{\partial C}{\partial z} - ik_0 C = -2ik_0 E_i(\rho, t), \quad z = 0; \frac{\partial C}{\partial z} + ik_0 C = 0, \; z = L_z; \; k_0 = \frac{\omega}{c} \tag{10.14}$$

In (10.14), $E_i(\rho,t)$ is the amplitude of the incident wave at the input. The method used is a further development of the methods we proposed earlier for solitons of the magnetostatic waves (Buttner *et al* 2000a, 2000b, Slavin *et al* 2003), solitons and waves in nonlinear MMs and MM concentrators of the electromagnetic field (Rapoport *et al* 2014b).

An interesting phenomenon is the nonlinear switching of the pass-reflection modes for short THz pulses in the DG structure.

In this case, the incident wave has the form:

$$E_i(\rho, t) = E_{i0} \exp\left\{-[(t - t_1)/t_0]^4\right\}\exp[-(\rho/\rho_0)^4] \tag{10.15}$$

The structure under consideration includes seven dielectric layers ($\varepsilon = \varepsilon^{(1)} = 4$), the thickness of each of which is equal to $h = 5 \times 10^{-3}$ cm. The system includes eight graphene layers located perpendicular to the Z axis with coordinates $z = 0, h, 2h$ etc (figure 10.5(a)). The concentration of the 2D layer of electrons is equal to $n_{20} = 2.5 \times 10^{12}$ cm^{-2}. The half-width of the input pulse included in the relation (10.15), $\rho_0 = 0.025$ cm. THz pulses with a carrier frequency $\omega = 1 \times 10^{13}$ s^{-1} are considered. The half-width of the input pulse is equal to $\rho_0 = 0.025$ cm.

The following conclusions can be pointed out. The electromagnetic–electronic nonlinearity can provide nonlinear switching from the mode of reflection to the mode of transmission (figures 10.6(a)–(c)) of electromagnetic waves of the THz range in layered dielectric–graphene structures when the amplitude of the incident pulse exceeds the threshold level of about 20 kV cm^{-1}, see figure 10.6(c). In this case, the maximum intensity of the transmitted pulse exceeds the maximum value of the intensity of the incident pulse (figure 10.6(c)).

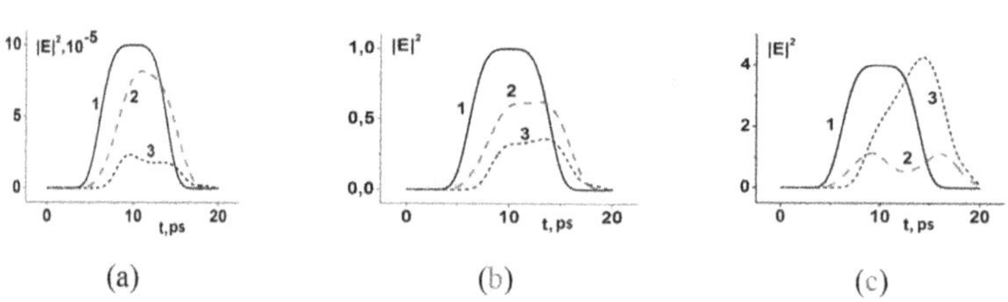

Figure 10.6. The propagation of a relatively short THz pulse through the multilayer structure of the DG (Rapoport *et al* 2013a, 2013b). Curves 1, 2 and 3 correspond to incident, reflected and passed through the layered structure, pulses, respectively. The dissipation parameter is equal to $\nu = 3 \times 10^{11}$ s^{-1}; (a) is the linear mode with a significant reflection of the pulse incident on the system; (b) is the quasi-linear mode; (c) is the passage of the pulse in the mode of strong nonlinearity. The amplitude $|E|$ is normalized to 20 kV cm^{-1}. Curves 1, 2 and 3 correspond to the incident, reflected, and transmitted pulses, respectively. Reprinted from Rapoport *et al* (2013a), copyright (2013) with permission of Springer.

10.4 Conclusions

The following main results are obtained.

(1) The conditions of transition to the mode of strong nonlinearity at the giant generation of the localized surface second harmonic of surface plasmons in the mode of temporal resonance are estimated.

(2) For the first time, a nonlinear MM field concentrator/a 'nonlinear electromagnetic black hole' was proposed. To simulate the effects in a mesoscale linear–nonlinear field concentrator, a new method for simulating the phenomena of the nonlinear transformation optics with strong nonlinearity was put forward. This method includes two related approaches, namely the CGO in the (linear) domain, where the wavelength is much less than the characteristic inhomogeneity length, and the full-wave solution of the Maxwell equations in the (strong) nonlinearity domain, and the boundary conditions for sewing together both solutions borders between the respective areas. A theoretically new nonlinear phenomenon of jumping the focus point and hot spot formation when the input amplitude exceeds a certain threshold value in a linear–nonlinear mesoscale active isotropic MM field concentrator is found. The jump of the focus point and the formation of the hot spot occurs regardless of the possible way to achieve the threshold amplitude by increasing the input amplitude, increasing the linear gain in an active nonlinear plasma-like MM medium, reducing nonlinear energy dissipation, etc. The above-threshold field concentration is also obtained in a nonlinear active field concentrator on a hyperbolic MM. The proposed method for a nonlinear field concentration can be useful for solar panels, subwavelength images, nonlinear antennas, generation of higher harmonics, etc.

(3) For the first time, the possibility of controlling the transition to the formation of hotspots in the process of strong nonlinear focusing, which includes two stages, namely sequential linear and nonlinear focusing in the outer linear and inner nonlinear regions of the nonlinear concentrator. In particular, the control of a nonlinear signal beam with an additional beam of relatively small amplitude, which simultaneously with the signal is launched into the system, and at a certain amplitude of the control beam there is a transition to the formation of a hot spot in the system. This effect of nonlinear control was obtained for concentrators based on both isotropic (Boardman *et al* 2011, Rapoport and Grimalsky 2011, Rapoport *et al* 2012a, 2012b, 2014b) and hyperbolic (Rapoport *et al* 2014a) MMs.

(4) Electromagnetic–electronic nonlinearity can provide nonlinear switching from the mode of reflection to the mode of transmission of electromagnetic waves of the THz range in layered dielectric–graphene structures when the amplitude of the incident pulse of the threshold level, which is ~20 kV cm^{-1}, is exceeded. In this case, the maximum intensity of the transmitted pulse exceeds the maximum value of the intensity of the incident pulse.

Problems to chapter 10

Problem 10.1

Show that the asymptotic solution of equation (10.7), finite at $r \to \infty$, corresponds to the condition (10.10) and, respectively, to (10.8).

We are looking for a solution for all $A_m(r)$ (and, accordingly, $F_m(r)$), finite at $r \to 0$. Such an asymptotics is reduced to solving the problem

$$A_m \approx \zeta_m(r), \; r^2 \frac{\mathrm{d}^2 \zeta_m}{\mathrm{d}r^2} + r \frac{\mathrm{d}\zeta_m}{\mathrm{d}r} - m^2 \zeta_m = 0. \tag{P10.1}$$

Note that this equation coincides with the corresponding equation, which arises in the asymptotic solution for cylindrical functions in the limiting case (10.10).

An analysis of the asymptotic solution corresponding to (P10.1) leads to the conclusion that the conditions (10.10), and, respectively, (10.8) are valid.

Problem 10.2

Obtain the boundary condition in the form of the third relation from (10.4) for the electromagnetic waves propagating in the structure the half-infinite homogeneous dielectric or vacuum in the region $z \geqslant L_z$ (figure 10.5(a)).

Take into account that the sources of electromagnetic waves are absent at $z > L_z$. Accounting for that the electromagnetic wave is incident onto the system shown in figure 10.5(a) from the left side and then, after the wave transition through this system propagates in the positive direction of the axis z, only the corresponding propagating wave (in the absence of reflecting one) exists at $z > L_z$. Write down the condition of continuity for tangential components of electric and magnetic fields at the boundary $z = L_z$. Exclude from the set of the corresponding two equations for the plane electromagnetic fields an amplitude of electric field in the infinitesimally close neighbored of the boundary $z = L_z$ at the right side of this boundary. As a result, the third relation from (10.4) should be obtained.

Problem 10.3

Obtain the boundary condition in the form of the first relation from (10.4) the plane electromagnetic wave incident from the left side (a half-infinite homogeneous dielectric or vacuum), $z < 0$ (see figure 10.5(a)).

Take into account that the region $z < 0$ is homogeneous half-space dielectric or vacuum and both incident reflected plane electromagnetic waves exist at $z < 0$, write down the condition of continuity for the tangential components of electric and magnetic fields at the boundary $z = 0$. Exclude from the obtained set of two equations the amplitude of the reflected electromagnetic wave at $z \leqslant 0$, obtain, as a result, the first relation from (10.4) (compare the analytical calculations with these made for the solution of problem 10.2).

List of abbreviation

DG Dielectric-graphene

References

Apostolov S S, Maizelis Z A and Yampol'skii V A *et al* 2016 Transmission of THz waves through layered superconductors controlled by a dc magnetic field *Phys. Rev.* B **94** 024513

Berczynski P 2011 Complex geometrical optics of nonlinear inhomogeneous fibres *J. Opt.* **13** 035707

Bergman L, Alitalo P and Tretyakov S 2011 Nonlinear transformation optics and engineering of the Kerr effect *Phys. Rev.* B **84** 205103

Bhardwaj A, Pratap D, Semple M, Iyer A K, Jayannavar A M and Ramakrishna S A 2020 Properties of waveguides filled with anisotropic metamaterials *C. R. Phys., Tome* **21** 677–711

Boardman A D, Grimalsky V V and Rapoport Y G 2011 Nonlinear transformational optics and electromagnetic and acoustic fields concentrators *AIP Conf. Proc. The Fourth Int. Workshop on Theoretical and Computational Nanophotonics* vol 1398 pp 120–2

Boardman A D, Grimalsky V V, Rapoport Y G and Kalinich N A 2012 Wave processes and new effects in hyperbolic and chiral nonlinear and active metamaterials *Progress in Electromagnetics Research Symp., (PIERS) (Moscow, August 19–23)* Abstracts p 20

Bravo-Ortega A and Glassor A H 1991 Theory and application of complex geometric optics in inhomogeneous magnetized plasmas *Phys. Fluids* B **3** 529–35

Buttner O, Bauer M, Demokritov S O, Hillebrands B, Kivshar Yu S, Grimalsky V, Rapoport Y, Kostylev M P, Kalinikos B A and Slavin A N 2000a Spatial and spatiotemporal self-focusing of spin waves in garnet films observed by space- and time-resolved Brillouin light scattering *J. Appl. Phys.* **87** 5088–90

Buttner O, Bauer M, Demokritov S O, Hillebrands B, Kivshar Yu S, Grimalsky V V, Rapoport Y and Slavin A N 2000b Linear and nonlinear diffraction of dipolar spin waves in yttrium iron garnet films observed by space- and time-resolved Brillouin light scattering *Phys. Rev.* B **61** 11576–87

Dong H and Conti C *et al* 2013 Terahertz relativistic spatial solitons in doped graphene metamaterials *J. Phys. B: At. Mol. Opt. Phys.* **46** 155401

Gershman B N, Erukhimov L M and Yashin Y Y 1984 *Wave Phenomena in the Ionosphere and Space Plasma (in Russian)* (Moscow: Nauka)

Ginzburg V L 1961 *Propagation of Electromagnetic Waves in Plasmas* (London: Gordon and Breach)

Krasavin A V, Ginzburg P and Zayats A V 2019 Nonlinear nanoplasmonics *Quantum Photonics: Pioneering Advances and Emerging Applications* ed R Boyd, S Lukishova and V Zadkov (Springer Series in Optical Sciences vol 217) (Berlin: Springer)

Narimanov E E and Kildishev A V 2009 Optical black hole: Broadband omnidirectional light absorber *Appl. Phys. Lett.* **9** 041106

Pertsch T and Khivshar Yu 2020 Nonlinear optics with resonant metasurfaces *MRS Bull.* **45** 210–20

Qu Z, Xu Y, Zhang B, Duan J and Tian Y 2020 Terahertz dual-band polarization insensitive electromagnetically induced transparency-like metamaterials *Plasmonics* **15** 301–8

Rapoport Y G and Grimalsky V V 2011 Transformational optics, complex geometrical optics and nonlinear electromagnetic energy concentrator *Proc. of the Int. Conf. Days on Diffraction (St. Petersburg, Russia, May 30–June 3)* pp 160–1

Rapoport Y G, Boardman A D and Grimalsky V V *et al* 2012a Metamaterial based electromagnetic and acoustic field concentrators and new physical phenomena: nonlinear focusing switching *Proc. XXXII Int. Science Conf. on Electronics and Nanotechnology (ELNANO) (Kyiv, Ukraine, 10–12 April)* pp 84–5

Rapoport Y, Boardman A, Grimalsky V, Selivanov Y and Kalinich N 2012b Metamaterials for space physics and the new method for modeling isotropic and hyperbolic nonlinear concentrators *Proc. of Int. Conf. on Mathematical Methods in Electromagnetic Theory (MMET) (Kharkiv, Ukraine, 28–30 August)* pp 76–9

Rapoport Y, Kalinich N, Grimalsky V and Boardman A 2013d Nonlinear beams and active controllable field concentrator with isotropic metamaterials *Proc. of IEEE 33rd Int. Scientific Conf. Electronics and Nanotechnology (ELNANO) (Kyiv, Ukraine, 16–19 April)* pp 165–8

Rapoport Y, Grimalsky Y, Iorsh I, Kalinich N, Koshevaya S, Castrejon-Martinez C and Kivshar Yu S 2013a Nonlinear reshaping of terahertz pulses with graphene metamaterials *JETP Lett.* **98** 503–6

Rapoport Y, Grimalsky V, Kivshar Y, Koshevaya S and Castrejon-M C 2013b Nonlinear switching of terahertz pulses in the structures with graphene layers *Proc. of Int. Kharkov Symp. on Physics and Engineering of Microwaves, Millimeter and Submillimeter Waves (MSMW) (Kharkov, Ukraine, 23–28 June)* pp 253–5

Rapoport Y, Kalinich N, Grimalsky V V, Nefedov I and Malnev V N 2013c Three-level approach to graphene metamaterials: Electron density waves and linear and nonlinear electrodynamics *Conf. Proc. IEEE 33rd Int. Scientific Conf. Electronics and Nanotechnology (ELNANO) Kyiv, Ukraine* pp 169–71

Rapoport Y G, Grimalsky V V, Koshevaya S V, Boardman A D and Malnev V N 2014a New method for modeling nonlinear hyperbolic concentrators *Proc. of IEEE 34th Int. Scientific Conf. on Electronics and Nanotechnology (ELNANO) (Kyiv, Ukraine)* pp 35–8

Rapoport Y G, Boardman A D, Grimalsky V V, Ivchenko V M and Kalinich N 2014b Strong nonlinear focusing of light in nonlinearly controlled electromagnetic active metamaterial field concentrators *J. Opt.* **16** 0552029–38

Rokhmanova T N, Apostolov S S, Maizelis Z A, Yampol'skii V A and Nori F 2013 Self-induced terahertz-wave transmissivity of waveguides with finite-length layered superconductors *Phys. Rev.* B **88** 014506

Schalch J S, Pos K and Guangwu D *et al* 2019 Strong metasurface–Josephson plasma resonance coupling in superconducting $La_{2-x}Sr_xCuO_4$ *Adv. Opt. Mater.* **7** 1900712

Siddiqui O F and Mohra A S 2017 Microwave dielectric sensing in hyperbolically dispersive media *IEEE Sens. Lett.* **1** 1500804

Slavin A N, Büttner O, Bauer M, Demokritov S O, Hillebrands B, Kostylev M P, Kalinikos B A, Grimalsky V V and Rapoport Y 2003 Collision properties of quasi-one-dimensional spin wave solitons and two-dimensional spin wave bullets *Chaos* **13** 693–701

Smirnova D and Kivshar Y S 2014 Second-harmonic generation in subwavelength graphene waveguides *Phys Rev* B **90** 165433

Sutherland R L 2003 *Handbook of Nonlinear Optics* (New York: Marcel Dekker)

Thomas P and Yuri K 2020 Nonlinear optics with resonant metasurfaces *MRS Bull.* **45** 210–20

Xie P, Sun W, Liu Y, Du A, Zhang Z, Wu G and Fan R 2018 Carbon aerogels towards new candidates for double negative metamaterials of low density *Carbon* 129 598–606

Yang R, Xu J, Shen N-H, Zhang F, Fu Q, Li J, Li H and Fan Y 2021 Subwavelength optical localization with toroidal excitations in plasmonic and Mie metamaterials *InfoMat* (Review article) **3** 577–97

IOP Publishing

Yuriy Rapoport and Vladimir Grimalsky

Chapter 11

Nonlinear stationary and non-stationary diffraction in active planar anisotropic hyperbolic metamaterial

11.1 Introduction

Using nonlinear active hyperbolic metamaterials (MMs) with tunable dispersion and remarkable Purcell effect (Rapoport *et al* 2012, 2014, Poddubny *et al* 2013, Shalaginov *et al* 2017) provides very effective functionality and controllability, and paves the way to quantum nanophotonics (Jacob and Shalaev 2011) devices including effective single photon sources (Shalaginov *et al* 2017), emerging nano-scale plasmonic amplifiers and lasers (Smalley *et al* 2016) photovoltaic devices (Simovski *et al* 2013) ultrasensitive sensors etc. Therefore, the contemporary nanophotonics/nanoplasmonics includes, as an important component, the physics of nonlinear active controllable hyperbolic MMs (Urbas *et al* 2016, Krasavin *et al* 2019, Kuznetsov *et al* 2016, Kauranen *et al* 2017).

In the present chapter, we consider nonlinear non-stationary diffraction in active planar anisotropic hyperbolic MMs (Smolyaninov and Hung 2011, Shekhar *et al* 2014, Boardman *et al* 2015, Alberucci and Assanto 2011, Rapoport *et al* 2012, 2014). Such an MM is definitely periodical and multilayered. In the present chapter each layer from which an MM consists (i.e. elementary layers), is considered to be as isotropic. The anisotropy of an MM as a whole will follow from the multilayered structure only and not from properties of each medium included in separate layers. In this section, we will emphasize the following questions. (i) Rather general nonlinear evolution equation is considered where only the temporal dependence of an amplitude is supposed to be slow, and not the spatial dependence. (ii) The hyperbolic MM is considered as a periodical medium with repeating elementary cells with three layers inside each. In the coordinate frame the OZ axis is aligned

doi:10.1088/978-0-7503-2336-9ch11

perpendicularly to the boundaries of layers, the OX one is along the boundaries. Therefore, the tensor of the effective permittivity possesses only diagonal components. One of these layers (namely the third one) is supposed to be active, and for the simplification, the approximation of the total compensation between the losses and gain is used here. (iii) We compare two approaches. Namely, one of them is based on the averaging over a unit cell, while the other one does not include such an averaging. The first one is, in fact, the MM approach. The second one is more accurate than the MM approach and is applicable for the mesoscale medium with the scales intermediate between MMs and photonic crystals. We would like to emphasize that these two approaches are complementary, rather than contradictive to each other! In fact, the more accurate approach without an averaging will be used, first of all, to justify the possibility of an application of the method of the averaging for the nonlinear active hyperbolic (MM) media. Again, we emphasize that such a possibility is not evident as such for nonlinear hyperbolic media even in the case when the MM approximation is valid in the linear limit (or in the case of relatively small amplitudes). It is very important in the context that the MM is nonlinear and active. In this chapter, we will consider only the case of total net gain/losses, in other words their exact mutual compensation. There is also another aspect of the comparison between the results of the above mentioned two approaches to the modeling of the nonlinear wave processes in the (layered) hyperbolic active MM media: with and without an averaging. The point is that, in the active nonlinear media, a stationary solution could not exist at all, at least for some specific set of parameters. Therefore, generally speaking, even an existence of a stationary solution (s) for some specific conditions pointed out below, will be investigated. To provide a possibility of such an investigation, the creation of the corresponding algorithms will be necessary. (iv) In the moderately nonlinear regime, a possibility of formation of hot spots is investigated. This is exactly the case, when the investigation included a possibility of an existence of a stationary regime. To provide this, a proper algorithm has been developed, based on the method of establishing stationary solution (asymptotically, as an evolution of the corresponding non-stationary solution) (Samarskii 2001, Marchuk 1988). When we consider a propagation of the electromagnetic waves through the layer of a hyperbolic active nonlinear MM with the finite width, such propagation includes the reflections of two types. The first one is the reflections at the boundaries between the elementary layers. The second one is the reflection at the (two) boundaries between the MM and the surrounded media, and the last will be considered as to be the semi-infinite and homogeneous, for the simplification. The following physical and methodological questions are addressed. (1) What should be the physical approximations and the form of the corresponding equations, necessary to describe such wave propagation and search the corresponding physical effects? (2) Which interesting nonlinear effects and under which conditions are possible in a real active nonlinear MM under a condition of the moderate nonlinearity (corresponding criteria of the moderate character of the nonlinearity will be clarified later)? The thicknesses of the elementary layers should be specified, when the results coincide, which are obtained from two approaches pointed above. Any real MM includes finite losses or gain. In the present chapter we will only start the consideration of these problems for the active nonlinear hyperbolic medium, and use an approximation of total compensation between gain and losses: the case of a medium with net zero losses/gain.

11.2 Basic equations. Two approaches: with and without an averaging

Consider the nonlinear propagation of electromagnetic waves in a bounded layered medium. There are three alternating elementary layers of thicknesses d_1, d_2, d_3, as seen in figure 11.1.

Thus, the thickness of the elementary cell is $d = d_1 + d_2 + d_3$. Along OY axis the system is uniform. The propagation of nonlinear modulated waves with the components E_x, H_y, E_z is investigated. The positive frequency components are ~$\exp(-i\omega t)$, where t is time, $\omega > 0$ is a circular frequency. Within each isotropic layer the dependence of the permittivity of the electric field is (Kivshar and Agrawal 2003):

$$\varepsilon(z, x) = \varepsilon_L(z) + \frac{\alpha(z)|E|^2}{1 + \gamma(z)|E|^2}; \qquad (11.1)$$
$$|E|^2 \equiv |E_x|^2 + |E_z|^2$$

The model of the local saturating nonlinearity is used here (Kivshar and Agrawal 2003). Each elementary layer is assumed as uniform, so ε_L, α, γ are step-like functions of z. Generally ε_L and α are complex, because the cases of possible dissipation and active media are investigated. The coefficient of the saturation of the nonlinearity is assumed as real and positive: $\gamma > 0$. The nonlinearity is moderate when $\gamma|E|^2 < 1$.

The Maxwell equations are:

$$\nabla \times \vec{H} = \frac{1}{c}\frac{\partial \vec{D}}{\partial t}; \quad \nabla \times \vec{E} = -\frac{1}{c}\frac{\partial \vec{H}}{\partial t} \qquad (11.2)$$

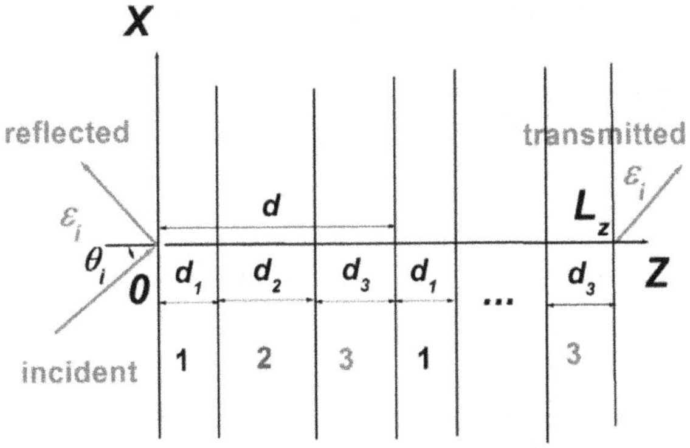

Figure 11.1. Geometry of the problem. The elementary cell includes three layers. Layer 1 possesses $\varepsilon_{L1} < 0$, layers 2, 3 possess $\varepsilon_{L2,3} > 0$. Reprinted from Boardman *et al* (2017), copyright IOP Publishing, all rights reserved.

Below the lengths are normalized to $l_n = 1$ μm, time is normalized to $t_n = l_n/c$. The relations between the components of the electromagnetic field are:

$$E_x = -\frac{i\beta}{\omega}\frac{\partial H}{\partial z}\mathrm{e}^{-i\omega t},\ H_y = H\mathrm{e}^{-i\omega t},\ E_z = \frac{i\beta}{\omega}\frac{\partial H}{\partial x}\mathrm{e}^{-i\omega t};\ \beta(z, x) \equiv \frac{1}{\varepsilon(z, x)} \quad (11.3)$$

(1) Let us consider the first approach. The equation for the slowly varying amplitude of the magnetic field H is:

$$\frac{\partial H}{\partial t} - \frac{i}{2\omega}\frac{\partial}{\partial x}\left(\beta\frac{\partial H}{\partial x}\right) - \frac{i}{2\omega}\frac{\partial}{\partial z}\left(\beta\frac{\partial H}{\partial z}\right) - \frac{i\omega}{2}H = 0 \quad (11.4)$$

Note that equation (11.4) is written for each elementary layer, and each of them is isotropic as such for the used model of a hyperbolic MM. The standard boundary conditions of the continuity of E_x, H_y at the boundaries $z = 0$ and $z = L$ between the layer of the MM and the surrounding medium and between all elementary layers included in the MM (figure 11.1) are applied. Generally, $H(z,x,t)$ is slowly varying in time only, whereas the dependences on z and x are arbitrary. Both direct and reflected waves in each layer are taken into account here. In equation (11.4) the inverse permittivities $\beta(z, x)$ depend on both coordinates, especially on x due to nonlinearity, and are step-like functions on z.

The thickness of layered medium is L_z, so it is localized within the interval $0 < z < L_z$. At $z < 0$ and $z > L_z$ there are linear media with the real permittivity ε_i. At $z < 0$ there are an incident wave with the incidence angle θ_i and a reflected one. At $z > L_z$ there exists the transmitted wave only.

At $z < 0$ the magnetic field is the sum of the incident wave and reflected one:

$$H = A_i(z, x, t)\exp(ik_{ix}x + ik_{iz}z) + A_r(x)\exp(ik_{ix}x - ik_{iz}z);$$
$$k_{ix} = k_i \sin\theta_i; \quad k_{iz} = k_i \cos\theta_i; \quad k_i \equiv \omega\varepsilon_i^{1/2} \quad (11.5)$$

The slowly varying amplitudes $A_{i,r}$ satisfy the following equations:

$$\frac{\partial A_i}{\partial t} + v_z\frac{\partial A_i}{\partial z} + v_x\frac{\partial A_i}{\partial x} \approx 0; \quad \frac{\partial A_r}{\partial t} - v_z\frac{\partial A_r}{\partial z} + v_x\frac{\partial A_r}{\partial x} \approx 0;$$
$$v_x \equiv \beta_i \sin\theta_i; \quad v_x \equiv \beta_i \cos\theta_i; \quad \beta_i \equiv \frac{1}{\varepsilon_i} \quad (11.6)$$

The boundary conditions at $z = 0$ can be written down as:

$$(A_i + A_r)|_{(z=-0)}\exp(ik_{ix}x) = H|_{(z=+0)};$$
$$\beta_i\left(ik_{iz}(A_i - A_r) + \frac{\partial A_i}{\partial z} + \frac{\partial A_r}{\partial z}\right)\Bigg|_{(z=-0)}\exp(ik_{ix}x) = \beta_1\frac{\partial H}{\partial z}\Bigg|_{(z=+0)}. \quad (11.7)$$

After expression of the derivatives $\partial A_i/\partial z$, $\partial A_r/\partial z$ from (11.6), one can get the following approximate boundary condition for H at $z = 0$:

$$\left(1 + \frac{k_{ix}v_x}{k_{iz}v_z}\right) \cdot H - \frac{i\beta_1}{\beta_i k_{iz}} \frac{\partial H}{\partial z} + \frac{i}{k_{iz}v_z}\left(\frac{\partial H}{\partial t} + v_x \frac{\partial H}{\partial x}\right)\Bigg|_{(z=+0)}$$
$$\approx 2\left[A_i + \frac{i}{k_{iz}v_z}\left(\frac{\partial A_i}{\partial t} + v_x \frac{\partial A_i}{\partial x}\right)\right]\Bigg|_{(z=-0)} \exp(ik_{ix}x) \tag{11.8}$$

Because at $z > L_z$ the outgoing wave exists only, at $z = L_z$ the corresponding boundary condition is

$$\left(1 + \frac{k_{ix}v_x}{k_{iz}v_z}\right) \cdot H + \frac{i\beta_3}{\beta_i k_{iz}} \frac{\partial H}{\partial z} + \frac{i}{k_{iz}v_z}\left(\frac{\partial H}{\partial t} + v_x \frac{\partial H}{\partial x}\right)\Bigg|_{(z=L_z-0)} \approx 0 \tag{11.9}$$

(2) Also the second approach is applied where the averaged values of the permittivity and therefore a consideration of the hyperbolic MM as effectively continuous medium, are used. Accurately speaking, this approach is applicable when the typical widths of the layers are much less than the wavelength in the medium. Nevertheless we will compare the results of both approaches for the values of the widths of the layers lying in the range from 0.1 to 0.25 of the wavelength. It is assumed that the elementary layers 2, 3 are nonlinear, whereas the first one is linear. The non-dimensional units are used with $c = 1$. Within the averaged medium we have in each point of it:

$$E_x = -\frac{i\beta_x}{\omega}\frac{\partial H}{\partial z}; \quad E_z = \frac{i\beta_z}{\omega}\frac{\partial H}{\partial x} \tag{11.10}$$

Here E_x, E_z are averaged values of the components of the electric field: $E_x \equiv \langle E_x \rangle$, $E_z \equiv \langle E_z \rangle$. In this approach the equation for the slowly varying amplitude $H(z,x,t)$ is:

$$\frac{\partial H}{\partial t} - \frac{i}{2\omega}\frac{\partial}{\partial x}\left(\beta_z \frac{\partial H}{\partial x}\right) - \frac{i}{2\omega}\frac{\partial}{\partial z}\left(\beta_x \frac{\partial H}{\partial z}\right) - \frac{i\omega}{2}H = 0 \tag{11.10b}$$

The amplitude $H(z,x,t)$ is slowly varying with respect to time t only, as in equation (11.4). The formulas for the averaged components of the inverse permittivity are:

$$\beta_z = \beta_1\frac{d_1}{d} + \beta_2\frac{d_2}{d} + \beta_3\frac{d_3}{d}, \quad \beta_{1,2,3} = \frac{1}{\varepsilon_{1,2,3}};$$
$$\varepsilon_x = \varepsilon_1\frac{d_1}{d} + \varepsilon_2\frac{d_2}{d} + \varepsilon_3\frac{d_3}{d}, \quad \beta_x = \frac{1}{\varepsilon_x}. \tag{11.11}$$

These formulas are applied in each point of the averaged medium, where the values of E_x, E_z are computed. Here $d_{1,2,3}$ are the thicknesses of the elementary layers, $d = d_1 + d_2 + d_3$. In (11.11) only the ratios $d_{1,2,3}/d$ are important but not their absolute values. Respectively, it is supposed that the change of the field components $E_{x,z}$, H_y, as well as of the values $\beta_{x,z}$ along the layers (in the direction x) happen on the lengths, much larger than the thicknesses of the layers $d_{1,2,3}$. The nonlinear permittivities of the elementary layers 2, 3 are calculated as:

$$\varepsilon_2 = \varepsilon_{2L} + \frac{\alpha_2 \cdot (|E_{2x}|^2 + |E_{2z}|^2)}{1 + \gamma_2 \cdot (|E_{2x}|^2 + |E_{2z}|^2)}; \qquad (11.12)$$
$$\varepsilon_3 = \varepsilon_{3L} + \frac{\alpha_3 \cdot (|E_{3x}|^2 + |E_{3z}|^2)}{1 + \gamma_3 \cdot (|E_{3x}|^2 + |E_{3z}|^2)}$$

Here E_{2x} etc are the components of the electric field in the corresponding elementary layers near the specific point within the averaged medium. There is the problem of the correspondence between the averaged values E_x, E_z and E_{2x}, E_{3x}, E_{2z}, E_{3z} within the elementary layers. Because the tangential component of the electric field is continuous in the layered medium, one can write down:

$$E_{2x} = E_{3x} = E_x \qquad (11.13)$$

The normal component of the electric induction is also continuous:

$$D_{2z} = D_{3z} = \langle D_z \rangle \equiv \frac{E_z}{\beta_z} \qquad (11.14)$$

It is follows from (11.4) that

$$E_{2z} = \beta_2 D_{2z} = \frac{\beta_2}{\beta_z} E_z; \; E_{3z} = \beta_3 D_{3z} = \frac{\beta_3}{\beta_z} E_z \qquad (11.15)$$

The formulas (11.10)–(11.15) are used jointly, so several iterations should be applied. Note that in distinction to equation (11.4) used in the first considered approach, equation (11.10b) describes the whole MM as a continuous medium. Therefore, the boundary conditions between the elementary layers included in the MM (figure 11.1) are not applied for the equation (11.4). Nevertheless the boundary conditions at the boundaries $z = 0$ and $z = L$ between the layer of the MM and surrounding medium are still necessary and applied. The last boundary conditions have the form similar to (11.9).

Note the following concerning the physical sense of the averaging, namely formulas (11.10)–(11.15). The presence in the MM, consisting of the periodical alternation of isotropic layers, of the anisotropy, which is evident from the formulas (11.11), has a clear physical sense. Let us consider first a linear medium. In this case formulas (11.11) have been derived accurately (this derivation is not presented here) using the consideration of the periodical media with the further approximation of thin layers. Qualitatively, the description of the anisotropy of the hyperbolic MM may be drawn from the equivalent transmission line, describing a planar multilayered MM (Felsen and Marcuvitz 1973, chapter 7, section 7.2, formulas (7a), (7b)). In accordance with this approach, averaged transverse field components (E_x, H_y in our case) would be determined through an average value of the effective impedance, which is proportional to $\langle\varepsilon_{xx}\rangle$, where the brackets $\langle...\rangle$ mean a proper averaging (in our case by the period of the structure) of the value placed inside the brackets. On the other hand, as follows from Felsen and Marcuvitz (1973, chapter 7, section 7.2, formula (4)), an averaged longitudinal field component, in our case E_z, would be determined through the value proportional to $\langle\varepsilon_{zz}{}^{-1}\rangle$. For the hyperbolic MM, based on the isotropic alternating layers, in each, there are $\varepsilon_{xx} = \varepsilon_{zz}$, with the signs of ε_{xx} alternate from layer to layer belonging to each cell of the structure. Providing that $\langle\varepsilon_{xx}\rangle > 0$, it is easy to see that $\langle\varepsilon_{xx}{}^{-1}\rangle = \langle\varepsilon_{zz}{}^{-1}\rangle < 0$. Therefore, we get hyperbolic uniaxial MM with the opposite signs of the diagonal tensors of the dielectric permittivity. In the nonlinear case, of course, the periodicity of the medium, which is the basis of an averaging, disappears. Nevertheless, we still use the relations (11.11) in the same form, as that for the linear case, phenomenologically, accounting for a contribution of the nonlinearity in the values $\varepsilon_{2,\,3}$ (formulas (11.12)) and, respectively, to $\beta_{1,2,3}$, $\varepsilon_{x,\,z}$ (formulas (11.11)), as described above. This is one of the reasons why the comparison between the accurate approach without an averaging (see equation (11.4)) with the MM approximation of the continuous media (equation (11.10b)) is really necessary and important.

Equation (11.4) added by boundary conditions has been solved by the finite differences method where the operator factorization, or the method of Douglas–Rachford (Samarskii 2001, Marchuk 1988, Douglas and Rachford 1956), has been applied. Namely, equation (11.4) is rewritten symbolically as:

$$\frac{\partial H}{\partial t} + \hat{L}_1 H + \hat{L}_2 H = 0, \quad \text{where } \hat{L}_1 H \equiv -\frac{i}{2\omega}\frac{\partial}{\partial z}\left(\beta\frac{\partial H}{\partial z}\right) - \frac{i\omega}{2}H,$$
$$\hat{L}_2 H \equiv -\frac{i}{2\omega}\frac{\partial}{\partial x}\left(\beta\frac{\partial H}{\partial x}\right) \tag{11.16}$$

The following notation is used:

$$\chi^{p+1} \equiv \frac{H^{p+1} - H^p}{\tau} \tag{11.17}$$

Here τ is the step for time t, $t^p \equiv p{\cdot}\tau$, $p = 0,1,2\ldots$; $H^p \equiv H(t^p)$. And equation (11.16) can be represented as:

$$(1 + \tau\hat{L}_2)\chi^{p+1/2} = -(\hat{L}_1 + \hat{L}_2)H^p;$$
$$(1 + \tau\hat{L}_1)\chi^{p+1} = \chi^{p+1/2}; \tag{11.18}$$
$$H^{p+1} = H^p + \tau\chi^{p+1}$$

The boundary conditions (11.8) and (11.9) are also rewritten in terms of χ^{p+1}. The operators $\hat{L}_{1,\,2}$ are approximated by finite differences (Samarskii 2001, Marchuk 1988, Felsen and Marcuvitz 1973). The inverse permittivity β is a piecewise-continuous function of z, so the values of β are approximated between the nodes for H: if the values of H in the nodes are $H_j \equiv H(z_j) \equiv H(j{\cdot}h_z)$, $j = 0,1,2\ldots$, then $\beta_{j+1/2} \equiv \beta((j+1/2){\cdot}h_z)$. Each distance $d_{1,2,3}$ includes integer numbers of spatial steps h_z. To take into account the nonlinearity, the iterations have been applied. This method is unconditionally stable.

The system is limited in X-direction, $0 \leqslant x \leqslant L_x$, so the boundary conditions at $x = 0$ and $x = L_x$ are $H = 0$.

The boundary conditions (11.8) and (11.9) have been approximated by the finite differences too. They are applied at the second fractional step in (11.18), to compute χ^{p+1}. The derivatives $\partial\chi^{p+1}/\partial x$ are calculated in the positive direction of x, because $\mathrm{v}_x > 0$. In the point $x_l \equiv l{\cdot}h_x$ the approximation is $\partial\chi^{p+1}/\partial x \approx (\chi_l{}^{p+1} - \chi_{l-1}{}^{p+1})/h_x$. This approximation makes possible to calculate χ^{p+1} from smaller values of x to higher ones: $l = 1,2,3,\ldots$, and to use for computing $\chi_l{}^{p+1}$ the value of $\chi_{l-1}{}^{p+1}$, which has been just calculated.

The implicit–explicit methods, like Peaceman–Rachford one (Samarskii 2001, Marchuk 1988, Felsen and Marcuvitz 1973), do not provide good stability in our nonlinear case, as our simulations demonstrated. The schemes like splitting with respect to physical factors, or the summatory approximation, require small temporal steps and are therefore practically unusable here.

The incident wave is assumed as a beam bounded in x-direction:

$$A_i(z = 0,\, x,\, t) = A_0 \exp\left(-\left(\frac{x - x_1}{x_0}\right)^2\right)\tanh\left(\frac{t}{t_0}\right) \tag{11.19}$$

The temporal dependence is tanh-like and the maximum amplitude of the incident wave tends to A_0 at the boundary $z = 0$. Below, the established values of the electromagnetic field are presented. As will be shown below, strongly nonlinear phenomena, namely hot spot formation, are possible for the present system with the corresponding parameters. This is the case, when a possibility of a stationary solution is not evident beforehand, and to prove such a possibility, the method of establishing (steady-state solution) and, respectively, the initial-boundary condition (11.19) are quite adequate.

11.3 Details of the structure and requirements for materials

11.3.1 Details of the structure

The simulations have been done for the hyperbolic media. For linear EM waves the parameters of the elementary layers are chosen to get $\varepsilon_x' > 0$, $\varepsilon_z' < 0$, where $\varepsilon' \equiv \mathrm{Re}(\varepsilon)$. The medium 1 is with $\varepsilon_{\mathrm{L1}}' < 0$, 2 and 3 are with $\varepsilon_{\mathrm{L1}}' > 0$. The hyperbolic media possess the properties, which are important both for theoretical and practical views. Below it is assumed that the real parts of the media 2, 3 are equal: $\varepsilon_{\mathrm{L2}}' = \varepsilon_{\mathrm{L3}}'$. The anisotropy is neglected.

11.3.2 Requirements for materials

The materials for the elementary layers should satisfy the following requirements: $| \varepsilon'_{L1} | < \varepsilon'_{L2}$; the dissipation within each layer should be as small as possible.

In the near-infrared and visible optical range (wavelengths $\lambda_0 = 0.5$–2 μm, $\omega = 5 \times 10^{14}$–$4 \times 10^{15}$ s^{-1}) medium 1 can be metallic of high conductivity or semi-metallic, like Ag, Au, Cu, Bi. The linear effective permittivity is in absolute units:

$$\varepsilon_{l1} = \varepsilon_{\text{lattice}} - \frac{\omega_p^2}{\omega(\omega + i\nu)}; \quad \omega_p^2 \equiv \frac{4\pi e^2 n_0}{m^*} \tag{11.20}$$

Here ω_p is the plasma frequency, n_0 is the electron concentration, m^* is the effective electron mass, ν is the electron collision frequency.

The critical parameters are the electron concentration n_0 and the collision frequency. It should preferably be $n_0 = 10^{21}$–2×10^{22} cm^{-3}, to provide the negative effective permittivity in the corresponding layer(s), as is required for the hyperbolic MM, in the optical range of **moderate** absolute values $\varepsilon'_{L1} = -3$ to -10. Namely, the effective permittivity is negative due to the electron plasma, thus, $0.2\omega_p < \omega < 0.5\omega_p$. The collision frequency should be relatively small $\nu \leqslant 10^{13}$ s^{-1}.

Layers 2, 3 should be dielectrics with high (with a positive real part) permittivity in the optical range and low losses (or even active, i.e. to provide amplification), like Hf_2O_5, Ta_2O_5, Al_2O_3, $\varepsilon'_{L2} \geqslant 10$.

In the THz range (wavelengths $\lambda_0 = 0.5$–0.03 mm, or $\omega = 10^{12}$–6×10^{13} s^{-1}) the narrow-gap semiconductors n-InSb, n-InAs, n-$Cd_{1-x}Hg_xTe$ are perspective. The frequency of EM wave should satisfy the inequality $\omega < E_g/\hbar$, where $E_g = 0.1 - 0.4$ eV is the forbidden gap for narrow-gap semiconductors. These semiconductors possess small effective electron masses $m^* = (0.002$–$0.02)m_e$ and low collision frequencies $\nu = (10^{11}$–$3 \times 10^{11})$ s^{-1} at moderate doping levels $n_0 = (10^{15}$–$10^{16})$ cm^{-3}. As the layers 2, 3, the dielectrics like TiO_2, MoO_2, $SrTiO_3$, $LiNbO_3$ with high permittivities $\varepsilon'_{L2} \geqslant 20$ can be used. Also $SrTiO_3$ and $LiNbO_3$ possess high dielectric cubic nonlinearity of the negative signs.

In the optical range the unity of unidimensional amplitude of the electric field can be estimated as 1 GW cm^{-2} for the intensity. In THz range the unity is estimated as 10 MW cm^{-2} for dielectrics of high nonlinearity.

It is demonstrated below that the nonlinear behavior of EM waves within the hyperbolic medium depends essentially on the sign of cubic nonlinearity.

The surrounding media at $z < 0$ and $z > L_z$ possess the permittivity ε_i matched to the hyperbolic medium in the linear case. Namely, the value of ε_i has been chosen to provide the zero reflection coefficient at $z = 0$ in the case of the incidence of the plane wave at the incidence angle θ_i.

In the experimental realization of hyperbolic media the essential problem is the linear dissipation, because usually the elementary layer 1 is metallic or semi-metallic. Therefore, to observe various wave phenomena there it is rather better to provide a compensation of this dissipation, and the elementary layer 3 should be active, where $\varepsilon'_{L3} < 0$. Below, namely, the case of the compensated dissipation is considered, where all linear permittivities ε_L and nonlinear coefficients α are assumed as real. The results of simulations are tolerant to changes of the lengths and the widths of the hyperbolic medium.

11.4 Results of simulations

In the linear case, i.e. absence of nonlinearity, the main attention is given to comparison of the simulations of wave beam propagation within the framework of direct simulations without averaging and for averaged media (figure 11.2). During the simulations without an averaging, it is used that the values in the brackets in the third terms in equations (11.4) and (11.10b) are continuous at the interface between the layers of the hyperbolic medium (because they are proportional to the tangential components of the corresponding electric fields, in accordance with equations (11.3) and (11.10b)). The results of simulations have been practically the same when the thicknesses d_j of the elementary layers should satisfy the inequalities $d_j{\cdot}\omega/c\ (|\varepsilon_{Lj}|)^{1/2} \leqslant 0.05$. When, for instance, the frequency of EM wave is $\omega = 2 \times 10^{15}\ s^{-1}$, the permittivities are $\varepsilon_{L1} = -10$, $\varepsilon_{L2,3} = 20$, the thicknesses of the elementary layers should be $d_{2,3} \leqslant 10$ nm, $d_1 \leqslant 20$ nm.

The used parameters are $\varepsilon_{L1} = -10$, $\varepsilon_{L2,3} = 20$, $L_z = 50$ μm, $L_x = 20$ μm, the half-widths of incidents beam are $x_0 = 1$–3 μm, $\omega = (1$–$30) \times 10^{14}\ s^{-1}$, incidence angles are $\theta_i = (0$–$30)°$. The thicknesses of the elementary layers are chosen as $d_{1,2,3} = 5$–20 nm. The typical results are given in figure 11.2. It is seen that the results of non-averaged simulations, equation (11.4), and within the framework of the averaged medium, equation (11.10b), coincide. In figure 11.2 the incident beam possesses the incidence angle $\theta_i = 30°$, the x-component of the group velocity of the incident beam is positive and directed **upwards**. The x-component of the group velocity of the refracted wave within the hyperbolic medium at $x > 0$ is negative and is directed **downwards**. Thus, the hyperbolic medium possesses the negative refraction.

In the nonlinear case, layers 2, 3 are assumed to be nonlinear, whereas 1 is linear. Two cases are considered: $\alpha > 0$ (figure 11.3) and $\alpha < 0$ (figures 11.4 and 11.5). In the first case the input beam is subject to essential nonlinear diffraction. The second case is much more interesting. Within the hyperbolic media the hot spots are formed (near $x \approx 13$ μm, $z \approx 27$ μm, figure 11.4 and $x \approx 12$ μm, $z \approx 35$ μm, figure 11.5) where the EM energy concentrates.

Note that: (1) while the beam width increases two times, the intensity in the peak of the hot spot increases ~1.5 times (compare figure 11.5 with figure 11.4); and (2) for the input beam width equal to 2 μm, the areas of the hot spots are of order of 2 μm^2 (figure 11.5).

From figures 11.3 and 11.4 one can see that the nonlinearity is moderate there and the results of simulations are practically the same both in the direct simulations, equation (11.4), and with the averaged permittivities, equation (11.10b). But even when the nonlinearity ceases to be moderate in figure 11.5, $\gamma\,|\,E\,|^2_{max} \geqslant 1$, the difference between these two approaches is also unessential. The sizes of the hot spots in figures 11.4 and 11.5 are 2–3 μm, i.e. 3–5 wavelengths calculated for the averaged permittivity $\lambda = \lambda_0/(\varepsilon_x')^{1/2} \approx 0.4\ \mu m$. The moderate character of the

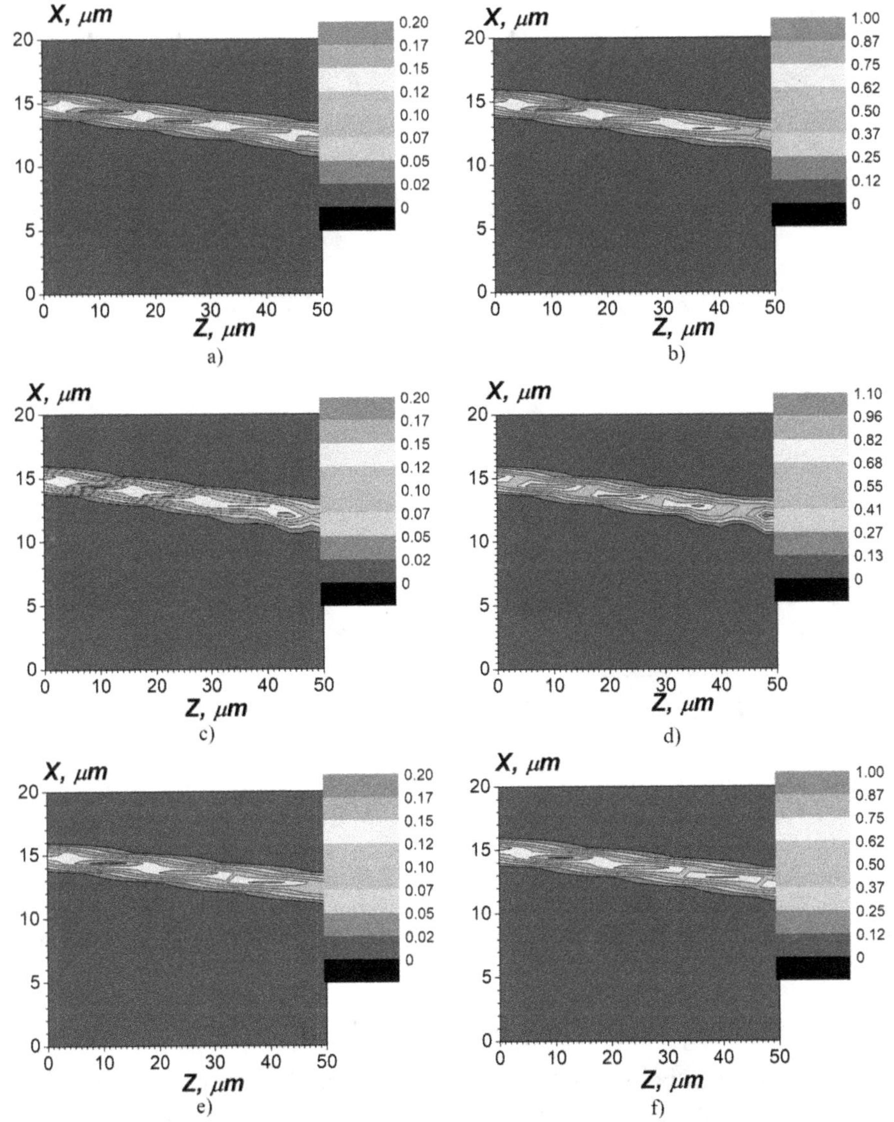

Figure 11.2. Propagation of linear waves. The circular frequency is $\omega = 6\pi \cdot 10^{14}$ s^{-1} ($\lambda_0 = 1$ μm). Parts (a) and (b) are $|E|^2$ and $|H|^2$ for the thicknesses of elementary layers $d_1 = 7.5$ nm, $d_2 = d_3 = 3.75$ nm. Parts (c) and (d) are the same, but $d_1 = 15$ nm, $d_2 = d_3 = 7.5$ nm. Parts (a)–(d) are simulated without averaging. Parts (e) and (f) are the same, but the simulations are within the framework of averaged permittivities. For the chosen parameters the averaged permittivities are $\varepsilon_x = 5$, $\varepsilon_z = -40$; the permittivity of the contacting medium is $\varepsilon_I = 4.203$. The half-width of the incident beam is $x_0 = 1$ μm. Reprinted from Baordman *et al* (2017), copyright IOP Publishing, all rights reserved.

nonlinearity is also determined by the following extra requirement, if the hot spots are under consideration. Namely, a typical size of the hot spots should be of the order of at least several, say five thicknesses of the elementary layers included in the MM. Such a condition is satisfied for the present simulations.

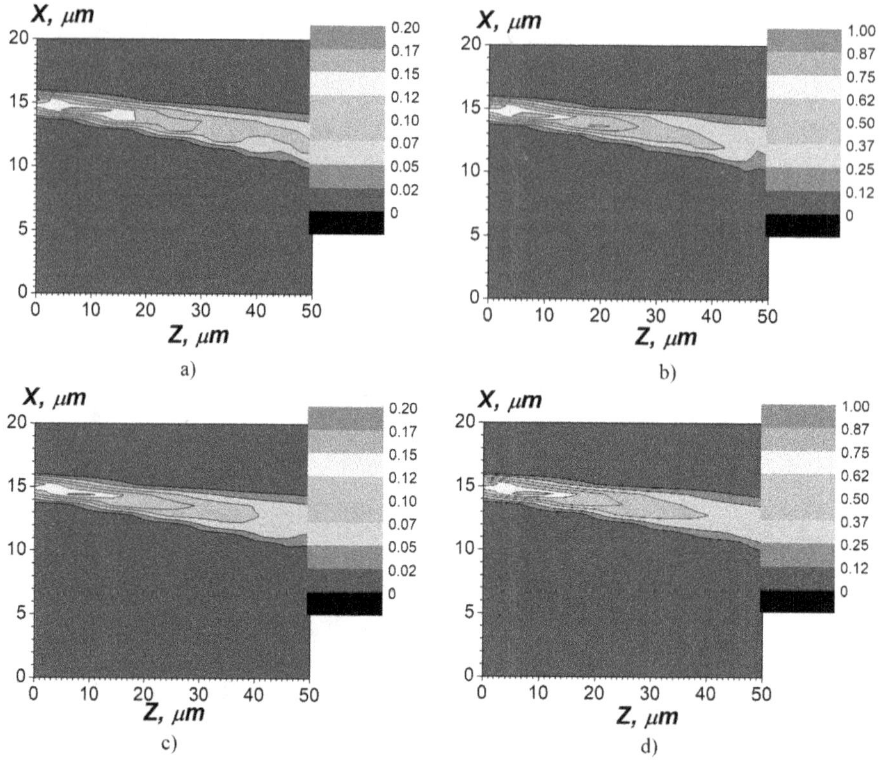

Figure 11.3. Propagation of nonlinear waves. The case of moderate nonlinearity. The circular frequency is $\omega = 6\pi \cdot 10^{14}\ \mathrm{s}^{-1}$ ($\lambda_0 = 1\ \mu\mathrm{m}$). The nonlinear coefficient for elementary layers 2, 3 is $\alpha = 1$, $\gamma = 0.5$. Parts (a) and (b) are $|E|^2$ and $|H|^2$ for the thicknesses of elementary layers $d_1 = 7.5$ nm, $d_2 = d_3 = 3.75$ nm. Parts (a) and (b) are simulated without averaging. Parts (c) and (d) are the same, but the simulations are within the framework of averaged permittivities. For the chosen parameters, the permittivity of the contacting medium is $\varepsilon_l = 4.203$. The half-width of the incident beam is $x_0 = 1\ \mu\mathrm{m}$. Reprinted from Baordman *et al* (2017), copyright IOP Publishing, all rights reserved.

11.5 The limiting case of the stationary NSE

Equation (11.11) can be reduced to the stationary nonlinear Schrödinger equation (NSE) (Kivshar and Agrawal 2003) in the case of the moderate nonlinearity. NSE has the structure:

$$K_z \frac{\partial U}{\partial \tilde{z}} + K_x \frac{\partial U}{\partial \tilde{x}} + iG \frac{\partial^2 U}{\partial \tilde{x}^2} + iNF(|U|^2)U = 0 \tag{11.21}$$

The wave amplitude $U(\tilde{z}, \tilde{x})$ is slowly varying with respect to $\tilde{z}$. The coefficients $K_{x,z}$, N, and the function F can be expressed through the parameters of the hyperbolic medium and the frequency. Equation (11.21) has been written down in the rotated coordinate frame $\widetilde{X}O\widetilde{Z}$, where $O\widetilde{Z}$ axis is aligned along the group velocity within the hyperbolic medium. The non-dimensional units are used with $c = 1$. The group velocity V_g is obtained from the linear dispersion equation:

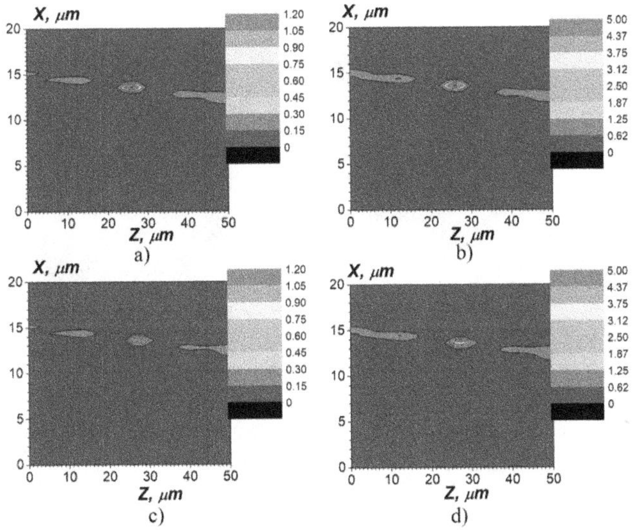

Figure 11.4. Propagation of nonlinear waves. The case of moderate nonlinearity. The circular frequency is $\omega = 6\pi\cdot 10^{14}\ s^{-1}$ ($\lambda_0 = 1$ μm). The nonlinear coefficient for elementary layers 2, 3 is $\alpha = -1$, $\gamma = 0.5$. Parts (a) and (b) are $|E|^2$ and $|H|^2$ for the thicknesses of elementary layers $d_1 = 7.5$ nm, $d_2 = d_3 = 3.75$ nm. Parts (a) and (b) are simulated without averaging. Parts (c) and (d) are the same, but the simulations are within the framework of averaged permittivities. For the chosen parameters, the permittivity of the contacting medium is $\varepsilon_I = 4.203$. The half-width of the incident beam is $x_0 = 1$ μm. Reprinted from Baordman *et al* (2017), copyright IOP Publishing, all rights reserved.

$$\begin{aligned} &D(k_z, k_x, \omega) = k_z^2\beta_x + k_x^2\beta_z - \omega^2 = 0,\\ &V_{gx} \equiv -\left(\frac{\partial D}{\partial k_x}\right)\cdot\left(\frac{\partial D}{\partial \omega}\right)^{-1} = \frac{k_x\beta_z}{\omega} < 0,\\ &V_{gz} \equiv -\left(\frac{\partial D}{\partial k_z}\right)\cdot\left(\frac{\partial D}{\partial \omega}\right)^{-1} = \frac{k_z\beta_x}{\omega} > 0,\\ &k_x, k_z, \omega, \beta_x > 0, \quad \beta_z < 0. \end{aligned} \tag{11.22}$$

Here the frequency dispersion of the components of the inverse permittivity $\beta_{x,\,z}$ is neglected. When the frequency dispersion is taken into account, the signs of the components of the group velocity preserve.

Therefore, under enough small thicknesses of the elementary layers in the case of the moderate nonlinearity the dynamics of the nonlinear beams in the hyperbolic medium can be described by NSE, where one-directional propagation of EM wave is considered.

In the case of the propagation along OZ, i.e. perpendicularly to the boundaries of the layers, the structure of stationary NSE is standard (Kivshar and Agrawal 2003):

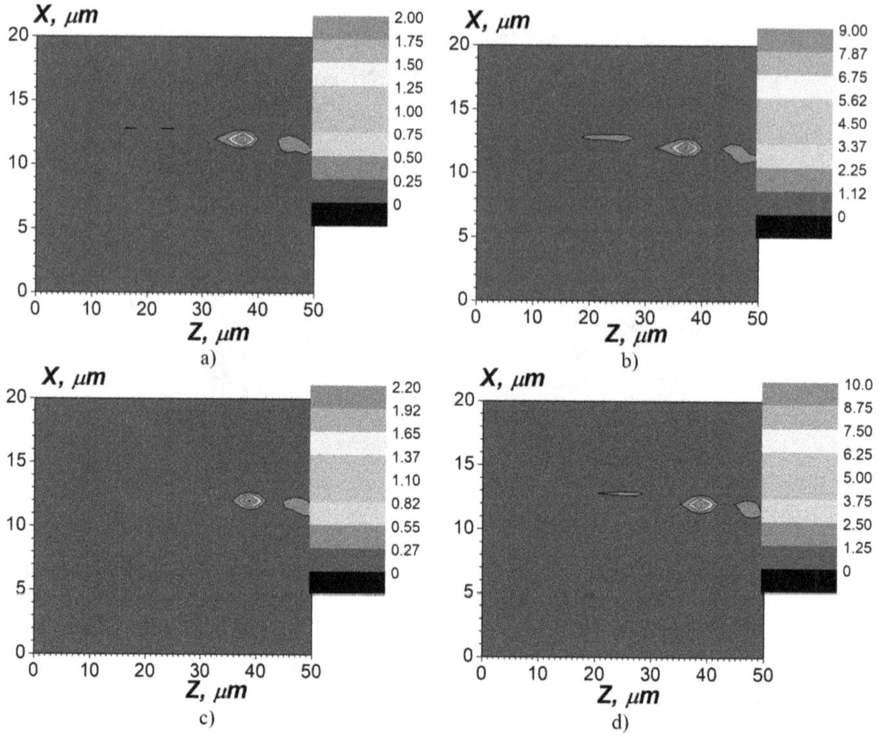

Figure 11.5. Propagation of nonlinear waves. The case is beyond the moderate nonlinearity. The circular frequency is $\omega = 6\pi \cdot 10^{14}\ \mathrm{s}^{-1}$ ($\lambda_0 = 1\ \mu\mathrm{m}$), the incidence angle is $\theta_i = 30°$, the amplitude of the incident beam is $A_0 = 1$. The nonlinear coefficient for elementary layers 2, 3 is **negative** $\alpha = -1$, $\gamma = 0.5$. Parts (a) and (b) are $|E|^2$ and $|H|^2$ for the thicknesses of elementary layers $d_1 = 7.5$ nm, $d_2 = d_3 = 3.75$ nm. Parts (a) and (b) are simulated without averaging. Parts (c) and (d) are the same, but the simulations are within the framework of averaged permittivities. For the chosen parameters, the permittivity of the contacting medium is $\varepsilon_1 = 4.203$. The half-width of the incident beam is $x_0 = 2\ \mu\mathrm{m}$, the center of the incident beam is $x_1 = 14\ \mu\mathrm{m}$. Reprinted from Baordman *et al* (2017), copyright IOP Publishing, all rights reserved.

$$\frac{\partial D}{\partial k_z}\frac{\partial U}{\partial z} - i\frac{\partial D}{\partial (k_x^2)}\frac{\partial^2 U}{\partial x^2} - i\frac{d_1 + d_2}{d}\omega^2 \alpha |U|^2 U = 0;$$
$$D(k_z, k_x, \omega) = k_z^2 \beta_x + k_x^2 \beta_z - \omega^2 = 0, \tag{11.23a}$$
$$\frac{\partial D}{\partial k_z} = 2k_z\beta_x > 0, \qquad \frac{\partial D}{\partial (k_x^2)} = \beta_z < 0.$$

Equation (11.23a) is written down in the case of moderate nonlinearity. Here $D = 0$ is the linear dispersion equation for EM waves in the hyperbolic medium. Equation (11.23a) can be rewritten in the equivalent manner:

$$\frac{\partial U}{\partial z} - i\frac{\beta_z}{2k_z\beta_x}\frac{\partial^2 U}{\partial x^2} - i\frac{d_1 + d_2}{d}\frac{\omega^2 \alpha}{2k_z\beta_x}|U|^2 U = 0. \tag{11.23b}$$

One can see that in the case of the **negative cubic nonlinearity** $\alpha < 0$ the signs of the diffraction and nonlinear coefficients coincide, and the bright spatial solitons can be formed.

11.6 The discussion and main results

In the near-infrared and visible optical range the metallic (or semi-metallic) layers with a high conductivity can be used as media with the negative permittivity, whereas the dielectrics with high values of permittivity can be used as another layer. In the THz range the narrow forbidden gap semiconductors, like *n*-InSb, can be used as media with the negative permittivity. The metallic layers possess the dissipation, even in metals with high conductivity. Therefore, the dissipation should be compensated, to observe the nonlinear wave phenomena. A mechanism of compensation of dissipation can be a creation of active dielectric layers, for instance, by means of inserting quantum dots with the inversion of energetic levels. In this chapter only the case of exact compensation is considered. The dynamics of nonlinear waves under moderate net amplification is of great interest, it is very non-trivial, in accordance with our preliminary evaluations, and is a subject of future work. The first approach described in section 11.2 including equation (11.4) is the appropriate one and will be applied for a solution of such a problem(s). In particular, the question of a possibility of stationary regimes as such, while the hot spots are formed in the nonlinear media with a wave amplification is, again, very non-trivial and will be considered elsewhere. The method of averaging can be applied when the thicknesses of elementary layers are small and the nonlinearity is moderate. When the following inequality is valid $d_i \leqslant 0.1\lambda_i$ where d_i is the thickness of each elementary layer, $\lambda_i \equiv \lambda_0 \cdot \varepsilon_i^{-1/2}$ is the wavelength of the electromagnetic wave in this medium, the results of simulations are the same both within the direct consideration of the layered medium and within the averaging approach for the hyperbolic medium. In this case it is possible to reduce the nonlinear equation for EM wave propagation to the standard NSE, where the wave amplitude is slowly varying with respect the longitudinal coordinate, i.e. along the direction of propagation. When the inequalities are valid $0.1\lambda_i \leqslant d_i \leqslant 0.25\lambda_i$, there are some quantitative differences within two approaches pointed above but the results are qualitatively similar. At larger thicknesses of elementary layers the differences between two approaches are qualitative and the averaging approach is not valid. The averaging in the MM approximation of the continuous media is based on the formulas (11.10b), (11.11). These formulas have been derived, accounting for the periodicity of the structure and with the farther application of the approximation of the thin layers (continuous media) in the linear limiting case. For the nonlinear media, the application of the formulas (11.11) in the same form, as that for the linear case, is only phenomenological. This is why it was necessary and important to verify the MM approximation for active nonlinear media (formulas (11.10b), (11.11)) by means of the comparison between the results of the corresponding modeling with these obtained using a more accurate approach without an averaging (based on equation (11.4)). We would like to emphasize that the positive result of such a comparison is obtained in the present work only for the case of the nonlinear active media with net zero gain (total compensation between gain and losses). A possibility of the MM approach to the hyperbolic nonlinear periodical active media with the nonzero net gain, as to the continuous media is questionable, even providing that the requirements of the MM approximation applicability are satisfied in the linear limiting case. These problems will be a subject of future papers.

For solving nonlinear problems various difference schemes have been applied. The implicit–explicit method of Peaceman–Rachford does not possess good stability. The method of the summatory approximation needs very small temporal steps and is practically not applicable there. It is very interesting that the method of the operator factorization, or the method of Douglas–Rachford (known more in the hydrodynamics (Marchuk 1988, Douglas and Rachford 1956) than in the nonlinear optics), seems the most appropriate.

The nonlinear effects are different for different signs of nonlinearity. In the case of the negative nonlinearity of the layers with the positive permittivity the hot spots can be formed within the hyperbolic medium. In the case of the positive nonlinearity of the layers with positive permittivity the nonlinear diffraction of the EM wave beams occurs.

Problems to chapter 11

Problem 11.1

Consider the system shown in figure P11.1. The structure is periodical in z direction and infinite in y direction.

The electromagnetic wave with field components $\{E_x, H_y, E_z\}$ propagate in the structure shown in figure P11.1. Note that the z components of the corresponding wave number in layers 1 and 2 of the element cell (figure P11.1) are equal to

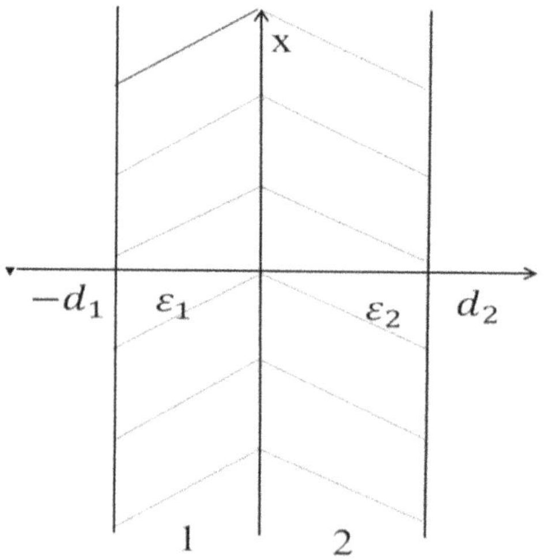

Figure P11.1. Elementary cell of hyperbolic MM; $d = d_1 + d_2$ is the period of structure.

$$k_{1,2}^2 = \frac{\omega^2}{c^2}\varepsilon_{1,2} - k_x^2 \tag{P11.1}$$

Note that it is supposed that

$$(H_x, H_y, H_z) \sim e^{-i\omega t + ik_x x + ik_{1,2} z}$$

Obtain the dispersion equation for the E-wave.

In the regions 1 ($-d_1 < z < 0$), 2 ($0 < z < d_2$) and $d_2 < z < d$, H_y component equal to:

$$-d_1 < z < 0: H_y = B_{11}\cos(k_1 z) + B_{12}\sin(k_1 z) \tag{P11.2}$$

$$0 < z < d_2: H_y = B_{21}\cos(k_2 z) + B_{22}\sin(k_2 z) \tag{P11.3}$$

$$d_2 < z < d: H_y = \left[B_{11}\cos(k_1(z-d)) + B_{12}\sin(k_1(z-d))\right]e^{ikd} \tag{P11.4}$$

Boundary conditions of continuity of H_y, E_x:

$$\text{at } z = 0: B_{11} = B_{21}; \frac{k_1}{\varepsilon_1}B_{12} = \frac{k_2}{\varepsilon_2}B_{22} \tag{P11.5}$$

As follows from (P11.5),

$$B_{21} = B_{11}; B_{12} = \frac{k_2\varepsilon_1}{k_1\varepsilon_2}B_{22} \tag{P11.6}$$

The boundary conditions at $z = d_2$ have the form:

$$\begin{gathered}(B_{11}\cos(k_1 d_1) - B_{12}\sin(k_1 d_1))e^{ikd} = B_{21}\cos(k_2 d_2) + B_{22}\sin(k_2 d_2)\\ \frac{k_1}{\varepsilon_1}(B_{11}\sin(k_1 d_1) + B_{12}\cos(k_1 d_1))e^{ikd} = \frac{k_2}{\varepsilon_2}(-B_{21}\sin(k_2 d_2) + B_{22}\cos(k_2 d_2))\end{gathered} \tag{P11.7}$$

Accounting for (P11.7) and (P11.6) the boundary conditions at $z = d_2$ take the form:

$$\begin{gathered}B_{11}\cos(k_1 d_1)e^{ikd} - \frac{k_2\varepsilon_1}{k_1\varepsilon_2}B_{22}\sin(k_1 d_1)e^{ikd} = B_{11}\cos(k_2 d_2) + B_{22}\sin(k_2 d_2)\\ \frac{k_1}{\varepsilon_1}B_{11}\sin(k_1 d_1)e^{ikd} + \frac{k_2}{\varepsilon_2}B_{22}\cos(k_1 d_1)e^{ikd} = -\frac{k_2}{\varepsilon_2}B_{11}\sin(k_2 d_2) + \frac{k_2}{\varepsilon_2}B_{22}\cos(k_2 d_2)\end{gathered} \tag{P11.8}$$

As follows from (P11.8)

$$B_{11}\left[\cos(k_2 d_2) - \cos(k_1 d_1)e^{ikd}\right] + B_{22}\left[\sin(k_2 d_2) + \frac{k_2\varepsilon_1}{k_1\varepsilon_2}\sin(k_1 d_1)e^{ikd}\right] = 0 \tag{P11.9}$$

$$B_{11}\left[\frac{k_1}{\varepsilon_1}\sin(k_1 d_1)e^{ikd} + \frac{k_2}{\varepsilon_2}\sin(k_2 d_2)\right] + B_{22}\frac{k_2}{\varepsilon_2}\left[\cos(k_1 d_1)e^{ikd} - \cos(k_2 d_2)\right] = 0$$

The corresponding dispersion relation is:

$$\frac{k_2}{\varepsilon_2}\left[\cos(k_2d_2) - \cos(k_1d_1)e^{ikd}\right]\left[\cos(k_1d_1)e^{ikd} - \cos(k_2d_2)e^{ikd}\right]$$
$$-\left[\sin(k_2d_2) + \frac{k_2\varepsilon_1}{k_1\varepsilon_2}\sin(k_1d_1)e^{ikd}\right]\frac{k_1}{\varepsilon_1}\sin(k_1d_1)e^{ikd} + \frac{k_2}{\varepsilon_2}\sin(k_2d_2) \qquad \text{(P11.10)}$$
$$= 0$$

Relation (P11.10) reduces to

$$-\frac{k_2}{\varepsilon_2}\cos^2(k_2d_2) - \frac{k_1}{\varepsilon_1}\sin(k_1d_1)\sin(k_2d_2)e^{ikd} - \frac{k_2}{\varepsilon_2}\sin^2(k_2d_2)$$
$$-\frac{k_2}{\varepsilon_2}\sin^2(k_1d_1)e^{ikd} - \frac{k_2^2\varepsilon_1}{k_1^2\varepsilon_2}\sin(k_1d_1)\sin(k_2d_2)e^{ikd} = 0 \qquad \text{(P11.11)}$$

Equation (11.11) can be presented, finally, as:

$$-\frac{k_2}{\varepsilon_2}(e^{2ikd} + 1) + 2\frac{k_2}{\varepsilon_2}\cos(k_1d_1)\cos(k_2d_2)e^{ikd}$$
$$-\left(\frac{k_1}{\varepsilon_1} + \frac{k_2^2\varepsilon_1}{k_1^2\varepsilon_2}\right)\sin(k_1d_1)\sin(k_2d_2)e^{ikd} = 0 \qquad \text{(P11.12)}$$

This is the dispersion equation for the E-mode propagating the structure shown in figure P11.1.

Problem 11.2

Obtain the dispersion equation for the E-mode propagating in the periodical structure, one period of which is shown in figure P11.1 under the conditions $kd \ll 1$, see the condition of problem 11.1 and caption to figure P11.1 and $k_x = 0$.

Use the dispersion relation for the E-mode in the periodical structure (figure P11.1) in the form P11.12 (problem P11.1). In the approximation $kd \ll 1$, one can get

$$1 - \frac{k^2d^2}{2} \simeq 1 - \frac{k_1^2d_1^2 + k_2^2d_2^2}{2} - \frac{1}{2}\left(\frac{k_1\varepsilon_2}{k_2\varepsilon_1} + \frac{k_2\varepsilon_1}{k_1\varepsilon_2}\right)k_1d_1 \cdot k_2d_2 \qquad \text{(P11.13)}$$

Note that here (see also equation (P11.1))

$$k \equiv k_z \qquad \text{(P11.14)}$$

Accounting (P11.13), it is possible to write:

$$k^2d^2 \simeq k_1^2d_1^2 + k_2^2d_2^2 + \left(\frac{k_1^2\varepsilon_2}{\varepsilon_1} + \frac{k_2^2\varepsilon_1}{\varepsilon_2}\right)d_1d_2 \qquad \text{(P11.15)}$$

Equation (P11.15) can be presented in the form

$$k^2d^2 \simeq k_1^2 d_1\left(d_1 + \frac{\varepsilon_2}{\varepsilon_1}d_2\right) + k_2^2 d_2\left(d_2 + \frac{\varepsilon_1}{\varepsilon_2}d_1\right)$$
$$= \left(\frac{k_1^2 d_1}{\varepsilon_1} + \frac{k_2^2 d_2}{\varepsilon_2}\right)(\varepsilon_1 d_1 + \varepsilon_2 d_2) \qquad \text{(P11.16)}$$

Equation (P11.6) can be rewritten as:

$$k^2d^2 \simeq \left(\frac{\omega^2}{c^2}d_1 - \frac{k_x^2 d_1}{\varepsilon_1} + \frac{\omega^2}{c^2}d_2 - \frac{k_x^2 d_2}{\varepsilon_2}\right)(\varepsilon_1 d_1 + \varepsilon_2 d_2)$$
$$= \left(\frac{k_1^2 d_1}{\varepsilon_1} + \frac{k_2^2 d_2}{\varepsilon_2}\right)(\varepsilon_1 d_1 + \varepsilon_2 d_2) \qquad \text{(P11.17)}$$

Relation (P11.17) is the dispersion equation for E-mode (see the condition of the problem 11.1) in the periodical structure one period of which is shown in figure P11.1 in the approximation $kd \ll 1$.

Problem 11.3

Obtain the components $\varepsilon_{1\text{eff}} \equiv \varepsilon_{x\text{eff}}$, $\varepsilon_{\varsigma\text{eff}} \equiv \varepsilon_{z\text{eff}}$ of effective tensor of electric permittivity $\hat{\varepsilon}_{\text{eff}}$ for the E-mode of electromagnetic wave in the periodical structure, one period of which is shown in figure P11.1. Suppose that the relation $kd \ll 1$ is valid.

Accounting for the condition $kd \ll 1$, relation (P11.17) in the form:

$$k_z^2 = \frac{w^2}{c^2}\varepsilon_{1\text{eff}} - \frac{k_x^2 \varepsilon_{1\text{eff}}}{\varepsilon_{\varsigma\text{eff}}} \qquad \text{(P11.18}a\text{)}$$

or

$$\frac{k_z^2}{\varepsilon_{1\text{eff}}} + \frac{k_x^2}{\varepsilon_{\varsigma\text{eff}}} = \frac{w^2}{c^2} \qquad \text{(P11.18}b\text{)}$$

where

$$\varepsilon_{1\text{eff}} = \frac{\varepsilon_1 d_1 + \varepsilon_2 d_2}{d} \equiv \varepsilon_{\chi\text{eff}} \qquad \text{(P11.19)}$$

$$\frac{1}{\varepsilon_{\varsigma\text{eff}}} = \frac{1}{d}\left(\frac{d_1}{\varepsilon_1} + \frac{d_2}{\varepsilon_2}\right) \equiv \varepsilon_{z\text{eff}} \qquad \text{(P11.20)}$$

Comparing the relations (P11.18a) and (P11.18b) with the corresponding relations from the paper (Poddubny *et al* 2013), one can see that the corresponding components of equivalent tensor $\hat{\varepsilon}$ describing the periodical hyperbolic MM structure, one period of which is shown in figure P11.1), really are described by equation (P11.20).

References

Alberucci A and Assanto G 2011 Nonparaxial (1+1)D spatial solitons in uniaxial media *Opt. Lett.* **36** 193–95

Boardman A D, Egan P and McCall M 2015 Optic axis-driven new horizons for hyperbolic metamaterials *EPJ Appl. Metamat.* **2** 1–7

Boardman A D, Alberucci A, Assanto G, Grimalsky V V, Kibler B, McNiff J, Nefedov I S, Rapoport Yu G and Valagiannopoulos C A 2017 Waves in hyperbolic and double negative metamaterials including rogues and solitons *Nanotechnology* **28** 444001

Douglas J and Rachford H 1956 On the numerical solution of heat conduction problems in two and three space variables *Trans. Am. Math. Soc.* **82** 421–39

Felsen L F and Marcuvitz M 1973 *Radiation and Scattering of Waves* vol 2 (Englewood Cliffs NJ: Prentice-Hall) p 551

Jacob Z and Shalaev V M 2011 *Plasmonics Goes Quantum. Sci.* **334** 463–4

Kauranen M, Linden S and Wegener M 2017 Nonlinear metamaterials *World Scientific Handbook of Metamaterials and Plasmonics (World Scientific Series in Nanoscience and Nanotechnology)* (Singapore: World Scientific) ch 3 pp 69–111

Kivshar Y S and Agrawal G P 2003 *Optical Solitons* (New York: Academic) p 539

Krasavin A V, Ginzburg P and Zayats A V 2019 *Nonlinear Nanoplasmonics. Quantum Photonics: Pioneering Advances and Emerging Applications* (Berlin: Springer) pp 267–316

Kuznetsov A I, Miroshnichenko A E, Brongersma M L, Kivshar Y S and Luk'yanchuk B 2016 Optically resonant dielectric nanostructures *Science* **354** 6314

Marchuk G I 1988 *Methods of Splitting* (Moscow: Nauka) p 264 (in Russian)

Poddubny A, Iorsh I, Belov P and Kivshar Y 2013 Hyperbolic metamaterials *Nat. Photonics* **7** 948–57

Rapoport Y U, Boardman A, Grimalsky V, Selivanov Y U and Kalinich N 2012 Metamaterials for space physics and the new method for modeling isotropic and hyperbolic nonlinear concentrators *Proc. 14th Intern. Conf. on Mathematical Methods in Electromagnetic Theory MMET (*Kharkiv Ukraine*)* Art. No. 6331154 pp 76–9

Rapoport Y G, Grimalsky V, Koshevaya S V, Boardman A D and Malnev V N 2014 New method for modeling nonlinear hyperbolic concentrators *Proc IEEE 34th Int. Sci. Conf. on Electronics and Nanotechnology ELNANO* Article number 6873975 pp 35–8

Samarskii A A 2001 The theory of difference *Economical Difference Schemes for Multidimensional Problems in Mathematical Physics* (New York: Schemes Marcel Dekker) ch 9 pp 543–642

Shalaginov M Y, Chandrasekar R, Bogdanov S, Wang Z, Meng X, Makarova O A, Lagutchev A, Kildishev A V, Boltasseva A and Shalaev V M 2017 Hyperbolic metamaterials for single-photon sources and nanolasers *Quantum Plasmonics* ed S Bozhevolnyi, L Martin-Moreno and F Garcia-Vidal (Springer Series in Solid-State Sciences) (Berlin: Springer) p 185

Shekhar P, Atkinson J and Jacob Z 2014 Hyperbolic metamaterials: Fundamentals and applications *Nano Convergence* **1** 1–17

Simovski C, Maslovski S, Nefedov I and Tretyakov S 2013 Optimization of radiative heat transfer in hyperbolic metamaterials for thermophotovoltaic applications *Opt. Express* **21** 14988–5013

Smalley J S T, Vallini F, Gu Q and Fainman Y 2016 Amplification and lasing of plasmonic modes *Proc. IEEE* **104** 2323–37

Smolyaninov S and Hung Y-J 2011 Modeling of time with metamaterials *J. Opt. Soc. Am.* B **28** 1591–5

Urbas A M *et al* 2016 Roadmap on optical metamaterials *J. Opt.* **18** 093005

Chapter 12

Analytical models of formation of nonlinear dissipative wave structures in active quantum hyperbolic planar resonant metamaterials in IR range

12.1 General description of the problem

Two approaches to moderately and strongly nonlinear active layered hyperbolic metamaterial (MM) are applied, with and without averaging over the cell. For the activity in the infrared (IR) range of the layers included in the unit cell MM, the active elements like quantum dots can be introduced into the material. In the parabolic approximation, new evolution equations of the amplitude of the envelope of nonlinear wave packets propagating in active hyperbolic media at an arbitrary angle to the optical axis are derived. In this case, the mixed derivative in the evolution equation disappears, and the equations are reduced to the nonlinear Schrödinger equation (NSE) for waves propagating along the optical axis. The nonlinear quantum-optical approach to waves in hyperbolic active MM is outlined.

12.1.1 Relevance of metamaterials, and in particular hyperbolic metamaterials, for modern photonics and the role of metamaterials in modern research projects

The most typical studies are the ones of advanced space agencies, as they require the miniaturization and optimization of mass and size characteristics, the minimization of the energy consumption and information capacity, the highest possible functionality and controllability. Therefore, the latest nanophotonic systems are used in this field, including optical MM, in particular, hyperbolic ones, from terahertz (THz) to the optical range.

doi:10.1088/978-0-7503-2336-9ch12

This is evidenced by the list of nanophotonics and MM projects for the space research and communication problems for which the use of modern MMs is critical. Such projects include, in particular, the following:

The recent US National Aeronautics and Space Administration projects (for instance, https://www.sbir.gov/node/1657429) on the MMs includes: MM antennas for space exploration; MM with a negative refractive index for high-temperature selective radiation for thermophotoelectrics; thermoelectric MM with high power density for energy collection; the miniature broadband antenna for communication and sounding; nanotechnologies for advanced images and for sensitive detectors.

The European Space Agency project (esa.int/gsp/ACT/ama/projects/metamaterials.html/) includes the MM-based development: the creation of the European consortium 'Metamorphose', which deals with MM research; materials with optical bands (bandgap materials); ideal lenses, antennas and waveguides; antenna pads; manipulation of electromagnetic fields and concentration of radiation in a certain space region; the invisible cloak, including reduction of radiation pressure; thermal insulators; equipment weight reduction.

12.1.2 The importance of nonlinear wave processes in active hyperbolic metamaterials

Studies of nonlinear hyperbolic, plasmonic and quantum MM (Boardman *et al* 2010, 2011, Shadrivov *et al* 2015, Smalley *et al* 2018, Savelev *et al* 2013), including the development of theory, physical and mathematical models and modeling, are promising for the development of methodological and technological bases for creating more efficient and highly sensitive sensors and solar panels; solving some problems of the electromagnetic compatibility of equipment, creating broadband communication systems and processing information with controlled parameters necessary, in particular, for communication between spacecraft, as well as between a spacecraft and the Earth in the presence of obstacles. The results of modeling the optical properties of MMs will be useful, in particular, for the development of a new high-speed controlled optical modulator that controls light, tiny antennas with controlled characteristics; improving the functional and weight characteristics of space research equipment and communication systems on a spacecraft; creation of ultra-high resolution image transmission systems. The study of plasmonic (Boardman *et al* 2017), hyperbolic nonlinear (Shekhar *et al* 2014) and active (Pustovit *et al* 2016, Schulz *et al* 2016) MMs will be promising for reducing energy consumption on board spacecraft, reducing their mass, increasing the service life of the device and information capacity in transmitting information, increasing efficiency and service life of spacecraft for research and communication in general.

12.1.3 Modern hyperbolic metamaterials

On the base of the hyperbolic MMs it is possible to create micro- and nanophotonic devices with fundamentally improved mass and size characteristics, controllability and functionality. Hyperbolic MMs: (1) are created by structuring layers of MMs, the thicknesses of which are less than the wavelength; (2) technologically the use of hyperbolic MMs is much easier than the use of many other MMs; (3) you can use the effect of the hyperbolic dispersion; (4) they have a negative refractive index; (5) active elements (e.g., quantum dots) can be easily incorporated into hyperbolic MMs; (6) they provide an increase in the quantum density and intensity of spontaneous EM radiation; (7) they are capable of enhanced superlensing; (8) they can be used to operate in a wide range, namely from radio frequencies to the optical range; (9) they have a significant increase in nonlinearity during the passage of the wave; (10) the linear and nonlinear characteristics of hyperbolic MMs can be controlled by external electric and magnetic fields.

In particular, this section will consider the geometry (figure 12.1),which is characteristic for hyperbolic MMs with a special layer in each cell, where active inclusions are made, which can be, depending on the frequency range, active diodes, quantum dots, dye molecules, etc (Poddubny *et al* 2013, Jacob and Shalaev 2011, Urbas *et al* 2016, Guo *et al* 2020, Boardman *et al* 2017, Shekhar *et al* 2014, Poddubny *et al* 2013, Makarova *et al* 2017).

When considering the propagation of electromagnetic waves (EMWs) in the structure (figure 12.1) the averaging of the parameters of the structure in the unit cell is realized (the approximation of a continuous medium). It is assumed that the length of each unit cell of the structure is significantly less than the wavelength for EMW in the structure. The electric permittivity of the structure is represented as a diagonal tensor. Another approach uses a direct solution of Maxwell equations without averaging the electrodynamic parameters over the unit cell.

When applying the MM approach, the dielectric permittivity tensor is used (see also figure 12.1)

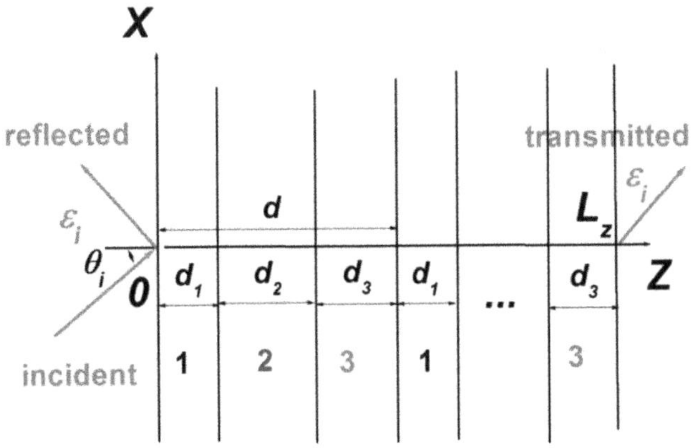

Figure 12.1. Geometry of the problem. It is assumed that the unit cell of the MM consists of three layers. Two approaches are used, namely the first one is the MM approximation, and it is considered that the length of an elementary cell is much less than wavelength. Layer 1 is a dielectric, layer 2 is a plasma-like medium (metal or semiconductor), layer 3 is a layer with active inclusions. θ_i is the angle of incidence of EMW onto the system. Reprinted from Boardman *et al* (2017), copyright IOP Publishing, all rights reserved.

$$\hat{\varepsilon} = \begin{pmatrix} \varepsilon_\perp & 0 & 0 \\ 0 & \varepsilon_\perp & 0 \\ 0 & 0 & \varepsilon_\parallel \end{pmatrix} \tag{12.1}$$

The elements of the averaged permittivity tensor included in formula (12.1) are defined as (Poddubny *et al* 2013)

$$\langle \varepsilon_\perp \rangle = \sum_{i=1}^{N} \varepsilon_i \frac{d_i}{d} (N = 3) \tag{12.2}$$

$$\langle \varepsilon_z \rangle = \frac{1}{\langle \varepsilon_z \rangle^{-1}} = \frac{1}{\sum_i \frac{1}{\varepsilon_i} \frac{d_i}{d}} \tag{12.3}$$

In the literature there are two types of MMs with the following relations of tensor signs included in (12.1)–(12.3):

$$\varepsilon_\perp \equiv \varepsilon_x = \varepsilon_y > 0;\ \varepsilon_\parallel \equiv \varepsilon'_z < 0 \tag{12.4}$$

$$\varepsilon_\perp \equiv \varepsilon_x = \varepsilon_y < 0;\ \varepsilon_\parallel \equiv \varepsilon_z > 0 \tag{12.5}$$

As a result, the corresponding dispersion equation describes the hyperbolic surface of a constant frequency (energy) in the space of wave vectors, taking into account the fact that the elements of the electric permittivity in (12.4), (12.5) have a different sign:

$$\frac{k_x^2 + k_y^2}{\varepsilon_z} + \frac{k_z^2}{\varepsilon_x} = \left(\frac{\omega}{c}\right)^2,\ \varepsilon_z \equiv \varepsilon_\parallel,\ \varepsilon_x \equiv \varepsilon_\perp \tag{12.6}$$

The conditions of opposite signs of the longitudinal and transverse components of the tensor ε included in the dispersion equation (12.6), as well as conditions of relatively small value of energy losses lead to rather strict conditions for choosing the parameters of plasma and metal layers (table 12.1).

Table 12.1. Materials for creating MM for frequency ranges from ultraviolet to THz EM radiation (Shekhar *et al* 2014).

Au/Al_2O_3 Ag/Al_2O_3	Au/TiO_2 Ag/TiO_2	TiN, ZrN, AZO, GZO, ITO	InGaAs, AlInAs, SiC, graphene
Ultraviolet	Visible light	Near IR	Middle IR
Plasmonic materials		Alternative plasmon materials	III–V semiconductors Phonon polaritons Two-dimensional materials

Such hyperbolic energy surfaces with two types of dispersion are considered, in particular, in (Poddubny *et al* 2013). For the use of hyperbolic materials, the Purcell effect is important, which is to increase the emission rate of the oscillator in the resonator compared to the rate of the spontaneous radiation in free space (Poddubny *et al* 2013). A significant increase in the radiation intensity of photonic sources, including single-photon ones, is used in modern systems for quantum cryptography. Physically, this effect is related to the theoretical infinity of the hyperbolic surface of constant energy or frequency in the space of wave vectors (the Fourier space). Another important application of such MMs is a new method of heat removal from heated bodies due to the effective Boltzmann constant increased by several orders of magnitude. A significant increase in the efficiency of single-photon sources placed in the middle of a cylindrical lens made of hyperbolic MM was proposed in (Makarova *et al* 2017).

12.2 Theoretical approach to modeling of modern nonlinear active hyperbolic metamaterials. Ginzburg–Landau equation

A nonlinear evolution equation for a wave with a moderate spectrum width and nonlinearity is a nonlinear parabolic equation, or NSE. Assume that a wave propagates in a nonlinear (for example, cubic nonlinearity) isotropic medium or along a magnetic field applied to an isotropic nonlinear medium. A wave packet with a narrow spectrum in the vicinity of a carrier wavenumber and a carrier frequency (k_0, ω_{k0}) is considered in the parabolic approximation, and we assume that $\Delta k \ll k$, $\Delta\omega \ll \omega$. Suppose that in the relation (12.7) $\varphi(z,t)$ is the amplitude of the wave packet, which slowly changes in space and time (Kadomtsev 1988):

$$E(z,\, t) = \varphi(z,\, t)\exp(ik_0 z - i\omega_{k0} t) \tag{12.7}$$

The relations for the spectral components $\omega = \omega_{k0} + \Delta\omega$, $k = k_0 + \Delta k$ are:

$$(\omega - \omega_k)\varphi_{\Delta\omega,\, \Delta k} = 0 \tag{12.8}$$

Equation (12.8) can be represented as

$$(\Delta\omega - \omega_k + \omega_{k0})\varphi_{\omega,\, \Delta k} = 0 \tag{12.9}$$

The group velocity and its spectral derivative, called the dispersion coefficient, are represented, respectively, in the form

$$V_{\mathrm{g}} = \partial\omega/\partial k;\; V'_{\mathrm{g}} \equiv \frac{\partial V_{\mathrm{g}}}{\partial k} = \frac{\partial^2\omega}{\partial k^2} \tag{12.10}$$

Using the relations (12.8)–(12.10), in the parabolic approximation we have the expansion

$$\omega_k \simeq \omega_{k0} + \omega_k\Delta k + \frac{1}{2}\omega_{kk}\Delta k^2 \tag{12.11}$$

We apply an inverse Fourier transform, using the operator method (Kadomtsev 1988):

$$i\Delta k \to \frac{\partial}{\partial z};\ i\Delta\omega \to -\frac{\partial}{\partial t} \tag{12.12}$$

Taking into account the nonlinear frequency shift $\Delta\omega_{NL} = \alpha \mid \varphi \mid^2$, where α is the nonlinearity coefficient, with using (12.7)–(12.12), we obtain a nonlinear parabolic Schrödinger-type equation (NSE), which describes, in this case, the evolution of the envelope wave packet in a medium without dissipation:

$$i\left(\frac{\partial\varphi}{\partial t} + V_g\frac{\partial\varphi}{\partial z}\right) + \frac{V'_g}{2}\frac{\partial^2\varphi}{\partial z^2} - \alpha \mid \varphi \mid^2 \varphi = 0 \tag{12.13}$$

In equation (12.13) α is the coefficient of nonlinearity, which determines the nonlinear frequency shift $\Delta\omega_{NL} = \alpha \mid \varphi \mid^2$.

Consider the general form of equations for a nonlinear medium. In contrast to equation (1.43), which describes a nonlinear dissipative medium, the Ginzburg–Landau equation allows us to study the propagation of a wave packet in an active (dissipative) medium. For such media, the dispersion relation is nonlinear and complex (Riskin and Trubetskov 2000).

As shown in figure 12.2, ν is a parameter of the order (figure 12.2), and when the value of this parameter exceeds some critical value, the dissipation of the medium changes its sign. At ν equal to the critical value of ν_C, ω'' is equal to zero,

$$\omega \equiv \omega'(k, \mid \psi \mid^2) + i\omega''(k, \nu), \tag{12.14}$$

where ψ is the amplitude of the envelope.

In the relation (12.14) ν is a parameter of the order that determines, in particular, the limit where the medium becomes active (figure 12.2).

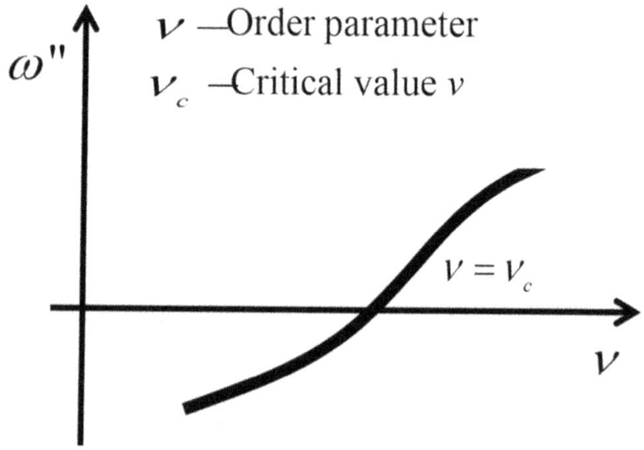

Figure 12.2. Dependence of the imaginary part of the frequency ω'' on a parameter of order ν.

The qualitative form of the dependence of ω'' on ν (figure 12.2) can be described by the following relations:

$$\begin{aligned} &\nu < \nu_c; \quad \omega''(k, \nu) < 0 \\ &\nu > \nu_c; \quad \omega''(k, \nu) > 0 \\ &\nu = \nu_c; \quad \omega''(k, \nu) = 0 \end{aligned} \tag{12.15}$$

Consider a wave packet in a nonlinear active medium:

$$E(x, t) = \psi(x, t)\exp[i(\Delta k \cdot x - \Delta\omega \cdot t)]\exp[i(k_0 x - \omega_0 t)] \tag{12.16}$$

We use the approximation of weak overcriticality. We apply in the dispersion relation the expansion in the vicinity of $\nu = \nu_c$; $\omega'' = 0$, taking into account the relations (12.15): $\Delta\omega = \omega - \omega_0$, $\Delta k = k - k_0$ where ω_0 and k_0 are, respectively, the carrier frequency and the wavenumber. We apply then the inverse Fourier transform with using the operator method (12.12). As a result, we obtain the amplitude of the envelope in a nonlinear medium:

$$\begin{aligned} &i(\psi_t + V_g\psi_x) + \beta\psi_{xx} + \gamma \mid \psi \mid^2 \psi = i\alpha\psi \\ &\alpha = (\nu - \nu_c)\partial\omega''/\partial\nu = \omega'' \end{aligned} \tag{12.17a}$$

Suppose that $\gamma'' \equiv \mathrm{Im}\gamma = 0$; $\tau = t$; $\xi = x - V_g t$; $\bar{\xi}^2 = \xi^2/\beta'$; $\psi = \psi_0\bar{\psi}$; $\psi_0^2 = \gamma'^{-2}$; $\gamma' \equiv \mathrm{Re}(\gamma)$;

After outlined normalization one can rewrite equation (12.17a) in the following form:

$$\bar{\psi}_\tau + i\bar{\psi}_{\xi\xi} + i|\bar{\psi}|^2\bar{\psi} = \nu_0\bar{\psi} - \mu\bar{\psi}_{\xi\xi} \tag{12.17b}$$

It is important that the coefficients in equation (12.17b), in particular, β and γ are complex, in the general case. In equation (12.17b) $\mu \equiv \beta''/|\beta'| \equiv \mathrm{Im}\beta/|\mathrm{Re}\,\beta|$; $\nu_0 \equiv \alpha'$; we suppose also that $\beta'|\equiv\mathrm{Re}(\beta) < 0$, $\beta'' \equiv \mathrm{Im}(\beta) > 0$, $\alpha'' \equiv \mathrm{Im}(\alpha) = 0$. Note that equation (12.17b) coincides with equation (7.40) from Guglielmi and Pokhotelov (1996). This equation is a particular case of the complex Ginzburg–Landau equation (Akhmediev and Ankiewitcz 2005, Boardman and Velasco 2006, García-Morales and Krischer 2012, Défi *et al* 2020). The second term in the left-hand side of equation (12.17b) describes the pulse dispersion, and the third one describes the nonlinear frequency shift; the first and the second terms in the right-hand side of this equation describe the effects of the linear gain and the gain restriction (including change sign) with the pulse spectrum broadening, respectively. As can be seen from this equation, small oscillations can be amplified in the active medium. Nevertheless the solution of equation (12.17b) does exist (Guglielmi and Pokhotelov 1996). The solution of equation (12.17b) in the form of the soliton in active media is presented in (Guglielmi and Pokhotelov 1996):

$$\bar{\psi}(\xi, \tau) = \bar{\psi}_0 \,\mathrm{sech}(\xi/\xi_0)\exp\left(-i\int_0^\tau q\mathrm{d}\tau'\right) \tag{12.17c}$$

With the amplitude corresponding to the stationary solution, equal to

$$\overline{\psi}_0 = \sqrt{6\nu_0/\mu}\,, \quad \xi_0 = \sqrt{\mu/3\nu_0}\,, \quad q = \overline{\psi}_0^2/2, \tag{12.17d}$$

Note that to consider the stability of the solution of Ginzburg–Landau equation it is necessary to include higher order terms, in particular (complex) quintic nonlinearity (Mihalache *et al* 2007, Boardman and Velasco 2006); the questions of stability of dissipative solitons will be not considered in the present chapter in any detail. Note that the Ginzburg–Landau equation (12.17) is related to the self-oscillation in nonlinear systems. The equation for the slowly varying amplitude A of the nonlinear oscillator near its resonant frequency in the presence of external harmonic force acting with the frequency close to the resonant frequency of the oscillator reduces to the form (Anisimov 2003, Rabinovich and Trubetskov 2000):

$$A_t = (\alpha_r + i\alpha_i)A - \gamma_i \,|\, A \,|^2\, A + i\gamma_r \,|\, A \,|^2\, A - iF \tag{12.18}$$

In equation (12.18), α_r describes a possibility of linear dissipation or amplification of small perturbations; α_i is proportional to the deviation of the frequency of the oscillations from the resonance frequency; the parameters γ_i and γ_r describe nonlinear energy losses and the nonlinear frequency shift, respectively; F is proportional to the external force. Equation (12.18), when the external force disappears, $F = 0$, can be considered as the simplest case of the Ginzburg–Landau equation (12.17). In this case, equations (12.17b) and (12.18) describe the oscillatory processes that develop in space and time and in time, respectively. The methods for modeling active controlled nonlinear hyperbolic MMs for the IR range are described in chapter 11.

12.3 Details of the structure of the active hyperbolic metamaterial

For a linear structure we require that the conditions are satisfied:

$$\varepsilon'_x > 0, \quad \varepsilon'_z < 0; \quad \varepsilon' \equiv \mathrm{Re}(\varepsilon), \tag{12.19}$$

moreover, in accordance with the condition (12.19), we require the satisfaction of the relevant conditions: for the medium forming the layer 1 we have $\varepsilon'_{\mathrm{L1}} < 0$; and for the media that make up layers 2, 3:

$$\varepsilon'_{\mathrm{L2,\,3}} > 0; \quad \varepsilon'_{\mathrm{L2}} = \varepsilon'_{\mathrm{L3}} \tag{12.20}$$

The corresponding requirements for materials that can be used to create layered active MMs are discussed in chapter 11, section 11.3.2. Note also that the necessary condition for the implementation (12.20) is:

$$|\,\varepsilon'_{\mathrm{L1}}| < |\,\varepsilon'_{\mathrm{L2}}|; \tag{12.21}$$

The expression for the dielectric permittivity is:

$$\varepsilon_{\mathrm{L1}} = \varepsilon_{\mathrm{lattice}} - \frac{\omega_p^2}{\omega(\omega + i\nu)}, \quad \omega_{\mathrm{p}}^2 = \frac{4\pi e^2 n_0}{m^*}, \tag{12.22}$$

Here $\varepsilon_{\text{lattice}}$ is the dielectric permittivity of the crystal lattice. In the expression (12.22) ω_p is the plasma oscillation frequency, n_0 is the equilibrium electron concentration, m^* is the effective mass of the electron in the medium, ν is the collision frequency, and $n_0 \sim (10^{21} \div 2 \times 10^{22}) cm^{-3}$; $\nu \leqslant 10^{13}$ s^{-1}.

12.3.1 Evolutionary equation for nonlinear momentum propagation at an arbitrary angle to the optical axis in the laboratory coordinate frame in the 'time representation' and coordinate representation

Consider two different coordinate frames: (1) The laboratory frame XYZ, shown in figure 12.1 (with the geometry of the problem) with the Z axis passing along the optical axis; and (2) the rotated coordinate frame $X'Y'Z'$, where: (z' is directed along $\vec{V}_g$, group speed EMW):

$$\vec{e}_{z'} \uparrow\uparrow \vec{V}_g \tag{12.23}$$

12.3.2 Evolution equation written with respect to the laboratory coordinate frame XYZ

Assume the following dependence of the transverse component of the EMW magnetic field:

$$H_y = H(x, y, z)\exp i(\vec{k}_c\vec{r} - \omega_c t) \tag{12.24}$$

$$\vec{k}_c = \vec{k}(\omega = \omega_c),\ \vec{k} = (k_x, 0, k_z) \tag{12.25}$$

$$\omega = \omega_c + \Delta\omega \tag{12.26}$$

$$\vec{k} = \vec{k}_c + \Delta\vec{k};\ \ \Delta\vec{k} = (\Delta k_x, 0, \Delta k_z);\ \frac{\partial}{\partial y} = 0 \tag{12.27}$$

In the expressions (12.24)–(12.27) ω, $\vec{k}_c$ are, respectively, the carrier frequency and wave vector of a wave packet with a spectrum of a moderate width.

Taking into account the relations (12.24)–(12.27), we derive a nonlinear dispersion equation and an evolution equation for the amplitude of the EMW wave packet. The nonlinear dispersion equation has the form:

$$D(\omega, k_x, k_z, H^2) = -k_x^2\beta_z(\omega, |H|^2) - k_z^2\beta_z(\omega, |H|^2) + k_0^2 = 0 \tag{12.28}$$

We represent the deviation of frequency $\Delta\omega$ from the carrier frequency in the form:

$$\Delta\omega = \frac{\partial\omega}{\partial k_x}\Delta k_x + \frac{\partial\omega}{\partial k_z}\Delta k_z + \frac{1}{2}\frac{\partial^2\omega}{\partial k_x^2}\Delta k_x^2 + \frac{1}{2}\frac{\partial^2\omega}{\partial k_x\partial k_z}\Delta k_x\Delta k_z + \frac{1}{2}\frac{\partial^2\omega}{\partial k_z^2}\Delta k_z^2 + \frac{\partial\omega}{\partial|H|^2}|H|^2 \tag{12.29}$$

To obtain the evolution equation, replace in (12.29):

$$-i\Delta\omega \to \frac{\partial}{\partial t};\ i\Delta\vec{k} \to \frac{\partial}{\partial\vec{r}} \tag{12.30}$$

Using (12.29) and (12.30) we obtain:

$$\begin{aligned} &i\frac{\partial H}{\partial t} + i\frac{\partial\omega}{\partial k_x}\frac{\partial H}{\partial x} + i\frac{\partial\omega}{\partial k_z}\frac{\partial H}{\partial z} + \frac{1}{2}\frac{\partial^2\omega}{\partial k_x\partial k_z}\frac{\partial^2 H}{\partial x\partial z} \\ &+ \frac{1}{2}\frac{\partial^2\omega}{\partial k_x^2}\frac{\partial^2 H}{\partial x^2} + \frac{1}{2}\frac{\partial^2\omega}{\partial k_z^2}\frac{\partial^2 H}{\partial z^2} - \frac{\partial\omega}{\partial|H|^2}|H|^2 H = 0 \end{aligned} \tag{12.31}$$

Note that the mixed derivative in (12.31) is present due to the fact that the EMW propagates in a hyperbolic MM medium at a finite angle to the optical axis.

We write down the following coefficients included in the left part of equation (12.31) (all values are taken as $\omega = \omega_c$, $\vec{k} = \vec{k}_c$):

$$\frac{\partial\omega}{\partial k_z} = D_{00}^{-1}2k_x\beta_z;\quad D_{00} = 2k_0\frac{1}{c} - k_x^2\frac{\partial\beta_z}{\partial\omega} - k_z^2\frac{\partial\beta_x}{\partial\omega};\ k_0 = \frac{\omega}{c} \tag{12.32}$$

$$\begin{aligned} \frac{\partial^2\omega}{\partial k_x\partial k_z} &= D_{00}^{-1}\left[2\left(k_x\frac{\partial\beta_z}{\partial\omega}\frac{\partial\omega}{\partial k_z} + k_z\frac{\partial\beta_x}{\partial\omega}\frac{\partial\omega}{\partial k_x}\right) + \right. \\ &\left. + \left(k_x^2\frac{\partial^2\beta_z}{\partial\omega^2} + k_z^2\frac{\partial^2\beta_x}{\partial\omega^2} - \frac{2}{c^2}\right)\frac{\partial\omega}{\partial k_z}\frac{\partial\omega}{\partial k_x}\right] \end{aligned} \tag{12.33}$$

$$\frac{\partial^2\omega}{\partial k_x^2} = D_{00}^{-1}\left[\left(\frac{\partial\omega}{\partial k_x}\right)^2\left(k_x^2\frac{\partial^2\beta_z}{\partial\omega^2} + k_z^2\frac{\partial^2\beta_x}{\partial\omega^2} - \frac{2}{c^2}\right) + 2k_x\frac{\partial\beta_z}{\partial\omega}\frac{\partial\omega}{\partial k_x} + 2\beta_z\right] \tag{12.34}$$

$$\frac{\partial^2\omega}{\partial k_z^2} = D_{00}^{-1}\left[\left(\frac{\partial\omega}{\partial k_z}\right)^2\left(k_x^2\frac{\partial^2\beta_z}{\partial\omega^2} + k_z^2\frac{\partial^2\beta_x}{\partial\omega^2} - \frac{2}{c^2}\right) + 2k_z\frac{\partial\beta_x}{\partial\omega}\frac{\partial\omega}{\partial k_z} + 2\beta_x\right] \tag{12.35}$$

$$\frac{\partial\omega}{\partial|H|^2} = D_{00}^{-1}\left(k_x^2\left(\frac{\partial\beta_z}{\partial|H|^2}\right) + k_z^2\left(\frac{\partial\beta_x}{\partial|H|^2}\right)\right) \tag{12.36}$$

12.3.3 Evolutionary equation for nonlinear propagation of a pulse at an arbitrary angle to the optical axis in the coordinate frame $X'Y'Z'$, where the group velocity is directed along z', in the spatial representation

$$\begin{aligned} \Delta k_z' &= \frac{\partial k_z'}{\partial k_x'}\Delta k_x' + \frac{\partial k_z'}{\partial\omega}\Delta\omega + \frac{1}{2}\frac{\partial^2 k_z'}{\partial k_x'}\Delta k_x'2 + \frac{1}{2}\frac{\partial^2 k_z'}{\partial k_x'\partial\omega}\Delta k_x'\Delta\omega \\ &+ \frac{\partial^2 k_z'}{\partial\omega^2}\Delta\omega^2 + \frac{\partial k_z'}{\partial|H|^2}\Delta|H|^2 \end{aligned} \tag{12.37}$$

Again, the replacement is

$$\Delta\omega \to i\frac{\partial}{\partial t};\ \ \Delta\vec{k}' \to i\frac{\partial}{\partial \vec{r}} \tag{12.38}$$

Equation (12.30) is made now in (12.37). Taking into account (12.37) and (12.38) we obtain the evolution equation in the form

$$\begin{aligned} &i\frac{\partial H}{\partial z'} + i\frac{\partial k_z'}{\partial\omega}\frac{\partial H}{\partial t} - i\frac{\partial k_z'}{\partial k_x'}\frac{\partial H}{\partial x'} + \frac{1}{2}\frac{\partial^2 k_z'}{\partial k_x'\partial\omega}\frac{\partial^2 H}{\partial x'\partial t} + \frac{1}{2}\frac{\partial^2 k_z'}{\partial k_x'2}\frac{\partial^2 H}{\partial x'^2} \\ &+ \frac{\partial^2\omega}{\partial k_z'^2}\frac{\partial^2 H}{\partial t^2} - \frac{\partial k_z'}{\partial|H|^2}|H|^2 H = 0. \end{aligned} \tag{12.39}$$

Here are the expressions of some coefficients included in equation (12.39):

$$\begin{aligned} &\frac{\partial k_z'}{\partial\omega} = (2D_{11})^{-1}D_{00}, \\ &D_{11} = k_x\frac{\partial k_x}{\partial k_z'}\beta_z + k_z\frac{\partial k_z}{\partial k_z'}\beta_x. \end{aligned} \tag{12.40}$$

$$\frac{\partial k_z'}{\partial|H|^2} = \left[2\left(k_x\frac{\partial k_x}{\partial k_z'} + k_z\frac{\partial k_z}{\partial k_z'}\right)\right]^{-1}\left(k_x^2\frac{\partial\beta_z}{\partial|H|^2} + k_z^2\frac{\partial\beta_x}{\partial|H|^2}\right) \tag{12.41}$$

$$\begin{aligned} \frac{\partial^2 k_z'}{\partial\omega\partial k_x'} = &-D_{11}\left\{\left[\frac{\partial\beta_x}{\partial\omega}k_z\left(\frac{\partial k_y}{\partial k_z'}\frac{\partial k_z'}{\partial k_x'} + \frac{\partial k_z}{\partial k_x'}\right) + \frac{\partial\beta_z}{\partial\omega}k_x\left(\frac{\partial k_x}{\partial k_z'}\frac{\partial k_z'}{\partial k_x'} + \frac{\partial k_z}{\partial k_x'}\right)\right] + \right. \\ &\left. + \frac{\partial k_z'}{\partial\omega}\left[\beta_z\frac{\partial k_x}{\partial k_z'}\left(\frac{\partial k_z}{\partial k_z'}\frac{\partial k_z'}{\partial k_x'} + \frac{\partial k_x}{\partial k_x'}\right) + \beta_x\frac{\partial k_x}{\partial k_z'}\left(\frac{\partial k_z}{\partial k_z'}\frac{\partial k_z'}{\partial k_x'} + \frac{\partial k_z}{\partial k_x'}\right)\right]\right\} \end{aligned} \tag{12.42}$$

The coefficients included in the left part of equation (12.39) are also expressed similarly. $\alpha = \angle(\vec{V}_g, \vec{e}_z)$ is the angle between the directions of the vector $\vec{V}_g$ and axis $\widehat{O}z$. $\vec{k}$ and $\vec{k}'$ are related by the following relationship:

$$\begin{aligned} &k_z = k_z(k_x', k_z') = k_z'\cos\alpha - k_x'\sin\alpha;\ k_x = k_x(k_x', k_z') \\ &= k_z'\sin\alpha + k_x'\cos\alpha \end{aligned} \tag{12.43}$$

Taking into account (12.43), one can express the derivatives $(\partial k_i/\partial k_j)$, which are mentioned in (12.41)–(12.43):

$$\frac{\partial k_x}{\partial k_x'} = \cos\alpha;\ \frac{\partial k_x}{\partial k_z'} = \sin\alpha;\ \frac{\partial k_z}{\partial k_x'} = -\sin\alpha;\ \frac{\partial k_z}{\partial k_z'} = \cos\alpha \tag{12.44}$$

Taking into account relations (12.43), (12.44) and (12.40)–(12.42), the coefficients in the evolution equation (12.39) can be completely definite.

12.4 The model of a two-level active medium and equations for nonlinear EMW in planar active resonant hyperbolic medium

A two-level active medium with quantum dots can be represented as an effective medium with an effective complex electric permittivity and a negative dissipation using the semiclassical Maxwell–Bloch equations (Zyablovsky *et al* 2011). To determine the complex electric permittivity in each of the isotropic elementary active layers, it is possible to simplify to consider for this system the one-dimensional Maxwell–Bloch model:

$$\begin{aligned} &\frac{\partial^2 E}{\partial z^2} - \frac{\varepsilon_0(z)}{c^2}\frac{\partial^2 E}{\partial t^2} = \frac{4\pi}{c^2}\frac{\partial^2 P}{\partial t^2}; \\ &\frac{\partial^2 P}{\partial t^2} + \frac{2}{\tau_p}\frac{\partial P}{\partial t} + {\omega_t}^2 P = \frac{2\omega_t|\mu|^2 nE}{\hbar}; \\ &\frac{\partial n}{\partial t} + \frac{1}{\tau_n}(n - n_0) = \frac{1}{\hbar\omega_t}\,\mathrm{Re}\left(E^*\frac{\partial P}{\partial t}\right). \end{aligned} \tag{12.45}$$

In (12.45) E, P are the electric field and the microscopic polarization, respectively, ω_t is the resonant frequency that corresponds to the energy difference between the energy levels, ε_0 is the permittivity (isotropic) of the host medium, including active centers (in particular, quantum dots) with a difference n of the populations of the upper and lower levels (the population inversion), μ is the off-diagonal matrix element of the dipole moment of a quantum dot, and τ_p, τ_n are typical values of relaxation times for the polarization and population inversion. An external excitation and the induced radiation cause n to tend to n_0 and, accordingly, n decreases. In the static case, the complex electric permittivity, which can be obtained on the base of the set of equation (12.45), takes the form (Zyablovsky *et al* 2011):

$$\varepsilon = \varepsilon_0 + q\frac{\omega_t}{\omega}\frac{-i + \dfrac{\omega^2 - \omega_t^2}{2\omega/\tau_p}}{1 + f|E|^2 + \left(\dfrac{\omega^2 - \omega_t^2}{2\omega/\tau_p}\right)^2} \tag{12.46}$$

$$q = 4\pi\mu\tau_p n_0/\hbar,\ f = \hbar^{-2}\mu\tau_n\tau_p \tag{12.47}$$

According to (12.46), the imaginary part of the electric permittivity, which determines the gain within the medium, under stationary conditions is equal to

$$\varepsilon'' = -q\frac{\omega_t}{\omega}\frac{1}{1 + f|E|^2 + \left(\dfrac{\omega^2 - \omega_t^2}{2\omega/\tau_p}\right)^2} \tag{12.48}$$

Thus, the way of studying the propagation of EMW in nonlinear hyperbolic MM at an arbitrary angle to the optical axis, taking into account the active processes in MMs from the first principles, is determined.

If

$$f \mid E \mid^2; \left(\frac{\omega^2 - \omega_e^2}{2\omega/\tau_P} \right) \ll 1 \tag{12.49}$$

then we obtain when taking into account (12.46):

$$\varepsilon \simeq \varepsilon_0 + q\frac{\omega_e}{\omega}\left[1 - f \mid E \mid^2 - \left(\frac{\omega^2 - \omega_e^2}{2\omega/\tau_P}\right)^2\right]\left[-i + \frac{\omega^2 - \omega_e^2}{2\omega/\tau_P}\right] \tag{12.50}$$

$$= \varepsilon_{\text{Lin}} + \alpha \mid E \mid^2$$

$$\varepsilon_{\text{Lin}} = \varepsilon_0 + q\frac{\omega_e}{\omega}\left[1 - \left(\frac{\omega^2 - \omega_e^2}{2\omega/\tau_P}\right)^2\right]\left[-i + \frac{\omega^2 - \omega_e^2}{2\omega/\tau_P}\right] \tag{12.51}$$

The possibility of EMW amplification follows from the presence of the imaginary part of the electric permittivity with the appropriate sign:

$$\varepsilon_{\text{NL}} = \alpha \mid E \mid^2; \alpha = -qf\frac{\omega_e}{\omega}\left[-i + \frac{\omega^2 - \omega_e^2}{2\omega/\tau_P}\right] \tag{12.52}$$

According to relations (12.47), we have:

$$qf = \frac{4\pi\tau n_0 \mid \mu \mid^2}{\hbar}\frac{\tau_n \tau_P}{\hbar^2}; \Omega(t) = 2 \mid \mu \mid E/\hbar \tag{12.53}$$

In relation (12.53), $\Omega(t)$ is the Rabi frequency (Makarova *et al* 2017). Taking into account (12.51), (12.52), (12.53) we get:

$$\Delta\omega_{\text{NL}} \sim \varepsilon_{\text{NL}} \sim \alpha E^2 \sim qf \mid E \mid^2 \sim \Omega^2 \tau_n \tau_P \tag{12.54}$$

The medium shown in figure 12.1, is active because the conditions for the layers in the unit cell are satisfied, namely:

The first layer is metallic

$$\text{Re}\, \varepsilon_1 = \varepsilon_1' < 0 \tag{12.55}$$

The second and third layers are dielectric

$$\text{Re}\varepsilon_{2,3} = \varepsilon_{2,3}' > 0 \tag{12.56}$$

A possibility of amplification of EMW in active hyperbolic MM and other features of the activity of the medium are presented by the properties of the coefficients of the corresponding evolution equation for the amplitude of the wave packet A:

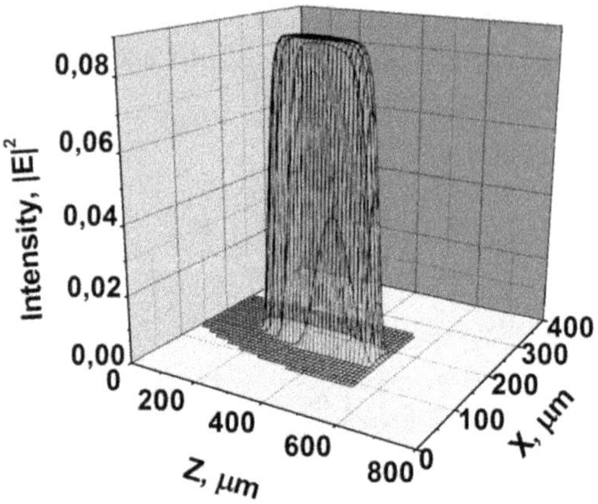

Figure 12.3. Example of propagation of pulse in active resonant hyperbolic MM: regime of saturation. The distributions of $|E|^2$ at time moments $t/t_n = 5000$ is shown.

$$\frac{\partial A}{\partial t} + V_x \frac{\partial A}{\partial x} + V_z \frac{\partial A}{\partial z} + ig_{11} \frac{\partial^2 A}{\partial x^2} + ig_{33} \frac{\partial^2 A}{\partial z^2} + ig_{13} \frac{\partial^2 A}{\partial x \partial z} + iN \mid A \mid^2 A + \Gamma A = 0 \quad (12.57)$$

Note that if the group velocity is directed along the axis z ($v_{gx} = 0$), then the coefficient $g_{13} = 0$. The 'activity' of the medium leads to the fact that all the coefficients $g_{11}, g_{33}, g_{13}, N$, as well as components of the effective group velocity, $V_{x,\,z}$ are complex. The linear gain mode corresponds to the condition $\Gamma < 0$.

12.5 Example of numerical modeling

An example of propagation of pulse in active resonant hyperbolic MM: regime of saturation is shown in figure 12.3. At the input there is a single pulse, the incidence angle is 30°. Distances are normalized to $l_n = 1$ μm, time is normalized to $t_n = l_n/c \approx$ 3.33 fs. The maximum amplitude of the input pulse is 0.07 at $t/t_n = 400$, its duration is

$t_{01}/t_n = 25$, the initial width is $x_{01} = 10$ μm. Under the propagation a small pulse is subject to amplification, but then the saturation of amplification occurs as well as the pulse broadening. Note that the pulse amplitude saturation happens at $t/t_n \sim 3000$ and pulse intensity of order $|E_{ma}|^2 \sim 0.09$.

12.6 Conclusions

A model of the nonlinear wave propagation in hyperbolic nonlinear active resonant MM in the IR range is developed. The nonstationary and stationary nonlinear beam propagation, the averaging method, the material and photonic-crystalline approximation are used. These approaches allow us to study the phenomena of arbitrary, or non-moderate, nonlinearity and the hot spot formation. Hyperbolic MMs are very useful for the development of controlled optical micro- and nanophotonic devices.

Hyperbolic MMs: (1) can be created by subwave structuring; (2) are technologically simpler than MMs of many other types; (3) demonstrate hyperbolic dispersion; (4) have the property of negative refraction; (5) active elements can be easily incorporated into hyperbolic MMs; (6) they provide an increase in the density of quantum states and spontaneous emission of light from a source located near MMs; (7) hyperbolic MMs are promising for the development of superlens with super-resolution; (8) they can be implemented from radio frequencies to the optical range; (9) provide a significant increase in resonant nonlinearity; (10) their linear and nonlinear characteristics can be controlled by external fields, the wave characteristics in hyperbolic MMs can also be controlled by the principle of 'light by light'. The models proposed in this chapter can be used to model all these phenomena and characteristics.

Problems to chapter 12

Problem 12.1

Suppose the dispersion equation for EMW in a hyperbolic MM is given as (see equation (12.28)):

$$D(\omega, k_x, k_z, H^2) = 0 \tag{P12.1}$$

The values which are included in equation (P12.1), are determined in the text of this chapter (see, in particular, equations (12.25), (12.24)). Determine, in general form, the relation for $\partial^2\omega/\partial k_x^2$, see also equation (12.34).

Differentiating equation (P12.1) or (12.28) with respect to k_x under constant k_z, obtain

$$D_\omega \omega_{k_x} + D_{k_x} = 0; \quad V_x \equiv \omega_{k_x} = -\frac{D_{k_x}}{D_\omega} \tag{P12.2}$$

In equation (12.2) the notations like $D_\omega \equiv \partial D/\partial\omega\ |_{k_x=\text{const},\ k_z=\text{const},\ H^e=\text{const}}$ etc are used; then the one can obtain the following relations:

$$D_\omega \equiv 2\omega - \frac{\partial}{\partial\omega}\left[k_x^2\beta_z(\omega) + k_z^2\beta_x(\omega)\right] \tag{P12.3}$$

$$D_{\omega\omega} = 2 - \frac{\partial^2}{\partial\omega^2}\left[k_x^2\beta_z(\omega) + k_z^2\beta_x(\omega)\right] \tag{P12.4}$$

$$D_{k_x} = -2k_x\beta_z(\omega); \ D_{k_xk_x} = -2\beta_z(\omega) \tag{P12.5}$$

$$D_{k_z} = -2k_z\beta_x(\omega); \ D_{k_zk_z} = -2\beta_x(\omega) \tag{P12.6}$$

Differentiate the first relation from (P12.2) with respect to k_x and obtain:

$$D_{\omega\omega}\omega_{kx}^2 + D_{kx\omega}\omega_{kx} + D_\omega\omega_{kxkx} + D_{kx\omega}\omega_{kx} + D_{kxkx} = 0 \tag{P12.7}$$

As follows from (P12.7),

$$\begin{aligned}\omega_{k_x k_x} &= -\frac{1}{D_\omega}\left\{D_{\omega\omega}\omega_{k_x}^2 + 2D_{k_x\omega}\omega_{kx} + D_{k_x k_x}\right\} \\ &\equiv -\frac{1}{D_\omega}\left\{D_{\omega\omega}v_x^2 + 2D_{k_x\omega}v_x + D_{k_x k_x}\right\}\end{aligned} \tag{P12.8}$$

In equation (P12.8), the denotation $v_x \equiv \omega_{k_x}$ (x- component of the group velocity of EMW) is used.

Problem 12.2

Determine, in general form, the relation for $\partial^2\omega/\partial k_z^2$ (see also equation (12.31)).
Similarly to relations (P12.2)–(P12.8) (see problem 12.1), one can obtain:

$$\omega_{k_z k_z} = -\frac{1}{D_\omega}\left\{D_{\omega\omega}\omega^2 + 2D_{k_z\omega}v_z + D_{k_z k_z}\right\};\ v_z \equiv \partial\omega/\partial k_z \equiv \omega_{k_z} \tag{P12.9}$$

Problem 12.3

Obtain, in general form, the relation for $\omega_{k_x k_z}$ (see also equation (12.31))
Similarly to relations (P12.2)–(P12.8) (see problem 12.1), one can obtain:

$$D_{k_x\omega} = -2k_x\frac{\partial}{\partial\omega}\beta_z(\omega);\ D_{k_z\omega} = -2k_x\frac{\partial}{\partial\omega}\beta_x(\omega) \tag{P12.10}$$

Differentiate the first relation from (P12.2) with respect to k_z and obtain:

$$D\omega_{k_z}\omega_{k_x} + D_{\omega\omega}\omega_{k_x}\omega_{k_z} + D_\omega\omega_{k_z k_z} + D_{\omega k_x}\omega_{k_z} + D_{k_x k_z} = 0 \tag{P12.11}$$

As seen from equation (P12.11),

$$\omega_{k_x k_z} = -\frac{1}{D_\omega}\left\{D_{\omega\omega}v_x v_z + D_{\omega k_z}v_z + D_{\omega k_z}v_x\right\}$$

Problem 12.4

Obtain the relation for $\omega_{k_x k_z}$ (see also equation (12.31)) using an approach, alternative to one proposed in the explanation to problem (12.3) using the dispersion equation in the form, alternative to (P12.1), namely:

$$k_z = k_z(k_x, \omega) \tag{P12.12}$$

Accounting for (P12.12), one can obtain

$$\omega_{k_z|k_x} = \frac{1}{k_{z\omega|k_x}} \tag{P12.13}$$

In equation (P12.13), the following denotation is used:

$$\omega_{k_z|_{k_x}} \equiv \left(\frac{\partial\omega}{\partial k_z}\right)_{k_x=\text{const}} \quad \text{(P12.14)}$$

Accounting (P12.13) and (P12.14), it is possible to get

$$\begin{aligned}\frac{\partial}{\partial k_x}\,\omega_{k_z}\big|_{k_z=\text{const}} &= -\frac{1}{k_z\omega^2}\frac{\partial}{\partial k_x}\,k_{z\omega}\big|_{k_z=\text{const}}\\ &= -\frac{1}{k_z\omega^2}\left\{\frac{\partial}{\partial k_x}\,k_{z_\omega}\big|_{\omega=\text{const}}\left(\frac{\partial k_x}{\partial k_x}\right)_{k_z}\right.\\ &\left.+\frac{\partial}{\partial\omega}\,k_{z\omega}\big|_{k_x=\text{const}}\left(\frac{\partial\omega}{\partial k_x}\right)_{k_z=\text{const}}\right\} = -\frac{1}{k_z\omega^2}\left\{\frac{\partial^2 k_z}{\partial\omega\partial k_x}+\frac{\partial^2 k_z}{\partial\omega^2}v_x\right\}\end{aligned} \quad \text{(P12.15)}$$

To find the second derivative included in the last of the relations from (P12.15), note that

$$\begin{aligned}\frac{\partial^2 k_z}{\partial\omega\partial k_x} &= \frac{\partial}{\partial\omega}\left(\frac{\partial k_z}{\partial k_x}\bigg|_{\omega=\text{const}}\right)_{k_x=\text{const}} = -\frac{\partial}{\partial\omega}\left\{\frac{k_x}{k_z}\frac{\beta_z(\omega)}{\beta_x(\omega)}\right\}\bigg|_{k_x=\text{const}}\\ &= -k_x\frac{\partial}{\partial\omega}\left\{\frac{1}{k_z}\frac{\beta_z(\omega)}{\beta_x(\omega)}\right\}\bigg|_{k_x=\text{const}} = k_x\left\{\frac{k_{z\omega}}{k_z^2}\frac{\beta_z(\omega)}{\beta_x(\omega)}-\frac{1}{k_z}\right.\\ &\left.\times\frac{\partial}{\partial\omega}\left(\frac{\beta_z(\omega)}{\beta_x(\omega)}\right)\right\}\bigg|_{k_x=\text{const}} = \frac{k_x}{k_z}\left\{\frac{k_{z\omega}}{k_z}\frac{\beta_z(\omega)}{\beta_x(\omega)}-\frac{\partial}{\partial\omega}\frac{\beta_z(\omega)}{\beta_x(\omega)}\right\}\end{aligned} \quad \text{(P12.16)}$$

Accounting for (P12.16) and (P12.15), the corresponding coefficient from equation (12.31) is obtained, namely

$$g_{13} \equiv -\frac{\partial\omega_{k_z}}{\partial k_x} \equiv -\frac{\partial^2\omega}{\partial k_x\partial k_z} \equiv -\frac{\partial}{\partial k_x}\omega_{k_z} \quad \text{(P12.17)}$$

List of abbreviations

IR	Infrared
EMWs	Electromagnetic waves
MM	Metamaterial
NSE	Nonlinear Schrödinger equation
THz	Terahertz

References

Akhmediev N and Ankiewitcz A 2005 *Dissipative Solitons* (Berlin: Springer)

Anisimov I O 2003 *Oscillations and Waves* (Kyiv: Akadempres) p 280 (in Ukrainian)

Boardman A D, King N and Rapoport Y 2010 Circuit model of gain in metamaterials *Nonlinearities in Periodic Structures and Metamaterials* (Springer Series in Optical Sciences Series in Optical Sciences vol 150) (Berlin: Springer) pp 259–71

Boardman A D, Grimalsky V V, Kivshar Y, Koshevaya S V, Lapine M, Litchinitser M, Malnev V N, Noginov M, Rapoport Yu G and Shalaev V M 2011 Active and tunable metamaterials *Laser Photonics Rev.* **5** 287–307

Boardman A, Alberucci A, Assanto G, Grimalsky V, Kibler B, McNiff J, Nefedov I S, Rapoport Y G and Valagiannopoulos C A 2017 Waves in hyperbolic and double negative metamaterials including rogues and solitons *Nanotechnology* **28** 1–41

Boardman A D and Velasko L 2006 Gyroelectric cubic—quintic dissipative solitons *IEEE J. Select. Top. Quant. Electron.* **12** 388–97

Boardman A D, Alberucci A, Assanto G, Rapoport Y G, Grimalsky V, Ivchenko V and Tkachenko E 2017 Spatial solitonic and nonlinear plasmonic aspects of metamaterials *World Scientific Handbook Metamaterials and Plasmonics* vol 4 (Singapore: World Scientific) pp 419–69

Défi F J Jr, Dikandé A M and Sunda-Meya A 2020 Continuous-wave stability and multipulse structures in a universal complex Ginzburg–Landau model for passively mode-locked lasers with a saturable absorber *J. Opt. Soc. Am.* B **37** A175–83

García-Morales V and Krischer K 2012 The complex Ginzburg–Landau equation: An introduction *Contemp. Phys.* **53** 79–95

Guglielmi A V and Pokhotelov O A 1996 *Geoelectromagnetic Waves* (Bristol and Philadelphia, PA: Institute of Physics Publishing) p 397

Guo Z, Jiang H and Chen H 2020 Hyperbolic metamaterials: From dispersion manipulation to applications *J. Appl. Phys.* **127** 071101

Jacob Z and Shalaev V M 2011 Plasmonics goes quantum *Science* **334** 463–4

Kadomtsev B B 1988 *Collective Phenomena in Plasma* (Moscow: Nauka) p 304 (in Russian)

Makarova O A, Shalaginov M Y and Bogdanov S *et al* 2017 Patterned multilayer metamaterial for fast and efficient photon collection from dipolar emitters *Opt. Lett.* **42** 3968

Makarova O, Shalaginov M, Bogdanov S, Guler U, Boltasseva A, Kildishev A and Shalaev V 2017 Patterning metamaterials for fast and efficient single-photon sources *Proc. Soc. Photo-Opt. Instrumentation Eng. Photonic Phononic Properties Eng. Nanostructures* 1011208

Mihalache D, Mazilu D, Lederer F, Leblond H and Malomed B A 2007 Stability of dissipative optical solitons in the three-dimensional cubic-quintic Ginzburg–Landau equation *Phys. Rev.* A **75** 033811

Poddubny A, Iorsh I, Belov P and Kivshar Y 2013 Hyperbolic metamaterials *Nat. Photonics* **7** 948–57

Pustovit V N, Urbas A M and Zelmon D E 2016 Surface plasmon amplification by stimulated emission of radiation in hyperbolic metamaterials *Phys. Rev.* B **94** 235445

Rabinovich M I and Trubetskov D I 2000 *Introduction to the Theory of Oscillations and Waves* (Izhevsk, Russia: Research Center 'Regular and Chaotic Dynamics') p 560

Savelev R S, Shadrivov I V and Belov P A *et al* 2013 Loss compensation in metal-dielectric layered metamaterials *Phys. Rev.* B **87** 115139

Schulz K M, Vu H and Schwaiger S *et al* 2016 Controlling the spontaneous emission rate of quantum wells in rolled-up hyperbolic metamaterials *Phys. Rev. Lett.* **117** 085503

Shadrivov I V, Lapine M and Kivshar Y S 2015 *Nonlinear, Tunable and Active Metamaterials* (Springer Series in Materials Science vol 200) (New York: Springer) p 253

Shekhar P, Atkinson J and Jacob Z 2014 Hyperbolic metamaterials: fundamentals and applications *Nano Converg.* **1** 14

Smalley J S T, Vallini F, Zhang X and Fainman Y 2018 Dynamically tunable and active hyperbolic metamaterials *Adv. Opt. Photonics* **10** 354–408

Urbas A M, Jacob Z and Negro L D *et al* 2016 Roadmap on optical metamaterials *J. Opt.* **18** 093005

Riskin N M and Trubetskov D I 2000 *Nonlinear Waves* (Moscow: Nauka Physmatlit) p 272 (in Russian)

Zyablovsky A A, Dorofeenko A V, Pukhov A A and Vinogradov A P 2011 Lasing in a gain slab as a consequence of the causality principle *J. Commun. Technol. Electron.* **56** 1139–45

IOP Publishing

Yurly Rapoport and Vladimir Grimalsky

Chapter 13

Rogue waves in metamaterial waveguides

13.1 Introduction

As the basis of consideration of various nonlinear wave phenomena with envelope waves, the nonlinear Schrödinger equation (NSE) is used. But in most practical cases the standard NSE should be modified. For instance, some additional terms that measure the self-steepening of the pulse may be required, i.e. the nonlinearity dispersion or diffraction. Also, when a pulse is short in the time domain, it possesses a broad frequency spectrum, and the nonlinear interaction between different frequency parts occurs, i.e. the Raman self-scattering. Under the pulse propagation in metamaterials (MMs), the important problem is the influence of the used model on the pulses self-steepening. Also the influence of applied frequency dispersion of the permittivity and the permeability of the pulse propagation can be critical. This frequency dependence can be introduced through a variety of Drude models.

The topic of this chapter is the optical rogue waves in the MM waveguides. The nonlinear envelope waves are subject to an MM environment and also a magneto-optic influence. The optical rogue waves were widely discussed in the literature because of the theoretical and practical interest. The rogue waves are interesting for theory because they unify the soliton pulse propagation and the development of the modulation instability (MI) of long pulses of relative small initial amplitudes. In practice they are used to excite short pulses of extremely high amplitudes.

Both nonlinear electromagnetic MMs and optical rogue waves became of a focus in recent years and they are separately large areas of research to study. Here a connection between these subjects is considered to generate new possibilities. A rogue generally is a new phenomenon that appears inexpectably, or unpredictably. The rogue wave is a sharp strong peak or several peaks that appears from flat almost smooth initial background. Rogue waves are known in hydrodynamics, nonlinear optics, plasma physics etc. Both creation and disappearance are with equal rapidity.

doi:10.1088/978-0-7503-2336-9ch13

An example of such a scenario is the famous case of a single peak appearing and disappearing, known as a Peregrine soliton. Also a multi-peak solution is discussed.

The rogue waves in hydrodynamics were associated with so-called wave-killers in the outer ocean. Rogue waves have been observed in nature, but they are a comparatively rare event. The reason is that in a natural physical system the boundary or initial conditions should correspond to rogue waves with a great accuracy. The natural sea measurements were complicated for observing hydrodynamic rogue waves directly. But it is possible to generate them in laboratory experiments, both in the nonlinear optics and in hydrodynamics. The confirmation of the existence now is complete (Solli *et al* 2007, Akhmediev and Pelinovsky 2010, Akhmediev *et al* 2013, Onorato *et al* 2013, Kibler *et al* 2010, Chabchoub *et al* 2011). The mathematical structure of governing equations for the envelope nonlinear waves is similar in most nonlinear systems, including nonlinear optics and hydrodynamics, also including the waves on the surface of deep water. The analogies between hydrodynamics and nonlinear optics resulted in the discovery of both envelope solitons as the localized pulse solutions and the MI of long initial pulses between 1960–70. The discovery was within the framework of NSE, and since NSE is the basic equation for investigations of envelope nonlinear waves (Zakharov and Shabat 1971, Lighthill 1965, Whitham 1965, Bespalov and Talanov 1966, Benjamin and Feir 1967, Benjamin 1967, Zakharov 1968, Zakharov and Ostrovsky 2009).

Thus, wave dynamics in weakly nonlinear dispersive media in a general case, including the optical Kerr media and waves on the surface of deep water, can be described by the NSE in common. It was recently shown that this correspondence may be applied even in the limit of extreme nonlinear wave localization described by the common mathematical model (Kibler *et al* 2010, Chabchoub *et al* 2011). Note that the NSE admits general breather solutions on a finite background, i.e. pulsating envelopes that imitate the dynamics of rogue waves that may appear from nowhere and disappear without leaving a trace. As a consequence, we can first address the issue of rogue waves in (almost-) conservative systems in terms of NSE breathers whose entire space–time evolution is analytically described (Akhmediev and Ankiewicz 1997). Their pulsating and localization properties make the mathematical solutions like the simplest nonlinear prototypes of the hydrodynamic rogue waves (Osborne 2010), in particular the doubly localized (in space and time) breather solutions (i.e. Peregrine soliton and higher orders) (Peregrine 1983). These pulsating solutions also include solutions that are either periodic in space and localized in time or periodic in time and localized in space, which are referred to as Kuznetsov-Ma breathers and Akhmediev breathers, respectively. Taking the period of both of these latter solutions to infinity results in the Peregrine solution.

If we now recall that breather solutions describe localized carrier perturbations with a strong amplification, they provide support to the nonlinear stage of the universal MI phenomenon (Akhmediev and Korneev 1986). We distinguish two MI regimes: on the one hand, the noise-driven MI that refers to the amplification of initial noise superposed to the plane wave that leads to spontaneous pattern formation from stochastic fluctuations; on the other hand, the coherent seeded MI (or coherent driving of MI) that refers to the preferential amplification of a specific perturbation (i.e. leading to a particular breather solution) relative to any broad-band noise. In either case, the wave dynamics can be interpreted in terms of breathers and competitive interactions. Most importantly, the coherent seeded MI can be used to efficiently stabilize and manipulate the output wave, thus allowing it to generate and quantitatively measure NSE breather properties in optical fibers.

It is also important to say that breather dynamics appear even with initial conditions that do not fulfil the mathematical ideal and optical studies have strongly contributed to that end. In fact it can be said any kind of disturbance of the ideal NSE propagation induces a deviation from the expected theoretical solutions (specific to each breather on a finite background), but most of the features related to their pulsating dynamics remain clearly observable. Breathers on a finite background may be considered as ‘robust solutions’ (but unstable solutions from the mathematical point of view) (Ankiewicz *et al* 2009), in the sense that they can be excited or propagated even with non-ideal conditions, and the main features of a localized high amplitude event (i.e. rogue wave) still occur. In general, fiber characteristics are chosen in accordance with both spectral bandwidth and peak power of the wave evolving into the fiber in order to avoid the potential impact of higher-order dispersive or nonlinear effects (Agrawal 2013). Nevertheless, besides fiber losses, higher order effects linked to pulse propagation such as third-order dispersion, self-steepening, and the Raman effect can be considered, even theoretically by extending the NSE (Ankiewicz *et al* 2009). Exact rogue wave solutions were even found in such complex equations that are integrable in special cases, such as the Sasa–Satsuma or the Hirota equations (Ankiewicz *et al* 2010, 2013, Bandelow and Akhmediev 2012, Chen 2013).

From an MMs perspective the key work that is often considered to be the starting point is that of Veselago in 1967 (Veselago 1967). However, it was not until the 1990s with improved computing power and manufacturing techniques that these materials could be studied in great depth and in 2000 John Pendry (Pendry 2000) extended Veselago’s work to creating the perfect lens. Since then there has been an explosion of papers in the MMs field. MMs offer the potential to control a range of electromagnetic behavior in ways that cannot be achieved with materials found in nature and the ability to create purpose-built materials to fulfil new and novel functionality where these artificial structures, that exert influence over electromagnetic waves at the sub-wavelength level, are extensively being studied and designed. Here we contribute to this expansion.

The study of rogue waves in MMs is currently at an early stage, only a few recent numerical studies based on a transmission line model (with NSE reduction) or a cubic-quintic NSE can be found. In particular, it was shown there is the possibility of producing extreme waveform events, with strong similarities to NSE breather waves (Peregrine, but also Akhmediev or Kuznetsov-Ma breathers) (Essama *et al* 2014, Dontsop *et al* 2016, Shen *et al* 2017). In the following, we bring together both of these subjects by investigating the propagation behavior of wave excitation corresponding to rogue breather solutions in transparent MMs with the addition of magnetooptic properties; wherein higher-order dispersive or nonlinear effects are included. Hence the fascinating idea of rogue wave emergence in the form of Peregrine solitons (and near-Peregrines) within a nonlinear MM environment and where potentially magnetooptic control could be exerted is researched as a new pathway.

13.2 Simulations

The world is becoming very interested in hyperbolic MMs and breather type solutions, however, for this type of MM appropriate extensions to the dispersion type NSE are not yet available. As the best way to investigate rogue waves is based on the dispersion type of NSE the use of double-negative media is the way forward where appropriate extensions to the NSE have already been developed and could thus be readily adapted for the purpose of studying rogue waves. In this section with the dispersion based NSE we are looking at solutions of the type that have a flat background with a small perturbation. We assume the material is isotropic and it has a negative electric permittivity and negative magnetic permeability thus the form of the MMs considered here is transparently double-negative (Boardman *et al* 2010a) and our aim is to give some new physical insights on how the MM properties can affect rogue waves in the form of NSE breathers. More particularly, we reveal that their dynamical behavior can be influenced and controlled, by MM effects, namely the self-steepening, and magnetooptic effects. Here we restrict our analysis to temporal forms of rogue solutions, however, it is worth mentioning that both spatial and temporal waveform solutions can be considered in MMs, in a similar way to the standard soliton solution (Boardman *et al* 2010a, 2010b).

The general approach to the NSE is to assume that the components (both electric and magnetic fields) of an electromagnetic wave propagating in an MM can be factorized such that there is an amplitude that is slowly varying along the propagation direction, a linear modal field contribution and a fast plane wave variation that introduces a propagation wavenumber. Here a modified approach is adopted such that an appropriate extension to the NSE (Boardman *et al* 2010a) is utilized that takes into account the MM properties with self-steepening and magnetooptic effects.

Although both electric and magnetic nonlineafrities could be included for a given MM, it has been shown (Boardman *et al* 2010b) that the effective nonlinearity can be combined into a single nonlinear coefficient. Here, however, it is assumed that the nonlinear behavior discussed originates from an isotropic Kerr dielectric. A typical waveguide structure that is used here is shown in figure 13.1. It is a planar structure with an MM core that has boundaries in x and y directions and propagation is along the z-axis.

When part of the structure (as shown in figure 13.1), in the form of the substrate, is replaced with a magnetooptic material, it is possible to control solitons with the

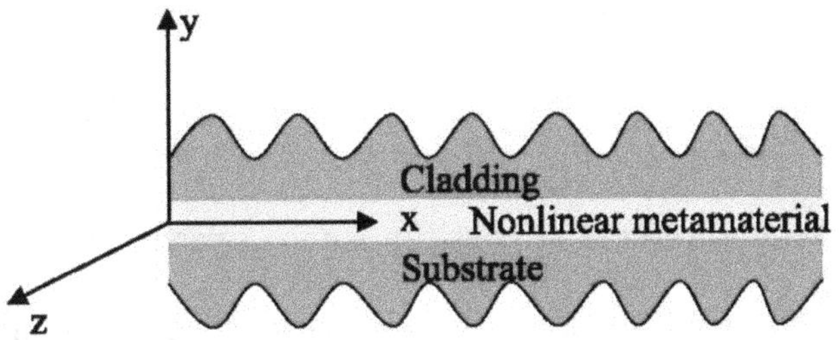

Figure 13.1. A diagram of a possible waveguide system with negative phase nonlinear MM. Reprinted from Boardman *et al* (2017), copyright IOP Publishing, all rights reserved.

application of a magnetic field. Here the magnetic field is applied in what is known as a Voigt configuration. This has an externally applied magnetic field in the plane of the guide perpendicular to the direction of optical propagation, which here can be applied along either the positive or negative x-direction.

The form of the extended NSE is given below in equation (13.1). It is noted that there is also the capability of adding higher order and Raman scattering effects, however, in the simulations discussed below these parameters are not invoked, but will be utilized in future work. This form of the NSE does not yet contain the magnetooptic parameters, which will be discussed further below.

$$i\frac{\partial \Psi}{\partial Z} - \frac{1}{2}\operatorname{sgn}(\beta_2)\frac{\partial^2 \Psi}{\partial \eta^2} - i\delta_3\frac{\partial^3 \Psi}{\partial \eta^3} + \operatorname{sgn}(\bar{\chi}^{(3)})(|\Psi|^2\Psi + iS\frac{\partial}{\partial \eta}(|\Psi|^2\Psi) - \tau_R \Psi\frac{\partial}{\partial \eta}(|\Psi|^2)) = 0. \tag{13.1}$$

Here β_2 is the group velocity dispersion, S is the self-steepening coefficient which is discussed below; $\delta_3 = \beta_3/6\beta_2{\cdot}(t_0)$, β_3 is the third-order dispersion, t_0 is interpreted as pulse width and τ_R is the Raman coefficient. $\chi^{(3)}$ is the third-order nonlinear Kerr coefficient in which the assumption of a weakly guiding system has been embedded. In the simulations presented later in this section the parameters that are considered are self-steepening with the use of a range of magnetooptic parameters that are used to adjust the effect of self-steepening in the MM. It should be noted that the self-steepening coefficient S in equation (13.1) involves very specific properties of the MMs, which has been pointed out by previous authors (Wen *et al* 2006, Scalora *et al* 2005). The expression is

$$S = \frac{1}{\omega_0}\left(2 + \frac{|v_p|}{v_g} + \frac{\omega}{\mu}\left(\frac{\partial \mu}{\partial \omega}\right)_{\omega_0}\right), \tag{13.2}$$

where ω is angular frequency, ω_0 is the operational frequency, v_p is phase velocity, v_g is group velocity and μ *is* the permeability, and:

$$\frac{|v_p|}{v_g} = -\frac{\beta_1\omega_1}{\beta_0}; \quad \beta_1 = \left(\frac{\partial k}{\partial \omega}\right)_{\omega_0} - \frac{1}{v_g}, \tag{13.3}$$

where β_0 is the wavenumber, it is further noted that the form of the NSE given here allows for the nonlinear coefficient to be dispersive which if included would give a form for the self-steepening coefficient:

$$S = \frac{1}{\omega_0 t_0}\left(2 + \frac{\beta_1\omega_0}{\beta_0} + \frac{\omega}{\mu}\left(\frac{\partial \mu}{\partial \omega}\right)_{\omega_0} + \left(\frac{\omega}{\chi^{(3)}}\frac{\partial \chi^{(3)}}{\partial \omega}\right)_{\omega_0}\right) \tag{13.4}$$

The approach here however is to consider a non-dispersive scenario for $\chi^{(3)}$ thus $\partial\chi^{(3)}/\partial\omega = 0$.

The important issue of whether the self-steepening coefficient is affected by MMs solutions is discussed in Boardman *et al* (2010a) (see figure 13.3 therein) and it is clearly demonstrated that this is the case for a judicious selection of S.

(In these early stage results similar values to those used in previous figure 13.2 temporal solitons simulations are adopted to allow comparisons of the different starting conditions used, see below.) The use of typical Drude models is adopted that are of the form:

$$\varepsilon(\varpi) = \varepsilon_{\rm D} - \frac{1}{\varpi^2}; \quad \mu(\varpi) = 1 - \frac{\omega_{\rm m}^2}{\omega_{\rm e}^2}\frac{1}{\varpi^2} \tag{13.5}$$

where $\omega_{\rm e}$ and $\omega_{\rm m}$ are plasma frequencies associated with the permittivity and permeability, respectively, and the dimensionless frequency $\varpi = \omega/\omega_{\rm e}$. For a specific application it is $\varpi = \varpi_0$, the operational frequency. It is then convenient to write (with a dispersionless $\bar{\chi}^{(3)}$):

$$\bar{S} = \frac{1}{\varpi}\frac{1}{\omega_0 t_0}\left(2 + \frac{\beta_1 \varpi}{|\bar{\beta}_0|} + \frac{\varpi}{\mu}\left(\frac{\partial \mu}{\partial \varpi}\right)_{\varpi_0}\right). \tag{13.6}$$

If the addition of the magnetooptic parameters are now considered it can be shown that:

$$i\frac{\partial \Psi}{\partial Z} - \frac{1}{2}\,\mathrm{sgn}(\beta_2)\frac{\partial^2 \Psi}{\partial \eta^2} + \mathrm{sgn}(\bar{\chi}^{(3)})|\Psi|^2\Psi + v\Psi = R;$$

$$\text{where } R = -iS \cdot \mathrm{sgn}(\bar{\chi}^{(3)})\frac{\partial}{\partial \eta}(|\Psi|^2\Psi) + i\delta_3\frac{\partial^3 \Psi}{\partial \eta^3} \tag{13.7}$$

$$+ \mathrm{sgn}(\bar{\chi}^{(3)})\tau_{\rm R}\Psi\frac{\partial}{\partial \eta}(|\Psi|^2).$$

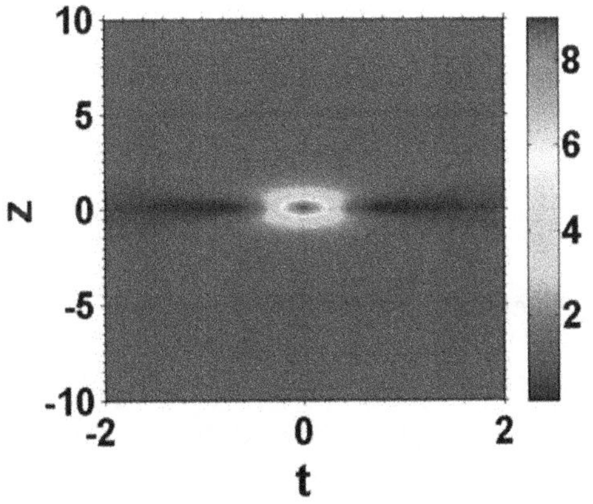

Figure 13.2. This numerical simulation uses the exact solution of the Peregrine soliton as a starting condition at normalized distance $z = -10$ (at this point none of the extensions to the NSE are being used). The Crank–Nicolson technique is then used to propagate along the z-direction. It can be seen that the result of this process is a single peak at position $z = 0$ that subsequently decays in line with the analytical result. It is noted that, although the important perturbation in the initial condition is included, at $z = -10$ the variation on the value of $|\Psi|^2$ is very close to 1 thus the plot appears almost flat at this point. The dark areas to either side of the main peak centered about $z = 0$ are due to the normalized value of $|\Psi|$ dropping below 1 as would be expected. Reprinted from Boardman *et al* (2017), copyright IOP Publishing, all rights reserved.

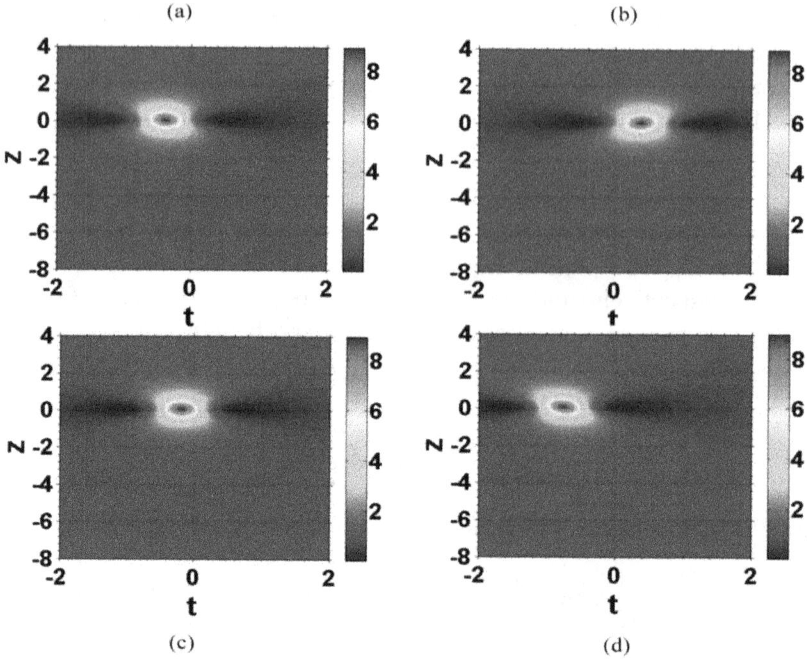

Figure 13.3. Effect of self-steepening on the excitation of the NSE Peregrine soliton using a value of (a) $S = -0.02$, (b) $S = +0.02$, (c) $S = -0.01$, and (d) $S = -0.04$. Note the opposite time shifting of the intensity peak of near-Peregrine soliton for opposite values of S. As in figure 13.2 the darker areas to the side represent the points where the value of $|\Psi|^2$ drops below 1. The initial starting condition also gives a very small deviation from 1 at $z = -8$, which accounts for the almost flat appearance of the initial starting condition. Reprinted from Boardman *et al* (2017), copyright IOP Publishing, all rights reserved.

where here the form of ν is selected similarly to equation (13.6) of Boardman *et al* (2010a), which is:

$$\nu = \frac{\nu_{\max}}{2}\left(1 - \left(\frac{\tanh\left(t - t_{\mathrm{v}} + (Z/v_{\mathrm{g}}) - 2\Delta t_{\mathrm{v}}\right)}{\Delta t_{\mathrm{v}}}\right)\right). \tag{13.8}$$

The magnetooptic parameters that are set in the simulations are thus: t_ν is the delay of the magnetization after excitation by the electromagnetic input, Δt_ν is the normalized time over which the magnetization takes place, v_g is the group velocity and $\nu_{\max}$ is the maximum value of magnetization which is defined in equation (3.4) in Boardman *et al* (2010a) as:

$$\nu_{\max} = \frac{\omega}{c} L_{\mathrm{D}} n_{\mathrm{m}}^2 Q_{\mathrm{sat}} \tag{13.9}$$

where L_{D} is the dispersion length, n_{m} is the refractive index of the magnetooptic material, and Q is the magnetooptic parameter that is usually taken to define the strength of its influence as Q_{sat} representing the saturation of this parameter. The values selected for n_{m} thus reflect the parameters in equation (13.9). For ordinary magnetooptic materials Q is typically of the order 10^{-4} but is not limited to this where suitably designed material for the substrate could enhance this by several orders of magnitude. Using an asymmetric structure for the waveguide also increases the influence of Q. Here in these initial results the various parameters are selected for the simulations (with the appropriate values given with individual simulations) in line with previous results obtained for temporal solitons.

In order to consider MM effects on the NSE breather type of input, the stable Crank–Nicolson method has been applied to numerically solve the propagation equation. The developed code is capable of solving the extended NSE as laid out in Boardman *et al* (2010a), which takes into account the appropriate effects of self-steepening and magnetooptic parameters. It requires an initial starting condition which then determines how this input condition will evolve as propagation occurs along z. Here in order to initiate a specific type of rogue wave condition we initially base our input on the solution to the standard NSE determined by Peregrine, which for the standard NSE gives rise to a solitary pulse on an almost flat background. Following this we go on to look at a non-ideal starting condition which is discussed in more detail below.

The Peregrine solution can of course be used to describe the evolution of a single peak analytically, however, by using a known solution as an input condition it allows the validity of the numerical solver to be checked for the case where there are no additional terms in the NSE. Further, having done this, it then allows us see how the evolution of such a well-known input is changed in the presence of MMs and magnetooptics by re-introducing the additional terms in the extended NSE.

The exact Peregrine solution to the NSE is

$$\psi = [1 - 4 \cdot (1 + 2iz)/(1 + 4z^2 + 4t^2)] \cdot \exp(iz), \tag{13.10}$$

which has the characteristic that it peaks at $z = 0$, both before and after this value the peak decays rapidly. If a specific value of z is now selected, for example $z = -10$, it can then be used as a starting condition for the numerical solver which in the absence of additional terms in the NSE should give a similar solution to the analytical approach.

Hence first we check the validity of our numerical solver for this new type of input and align it with previous work such as Kibler *et al* (2010) by selecting the exact solution at $z = -10$ along the z-axis prior to the peak of the breather and allowing it to propagate along z. It can be clearly seen from figure 13.2 that the result of this process is indeed a single growth/decay cycle centered at position $z = 0$ in agreement with the analytical solution. Note that all figures shown below are pseudocolour plots showing the evolution of wave intensity $|\Psi|^2$ as a function of normalized distance and time in order to highlight breather peak localization. The variation of $|\Psi|^2$ from a value of 1 is small at $z = -10$ and cannot easily be distinguished in the plots, however, it is still present and significant as without it there would be no peak at $z = 0$.

Subsequently, in order to consider MM effects the parameters that are adjusted here are as given above (see also Boardman *et al* 2010a), which are: S is the self-steepening coefficient, t_L is the delay of the magnetization after excitation by the electromagnetic input, Δt_ν is the normalized time over which the magnetization takes place, v_g is the group velocity and ν_{max} is the maximum value of magnetization.

Having established the Peregrine solution the next step is to introduce self-steepening. The self-steepening can be negative or positive, and it is shown that it is possible to arrange for the peak of the Peregrine to appear in different time slots depending on the values selected for the self-steepening coefficient.

As a starting point the values selected for the self-steepening coefficient S are in line with those of Boardman *et al* (2010a). As already stated above, a slight disturbance of the ideal NSE propagation induces a deviation from the expected theoretical solution, but as shown in figure 13.3 most of features related to the Peregrine soliton remain clearly observable, i.e. the main features of a localized high amplitude event (i.e. rogue wave) still occur with similar peak power. It is also interesting to note that by using $S = -0.02$ in the simulation the time shift on the near-Peregrine soliton is of the same order of the effect on the standard soliton pulse, see figure 8(a) in Boardman *et al* (2010a). Figure 13.3(a) illustrates this effect, the Peregrine soliton peak is shifted away from zero on the time axis by ~ -0.45 (for a distance $\Delta z = 8$) figure 13.3(b) shows the shift with $S = +0.02$ which is then shifted to the opposite side of zero-time line. Figure 13.3(c) shows a smaller shift from $S = -0.005$ and figure 13.3(d) has a larger shift resulting from a value for $S = -0.04$. This kind of tilted Peregrine soliton structure was also observed in the case of approximate polynomial rogue-wave solutions obtained beyond the integrable Sasa–Satsuma or Hirota equations (Ankiewicz *et al* 2013). This rogue wave structure makes an angle $\theta \neq 0$ with the z-axis, which corresponds to an effective velocity proportional to the self-steepening coefficient S (Ankiewicz *et al* 2013).

At this point magnetooptic effects are introduced. This, as discussed previously is enabled through the values given for t_ν, Δt_ν, and ν_{max}. It is also noted that with the parameters set, for the magnetooptic effect to operate on the central peak of the Peregrine soliton, then there needs to be sufficient propagation distance prior to where the peak manifests itself. Figure 13.4(a) shows the effect of using a value of $S = -0.02$ with magnetooptic parameters set as $\nu_{max} = -10$, $t_\nu = 200$, $\Delta t_\nu = 10$ and $v_g = 0.03$. It can be seen that the addition of the magnetooptic parameters moves the peak back to the position $t = 0$. It is noted that there is an overall clockwise rotation of the rogue wave structure. It is further noted that if a greater lead in distance is used then a lower value of ν_{max} can be used to achieve a similar response. Alternatively, a larger effect can be achieved with the same value of ν_{max}. Figure 13.4(b) shows the influence of the same magnetic parameters but with the propagation prior to the main peak being 12 dimensionless units along the z-axis rather than 8. It is thus demonstrated that a significant controllable effect can be brought about on the near-Peregrine soliton through MM effects, without cancelling the main features of the rogue wave solution.

As discussed above there are broader categories of solution to the NSE that can be studied; for example the Akhmediev breather solution is one of these. This category of solution has, for example, been studied in Kibler *et al* (2016) and gives multiple peaks in the time domain. However, of particular interest here are non-exact types of breather solutions mentioned above, which can reproduce mainly the exact solution for the Akhmediev breathers but also may yield additional features.

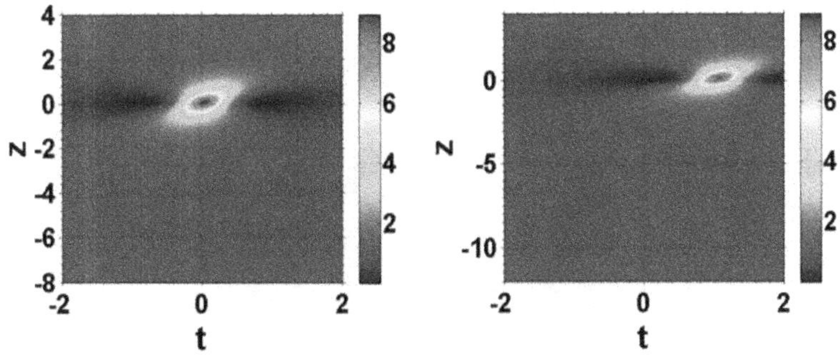

(a) (b)

Figure 13.4. (a) Effect of magnetooptic parameters set as $\nu_{\max} = -10$, $t_\nu = 200$, $\Delta t_\nu = 10$ and $v_g = 0.03$ on the self-steepening $S = -0.02$. The addition of the magnetooptics has moved the peak back towards the origin of the time axis (see figure 13.3(a)). (b) Same parameters but the difference lies in the distance propagated prior to the peak localization. Here the starting point for the input is at -12 rather than -8, it can be seen that the magnetooptic effect now moves the peak to the positive part of the time axis. The small deviation of the starting of $|\Psi|^2$ from 1 again gives the appearance of an almost flat initial condition. The dark areas either side of the main peak are again present where the value of $|\Psi|^2$ drops below 1. Reprinted from Boardman *et al* (2017), copyright IOP Publishing, all rights reserved.

Hence we now investigate the excitation of NSE breather solutions with non-ideal input conditions. Our objective is to point out the impact of negative phase nonlinear MMs in this area, in particular to extend the conclusions drawn for optical fiber platforms (Kibler *et al* 2010, Kibler *et al* 2016, Akhmediev *et al* 2016) in the presence of self-steepening and the addition of magnetooptic parameters. This is carried out in a similar manner to the process adopted for the exact solution simulations discussed above.

Here again the first step is to ensure correlation between the simulations previously carried out for the standard NSE with non-ideal inputs (Kibler *et al* 2016) and the new simulator with MM extensions. Figure 13.5 shows a test simulation carried out with the following input condition: $\Psi = (1 + 0.145\cos(\omega t))^{1/2}$, as originally given in figure 13.7 from Kibler *et al* (2016) where $\omega = 2\cdot(1 - 2a)^{1/2}$ and $a = 0.25$. This input condition simply corresponds to an intensity-modulated continuous wave with the angular frequency ω linked to the governing parameter a (recall that a determines the physical behavior of the excited breather solution, the excited breather is part of the family of Akhmediev breathers). At this point the simulation does not contain any extensions to the NSE. We observe that non-ideal initial conditions yield periodic evolution as a function of propagation in contrast to the exact Akhmediev-breather (AB) theory with a single growth/decay cycle of the temporal periodic pattern. However, each growth-return cycle remains well-described by the analytic AB solution.

Having established a correlation between this and previous work, the self-steepening effect is now added to the simulation ($S = -0.02$).

Figure 13.6 shows such a scenario where it can be seen that both the first and second set of peaks have been rotated and moved in a negative direction with respect to the time axis. The second set have moved by approximately -0.6 on the normalized time axis for a propagation length $\Delta z = 10.5$. This corresponds to a

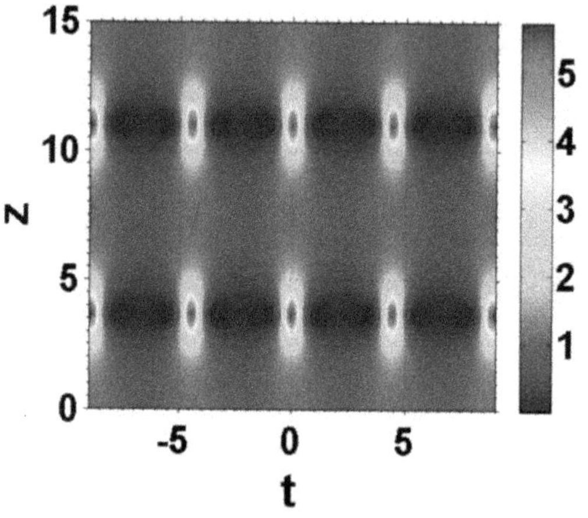

Figure 13.5. Simulation of the standard NSE with the following input condition $\Psi = (1 + 0.145\cos(\omega t))^{1/2}$ plots the darker areas to the sides of the peaks represent $|\Psi|$ dropping below a value of 1. Reprinted from Boardman *et al* (2017), copyright IOP Publishing, all rights reserved.

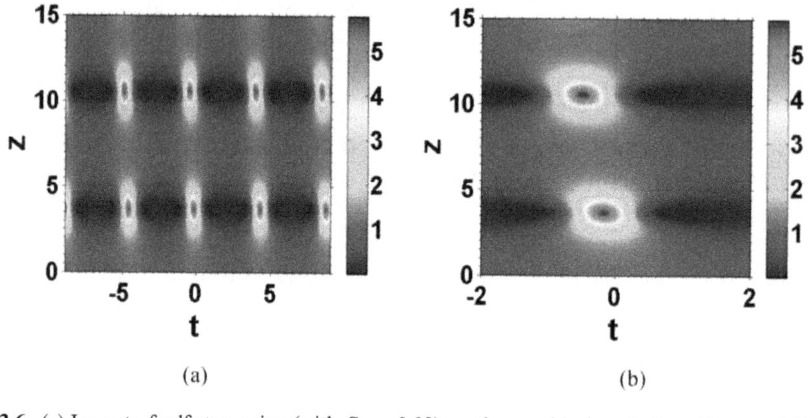

Figure 13.6. (a) Impact of self-steepening (with $S = -0.02$) on the non-ideal excitation of a near-Akhmediev breather ($a = 0.25$). The input condition is $\Psi = (1 + 0.145\cos(\omega t))^{1/2}$. It can be seen that both sets of peaks have moved along the time axis in a negative direction. (b) Expanded plot of the central portion of figure 13.6(a). On both plots the darker areas to the side of the main peaks show the value of $|\Psi|$ dropping below a value of 1. Reprinted from Boardman *et al* (2017), copyright IOP Publishing, all rights reserved.

slightly longer propagation distance from the input than that in figure 13.3(a) for the exact solution where there is a slightly smaller shift. The first set of peaks shifts by approximately −0.2 or less, for a significantly shorter propagation distance from the input ($\Delta z = 3.5$). It can thus be seen that the effective velocity induced by self-steepening correlates well with different input conditions. It is noted that the shift leaves both sets of peaks aligned in the same orientation and rotated counter-clockwise. Moreover, the spatial localization of the maximum intensity peak does not change significantly.

Figure 13.7 shows the effect of adding magnetooptic parameters with similar values to that used in figure 13.4 ($\nu_{\max} = -10$, $t_\nu = 200$, $\Delta t_\nu = 10$ and $v_g = 0.03$). It can be seen that the first set of peaks has not reached its original position prior to the addition of self-steepening and magnetooptic effects and still has counter-clockwise rotation, whereas the second set of peaks has moved well past the original position with a clockwise rotation. This again illustrates here that the magnitude of the response from the magnetooptics is position dependent.

Figure 13.8 has similar input parameters to figure 13.7 but importantly with $\nu_{\max} = -4$ rather than -10. This has the effect of moving the central point of the second set of peaks back to the starting position prior to the addition of self-steepening and magnetooptics. However, as with previous simulations there is a

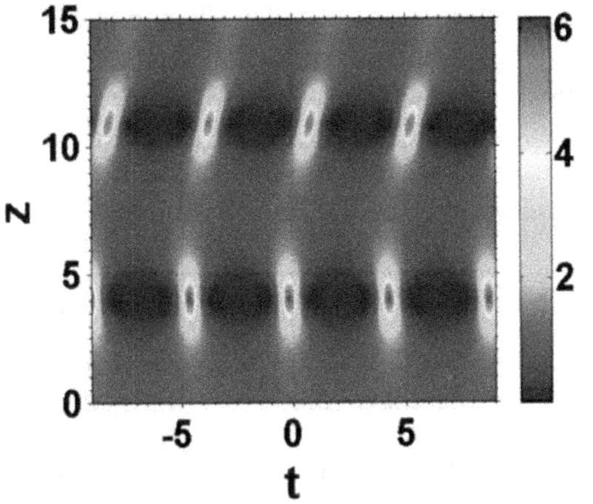

Figure 13.7. Impact of self-steepening and magnetooptic parameters on the non-ideal excitation of a near-Akhmediev breather ($a = 0.25$). The input condition is $\Psi = (1 + 0.145\cos(\omega t))^{1/2}$ continues to be darker areas on the plot where $|\Psi| < 1$. Reprinted from Boardman *et al* (2017), copyright IOP Publishing, all rights reserved.

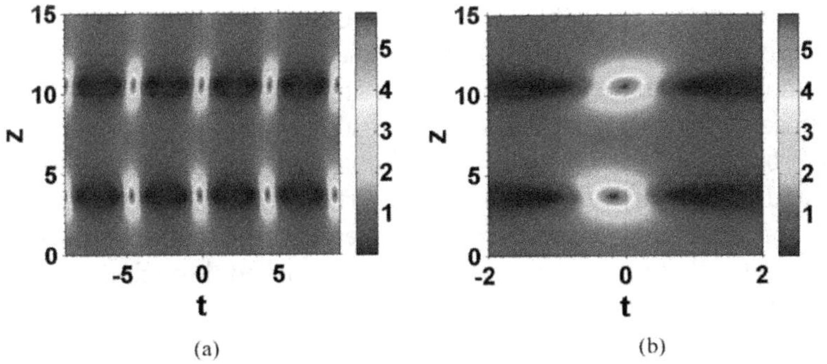

Figure 13.8. (a) Effect of magnetooptic parameters set as $\nu_{\max} = -4$, $t_\nu = 200$, $\Delta t_\nu = 10$ and $v_g = 0.03$ on the self-steepening $S = -0.02$. The second set of pulses is moved back toward the original starting position by the addition of magnetooptic parameters and after being moved away by the effect of self-steepening. (b) Expanded plot of the central portion of figure 13.8(a). On both plots it can be observed alongside the peaks there continues to be darker areas that represent $|\Psi|^2 < 1$. Reprinted from Boardman *et al* (2017), copyright IOP Publishing, all rights reserved.

small clockwise rotation as there is a greater effect of the magnetooptic addition the further along z. It also raises the possibility of having different magnetooptic parameters in different parts of the guide to allow for similar shifts on the time axis for different distances. Here we are able to almost cancel the effective velocity induced by self-steepening with an appropriate magnetooptic response.

These latter results with the use of non-exact input conditions thus show similar responses to the exact breather excitation. It is thus possible to see how a range of different input conditions could potentially be controlled. Although not yet fully optimized, the parameters used here indicate that reasonably accurate control can be achieved that would allow useful systems to be developed. There are now a wide range of possibilities in terms of self-steepening, magnetooptic and other parameters that are now being researched for this purpose. The fact that breathers propagating with a certain angle to the line $t = 0$ can also be induced by non-ideal input wave excitation, as well as propagation effects, is interesting to note (Kibler *et al* 2016). The resulting inclined trajectory is usually associated with an asymmetric spectrum or/and frequency detuning, with a distinct mean group velocity of the breather under study. Note that rogue wave solutions analyzed in the framework of the Sasa–Satsuma equation already revealed strong spectral asymmetries (Akhmediev *et al* 2015).

13.3 Conclusions to chapter 13 and future trends

To conclude this section, we investigated the propagation behavior of wave excitation corresponding to rogue NSE breather solutions in transparent double-negative MMs, wherein higher-order dispersive or nonlinear effects are included. An appropriate extended NSE was used which is not integrable. It takes into account typical higher order effects from MMs, such as the self-steepening and the magnetooptics parameters. We revealed the impact of the self-steepening on both Peregrine and Akhmediev breathers as the change of the mean group velocity of the evolving breather under study. Such time-shifting signatures can be also found in integrable systems such as the Sasa–Satsuma or the Hirota equations. With the application of magnetooptic influence/control over these phenomena, we demonstrated that the impact of self-steepening on breather waves can be cancelled or overcome, similarly to studies of the standard NSE soliton. Most importantly, we confirmed that the extra-terms studied here applied to the NSE do not prevent the emergence of rogue wave structures that are almost identical to rogue NSE breather solutions. Conversely, they offer a possible management of their unique pulsating dynamics and spatiotemporal localization properties. All of the investigations presented here offer some interesting perspectives on the interaction of electromagnetics and MMs. It is demonstrated that promising responses can be elicited through the propagation of rogue waves in MMs under magnetooptic influence.

Problems to chapter 13

13.1 Check directly that the Peregrine solution equation (13.10) satisfies the unperturbed NSE equation (13.1) without the perturbation terms. Then estimate an influence of the perturbation terms of the generalized NSE, with

$S \neq 0$, at the time moment of the maximum compression $t = 0$ near the pulse maximum.

13.2 As a generalization of the Peregrine solution, it is possible to consider the solutions of higher orders of NSE, see for instance (Boling Guo, Lixin Tian, Zhenya Yan, Liming Ling, Yu-Feng Wang 2017 Rogue waves *Mathematical Theory and Applications in Physics* (Berlin: De Gruyter) p 140 and after). Please compare the input distribution and the pulses under the maximum compression.

13.3 Please check numerically that the generation of the rogue wave from the Peregrine solution is stable with respect of the cut-off of the periphery of the initial condition for NSE. As a numerical method, the stable Crank–Nicolson scheme can be proposed (A A Samarskii 2001 *The Theory of Difference Schemes* (Boca Raton, FL: CRC Press). What is about the stability for the higher order solutions?

References

Agrawal G P 2013 *Nonlinear Fiber Optics* 5th edn (Oxford: Academic)

Akhmediev N and Ankiewicz A 1997 *Solitons, Nonlinear Pulses and Beams* (London: Chapman and Hall)

Akhmediev N and Korneev V I 1986 Modulation instability and periodic solutions of the nonlinear Schrödinger equation *Theor. Math. Phys.* **69** 1089–93

Akhmediev N and Pelinovsky E 2010 Discussion and debate: Rogue waves—towards a unifying concept? *Eur. Phys. J. Spec. Top.* **185** 1–258

Akhmediev N *et al* 2016 Roadmap on optical rogue waves and extreme events *J. Opt.* **18** 063001

Akhmediev N, Dudley J M, Solli D R and Turitsyn S K 2013 Recent progress in investigating optical rogue waves *J. Opt.* **15** 060201

Akhmediev N, Soto-Crespo J M, Devine N and Hoffmann N P 2015 Rogue wave spectra of the Sasa–Satsuma equation *Phys.* D **294** 37–42

Ankiewicz A, Devine N and Akhmediev N 2009 Are rogue waves robust against perturbations? *Phys. Lett.* A **373** 3997–4000

Ankiewicz A, Soto-Crespo J M and Akhmediev N 2010 Rogue waves and rational solutions of the Hirota equation *Phys. Rev.* E **81** 046602

Ankiewicz A, Soto-Crespo J M, Chowdhury M A and Akhmediev N 2013 Rogue waves in optical fibers in presence of third-order dispersion, self-steepening, and self-frequency shift *J. Opt. Soc. Am.* B **30** 87–94

Bandelow U and Akhmediev N 2012 Persistence of rogue waves in extended nonlinear Schrödinger equations: integrable Sasa–Satsuma case *Phys. Lett.* A **376** 1558–61

Benjamin T B 1967 Instability of periodic wavetrains in nonlinear dispersive systems *Proc. R. Soc.* A **299** 59–75

Benjamin T B and Feir J E 1967 The disintegration of wave trains on deep water: I. Theory *J. Fluid Mech.* **27** 417–30

Bespalov V I and Talanov V J 1966 Filamentary structure of light beams in nonlinear liquids *JETP Lett.* **3** 307–10

Boardman A D, Hess O, Mitchell-Thomas R C, Rapoport Y G and Velasco L 2010a Temporal solitons in magnetooptic and metamaterials waveguides *Photon. Nanostruct. – Fundam. Appl.* **8** 228–43

Boardman A D, Mitchell-Thomas R C, King N J and Rapoport Y G 2010b Bright spatial solitons in controlled negative phase metamaterials *Opt. Commun.* **283** 1585–97

Boardman A D, Alberucci A, Assanto G, Grimalsky V V, Kibler B, McNiff J, Nefedov I S, Rapoport Yu G and Valagiannopoulos C A 2017 Waves in hyperbolic and double negative metamaterials including rogues and solitons *Nanotechnology* **28** 444001

Chabchoub A, Hoffmann N P and Akhmediev N 2011 Rogue wave observation in a water wave tank *Phys. Rev. Lett.* **106** 204502

Chen S 2013 Twisted rogue-wave pairs in the Sasa–Satsuma equation *Phys. Rev.* E **88** 023202

Dontsop P Y G, Essama B G O, Dongo J M, Dedzo M M, Atangana J, Yemele D and Kofane T C 2016 Akhmediev–Peregrine rogue waves generation in a composite right/lefthanded transmission line *Opt. Quantum Electron.* **48** 59

Essama B G O, Atangana J, Biya-Motto F, Mokhtari B, Eddeqaqi N C and Kofane T C 2014 Optical rogue waves generation in a nonlinear metamaterial *Opt. Commun.* **331** 334–47

Kibler B, Fatome J, Finot C and Millot G 2016 *Rogue and Shock Waves* ed M Onorato (Lectures Notes in Physics vol 926) (Berlin: Springer) p 89

Kibler B, Fatome J, Finot C, Millot G, Dias F, Genty G, Akhmediev N and Dudley J M 2010 The Peregrine soliton in nonlinear fibre optics *Nat. Phys.* **6** 790–5

Lighthill M J 1965 Contribution to the theory of waves in nonlinear dispersive systems *J. Inst. Math. Appl.* **1** 269–306

Onorato M, Residori S, Bortolozzo U, Montina A and Arecchi F T 2013 Rogue waves and their generating mechanisms in different physical contexts *Phys. Rep.* **528** 47–89

Osborne A R 2010 *Nonlinear Ocean Waves and the Inverse Scattering Transform* (San Diego, CA: Academic)

Pendry J B 2000 Negative refraction makes a perfect lens *Phys. Rev. Lett.* **85** 3966–9

Peregrine D H 1983 Water waves, nonlinear Schrödinger equations and their solutions *J. Aust. Math. Soc.* B **25** 16–43

Samarskii A A 2001 The theory of difference *Economical Difference Schemes for Multidimensional Problems in Mathematical Physics* (New York: Schemes Marcel Dekker) ch 9 pp 543–642

Scalora M *et al* 2005 Generalized nonlinear Schrödinger equation for dispersive susceptibility and permeability: application to negative index materials *Phys. Rev. Lett.* **95** 013902

Shen Y, Kevrekidis P G, Veldes G P, Frantzeskakis D J, DiMarzio D, Lan X and Radisic V 2017 From solitons to rogue waves in nonlinear left-handed metamaterials *Phys. Rev.* E **95** 032223

Solli D R, Ropers C, Koonath P and Jalali B 2007 Optical rogue waves *Nature* **450** 1054–7

Veselago V G 1967 The electrodynamics and substances with simultaneously negative values of ε and μ *Sov. Phys. – Usp.* **10** 509–14

Wen S C, Xiang Y J, Su W H, Hu Y H, Fu X Q and Fan D Y 2006 Role of the anomalous self-steepening effect in modulation instability in negative-index material *Opt. Express* **14** 1568–75

Whitham G B 1965 A general approach to linear and nonlinear dispersive waves using a Lagrangian *J. Fluid Mech.* **22** 273–83

Zakharov V E 1968 Stability of periodic waves of finite amplitude on a surface of deep fluid *J. Appl. Mech. Tech. Phys.* **9** 190–4

Zakharov V E and Ostrovsky L A 2009 Modulation instability: The beginning *Phys.* D **238** 540–8

Zakharov V E and Shabat A B 1971 Exact theory of two-dimensional self-focusing and one-dimensional self-modulation of waves in nonlinear media *Zh. Eksp. Teor. Fiz.* 118–34 (in Russian, English transl. Zakharov V E and Shabat A B 1972 *Sov. Phys. - JETP* **34** 62–9

Yuriy Rapoport and Vladimir Grimalsky

Chapter 14

Waves in nonlinear layered metamaterials, gyrotropic and plasma media. The main results of the book and the proposed directions for future research

14.1 The main results obtained in the previous chapters

This monograph presents the results of theory development, detection, study, and generalization of regularities of wave processes in layered, complex, and active media of different physical nature, including metamaterials (MMs), gyrotropic and plasma media, in the presence of bulk and surface nonlinearities; detection of new wave structures and effects in the above-mentioned media, arising from resonant, moderate, and strong nonlinearities; searching opportunities to control wave processes in such media with the help of inhomogeneous and non-stationary external fields.

The main results and conclusions are formulated below:

1. It is shown for the first time that the frequency band where the corresponding elements are active must be finite to provide the spatial amplification of electromagnetic waves and negative phase behavior in MMs with metaparticles loaded on active elements. The conditions of the spatial amplification are found jointly with negative phase behavior of MMs with metaparticles loaded on active elements with frequency-dependent conductivity. A generalized theoretical scheme for constructing such an active MM that provides the spatial amplification of electromagnetic waves in a negatively phase MM medium is proposed. In the analysis of the absolute instability, three main types of potentially dangerous sources of the absolute instability in the infinite periodic system are considered.

   A model of a one-dimensional photonic crystal or a transmission line (TL) is used to determine the possibility of a new NPM with spatial amplification.

doi:10.1088/978-0-7503-2336-9ch14

The active frequency-dependent elements and dissipative resistances have been appropriately added. The used model is a good one for an active NPM on the basis of SRRs (split ring resonators) and metal rods. On the other hand, it can be used as a model of a real active NPM based on periodic TL. It is shown that the absolute instability is absent in such a material, while it is active, according to the chosen parameters and it is determined how the corresponding parameters should be selected.

The active medium with a negative phase has been proposed for the first time and is useful for many practical applications such as advanced lenses, special systems for linear and nonlinear signal processing, etc. Scaling the system and finding adequate active elements can lead to the application of the proposed principle of the new active medium in higher frequency bands.

A nonlinear homogenization method for a bi-anisotropic medium is developed. It is applied to a particular case of Ω-particles with a nonlinear load, arranged parallel to each other. For a case of GHz frequencies, a single metaparticle is analyzed on the basis of the Kirchhoff rules for currents and voltages, taking into account nonlinear elements excited by the total action of local electromagnetic fields including the action of other metaparticles in the Lorentz–Lorenz approximation.

2. The NEELS method has been developed in a very general way for the study of waves in layered and complex media of different physical nature, with spatial and temporal scales that differ by many orders of magnitude, namely in MMs, plasma, gyrotropic and chiral media, as well as in the LAI system. It is proven that the developed method allows one to study nonlinear wave effects in active media, taking into account surface and volume nonlinearities, higher harmonics, parametric interaction, NEHO (nonlinear effects of high order). The methods are proposed to determine the contribution to the total nonlinearity coefficient from the surface nonlinearity included in the additional boundary conditions associated with the nonlocality, or the spatial dispersion, of dielectric-ferroelectric and dielectric-ferromagnetic media. The high efficiency of the NEELS method is proven: (1) by comparison with the analytical results of other authors; (2) comparing theoretical results with experimental ones; (3) obtaining new results, some of which have been confirmed experimentally and theoretically (by other authors and methods), including references to the works listed in the monograph.

   A method for deriving evolutionary macroscopic equations for the amplitudes of envelope waves in layered systems in the presence of bulk and surface nonlinearities has been developed for media with bi-anisotropic layers.
3. Using the NEELS method, the possibility of formation of magnetic bullets is shown and their collapse in ferrite films is studied. New nonlinear effects arising during the formation of wave structures in gyrotropic waveguides like ferrite films have been identified, namely quasi-stationary diffraction of

a spin-wave beam with the concentration of wave energy at a single focal point. It is shown that for a gyrotropic nonlinear layer with dissipation, in some interval of propagation distances, the nonlinear collapse is stabilized by dissipation and a quasi-stable strongly localized two-dimensional wave packet is formed, the so-called magnetic bullet, i.e. (2 + 1) space-time 'soliton'; interactions of magnetic bullets at their collisions are described. It is shown that the length of formation of magnetic bullets in films of iron-yttrium garnet is of the order of several millimeters, which is only a few wavelengths. In the presence of parametric interaction and amplification, the possibility and features have been pointed out of formation of new nonlinear wave structures, including knife-shaped elongated phase-conjugate bullets (on the idle wave(s)) in the absence of bullets on an incident wave(s), and multibullet structures in narrow and wide gyrotropic layers, respectively. A new 'idler wave splitting' effect was found. It appeared on the basis of nonlinear phase degradation in the presence of a strong incident wave. The effect is achieved at relatively low values of the amplitude of the input waves and relatively small values of the parametric pump fields. It was found that for a relatively small value of the input pulse amplitude and insignificant values of the pump amplitude, the integral energies of the incident pulse and the idle pulse tend to collect after the end of the pumping action. The system 'forgets' the initial conditions, as well as the amplitudes of the counter propagating pulses (incident and idle), after the end of the parametric interaction, usually have magnitudes of the same order. It is shown that new nonlinear structures are formed by the amplitude of the input pulses and large pump fields in the 'wide' ferrite area. Such a structure can develop after finishing the pumping, consisting of a structure of four strong bullets (two bullets based on the incident wave and two based on the idler one). There were found the features of the transformation of the amplitude of the incident pulse during the passage through the gyrotropic layer under the action of the pump field; they appear in connection with the development of nonlinear wave structures in the gyrotropic layer. The vortex structures in normally magnetized gyrotropic layers have been investigated. Nonlinear vortex structures can be excited in a ferrite film on the basis of the interaction of three beams with relative phase differences between adjacent beams which equals $2\pi/3$ (between the second and the first beams and between the third and the second ones). For beams with non-shifted centers (with symmetric geometry), quasi-periodic wave structures with vortex singularities and a topological charge equal to 1 or -1 are formed. In the absence of nonlinearity, slightly different amplitudes of interacting beams do not, in general, change the symmetry of the forming wave structures, but nonlinearity, together with a small difference in the amplitudes of the beams, can change the symmetry. Possibilities of formation of non-stationary nonlinear vortex structures at interaction of three waves with corresponding mutual phase shifts in the gyrotropic layer (ferrite film) are shown.

The simulation results for nonlinear wave structures in ferrite films coincide qualitatively, and in a number of key parameters, and quantitatively with observations based on the Brillouin light scattering.

4. There were determined the characteristics of formation and the proposed method of control, using a non-stationary magnetic field, as well as the characteristics (including the delay time) of temporal magnetooptical solitons, in the presence of nonlinear dispersion and the Raman interaction, in MM layered structures. Evolutionary equations are derived taking into account the effects of magnetic and electrical nonlinearities, on the base of which the mechanism of effective control of spatial soliton characteristics in layered MMs in the presence of NEHO is theoretically substantiated. Surface magnetic polaritons with a new type of dispersion have been found, for which both direct and inverse waves can be excited in the same frequency range in the system 'half-infinite MMs–half-infinite ferromagnetic'.
5. It is shown that to provide the spatial amplification of electromagnetic waves and negative phase behavior in MMs with metaparticles loaded on active elements, the frequency band should be finite where the corresponding elements are active must be finite. Based on the analysis of three main possible sources of the absolute instability, the conditions of spatial amplification with simultaneous negative phase behavior of MMs with metaparticles loaded on active elements with frequency-dependent conductivities are found. A generalized theoretical scheme is proposed for constructing such an active MM that provides spatial amplification of electromagnetic waves in a negative phase MM environment.

   An analysis of (2 + 1) (space-time) solitons/quasisolitons in MM waveguides shows that bullets are possible with the proper set of signs of the coefficients of nonlinear equations.

   The possibility of significant and non-trivial influence of NEHO (both dispersion and diffraction) on the instability of bullets in a nonlinear waveguide is shown, in particular, as a factor of bullet splitting in the direction of propagation. It is shown that with the use of diffraction management, the bullets in the active waveguide in the presence of NEHO can be amplified with the preservation of the shape at a greater distance than without diffraction management. The following conclusion is made, namely the effects of NEHO are non-trivial and significant for the formation, propagation and control, with the help of external fields, nonlinear wave structures in layered complex media of different physical nature.

   The evolution of magnetic bullets/(2 + 1) space-time solitons in MMs is determined by the competition between two types of nonlinear instabilities and dissipative losses. The instability of the first type is that which causes the formation of a bullet and tends to lead to an increase in the amplitude and decrease in the cross section of the bullet. In the absence of diffraction control and in a certain range of parameters, the instability of the second type tends to destroy the bullets and is manifested primarily in the splitting of the bullets in the longitudinal direction. Then, unlike the situation where higher nonlinearities are absent, the larger of the forming peaks extends over a relatively long distance, as a pronounced quasi-stable bullet. Nonlinear diffraction has a non-trivial and significant influence on the formation of nonlinear structures in the longitudinal direction. The presence of self-compression of different signs causes in the nonlinear waveguide the arrangement of peaks of different intensity in different order when a multi-soliton structure with many peaks is formed.

6. It is determined that almost reflectiveless phase conjugation is achieved at normal signal wave incidence and non-degenerate three-wave interaction due to the use of an active nonlinear Huygens metasurface, which is made on the base of chiral metaparticles. Qualitative estimates for the microwave wave range were performed. The possibility of the almost ideal phase conjugation is shown. The almost ideal phase conjugation is achieved with the use of an active metasurface with a normal incidence of a signal wave and a non-degenerate three-wave interaction, with a small inequality of the frequencies of the incident and phase-conjugate waves. The use of linear and nonlinear Huygens sources for both linear (signal) and nonlinear (phase-coupled) waves causes zero or very small values of the reflection coefficient for both waves simultaneously. It is possible to achieve the value of energy transformation of an incident to a phase-conjugate EMW of order 10^{-5} that is sufficient for experimental demonstration.

On the basis of the derived evolution equations, taking into account the effects of magnetic and electrical nonlinearities and diffraction management, a qualitative explanation of the mechanism of effective control of spatial solitons, including breathers, is given by means of a magnetooptical effect in a transversely inhomogeneous magnetic field in magnetooptical waveguides.

Temporal (pulse) solitons in MMs in the presence of external stationary and transversely inhomogeneous magnetic fields are investigated. A frequency window with negative magnetic permeability and electric permittivity and different signs of group and phase velocities was found. The frequency dependences of the coefficients of the modified normalized nonlinear Schrödinger equation (NSE) are found. The model includes nonlinear dispersion, diffraction (for two-dimensional pulses), Raman interaction, and higher-order (including third order) linear dispersion. It is the frequency dispersion of these normalized coefficients that represent the essential properties of solitons in MMs. In particular, the standard Drude model for the electric permittivity and magnetic permeability is considered. MO interaction is used as a practical mechanism for controlling the velocity of nonlinear (including one-dimensional) pulses. Practically, this effect can be used for data processing systems.

The possibility of effective magnetooptical control of bullets in an MM waveguide with a transverse magnetic field (due to the Voigt effect) is shown.

The magnetooptical interaction leads to a quasi-periodic dependence of the intensity of the bullets on the coordinate along the propagation direction in the waveguide. The most effective stabilization of the bullets in the MM magnetooptical waveguide is achieved at the parameters of the waveguide providing a zero (or very small) value of the self-steepening coefficient (nonlinear dispersion).

It is shown that in the 'ferromagnet/NPM' system there are frequency regions where forward and backward waves can simultaneously propagate, unlike the 'dielectric-ferromagnet' system. The coefficients of reflection of electromagnetic waves from the system 'dielectric-ferromagnet-MM (NPM)' were obtained. These coefficients are necessary for the study of non-trivial Goos–Hänchen shift, when reflecting electromagnetic pulses from the system 'dielectric-ferromagnet-MM (NPM)' (Boardman *et al* 2005). A possibility of nonlinear control of Goos–Hänchen shift in non-linear dielectric-MM layered media has been proposed.

7. The end-to-end method of modeling nonlinear waves in active layered bi-anisotropic MM structures with Ω-particles is proposed, which includes: (1) modeling and description of a single-particle with nonlinear loading; (2) a nonlinear homogenization of such a bi-anisotropic medium; (3) simulation of nonlinear wave propagation using the NEELS method for bi-anisotropic MMs. For a special case of a layered bi-anisotropic medium with metaparticles containing nonlinear loads that allow both self-action and parametric interaction of nonlinear counter propagating waves is considered. The possibilities of self-action, cross-modulation, the second harmonic generation, and the parametric interaction for counter propagating waves in layered bi-anisotropic structure are accounted for. The surface nonlinearity and nonlinear losses or amplification are taken into account as well. The physical reason of the effects mentioned above could be, in particular the presence of the effective negative conductivity in the loading of metaparticles.
8. A NEELS method for layered dielectric-plasma structures with bulk and surface nonlinearities arising as a result of nonlinear motion of surface charges on the plasma-dielectric surface has been developed. The defining role of the surface nonlinearity (compared to the volume one) for these structures is established for the first time. The possibility of generating a giant plasmon localized second harmonic, the amplitude of which exceeds the amplitude of the main harmonic, is shown; the conditions for the transition of the process of resonant generation of a giant second harmonic to the regime of strong nonlinearity are found.
9. The existence of strong nonlinearity effects in a multilayer MM dielectric-graphene structure is shown, in particular, the nonlinear threshold switching of THz pulses from the reflection mode to the transmission mode. It is determined that to achieve the effect of exceeding the amplitude of the transmitted pulse over the amplitude of the incident pulse, the amplitude of

the incident beam must exceed a threshold value of about 20 KV cm^{-1}, and the effective collision frequency should not exceed 1.5×10^{12} s^{-1}.

10. A nonlinear MM field concentrator (nonlinear electromagnetic 'black hole') is proposed. For the case of strong nonlinearity in the field concentrator (beyond the use of the NEELS method), a new method is proposed and developed. This method consists in using a combination of a new variant of complex geometric optics in the linear domain, where the wavelength is much less than the characteristic inhomogeneity, and a full-wave solving of Maxwell equations in the domain with nonlinearity, according to the proposed boundary conditions for matching both solutions. This approach can be used for wave processes in both mesoscale MM and the LAIM system, despite the fact that the values of spatial and temporal scales for various media differ in many orders of magnitude. This method theoretically revealed a new physical effect associated with strong nonlinearity in active MM in the IR range, namely the threshold jump of the focusing point and the formation of a self-consistent highly localized nonlinear resonator (the hot spot) at the boundary of linear and nonlinear regions in the field concentrator. It is proven that the jump of the focusing point is achieved when the amplitude of the field exceeds the threshold value, regardless of the physical method of such exceeding (either increasing the amplitude of the incident pulse, or linear gain, or by using an additional weak beam). The numerical estimations adopt the doped *n*-Si. The electromagnetic wave (with intensity of order ~1–5 GW cm^{-2}) is necessary for observing the nonlinear phenomena in the field concentrator with the internal nonlinear region of the radius R_c = 10 μm. Respectively, the intensity of the electromagnetic wave incident on the external cylinder of the field concentrator should be of order (100–500) MW cm^{-2}. Applications of field concentrators based on electromagnetic or acoustic MM structures may be perspective for nonlinear versions of energy harvesting systems, such as solar cells, subwavelength imaging, antenna-based sensing including field-enhanced microscopy, systems for feeding optical fibers, frequency mixing and high-harmonic generation, new types of optical nonlinear switches, reduction of strong acoustic noise. The proposed method which includes the matching of full-wave nonlinear solution and complex geometrical optics is rather general and, besides electromagnetic and acoustic MMs, can be used for modeling natural inhomogeneous, active, and strongly nonlinear media, such as, for example, laboratory and space plasmas (Tao and Bortnik 2010a, Tao *et al* 2010b, Demekhov *et al* 2002). The developed technique may be useful for active space experiments (Luhmann *et al* 2005), where the processes of wave coupling in the nonlinear system 'magnetosphere–ionosphere' will be studied, while spatial scales in different regions of a system differ in many orders of magnitude and the matching of complex geometrical optics and full-wave electromagnetic solutions is promising. The effect of nonlinear control was obtained for concentrators based on both isotropic (Boardman *et al* 2011,

Rapoport and Grimalsky 2011, Rapoport *et al* 2012a, 2012b, 2014b) and hyperbolic (Rapoport *et al* 2014a) MMs. Details concerning nonlinear superfocusing in acoustic MMs and anisotropic MMs are the subject of strong interest for future investigation.

11. The propagation behavior of wave excitation corresponding to rogue NSE breather solutions in transparent double-negative MMs, wherein higher-order dispersive or nonlinear effects are included, has been investigated. An appropriate extended NSE was used which is not integrable. It takes into account typical higher order effects from MMs, such as the self-steepening and the magnetooptics parameters. The impact of the self-steepening has been revealed on both Peregrine and Akhmediev breathers as the change of the mean group velocity of the evolving breather under study. Such time-shifting signatures can be also found in integrable systems such as the Sasa–Satsuma (Chen 2013) or the Hirota equations (Ankiewicz *et al* 2010). With the application of magnetooptic influence/control over these phenomena, it has been demonstrated that the impact of self-steepening on breather waves can be cancelled or overcome, similarly to studies of the standard NSE soliton. Most importantly, it has been confirmed that the extra-terms studied here applied to the NSE do not prevent the emergence of rogue wave structures that are almost identical to rogue NSE breather solutions. Conversely, they offer a possible management of their unique pulsating dynamics and spatiotemporal localization properties. All of the investigations presented here offer some interesting perspectives on the interaction of electromagnetics and MMs. It is demonstrated that promising responses can be elicited through the propagation of rogue waves in MMs under magnetooptic influence.
12. The method of averaging can be applied for hyperbolic MMs when the thicknesses of elementary layers are small and the nonlinearity is moderate. When the following inequality is valid $d_i \leqslant 0.1\lambda_i$ where d_i is the thickness of each elementary layer, $\lambda_i \equiv \lambda_0 \cdot \varepsilon_i^{-1/2}$ is the wavelength of the electromagnetic wave in this medium, the results of simulations are the same both within the direct consideration of the layered medium and within the averaging approach for the hyperbolic medium. In this case it is possible to reduce the nonlinear equation for EM wave propagation to the standard nonlinear Schrödinger equation, where the wave amplitude is slowly varying with respect to the longitudinal coordinate, i.e. along the direction of propagation. When the inequalities are valid $0.1\lambda_i \leqslant d_i \leqslant 0.25\lambda_i$, there are some quantitative differences within the two approaches indicated above but the results are qualitatively similar. At larger thicknesses of elementary layers the differences between two approaches are qualitative and the averaging approach is not valid. The averaging in the MM approximation of the continuous media has been provided. The formulas for averaging have been derived, accounting for the periodicity of the structure and with the farther application of the approximation of the thin layers (continuous media) in the linear limiting case. For the nonlinear media, the application of the

averaging formulas is only phenomenological. This is why it was necessary and important to verify the MM approximation for active nonlinear media by means of the comparison between the results of the corresponding modeling with these obtained using a more accurate approach without an averaging. We would like to emphasize that the positive result of such a comparison is obtained in the present work only for the case of nonlinear active media with net zero gain (total compensation between gain and losses). A possibility of the MM approach to the hyperbolic nonlinear periodical active media with the nonzero net gain, as to the continuous media is questionable, even providing that the requirements of the MM approximation applicability are satisfied in the linear limiting case. These problems will be a subject of future work.

For solving nonlinear problems various different schemes have been applied. The implicit–explicit method of Peaceman–Rachford does not possess good stability. The method of the summatory approximation needs very small temporal steps and is practically not applicable there. It is very interesting that the method of the operator factorization, or the method of Douglas–Rachford (known more in hydrodynamics (Boardman *et al* 2017, Marchuk 1990, Douglas and Rachford 1956) than in nonlinear optics), seems the most appropriate.

The nonlinear effects are different for different signs of nonlinearity. In the case of the negative nonlinearity of the layers with the positive permittivity the hotspots can be formed within the hyperbolic medium. In the case of the positive nonlinearity of the layers with positive permittivity the nonlinear diffraction of the EM wave beams occurs.

In the near-infrared and visible optical range metallic (or semi-metallic) layers with high conductivity can be used as media with the negative permittivity, whereas the dielectrics with high values of permittivity can be used as other layers. In the THz range the narrow forbidden gap semi-conductors, like *n*-InSb, can be used as media with the negative permittivity. The metallic layers possess dissipation, even in metals with high conductivity. Therefore, the dissipation should be compensated, to observe the nonlinear wave phenomena. A mechanism of compensation of dissipation can be a creation of active dielectric layers, for instance, by means of inserting quantum dots with the inversion of energetic levels.

13. A model of the nonlinear wave propagation in the hyperbolic nonlinear active resonant planar MM in the IR range is developed. The Maxwell–Bloch equations have been used for EMW propagation in quantum active media. The non-stationary and stationary nonlinear beam propagation, the averaging method, the material and photonic-crystalline approximation are used. These approaches allow us to study the phenomena of arbitrary, or non-moderate, nonlinearity and hot spot formation. Hyperbolic MMs are very useful for the development of controlled optical micro- and nano-photonic devices. Hyperbolic MMs: (1) can be created by subwave structuring; (2) are technologically simpler than MMs of many other types;

(3) demonstrate hyperbolic dispersion; (4) have the property of negative refraction; (5) active elements can be easily incorporated into hyperbolic MMs; (6) they provide an increase in the density of quantum states and spontaneous emission of light from a source located near MMs; (7) hyperbolic MMs are promising for the development of superlenses with super-resolution; (8) they can be implemented from radio frequencies to the optical range; (9) provide a significant increase in resonant nonlinearity; (10) their linear and nonlinear characteristics can be controlled by external fields, the wave characteristics in hyperbolic MMs can also be controlled by the principle of 'light by light'.

14. The results presented in the monograph can be used for basic research of wave processes in layered nonlinear and active artificial and natural media of different physical nature. These media can be artificially created MMs, plasmas and gyromagnetic structures, and natural environments (in particular, in the system 'lithosphere–atmosphere–ionosphere–magnetosphere' (LAIM)) (Boardman *et al* 2007, Boardman *et al* 2018, Ferrari *et al* 2015, Grimalsky *et al* 1997, Grimalsky and Rapoport 1998, Grimalsky *et al* 1999, Pitilakis *et al* 2016). The developed theory, physical–mathematical apparatus, and the MM approach to detection, study, and generalization of regularities of wave processes in layered, complex and active environments of different physical nature are a perspective for designing qualitatively improved signal processing systems, MMs and active media, creation of optical processors, superlenses, transformational optics devices, construction of active devices based on quantum plasmonics, the development of highly sensitive sensors for biomedical technologies, study of the mechanism of influence of powerful phenomena in the lower atmosphere and lithosphere (cyclones, earthquakes) on the ionosphere for use of ionospheric monitoring. improvement of space communication systems, geosurveillance and navigation, etc.

As a rule in monographs devoted to some fields of physics, the list of the future tasks is a subject of personal preferences of the author(s). The present monograph and the corresponding list of tasks are not an exception from this rule. We would like to ask readers to take this into account when treating the list of possible future problems mentioned below.

The general directions are

Wave processes and structures in active nonlinear, resonant and focusing controllable MM media, including singular optics and variable topology. Investigation of wave synergetic processes with extremely high concentration of EM field in nonlinear controlled active quantum and resonant nanomaterials in THz and infrared ranges.

Some specified problems are:

I. Wave processes and strongly localized spatiotemporal structures in active MM waveguides with higher-order nonlinearities and magnetooptic control. Wave processes and structures in active quantum nonlinear hyperbolic MMs.

II. Modeling topological properties of strongly localized nonlinear waves and wave structures. The models of magnetooptical vortex structures with phase defects. *Ab initio* models of variable topology wave structures on the base of hyperbolic nonlinear quantum active controlled MMs. Rogue waves in variable topologically controllable active MMs.
III. Generation of pulses at metasurfaces, in MMs, and photonic crystals with quasi-periodic distributions of metaparticles, with the using of nonlinearity, possible active elements (for instance quantum points, quantum wells etc), singularity and topologic control. Also, the use of lattice, Fano (Ranjan Singh et al 2011), and (BIC) bound states in the continuum (Marinica and Borisov 2008) resonances for focusing, harmonic generation, laser generation, including ultrashort pulses, with very low thresholds.
IV. Formation of nonlinear active controlled singular vortex-type nonlinear structures, like bullets. Rogue waves controllable orbital moment vortices etc, using nonlinear resonant metasurfaces.
V. Developing the methods of topological control over nonlinear singular structures including vortices.
VI. Light focusing and nonlinear effects basing on nonlinear nanoparticles.
VII. Searching dynamic and chaotic behavior of above-threshold EMWs in nonlinear active isotropic and hyperbolic field concentrators in different geometries (cylindrical, spherical) and planar nonlinear active hyperbolic structures.
VIII. Searching controllable strongly nonlinear regimes in multilayered dielectric–plasma of narrow gap semiconductors/graphene structures.
IX. Searching ways of manipulation of quantum waves of electron states (QWES) by external electric and magnetic fields in graphene including electron resonators and diffraction gratings and developing graphene/metamaterial electron optics (GMEO). Searching effects of interaction between EMWs of optical range and QWES in advanced signal processing.
X. Searching advanced ways of terahertz generation in the layered dielectric–plasma structures with narrow gap semiconductors/graphene and electron drift. Using ultra-high quality resonances (BIC, Fano, grid, and other types) in periodic MM structures.
XI. Using formed vortex and singular structures for signal transmission through the system ‘atmosphere–ionosphere’.

List of abbreviations

AGW	Acoustic-gravity waves
BIC	Bound states in the continuum
EMW	Electromagnetic wave
GMEO	Graphene/metamaterial electron optics
LAI	Lithosphere–atmosphere–ionosphere
LAIM	Lithosphere–atmosphere–ionosphere–magnitosphere

MMs	Metamaterials
NEELS	Nonlinear evolution equation for layered structures
NEHO	Nonlinear effects of high order
NSE	Nonlinear Schrödinger equation
QWES	Quantum waves of electron states
SRRs	Split ring resonators

References

Ankiewicz A, Soto-Crespo J M and Akhmediev N 2010 Rogue waves and rational solutions of the Hirota equation *Phys. Rev.* E **81** 046602

Boardman A D, Grimalsky V V, Ivanov B, Koshevaya S V, Velasko L, Zaspel C and Rapoport Y G 2005 Excitation of vortices using linear and nonlinear magnetostatic waves *Phys. Rev.* E **71** 026614–24

Boardman A D, King N and Rapoport Y 2007 Metamaterials driven by gain and special configurations *SPIE Proc. of Metamaterials II* **6581** 1–10

Boardman A D, Grimalsky V and Rapoport Y 2011 Nonlinear transformational optics and electromagnetic and acoustic fields concentrators *AIP Conf. Proc. Fourth Int. Workshop Theoretical Comput. Nanophotonics* **1398** 120–2

Boardman A D, Alberucci A, Assanto G, Grimalsky V V, Kibler B, McNiff J, Nefedov I, Rapoport Y G and Valagiannopoulos C A 2017 Waves in hyperbolic and double negative metamaterials, including rogues and solitons *Nanotechnology* **28** 444001

Boardman A D, Alberucci A, Assanto G, Rapoport Y G, Grimalsky V V, Ivchenko V M and Tkachenko E N 2018 Spatial solitonic and nonlinear plasmonic aspects of metamaterials *World Scientific Handbook of Metamaterials and Plasmonics* vol 16 ed E Shamonina and S A Maier (Singapore: World Scientific) 419–69

Chen S 2013 Twisted rogue-wave pairs in the Sasa–Satsuma equation *Phys. Rev.* E **88** 023202

Demekhov A G, Nunn D and Trakhtengerts V Y 2002 Backward wave oscillator regime of whistler cyclotron instability in an inhomogeneous magnetic field *Physics of Auroral Phenomena Proc. XXV Annual Seminar Apatity* pp 65–8

Douglas J and Rachford H H 1956 On the numerical solution of heat conduction problems in two and three space variables *Trans. Am. Math. Soc.* **82** 421–39

Ferrari L, Wu C, Lepage D, Zhang X and Liu Z 2015 Hyperbolic metamaterials and their applications *Prog. Quantum Electron.* **40** 1–40

Grimalsky V V, Rapoport Y G and Slavin A N 1997 Nonlinear diffraction of magnetostatic waves in ferrite films *J. Phys. IV Colloque* **7** 393–4

Grimalsky V V and Rapoport Y G 1998 Modulational instability of surface plasma waves in the second-harmonic resonance region *Plasma Phys. Rep.* **24** 980–2

Grimalsky V V, Kremenetsky I A and Rapoport Y G 1999 Excitation of electromagnetic waves in the lithosphere and their penetration into ionosphere and magnetosphere *J. Atmos. Electr.* **19** 101–17

Luhmann J G, Curtis D W and Lin R P *et al* 2005 IMPACT: Science goals and firsts with STEREO *Adv. Space Res.* **36** 1534–543

Marchuk G I 1990 Splitting and alternating direction methods *Handbook of Numerical Analysis* vol 1 ed P G Ciarlet and J L Lions (Amsterdam: Elsevier Science Publishers) pp 197–462

Marinica D C and Borisov A G 2008 Bound states in the continuum in photonics *Phys. Rev. Lett.* **100** 183902

Pitilakis A, Chatzidimitriou D and Kriezis E E 2016 Theoretical and numerical modeling of linear and nonlinear propagation in graphene waveguides *Opt. Quantum Electron.* **48** 1–22

Rapoport Y G and Grimalsky V V 2011 Transformational optics, complex geometrical optics and nonlinear electromagnetic energy concentrator *Proc. of the Int. Conf. Days on Diffraction (St. Petersburg, Russia)* pp 160–61

Rapoport Y, Boardman A, Grimalsky V, Selivanov Y and Kalinich N 2012a Metamaterials for space physics and the new method for modeling isotropic and hyperbolic nonlinear concentrators *Proc. Int. Conf. on Mathematical Methods in Electromagnetic Theory MMET (Kharkiv Ukraine)* Art. No. 6331154 pp 76–9

Rapoport Y, Boardman A and Grimalsky V 2012b Metamaterial based electromagnetic and acoustic field concentrators and new physical phenomena: Nonlinear focusing switching *Proc. XXXII Int. Science Conf. on Electronics and Nanotechnology (ELNANO) (Kyiv Ukraine)* pp 84–5

Rapoport Y G, Grimalsky V V, Koshevaya S V, Boardman A D and Malnev V N 2014a New method for modeling nonlinear hyperbolic concentrators *Proc. of IEEE 34th Int. Scientific Conf. on Electronics and Nanotechnology (ELNANO) (Kyiv Ukraine)* pp 35–8

Rapoport Y G, Boardman A D, Grimalsky V V, Ivchenko V M and Kalinich N 2014b Strong nonlinear focusing of light in nonlinearly controlled electromagnetic active metamaterial field concentrators *J. Opt.* **16** 0552029

Singh R, Al-Naib I A I, Koch M and Zhang W 2011 Sharp Fano resonances in THz metamaterials *Opt. Express* **19** 6314

Tao X and Bortnik J 2010a Nonlinear interactions between relativistic radiation belt electrons and oblique whistler mode waves *Nonlinear Process. Geophys.* **17** 599–604

Tao X, Bortnik J and Friedrich M 2010b Variance of transionospheric VLF wave power absorption *J. Geophys. Res.* **115** A07303

www.ingramcontent.com/pod-product-compliance
Lightning Source LLC
Chambersburg PA
CBHW082130210326
41599CB00031B/5937